**Books are to be returned on or before
the last date below**

Differential Equations with Applications

International series in pure and applied mathematics

William Ted Martin and E. H. Spanier
CONSULTING EDITORS

Differential Equations with Applications

Paul D. Ritger

Bell Telephone Laboratories

and

Nicholas J. Rose

North Carolina State University

McGraw-Hill Book Company

New York   St. Louis   San Francisco   Toronto   London   Sydney

Differential Equations with Applications

Library of Congress Catalog Card Number 67-13202

ISBN 07-052945-0

6 7 8 9 10 11 12 – MAMM – 7 5 6 4 3 2

Preface

This book is intended as a textbook for a first course in the theory and applications of differential equations. It is an outgrowth of courses given by us over the last 10 years to students of mathematics, science, and engineering at the Stevens Institute of Technology.

Elementary books on differential equations generally fall into one of the following three categories: (1) those that concentrate on the theory and ignore the applications, (2) those that emphasize the applications and slight the theory, and (3) those that are so concerned with techniques of solution that both theory and application receive minimal attention. This book attempts to avoid fitting into any of these categories by providing a balanced presentation of theory, applications, and techniques.

We believe that a knowledge and appreciation of the basic theory of differential equations are important for the scientist and engineer as well as the mathematician. Accordingly we have tried to present this theory in a careful and straightforward manner. The meaning of a solution of a differential equation and the conditions for existence and uniqueness are discussed more thoroughly than is customary in an elementary text. Considerable emphasis is also given to initial- and boundary-value problems, the general properties of linear equations, and the differences between linear and nonlinear systems. Although techniques of solution are certainly important, we have not over-emphasized them. Necessary techniques are well motivated and explained, with full attention given to the conditions for their applicability. Methods of general use are given preference over particular methods of limited use.

As we have stated, applications are not neglected in this book. We feel that differential equations and the physical sciences are so closely related that it is highly unnatural and artificial to separate them. Applications to the physical sciences and engineering are treated as an integral part of the book; they serve to motivate and illustrate the theory, to suggest new problems, and to give meaning and substance to the methods. We have provided sufficient background material and enough detail in the derivations so that the applications can be followed by a reader of limited experience. Moreover, we have attempted to make the discussions of applications complete enough so that the reader can achieve some real understanding of the physical phenomena involved.

We have included much more material than is usual for a book on this level. This has been done to allow greater flexibility in the use of the book for students of varying backgrounds and interests. Furthermore, the value of the book as a reference for students in later work and for practicing scientists and engineers is enhanced.

Considerable care has been taken to make our presentation clear and understandable. Wherever possible the student's previous experience is drawn on to motivate new concepts and methods. We have not hesitated to present topics from more than one point of view when we thought this would increase understanding. On occasion, in order to avoid interrupting the development of ideas, we have deferred some of the longer or more difficult proofs to the appendixes to the chapters.

Although the ability to work through a particular example is no substitute for understanding general principles, it is often a prelude to such understanding. Therefore we have included an unusually large number of illustrative examples worked out in detail. In addition, rather extensive sets of problems are provided; these include routine exercises, theoretical questions, completion of proofs, and extensions of the material in the text. Answers or hints to most of the problems are collected at the end of the book.

The book may be divided roughly into four parts. The first four chapters are concerned mainly with first-order equations. Linear equations and linear systems are discussed in Chaps. 5 through 9. Chapters 10 through 12 cover approximate methods of solution, and Chaps. 13 through 15 provide an introduction to boundary-value problems and the partial differential equations of mathematical physics. A detailed description of the chapters follows.

Chapter 1 covers definitions, classifications, and simple illustrations of differential equations. Chapter 2 takes up standard methods of solving first-order differential equations which are used in the applications in Chap. 3. The fourth chapter is somewhat more theoretical and deals with existence and uniqueness of solutions and methods of approximating solutions. A proof of existence and uniqueness is given in the appendix to this chapter.

Chapter 5, which assumes only the fundamental existence theorem, develops the basic theory of linear differential equations and the methods of solution for equations with constant coefficients. Physical applications of linear equations are presented in Chap. 6. Chapter 7 deals with the series solution of linear equations with variable coefficients. Chapter 8 considers systems of first-order linear equations; the basic theory is developed in close analogy with the theory of a single equation. Matrix methods for finding solutions of linear systems with constant coefficients are discussed, including an elementary treatment of the matrix exponential.   Chapter 9 is a self-contained development of the Laplace-transform method; the last section is devoted to the basic concepts of linear-systems analysis.

Chapter 10 provides an introduction to some of the methods of obtaining qualitative information about the solutions of nonlinear equations.   Chapter 11 discusses linear difference equations.   Here again the basic theory is developed in close analogy with the theory of linear differential equations.   Chapter 12 covers numerical methods; it emphasizes the predictor-corrector methods and the Runge-Kutta method and presents a fairly detailed treatment of error analysis and stability.

Chapter 13 takes up linear two-point boundary-value problems and introduces the concepts of eigenfunction expansions and Fourier series with a brief introduction to Green's functions.   Chapters 14 and 15 include the derivations of some important partial differential equations and their solution by the method of separation of variables.

The book is designed to serve as a text for either a one- or a two-semester course.   It should be possible to cover most of the material in the book in a course meeting three hours per week for two semesters.

A great variety of one-semester courses can be designed by a selection of material.   The following one-semester course has been given successfully by us and our colleagues at Stevens: Chaps. 1 and 2; selections from Chap. 3; Chap. 4, Secs. 1 through 4; most of Chaps 5 and 6; Chap. 7, Secs. 1 though 4; Chap. 8; most of Chap. 9; Chap. 13, Secs. 3, 4, 5, and 9; Chap. 14, Secs. 1 through 4.

It is a pleasure to express our gratitude to many friends and colleagues and to "generations" of students at Stevens for their valuable criticism of a preliminary version of this book.   We would particularly like to thank Richard Bronson and Charles Giardina, who checked most of the solutions to the problems.   We also owe a special debt of gratitude to Mrs. Katherine Melis for an exceptionally accurate job of typing the entire manuscript.   Any errors in the manuscript are most certainly the fault of the authors.

Paul D. Ritger

Nicholas J. Rose

Contents

Basic Concepts

1

1-1 Introduction

A differential equation is an equation involving an unknown function and one or more of its derivatives. The importance of differential equations is attested to by the frequency with which they occur in scientific phenomena; in fact, many of the fundamental laws of science are formulated in terms of differential equations. We will study the basic theory and methods of solution of many of the differential equations that arise in applications.

This first chapter is devoted to definitions, classifications, and illustrations of differential equations. We must assume that the reader has a knowledge of the techniques and concepts of elementary calculus; in this chapter, however, many opportunities will arise for the reader to review and extend his knowledge of these matters.

1-2 Classifications and examples of differential equations

The derivatives appearing in a differential equation must be derivatives of *functions*. If these functions depend on only one independent variable, the derivatives are *ordinary* derivatives, and the equation is called an *ordinary differential equation*. If these functions depend on two or more independent variables, the derivatives are *partial* derivatives, and the equation is called a *partial differential equation*. Differential equations are also classified according to their *order*, the order of a differential equation being the order of the highest derivative appearing in the equation.

The simplest example of a differential equation is

$$\frac{dy}{dx} = f(x) \qquad \text{or} \qquad y' = f(x)\dagger \tag{1}$$

where $f(x)$ is a given function. This is a first-order ordinary differential equation which should be familiar to the reader from elementary calculus. In this equation the independent variable is x and the dependent variable or unknown function is y. To solve this equation, we must find a function $y(x)$ whose derivative is the given function $f(x)$. The solution is therefore the antiderivative or indefinite integral of $f(x)$. We will study this equation in Sec. 1-6.

The equation

$$y' = f(x,y) \tag{2}$$

is a more general ordinary differential equation of the first order. The presence of the unknown function on the right-hand side means that the solution cannot be obtained by integrating both sides of the equation. Actually there is no simple way of solving this equation for arbitrary functions f; various special cases must be considered. This will be done in Chap. 2.

A very important second-order ordinary differential equation is

$$a(x)y'' + b(x)y' + c(x)y = f(x) \tag{3}$$

where a, b, c, and f are given functions of x. In this equation the unknown function y and its derivatives appear linearly;‡ consequently the equation is called a *linear* differential equation. This equation will be extensively studied in this book.

If $f(x)$, the term independent of y and its derivatives, is identically zero, the equation is called a *homogeneous* linear equation; otherwise it is called *nonhomogeneous*. Clearly this definition can be extended to ordinary and partial differential equations of any order. For instance, the equation

$$\frac{\partial^2 u}{\partial x^2} = \frac{1}{c^2}\frac{\partial^2 u}{\partial t^2} \qquad \text{or} \qquad u_{xx} = \frac{1}{c^2}u_{tt}\S \tag{4}$$

is an example of a second-order linear homogeneous partial differential

† If it is clear (or does not matter) what the independent variable is, the abbreviation y' is used for dy/dx. Similarly, y'' is used for d^2y/dx^2, etc.

‡ We say a set of variables u, v, and w appear "linearly" if they appear in the form: $au + bv + cw + d$, where a, b, c, and d are independent of any of the variables u, v, and w.

§ Subscripts are often used to indicate partial differentiation. Thus

$$u_x = \frac{\partial u}{\partial x} \qquad u_{xx} = \frac{\partial^2 u}{\partial x^2} \qquad u_{xt} = \frac{\partial^2 u}{\partial t\,\partial x} = \frac{\partial}{\partial t}\left(\frac{\partial u}{\partial x}\right) \qquad \text{etc.}$$

equation. In this equation the independent variables are x and t and the unknown function is $u(x,t)$.

A *nonlinear* differential equation is simply one that is not linear. The equation

$$y'' + k \sin y = 0 \tag{5}$$

is nonlinear because of the $\sin y$ term and the equation

$$yy' = ay + b \tag{6}$$

is nonlinear because of the yy' term.

In certain cases differential equations are classified according to their *degree*. The degree of a differential equation is defined only if the equation is a polynomial in the unknown function and its derivatives, and the degree is then the power to which the *highest*-order derivative is raised. All linear differential equations are of the first degree. The nonlinear equation (6) above is also of the first degree, but the degree of (5) is not defined.

The equation

$$(y'')^3 + 5x(y')^5 = xy \tag{7}$$

is of the third degree, but

$$\sin\left(\frac{dy}{dx}\right) = y + x \tag{8}$$

has no degree assigned to it since it is not a polynomial equation in y and y'. Most of this book will deal with first-degree equations. On occasion, however, we will study some of the simpler equations of higher degree.

All of the above examples of differential equations give some functional relation between the independent variables, the unknown function, and the derivatives of the unknown function. An ordinary differential equation of the nth order can therefore be described as an equation in the form

$$f\left(x, y, \frac{dy}{dx}, \ldots, \frac{d^n y}{dx^n}\right) = 0 \tag{9}$$

where f is a well-defined function of its various arguments. A similar expression could be written for a partial differential equation of the nth order.

We mention, finally, that we must also consider *systems* of differential equations, that is, several differential equations in several unknown functions. An example of a system of differential equations is

$$\begin{aligned} y' &= ay + bz \\ z' &= cy + dz \end{aligned} \tag{10}$$

In these equations there is only one independent variable, say x, and two unknown functions $y(x)$ and $z(x)$ which must satisfy both equations simultaneously.

Problems 1-2

1. Classify the following differential equations as to order, degree, and linearity. State the independent variables and unknown functions.

a. $y' + P(x)y = Q(x)$ **b.** $4y'' + 3y' + 6y = 0$

c. $\left(\dfrac{d^3s}{dt^3}\right)^2 + 5s^4t^3 = 0$ **d.** $y'' = x^3$

e. $\displaystyle\sum_{i=0}^{n} a_i(x)\dfrac{d^iy}{dx^i} = f(x)$, where $\dfrac{d^0y}{dx^0} = y$ **f.** $\dfrac{\partial^2 u}{\partial x^2} + \dfrac{\partial^2 u}{\partial y^2} = 0$

g. $\begin{cases} 3xy' + xz'' - yz = 0 \\ y - z' + y'' = 0 \end{cases}$ **h.** $\dfrac{dx}{dy} + y^4x = \sin y$

2. Write the most general second-order linear partial differential equation in two independent variables.

3. Prove that if $y_1(x)$ is a solution of the linear equation

$$a(x)y'' + b(x)y' + c(x)y = f_1(x)$$

and $y_2(x)$ is a solution of

$$a(x)y'' + b(x)y' + c(x)y = f_2(x)$$

then $y = y_1(x) + y_2(x)$ is a solution of

$$a(x)y'' + b(x)y' + c(x)y = f_1(x) + f_2(x)$$

This additive property of the solutions of linear differential equations is known as the *principle of superposition*.

4. Show that the principle of superposition (see Prob. 3) does not apply to the nonlinear differential equation

$$y'' + y^2 = f(x)$$

1-3   The motion of a particle

The problems of planetary motion and of motion in general have always intrigued mankind. These problems, more than anything else, forced the development of calculus and differential equations. One of the first, and most striking, applications of differential equations was Newton's derivation of Kepler's three laws of planetary motion. Previously these laws were empirical laws; Newton was able to derive them from the universal law of gravitational attraction† by solving the differential equations of motion.

† This law states that the magnitude of the force of attraction between two particles is directly proportional to the product of their masses and inversely proportional to the square of the distance between the particles.

Planetary motion will be considered in Sec. 6-5.   In this section we consider the general problem of motion of a particle to illustrate how physical problems lead to differential equations.   The reader who is unfamiliar with the basic concepts of mechanics should consult a book† on physics or mechanics for more details than we can present here.

Consider a particle of mass m moving along the x axis and located at position x (at time t).   The velocity and acceleration of the mass are defined by

$$v = \frac{dx}{dt} = \dot{x}\ddagger \tag{11}$$

and

$$a = \frac{d^2x}{dt^2} = \ddot{x}\ddagger \tag{12}$$

respectively.   We assume a force F, directed along the x axis, acts on the mass (F is positive if it acts in the direction of the positive x axis and negative if it acts in the direction of the negative x axis).   Newton's second law states that "force equals mass times acceleration" or, in symbols,

$$F = m\ddot{x} \tag{13}$$

This is a second-order differential equation for the displacement $x(t)$.   If the force F is a function of time only, the equation can be solved by two integrations.   However, if F depends on the position x or the velocity $\dot{x}$, other methods must be used to solve the equation.

We mention briefly the *units* of the physical quantities involved in the above equations.   We can take the quantities *mass, length,* and *time* to be fundamental physical quantities whose units can be chosen arbitrarily.   The units of velocity, acceleration, and force are then defined by (11), (12), and (13), respectively.

The two sets of units which are in most common use are shown in the accompanying table.   To illustrate, a constant force of one pound acting on a

Quantity	Cgs system	Engineering system
Unit of mass	gram (g)	slug (slug)
Unit of length	centimeter (cm)	foot (ft)
Unit of time	second (sec)	second (sec)
Unit of velocity ($v = \dot{x}$)	centimeter per second (cm/sec)	foot per second (ft/sec)
Unit of acceleration ($a = \ddot{x}$)	centimeter per second per second (cm/sec²)	foot per second per second (ft/sec²)
Unit of force ($F = ma$)	dyne (dyne) = g cm/sec²	pound (lb) = slug ft/sec²

† E.g., see Sears [32] or Synge and Griffith [34].

‡ When the independent variable is time, a dot over a letter is often used to indicate differentiation with respect to time.

mass of one slug will produce an acceleration of one foot per second per second.

A familiar example of Newton's law of motion is the freely falling body with no air resistance.   In this case the only force on the body is the force of gravity.   This force is called the *weight* of the body and is equal to the product mg where g† is the acceleration due to gravity (g is approximately equal to 32 ft/sec² or 980 cm/sec²).   If the x axis is taken *positively downwards*, the differential equation of motion is simply

$$mg = m\ddot{x} \qquad \text{or} \qquad g = \ddot{x} \qquad\qquad (14)$$

If the body encounters air resistance, there is an additional force on the body which opposes the motion; often this force can be assumed to be proportional to some power of the velocity.   The differential equation of motion then becomes

$$mg - r(\dot{x})^n = m\ddot{x} \qquad\qquad (15)$$

where r is the constant of proportionality.   In case n is 0 or 1, the equation is linear; otherwise it is nonlinear.

As another example, consider the motion of a mass on a spring.   The spring is assumed to obey *Hooke's law*, which states that *the force required to stretch or compress the spring a distance d is proportional to d*.   The constant of proportionality is called the *spring constant* or *stiffness* and is denoted by k; its dimension is force per unit length (pounds per foot in engineering units). If a mass m is attached to the spring and lowered until equilibrium is reached, the spring is stretched some distance s, satisfying

$$mg = ks \qquad\qquad (16)$$

that is, the spring force is balanced by the weight of the mass (see Fig. 1*a* and *b*).   If the mass is now disturbed from its equilibrium position, it will

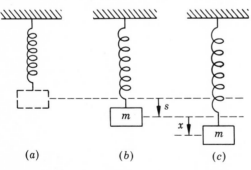

(a)   (b)   (c)

figure 1.   (*a*) *Unstretched spring*, (*b*) *equilibrium position*, (*c*) *general position*.

† g is not actually a constant, but it may be assumed to be constant near the surface of the earth.

oscillate up and down in some manner. (We assume the motion takes place along a vertical straight line.) Let the displacement of the mass from its equilibrium position be denoted by x measured positively *downwards*. If the mass is at a displacement x (see Fig. 1c), the spring force on it is

$$-k(x + s)$$

and the weight is mg. If air resistance and other effects are ignored, these are the only forces on the mass. Applying Newton's law, we obtain

$$-k(x + s) + mg = m\ddot{x} \tag{17}$$

Since $mg = ks$, the equation simplifies to

$$m\ddot{x} + kx = 0 \tag{18}$$

This equation is often called the equation of the *harmonic oscillator*; it is a second-order linear homogeneous differential equation, with constant coefficients, for the displacement $x(t)$. If an external force $F(t)$ is applied to the mass, the resulting differential equation is

$$m\ddot{x} + kx = F(t) \tag{19}$$

which is still a linear equation but is nonhomogeneous. Friction or damping forces, which are always present in physical systems, can also be taken into account. In the simplest case the damping force can be assumed to be proportional to the velocity, and the equation of motion becomes

$$m\ddot{x} + r\dot{x} + kx = F(t) \tag{20}$$

where $r > 0$ is the constant of proportionality. This is the differential equation for a damped, forced oscillation. This linear equation is the prototype for all vibration problems; it will be studied extensively in Chap. 6.

We consider briefly the more general problem of motion of a particle in space. Suppose the particle of mass m is located at the point with rectangular coordinates (x,y,z). It is convenient to denote this position by a *vector*† **r** from the origin to the point (x,y,z). We also denote the velocity vector, with *components*† $(\dot{x},\dot{y},\dot{z})$, by $\dot{\mathbf{r}}$ and the acceleration vector, with components $(\ddot{x},\ddot{y},\ddot{z})$, by $\ddot{\mathbf{r}}$. If the particle is subjected to a vector force **F**,

---

† For the reader who is unfamiliar with vectors, we briefly define what we mean; more details can be found in Chap. 8. See also Courant [12], vol. 2, p. 3.

A *vector* **V** is a directed line segment in space. A vector is therefore characterized by a length and a direction. The *components* of **V** are denoted by (V_1,V_2,V_3) and are the projections of **V** on the x, y, and z axes respectively. Two vectors **V** and **W** are called equal if $V_1 = W_1$, $V_2 = W_2$, $V_3 = W_3$. The product of **V** by a number α is defined by $\alpha\mathbf{V} = (\alpha V_1,\alpha V_2,\alpha V_3)$. The sum of two vectors **V** + **W** is defined by **V** + **W** = $(V_1 + W_1, V_2 + W_2, V_3 + W_3)$, the *parallelogram law*.

with components (F_1, F_2, F_3), Newton's law can be expressed by the vector equation

$$\mathbf{F} = m\ddot{\mathbf{r}} \tag{21}$$

In particular, this law states that the force and acceleration vectors at every instant must lie in the same straight line. The vector equation (21) may be considered a shorthand for the three equations

$$F_1 = m\ddot{x}$$
$$F_2 = m\ddot{y} \tag{22}$$
$$F_3 = m\ddot{z}$$

Thus, to determine the motion of a particle in space, we must solve this system of three equations for the functions $x(t)$, $y(t)$, and $z(t)$. One could not solve each of the equations in (22) individually because F_1, F_2, and F_3 are generally functions of x, y, z, and t.

If there are n particles instead of only one, three equations similar to (22) can be written for each particle; hence there is a total of $3n$ differential equations to determine the motion of n particles in space.

Problems of motion also lead to partial differential equations if, instead of discrete particles, a continuum of mass is postulated. For instance, the small motions of a tightly stretched string (such as a piano wire) are governed by the so-called *wave equation*

$$c^2 u_{xx} = u_{tt} \tag{23}$$

where $u(x,t)$ is the displacement of the string at the position x and at time t, and c depends on the tension and the mass density of the string. This equation will be studied in Chap. 14.

The above examples illustrate how problems of motion lead to differential equations. In the problems below, and in later chapters, many other examples from the various fields of science will be studied.

Problems 1-3

1. A projectile of mass m is shot vertically upwards from the surface of the earth. Neglecting all forces on the projectile except for the attraction of the earth given by Newton's law of gravitation, show that the equation of motion is

$$\frac{-kM}{x^2} = \ddot{x}$$

where M is the mass of the earth and x is the distance from the earth's center. How can the constant k be evaluated?

2. A particle of mass m is constrained to move along a fixed curve. The position of the particle is given by the length of the arc s measured from a fixed

point on the curve. Show that the differential equation of motion can be written

$$m\ddot{s} = f(s,t)$$

where $f(s,t)$ is the component of force acting on the particle in the direction of the curve.

3. A simple pendulum consists of a mass m supported by a weightless rod of length L pivoted at the top. The position of the mass is given by the angle θ that the rod makes with the vertical. Show that the differential equation of motion is

$$L\ddot{\theta} + g \sin \theta = 0$$

4. Suppose a particle of mass M is fixed at the origin, and another particle of mass m is located at (x,y,z). Assuming Newton's law of attraction, show that the differential equations of motion of the particle at (x,y,z) are for $r > 0$:

$$\ddot{x} = -kM \, \frac{x}{r^3}$$

$$\ddot{y} = -kM \, \frac{y}{r^3}$$

$$\ddot{z} = -kM \, \frac{z}{r^3}$$

where $r = (x^2 + y^2 + z^2)^{\frac{1}{2}}$

1-4 The solution of a differential equation

Before discussing methods of solving differential equations, it is important to understand what is meant by a solution of a differential equation. Consider a general ordinary differential equation of the nth order

$$f\left(x, y, \frac{dy}{dx}, \ldots, \frac{d^n y}{dx^n}\right) = 0$$

A solution of this equation is defined to be a real-valued† function $y(x)$ defined over some interval I‡ having the following two properties:

1. The function $y(x)$ and its first n derivatives§ exist for each x in I. This implies that $y(x)$ and its first $n - 1$ derivatives must be continuous for each x in I.

† We will have occasion later to allow $y(x)$ to take on complex values.

‡ The interval I can be $-\infty < x < \infty$, an infinite interval; $a \leq x \leq b$, a closed interval; $a < x < b$, an open interval; or $a \leq x < b$ or $a < x \leq b$, half-open intervals. In all cases the interval I will be assumed to be non-degenerate; i.e., I will contain more than one point.

§ In the case of a closed interval, the derivatives at the end points are assumed to be appropriate one-sided derivatives.

2. The substitution of $y(x)$ into the differential equation makes the equation an *identity* in x in the interval I; that is, the equation holds for each x in I.

Any single solution of a differential equation is called a *particular solution*, and the *set†* *of all solutions* is called the *general solution.‡* A particular solution can be represented as a curve in the xy plane called an *integral curve*. The general solution is the set of all integral curves, or the *family* of all integral curves.

We illustrate these definitions with several examples.

Example 1   A particular solution of the differential equation

$$y' = x^2 \qquad -\infty < x < \infty$$

is

$$y = \tfrac{1}{3}x^3 \qquad -\infty < x < \infty$$

since

$$\frac{d}{dx}\,(\tfrac{1}{3}x^3) = x^2 \qquad -\infty < x < \infty$$

Note that $\tfrac{1}{3}x^3$ is a continuous function and possesses a derivative for all x. In fact, since x^2 is continuous, the solution has a *continuous* derivative. It may be recalled from calculus that any two functions having the same derivative differ only by a constant. Therefore the general solution of the differential equation is

$$y = \tfrac{1}{3}x^3 + c$$

where c is any real number. For a particular value of c the solution can be represented by an integral curve as shown in Fig. 2. The general solution is represented by the family of all integral curves. Since there is one curve for each

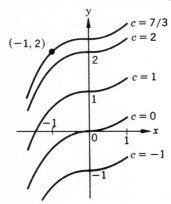

figure 2.   *Integral curves for* $y' = x^2$.

† The word *set* is used in its intuitive sense as "a collection of objects."

‡ This definition of general solution is different from that used in many books. See "Remarks about Solutions" at the end of this section.

value of the parameter c, the general solution is represented by a *one-parameter family of curves*.

The equations

$$y = \begin{cases} \frac{1}{3}x^3 + 1 & -\infty < x < 0 \\ \frac{1}{3}x^3 & 0 \le x < \infty \end{cases}$$

define a function for all x, the function being discontinuous at $x = 0$. This function is *not* a solution of the differential equation in the domain $-\infty < x < \infty$ because of the discontinuity at $x = 0$. The "piece" of the function $y = \frac{1}{3}x^3 + 1$ is a solution in the domain $-\infty < x < 0$, and the "piece" $y = \frac{1}{3}x^3$ is a solution in the domain $0 \le x < \infty$. The two pieces of the function do not fit together continuously at $x = 0$, and therefore the function is not a solution for all x.

Example 2   The differential equation

$$y \frac{dy}{dx} + x = 0$$

can be solved by separating the variables as follows:

$$y \, dy + x \, dx = 0$$

If $y(x)$ is any differentiable function satisfying this equation, we obtain by integration†

$$\tfrac{1}{2}y^2 + \tfrac{1}{2}x^2 = c_1$$

It is convenient to let c_1 equal $\frac{1}{2}c$ and thus obtain

$$y^2 + x^2 = c$$

This is a *tentative* solution of the differential equation. To see whether or not we have an *actual* solution, we must show two things: (1) that this equation determines y as a differentiable function of x and (2) that the function thus determined satisfies the differential equation. If we *assume* that the equation $y^2 + x^2 = c$ determines y as a differentiable function of x, it is easy to show that requirement 2 is satisfied. For, by implicit differentiation, we have

$$2yy' + 2x = 0$$

which is equivalent to the differential equation. We still must show that y is determined as a differentiable function of x. This is easily done in this particular case by actually solving the equation for y:

$$y = \pm\sqrt{c - x^2}$$

If $c > 0$, we have two differentiable functions, and therefore two solutions,‡

† See page 36.
‡ Note that the solutions possess not only a derivative but a *continuous* derivative.

defined in the interval

$$-\sqrt{c} < x < \sqrt{c}\,\dagger$$

(The end points $x = \pm\sqrt{c}$ are not included since $y' = \mp x/\sqrt{c - x^2}$, and hence y' does not exist for $x = \pm\sqrt{c}$.)   However, if $c < 0,\ddagger$ y is *not* determined as a (real-valued) function of x and no solution of the equation is obtained.

Remark   In the case where a solution of a differential equation is obtained as an implicit function of the form

$$f(x,y) = 0$$

we shall not always carry out the complete analysis given in the above example.   We shall usually *assume* that the implicit equation determines a function $y(x)$ and show only that this $y(x)$ must satisfy the differential equation.   This can be done by implicit differentiation, as in the above example.

The matter of showing that $y(x)$ exists, § and of finding the values of x for which it is defined, will usually be omitted.

Example 3   Consider the equation of the harmonic oscillator

$$m\ddot{x} + kx = 0 \qquad k, m > 0$$

It may be verified by direct substitution that the functions

$$x = \cos\sqrt{\frac{k}{m}}\,t \qquad \text{and} \qquad x = \sin\sqrt{\frac{k}{m}}\,t$$

are solutions for all values of t.   It can also be verified that the linear combination

$$x = c_1 \cos\sqrt{\frac{k}{m}}\,t + c_2 \sin\sqrt{\frac{k}{m}}\,t$$

is also a solution for all t, where c_1 and c_2 are arbitrary constants.   This is actually the general solution; every solution can be obtained by assigning suitable values to c_1 and c_2.   We note that the general solution of this second-order equation is a two-parameter family of functions.

† If $c > 0$, the symbol $\sqrt{c}$ will always be taken to mean the positive square root of c.
‡ For $c = 0$, $y(x)$ is defined for only the one point $x = 0$.   Clearly a function defined for only one point cannot possess a derivative.
§ An important theorem called the *implicit-function theorem* gives sufficient conditions for $f(x,y) = 0$ to determine y as a differentiable function of x. See Courant [12], vol. 2, pp. 111–116.

Example 4   Consider the first-order nonlinear equation of the second degree:

$$(y')^2 = 4y$$

It can easily be verified that

$$y = (x + c)^2$$

is a solution for all x and for every number c (this solution can be obtained by taking square roots and separating variables).

It might be expected that this one-parameter family of solutions is the general solution, that is, contains all solutions.   However, it is easily seen that $y = 0$ is an "extra" solution which cannot be obtained by assigning a particular value to c.   The solution $y = 0$ is called a *singular solution*.   We will return later to a discussion of singular solutions.   However, a detailed treatment of this subject is outside the scope of this book.

Remarks about solutions   So far, all of the differential equations considered have possessed solutions, in fact, infinitely many solutions; however, it is easy to give examples of differential equations that have only one solution or no solution at all.   For instance, the differential equation

$$(y')^2 + y^2 = 0 \tag{24}$$

has for its only solution $y = 0$.   For if $y(x)$ were different from zero for any x, then $(y(x))^2$ would be greater than zero, and the equation could not be satisfied.   Similarly, it can be seen that the differential equation

$$(y')^2 + y^2 = -1 \tag{25}$$

has no solution.   These examples indicate that it is not easy to make accurate general statements about the solutions of a differential equation.

Fortunately there are several important theorems that guarantee the existence of solutions for certain classes of differential equations.   Some of these theorems will be considered in the next section, and, in more detail, in Chap. 4.   The most complete information available is about linear differential equations.   For instance, consider the second-order homogeneous linear equation

$$a(x)y'' + b(x)y' + c(x)y = 0$$

If $a(x) \neq 0$ and if $a(x), b(x), c(x)$ are continuous† functions for some interval I, then there exist two solutions $y_1(x)$ and $y_2(x)$ that are defined in I.   Furthermore these two solutions can be so picked that the general solution of the differential equation is the linear combination

$$y = c_1 y_1(x) + c_2 y_2(x)$$

† See page 31 for a definition of continuity.

Note that the general solution contains two arbitrary constants or parameters. In a similar manner the general solution of the nth-order homogeneous linear equation contains n arbitrary constants which enter into the general solution in the same linear manner.

For nonlinear equations the situation is more complicated. We certainly cannot expect the general solution of an nth-order equation to be an n-parameter family of functions. Equation (25) shows that a nonlinear equation may have no solutions; even in those cases where an n-parameter family of solutions exists, there may also be extra (singular) solutions, as indicated in Example 4.

It is well to defer further consideration of these matters until Chap. 4. This will not deter us in our study for, in particular cases, it is possible to show that solutions exist by actually finding the solutions; it is sometimes even possible to show that all solutions have been obtained. For most of the differential equations that occur in this book, and for all linear equations, it actually turns out that the general solution is a certain n-parameter family of functions. The reader should expect this, but should not be unduly surprised if, in the case of nonlinear equations, certain exceptions to this statement arise.

Finally, we must warn the reader that in many books the phrase *general solution* is used in a different sense.† A general solution of an nth-order ordinary differential equation is often defined to be "a solution containing n essential arbitrary constants." This usage has two drawbacks: (1) it disguises the fact that the general solution (as defined in the previous sentence) need not contain all solutions, and (2) the phrase "containing n essential arbitrary constants" is not well-defined; how does one know when the constants are essential? We shall not adopt this usage; if a solution contains n parameters, we shall simply call it an n-parameter family of solutions; if the n-parameter family can be proved to contain *all* solutions, we shall then call it the general solution.

Problems 1-4

1. Verify that $y = c_1 e^{-ax} + c_2 e^{ax}$ is a solution of

$$y'' - a^2 y = 0$$

for all x and for all constant c_1 and c_2.

2. By integrating n times, find the general solution of

$$\frac{d^n y}{dx^n} = 0$$

3. For what values of x does $y^2 - 2x + 1 = 0$ determine y as a differentiable function of x which satisfies $yy' = 1$?

† See the excellent discussion in Agnew [1], pp. 116–117.

4. Find the values of a, if any, for which $y = ax^3$ is a solution of:

a. $x^2y'' + 6xy' + 5y = 0$
b. $x^2y'' + 6xy' + 5y = 2x^2$
c. $x^2y'' + 6xy' + 5y = x^3$

5. Find the values of λ for which $e^{\lambda x}$ is a solution of

$$y'' + 5y' + 6y = 0$$

6. Assume that the implicit function $e^{xy} = y$ determines y as a differentiable function of x in some interval. Without attempting to solve the equation for $y(x)$, show that $y(x)$ satisfies the differential equation

$$(1 - xy)y' - y^2 = 0$$

7. a. Define what should be meant by a solution of the partial differential equation

$$\frac{\partial^2 u}{\partial x^2} = \frac{1}{c^2} \frac{\partial^2 u}{\partial t^2}$$

b. Show that $u = f(x - ct) + g(x + ct)$ is a solution of this equation provided $f(v)$ and $g(v)$ are twice-differentiable functions. Hint: Use the chain rule:

$$\frac{\partial f}{\partial x} = \frac{df}{dv} \frac{\partial v}{\partial x}$$

8. Find the values of a, b, and w for which

$$x = a \sin wt$$
$$y = b \sin wt$$

is a solution of the system of equations

$$\ddot{x} + 8x - 4y = 0$$
$$\ddot{y} - 4y + 8x = 0$$

1-5 *Initial- and boundary-value problems*

In many problems involving a differential equation certain subsidiary conditions are given which must be satisfied by the solution of the differential equation. These subsidiary conditions often enable the arbitrary constants appearing in the general solution to be evaluated to yield a unique solution.

Example 1 It is required to find the solution of

$$y' = x^2$$

which also satisfies $y(-3) = 2$.

This is easily solved.   The general solution of the differential equation is

$$y = \tfrac{1}{3}x^3 + c$$

To satisfy $y(-3) = 2$, the constant c must satisfy

$$2 = \tfrac{1}{3}(-3)^3 + c$$

This yields $c = 11$, and the solution is

$$y = \tfrac{1}{3}x^3 + 11$$

Geometrically, we have picked out that integral curve which passes through the point $(-3,2)$ as shown in Fig. 2 (page 10).

Since the solution of a second-order differential equation generally involves *two* arbitrary constants, two subsidiary conditions can be specified. These can take a variety of forms.   The conditions may be given at the *same* value of x, such as

$$\begin{aligned} y(x_0) &= a \\ y'(x_0) &= b \end{aligned} \tag{26}$$

Such conditions are called *initial conditions* (IC).   An *initial-value problem* is a differential equation together with initial conditions.   The conditions may also be given at two *different* values of x, such as

$$\begin{aligned} y(x_0) &= a \\ y(x_1) &= b \end{aligned} \tag{27}$$

or

$$\begin{aligned} y(x_0) &= a \\ y'(x_1) &= b \end{aligned} \tag{28}$$

Such conditions are called *boundary conditions* (BC).   A *boundary-value problem* is a differential equation together with boundary conditions. Clearly these definitions can be extended to differential equations of any order.†

The following examples illustrate how an initial-value problem or a boundary-value problem can be solved when the general solution of the differential equation can be found.

Example 2   DE:‡ $y'' = 1$

$$\text{IC: } y(0) = 1$$
$$y'(0) = 2$$

---

† Boundary conditions may, of course, be given at more than two points. We shall consider only two-point boundary-value problems.   These include many problems of interest in physics and engineering.

‡ DE stands for differential equation.

The differential equation can be integrated twice to yield the general solution

$$y = \tfrac{1}{2}x^2 + c_1 x + c_2$$

From $y(0) = 1$ we obtain $c_2 = 1$. Since $y' = x + c_1$ the condition $y'(0) = 2$ yields $c_1 = 2$. The final solution is therefore

$$y = \tfrac{1}{2}x^2 + 2x + 1$$

Example 3   DE: $y'' = 1$
            BC: $y(0) = 1$
                $y(2) = 4$

The general solution of the differential equation is the same as in the previous example, and the condition $y(0) = 1$ again yields $c_2 = 1$. The second condition gives

$$4 = \tfrac{1}{2}(2^2) + 2c_1 + 1$$

The constant c_1 is therefore equal to $\tfrac{1}{2}$ and the solution is

$$y = \tfrac{1}{2}x^2 + \tfrac{1}{2}x + 1$$

Example 4   According to Newton's second law, a particle having no forces acting on it must satisfy the equation

$$\ddot{x} = 0$$

If the particle has an initial position x_0 and initial velocity v_0, that is, if

$$x(0) = x_0$$
$$\dot{x}(0) = v_0$$

then it can easily be shown that the *unique* solution for this problem is

$$x = x_0 + v_0 t$$

This is motion with a uniform velocity $\dot{x} = v_0$. Thus we see that a particle that is not acted on by external forces either must be at rest, if initially it is at rest, or else it must be in motion with uniform velocity equal to the initial velocity. These facts, which constitute Newton's first law of motion, are therefore simple consequences of Newton's second law.

Remarks about initial- and boundary-value problems   In the above examples the initial- and boundary-value problems all possessed solutions. We now investigate, using physical intuition as a guide, whether or not this can be generally expected.

From physical intuition we expect that the motion of a particle acted on by known forces is *uniquely determined* provided the initial position and initial velocity are known.   In mathematical language this states that the initial-value problem

DE: $m\ddot{x} = F(t,x,\dot{x})$

IC: $x(0) = x_0$

$\quad\ \dot{x}(0) = v_0$

can be expected to possess a unique solution.   Since the force F can be a rather general function of t, x, and $\dot{x}$, this covers a rather wide class of second-order differential equations.   In Chap. 4 we shall see that our intuition has not led us astray; with some restrictions on F to make the statement precise, *the initial-value problem generally possesses a unique solution*.

Now let us consider a boundary-value problem:

DE: $m\ddot{x} = F(t,x,\dot{x})$

BC: $x(t_0) = x_0$

$\quad\ \ x(t_1) = x_1$

In physical terms this means that a particle starts off at position x_0 at time t_0 (with *some* initial velocity), and we require that the particle be at position x_1 at time t_1.   We *cannot* generally expect this to be possible; the particle may have such a large force opposing the motion that, for any initial velocity, the particle may not reach position x_1 at time t_1, or at any other time for that matter.

Even if the particle reaches x_1 at time t_1, this may be possible with *more than one motion*.†   Mathematically this means that a boundary-value problem does not always possess a solution and even if a solution exists, it may not be unique (see Probs. 5 and 6 below).

In general the conclusions reached above for second-order equations also hold for higher-order equations.   Boundary-value problems are intrinsically more difficult than initial-value problems, although they have equal importance in applications.   The theory of initial-value problems is much more complete than that of boundary-value problems.   Only for linear differential equations has the boundary-value problem been thoroughly investigated.   We shall discuss some of the simpler boundary-value problems in Chap. 13.

Problems 1-5

1. Solve: DE: $y''' = x$

$\qquad\quad$ IC: $y(0) = 0 \qquad y'(0) = 1 \qquad y''(0) = 0$

† For example, in a free vibrating system (no external force), such as a mass-spring system (see Sec. 1-3), if $x(0) = \dot{x}(0) = 0$, the system remains at rest for all time.   However, if an initial velocity is given to the mass, it will return to $x = 0$ at *some* time t_1.   Thus there are two solutions satisfying $x(0) = 0$, $x(t_1) = 0$.

2. Solve: DE: $y'' - 4y = 0$

IC: $y(0) = 1$ $y'(0) = 1$

Assume that the general solution of the differential equation is $y = c_1 e^{-2x} + c_2 e^{2x}$.

3. The general solution of the differential equation

$$\ddot{x} + \lambda^2 x = 0 \qquad \lambda > 0$$

is

$$x = c_1 \sin \lambda t + c_2 \cos \lambda t$$

Find the solution satisfying the following sets of initial conditions,

a. $x(0) = x_0$ $\dot{x}(0) = 0$
b. $x(0) = 0$ $\dot{x}(0) = v_0$
c. $x(0) = x_0$ $\dot{x}(0) = v_0$

4. Assume $y(x)$ is a solution of

DE: $y' = x^2 + y^2$
IC: $y(0) = 1$

Without attempting to solve the equation, find $y'(0)$ and $y''(0)$. [It may be assumed that $y''(x)$ exists.]

5. Consider the boundary-value problem, where f is continuous,

DE: $y'' = f(x)$
BC: $y'(0) = 0$
     $y'(1) = 0$

Show that this problem has no solution unless $f(x)$ satisfies the condition

$$\int_0^1 f(x) \, dx = 0$$

which is certainly not satisfied for all functions f. Also show that when f does satisfy this condition, the solution is

$$y = \int_0^x \left[\int_0^t f(s) \, ds \right] dt + c$$

where c is any constant. The solution is therefore not unique. This problem illustrates both that a boundary-value problem may have no solution and that even if there is a solution, it may not be unique.

6. One solution of the boundary-value problem

DE: $\dfrac{d^2 y}{dx^2} + \lambda^2 y = 0 \qquad \lambda > 0$

BC: $y(0) = 0$
     $y(L) = 0$

is the trivial solution $y = 0$.   Show that a nontrivial solution exists only if λ is one of the numbers

$$\lambda_n = \frac{n\pi}{L} \qquad n = 1, 2, 3, \ldots$$

and that the solution in this case is

$$y = c_n \sin \frac{n\pi x}{L}$$

where c_n is an arbitrary constant.   (See Prob. 3.)

1-6   The differential equation $y' = f(x)$

The fundamental problem of integral calculus is to find all functions having a given derivative.   This is equivalent to finding all solutions of the differential equation

$$y' = f(x) \tag{29}$$

where $f(x)$ is a given function.   The solution can be written

$$y = \int f(x)\, dx \tag{30}$$

where the *indefinite integral* stands for all functions whose derivative is $f(x)$.   If $F(x)$ is one function whose derivative is $f(x)$, that is, if $F(x)$ is a particular solution of (29), then, by a basic theorem of calculus, *all* solutions are given by

$$y = F(x) + c \tag{31}$$

where c is an arbitrary constant.   We may take the function

$$F(x) = \int_a^x f(t)\, dt \tag{32}$$

as a particular solution, where a is an arbitrary real number.   The symbol on the right side of (32) is a *definite integral*.   The reader will recall that this definite integral is, by definition, a limit of a sum which represents the area under the curve $f(t)$ from $t = a$ to $t = x$.   It should be clear that the variable t is a dummy variable; that is, any other variable could be used in place of t without changing the value of the definite integral.

The fact that the definite integral in (32) is a solution of the differential equation (29) is really the content of the fundamental theorem of calculus. This theorem is often stated as a method of evaluating a definite integral, namely,

$$\int_a^x f(t)\, dt = G(x) - G(a) \tag{33}$$

where $G(t)$ is any function having the property that

$$G'(t) = f(t) \tag{34}$$

By differentiating both sides of (33) and using (34), we find

$$\frac{d}{dx} \int_a^x f(t)\, dt = f(x)\dagger \tag{35}$$

so that the definite integral in (32) actually provides a solution of (29).

We have assumed that the definite integral exists. The following theorem, which is usually proved in advanced calculus, gives a sufficient condition for the existence of the integral.‡

If $f(t)$ is a continuous function in the interval $\alpha \le t \le \beta$ and if a and x are any two points in this interval, then the integral

$$F(x) = \int_a^x f(t)\, dt$$

exists and is a continuous function with a continuous derivative.

This theorem and the preceding discussion enable us to state the following fundamental result:

If $f(x)$ is a continuous function in an interval $\alpha \le x \le \beta$, then all solutions of $y' = f(x)$ are given by

$$y = \int_a^x f(t)\, dt + c\S$$

where a and x are any points in the interval and c is an arbitrary real number.

We note that the solution y possesses not only a derivative but a continuous derivative.

It is easy to show from the above result that the *unique* solution of the initial-value problem

$$\begin{aligned} &\text{DE: } y' = f(x)\\ &\text{IC: } y(x_0) = y_0 \end{aligned} \tag{36}$$

† In other words, the definite integral, considered as a function of its upper limit, is also an indefinite integral or antiderivative.

‡ See Courant [12], vol. 1, p. 131.

§ Since, for different values of a, $\displaystyle\int_a^x f(t)\, dt$ differs by a constant, it might be expected that all solutions could be obtained by varying a and that the constant c is not needed. This, however, is not always the case: the equation $y' = \cos x$ has $y = \sin x + c$ for its general solution, but $\displaystyle\int_a^x \cos t\, dt = \sin x - \sin a$, and since $|\sin a| \le 1$, all solutions are not obtained.

is

$$y = y_0 + \int_{x_0}^{x} f(t)\, dt \tag{37}$$

provided $f(x)$ is a continuous function in the relevant interval.

A convenient way of obtaining (37) is to write the differential equation in the differential form

$$dy = f(x)\, dx$$

and integrate both sides from the initial point (x_0, y_0) to the point (x, y):

$$\int_{y_0}^{y} dy = \int_{x_0}^{x} f(x)\, dx = \int_{x_0}^{x} f(t)\, dt$$

or

$$y - y_0 = \int_{x_0}^{x} f(t)\, dt$$

which is equivalent to (37). This formal procedure is often used, and it gives the correct answer. However, it should not be considered a proof.

Example   DE: $y' = \sin x$

IC: $y(0) = 1$

Solution $\displaystyle \int_{1}^{y} dy = \int_{0}^{x} \sin x\, dx$

Therefore

$$y - 1 = -\cos x - (-1)$$

or

$$y = 2 - \cos x$$

Problems 1-6

1. Solve for $-\infty < x < \infty$:

a. DE: $y' = e^{-x}$ b. DE: $y' = |x|$

   IC: $y(0) = 3$ IC: $y(-1) = 2$

2. Find the *continuous* function $y(x)$ which satisfies the following initial-value problem except for $x = 0$.

DE: $y' = \begin{cases} 0 & x < 0 \\ 1 & x > 0 \end{cases}$

IC: $y(-5) = 2$

3. Verify that the solution of

DE: $y' + y \sin x = x$

IC: $y(0) = 0$

is

$$y = e^{\cos x} \int_0^x t e^{-\cos t} \, dt$$

4. Derive the solution $y = \int_0^x \left[\int_0^t f(s) \, ds \right] dt$ of the problem

DE: $y'' = f(x)$

IC: $y(0) = y'(0) = 0$

assuming $f(x)$ is a continuous function.

5. Show that

$$\frac{d}{dx} \int_{u(x)}^{v(x)} f(t) \, dt = f(v) \frac{dv}{dx} - f(u) \frac{du}{dx}$$

provided f is continuous and u and v are differentiable.   Hint: Use the chain rule

$$\frac{d}{dx} F(u,v) = \frac{\partial F}{\partial u} \frac{du}{dx} + \frac{\partial F}{\partial v} \frac{dv}{dx}$$

6. Show that there is no continuous function f which satisfies

$$x^4 = \int_a^x f(t) \, dt$$

unless $a = 0$, and find the solution in this case.

1-7   Integrals as functions of parameters

We shall have several occasions to consider integrals that are functions of parameters.   Consider the integral

$$F(x) = \int_a^b f(x,t) \, dt \tag{38}$$

which is a function of the parameter x appearing under the integral sign. We assume that $f(x,t)$ is continuous in the rectangle $R: \alpha \le x \le \beta, a \le t \le b$.   The integral then certainly exists for each x in the interval $\alpha \le x \le \beta$.   It is natural to inquire whether or not the integral, that is, $F(x)$, is a continuous function of x.   This question is answered by the following theorem which is proved in the appendix† to this chapter (page 33).

† Besides a proof of the theorems, the appendix also provides an opportunity to review some of the important definitions of calculus.

Theorem 1 *If $f(x,t)$ is continuous in R, then $F(x)$ is continuous for $\alpha \leq x \leq \beta$; that is, for any x_0 in this interval,†*

$$\lim_{x \to x_0} F(x) = F(x_0) \qquad or \qquad \lim_{x \to x_0} \int_a^b f(x,t)\, dt = \int_a^b f(x_0,t)\, dt$$

We note that this theorem gives conditions under which it is possible to interchange the order of the two operations: (1) taking the limit as $x \to x_0$ and (2) taking the definite integral from a to b. The problem of interchanging the order of two operations, particularly operations involving limiting processes, is of fundamental importance in mathematics; we shall meet this problem several times in later work.

We now consider the problem of finding $F'(x)$, the derivative of the integral, without first performing the integration and then differentiating the result. It is again natural to try to invert the order of these two operations, that is, first to differentiate with respect to x and then to integrate. In fact, the following theorem is true, the proof being deferred to the appendix to this chapter (page 34).

Theorem 2 *If $f(x,t)$ and its partial derivative $f_x(x,t)$ are continuous in R, then we may differentiate the integral with respect to x under the integral sign; that is, for $\alpha \leq x \leq \beta$,*

$$F'(x) = \frac{d}{dx} \int_a^b f(x,t)\, dt = \int_a^b f_x(x,t)\, dt$$

Before considering examples, we generalize the above theorem to the case where the limits of integration are also functions of x:

$$F(x) = \int_{u(x)}^{v(x)} f(x,t)\, dt \tag{39}$$

In order to compute $F'(x)$, we consider this integral to be a compound function

$$F(x) = \int_{u(x)}^{v(x)} f(x,t)\, dt = G(x,v,u) \tag{40}$$

where u and v are differentiable functions of x. According to the chain rule,

$$F'(x) = \frac{\partial G}{\partial x} + \frac{\partial G}{\partial v}\frac{dv}{dx} + \frac{\partial G}{\partial u}\frac{du}{dx} \tag{41}$$

However, assuming $f(x,t)$ satisfies the conditions of Theorem 2, we have

$$\frac{\partial G}{\partial x} = \int_u^v f_x(x,t)\, dt$$

† If $x_0 = \alpha$ or $x_0 = \beta$, the limits must be taken as the appropriate one-sided limits.

and according to the fundamental theorem of calculus,

$$\frac{\partial G}{\partial v} = f(x,v) \quad \text{and} \quad \frac{\partial G}{\partial u} = -f(x,u)$$

Therefore we have the important formula

$$\frac{d}{dx} \int_{u(x)}^{v(x)} f(x,t) \, dt = \int_{u(x)}^{v(x)} f_x(x,t) \, dt + f(x,v(x)) \frac{dv}{dx} - f(x,u(x)) \frac{du}{dx} \qquad (42)$$

Example 1   Consider the integral

$$F(x) = \int_{x}^{x^2} \frac{e^{-xt}}{t} \, dt \qquad x > 0$$

This integral is a non-elementary function (see Sec. 1-8), and it is not possible to perform the integration in terms of elementary functions.   However, $F'(x)$ can easily be computed by using formula (42):

$$F'(x) = -\int_{x}^{x^2} e^{-xt} \, dt + \frac{e^{-x^3}}{x^2} \, 2x - \frac{e^{-x^2}}{x} \cdot 1$$

Evaluating the integral and simplifying, we obtain the result

$$F'(x) = 3 \, \frac{e^{-x^3}}{x} - 2 \, \frac{e^{-x^2}}{x}$$

Example 2   Show that

$$y = \frac{1}{\lambda} \int_{0}^{x} f(t) \sin \lambda(x - t) \, dt \qquad \lambda \neq 0$$

is a solution of the initial-value problem

DE: $y'' + \lambda^2 y = f(x)$

IC: $y(0) = y'(0) = 0$

assuming $f(x)$ is a continuous function.

We see immediately that $y(0) = 0$.   According to formula (42), we have

$$y' = \int_{0}^{x} f(t) \cos \lambda(x - t) \, dt$$

and

$$y'' = -\lambda \int_{0}^{x} f(t) \sin \lambda(x - t) \, dt + f(x)$$

We see that $y'(0) = 0$.   Direct subsitution into the differential equation shows that it is identically satisfied.

Problems 1-7

1. Evaluate $F'(x)$ by two methods:

a. $F(x) = \int_0^x te^{-xt^2}\, dt$ b. $F(x) = \int_0^{x^2} (x - t)^2\, dt$

2. Show that $y = \int_0^x (x - t)f(t)\, dt$ is a solution of the initial-value problem

DE: $y'' = f(x)$

IC: $y(0) = y'(0) = 0$

assuming $f(x)$ is a continuous function.

3. The Bessel function of order zero can be defined by

$$J_0(x) = \frac{1}{\pi} \int_0^{\pi} \cos\,(x \cos \theta)\, d\theta$$

Show that this function satisfies the differential equation

$$y'' + \frac{1}{x} y' + y = 0$$

Hint: Find $J_0'(x)$ and integrate by parts.

4. Show that

$$u(x) = \int_0^1 g(x,t)f(t)\, dt \qquad \text{where } g(x,t) = \begin{cases} (1 - x)t & 0 \le t \le x \\ (1 - t)x & x \le t \le 1 \end{cases}$$

is a solution of the boundary-value problem

DE: $u'' = -f(x)$

BC: $u(0) = u(1) = 0$

where $f(x)$ is a continuous function.   Hint: $g(x,t)$ is a continuous function in the rectangle $R\colon 0 \le x \le 1,\ 0 \le t \le 1$. $g(x,t)$ has a continuous derivative in R except along the line $x = t$.   In order to apply the above theorem, it is necessary to split up the integral into two parts:

$$u(x) = \int_0^x (1 - x)tf(t)\, dt + \int_x^1 (1 - t)xf(t)\, dt$$

5. Combining the results of Prob. 2 above and Prob. 4 of Sec. 1-6, we have

$$\int_0^x (x - t)f(t)\, dt = \int_0^x \left[\int_0^t f(s)\, ds \right] dt$$

Prove this result directly.

6. Prove that if $g(x,y)$, $g_x(x,y)$, $g_y(x,y)$, and $g_{xy}(x,y)$ are continuous in the rectangle $R\colon \alpha \le x \le \beta,\ a \le y \le b$, then $g_{xy}(x,y) = g_{yx}(x,y)$ in R.   Hint: Set $f(x,y) = g_y(x,y)$; then $g(x,y) = g(x,a) + \int_a^y f(x,t)\, dt$.   Show that $g_{yx}(x,y) = f_x(x,y)$; similarly show that $g_{xy}(x,y) = f_x(x,y)$.

1-8   Elementary and non-elementary functions

In elementary calculus the reader learned to differentiate the so-called elementary functions. These functions are powers, roots, logarithms, exponentials, trigonometric and inverse trigonometric functions, and finite combinations of these functions. The derivative of such an elementary function is again an elementary function. In integration, the process inverse to differentiation, the situation is not so simple. The integral of an elementary function is *not* always an elementary function. In Sec. 1-6 we saw that the integral

$$F(x) = \int_0^x f(t)\,dt$$

exists if $f(x)$ is continuous and that $F'(x) = f(x)$. Therefore, since the elementary functions are all continuous (at least in *some* interval), there is no question that the integral of an elementary function exists; the point is that the integral of an elementary function may be a *new* function, that is, a function different from the rather arbitrarily defined elementary functions.

Certain non-elementary functions defined by integrals appear often in mathematics and science. Several of these are listed below:

Error function: $\text{Erf}(x) = \dfrac{2}{\sqrt{\pi}} \displaystyle\int_0^x e^{-t^2}\,dt$

Sine-integral function: $\text{Si}(x) = \displaystyle\int_0^x \dfrac{\sin t}{t}\,dt$

Exponential-integral function: $\text{Ei}(x) = \displaystyle\int_1^x \dfrac{e^{-t}}{t}\,dt$

Elliptic integral of the first kind: $K(k,\theta) = \displaystyle\int_0^\theta \dfrac{dt}{\sqrt{1 - k^2 \sin^2 t}}$

Elliptic integral of the second kind: $E(k,\theta) = \displaystyle\int_0^\theta \sqrt{1 - k^2 \sin^2 t}\,dt$

The values of all of these functions have been computed† and are available in tables‡ similar to tables of the trigonometric, logarithmic, and exponential functions. We shall have occasion to use some of these functions in our work.

It is by no means an easy task to determine whether or not a given integral can be expressed in terms of elementary functions; certain integrals

† These integrals are computed by numerical methods such as Simpson's rule, infinite series, or more sophisticated methods. In fact, the usual tables of $\sin x$, $\ln x$, etc., are computed in the same way.

‡ E.g., See Jahnke, Emde, and Lösch [20].

which appear often in important problems, such as the integrals shown above, have been thoroughly investigated, and it has been shown that they are non-elementary functions. However, little is known about this matter in general. There is an obvious advantage in knowing that an integral is a non-elementary function: it is not necessary to waste time trying to express the integral in terms of elementary functions.

We have seen that the process of integration, that is, the solution of the differential equation $y' = f(x)$, leads us outside the realm of elementary functions. Therefore it should not be surprising to learn that the solutions of more complicated differential equations also lead to non-elementary functions. Some of these non-elementary functions, such as the Bessel and Legendre functions, occur often in applied problems; we shall study them in Chap. 7.

Problems 1-8

1. Show that Erf $(-x) = -$Erf (x).

2. Show that $u(x,t) = 2$ Erf $(x/2\sqrt{t})$ is a solution of the partial differential equation $u_{xx} = u_t$.

3. Assume that the equation $t = 2K(k, \tfrac{1}{2}\theta)$ determines θ as a differentiable function of t. Show that this function satisfies the equation

$$\dot{\theta}^2 = 1 - \tfrac{1}{2}k^2(1 - \cos\theta)$$

4. The Bessel function of order zero can be defined by the infinite series

$$J_0(x) = 1 - \frac{x^2}{2^2} + \frac{x^4}{2^4(2!)^2} + \cdots + (-1)^k \frac{x^{2k}}{2^{2k}(k!)^2} + \cdots$$

This series converges for all x. Recalling that a power series can be differentiated term by term within its interval of convergence,† show that $J_0(x)$ is a solution of the differential equation

$$xy'' + y' + xy = 0$$

1-9   The gamma function

Another important non-elementary function which will be needed in our later work is the gamma function.‡ This function is defined by the *improper integral*

$$\Gamma(s) = \int_0^\infty x^{s-1}e^{-x}\,dx \tag{43}$$

---

† See Appendix A.

‡ An interesting fact is that the gamma function is not the solution of any differential equation whose coefficients are elementary functions.

This integral can be shown to converge† only for $s > 0$, so that the gamma function is defined by this formula only for positive values of its argument. Later we shall define the gamma function for negative values of its argument.

It is not difficult to compute the value of the gamma function when s is a positive integer. For instance, if $s = 1$,

$$\Gamma(1) = \int_0^\infty e^{-x}\,dx = -e^{-x}\,\Big|_0^\infty = 1 \tag{44}$$

The values of this function for other positive integers can be computed by integrating by parts. Let us proceed in this manner.

$$\Gamma(s) = \int_0^\infty x^{s-1}\,d(-e^{-x}) = -x^{s-1}e^{-x}\,\Big|_0^\infty + \int_0^\infty (s-1)x^{s-2}e^{-x}\,dx$$
$$= 0 + (s-1)\int_0^\infty x^{s-2}e^{-x}\,dx$$

Noting that the integral on the right is $\Gamma(s-1)$, we have the *recurrence formula* or *functional equation*

$$\Gamma(s) = (s-1)\Gamma(s-1) \qquad s > 1 \tag{45}$$

which holds for all values of s greater than 1 whether or not this value is an integer.

If $s = n$, a positive integer, repeated application of this formula yields

$$\Gamma(n) = (n-1)\Gamma(n-1)$$
$$= (n-1)(n-2)\Gamma(n-2)$$
$$\vdots$$
$$= (n-1)(n-2)\cdots 2.1\Gamma(1)$$

and therefore we have the important result

$$\Gamma(n) = (n-1)! \tag{46}$$

Because of this agreement with the factorial function for integral values of s, the gamma function can be thought of as a generalization of $(s-1)!$ when s is not an integer.

The recurrence formula (45) enables us to calculate the value of the gamma function for all $s > 0$ as soon as the values for $1 \le s \le 2$ are known. These values have been computed and are available in tables.‡

We have mentioned that the gamma function is not defined by (43) for negative values of the argument. We now extend the definition to include negative values. Although the integral definition (43) cannot be used for

† That is, $\lim\limits_{b \to \infty} \int_0^b x^{s-1}e^{-x}\,dx$ exists and is a finite number.

‡ E.g., See Jahnke, Emde, and Lösch [20].

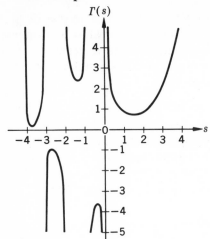

figure 3. *The gamma function* $\Gamma(s)$.

this purpose, the recurrence formula (45) can be used. We rewrite this formula in the form

$$\Gamma(s) = \frac{\Gamma(s+1)}{s} \tag{47}$$

which holds only for $s > 0$. However, we can use (47) to *define* $\Gamma(s)$ for the range

$$-1 < s < 0$$

since the right-hand side of (47) is well-defined for s in this range. For instance, for $s = -\frac{1}{2}$ we have

$$\Gamma(-\tfrac{1}{2}) = \frac{\Gamma(\tfrac{1}{2})}{-\tfrac{1}{2}}$$

Having defined $\Gamma(s)$ for $-1 < s < 0$, we can use (47) again to define $\Gamma(s)$ for $-2 < s < -1$, etc. Thus $\Gamma(s)$ is defined for all values of s except zero and negative integers. At $s = 0$ we have

$$\lim_{s \to 0^+} \Gamma(s) = \lim_{s \to 0^+} \frac{\Gamma(s+1)}{s} = +\infty \tag{48}$$

Similarly the limit as s approaches a negative integer is $\pm\infty$. A graph of the gamma function is shown in Fig. 3.

Problems 1-9

1. Given $\Gamma(1.6) = 0.8935$, evaluate $\Gamma(5.6)$ and $\Gamma(-2.4)$.

2. Evaluate $\Gamma(\tfrac{1}{2})$ by making the substitution $x = y^2$ in the defining equation and using the fact that $\displaystyle\int_0^\infty e^{-y^2}\, dy = \tfrac{1}{2}\sqrt{\pi}.$

3. Show that for n a positive integer,

$$\Gamma(n + \tfrac{1}{2}) = \frac{(2n - 1)(2n - 3) \cdots 3.1}{2^n} \sqrt{\pi}$$

4. Prove that $\displaystyle\int_0^\infty e^{-x^2} x^s \, dx = \tfrac{1}{2}\Gamma(\tfrac{1}{2}(s + 1))$.

5. The Bessel function $J_n(x)$ can be defined by

$$J_n(x) = \frac{x^n}{\sqrt{\pi}\, 2^n \, \Gamma(n + \tfrac{1}{2})} \int_0^\pi \cos(x \cos \theta) \sin^{2n} \theta \, d\theta$$

where $n \geq 0$ (n is not necessarily an integer). Show that:

a. $J_1 = -J_0'$

b. $J_{n+1} = J_{n-1} - 2J_n' \qquad n \geq 1$

c. $J_n'' + \dfrac{1}{x} J_n' + \left(1 - \dfrac{n^2}{x^2}\right) J_n = 0 \qquad n \geq 0$

6. Show that $\Gamma(s) \equiv \displaystyle\int_0^1 \left[\ln\left(\frac{1}{y}\right)\right]^{s-1} dy$.

Appendix

We shall prove the two theorems stated in Sec. 1-7 concerning the continuity and differentiability of an integral with respect to a parameter. Before proceeding with the proofs, we shall review some important definitions and theorems. Readers interested in more details than can be presented here are referred to books on advanced calculus.†

Continuity We consider a function $f(x)$ defined in a closed interval I: $a \leq x \leq b$.

Definition 1 *The function $f(x)$ is continuous at $x = x_0$ in I if for every $\epsilon > 0$, there exists a $\delta > 0$ so that*

$$|f(x) - f(x_0)| < \epsilon$$

provided that $|x - x_0| < \delta$ and $x \in I$.

In other words, $f(x)$ is arbitrarily "close" to $f(x_0)$ provided that x is "sufficiently close" to x_0. If a function $f(x)$ is continuous at every point of an interval I, it is said to be *continuous in the interval I*.

In the above definition we note that the number δ may depend on both ϵ

† See Buck [5] or Courant [12].

and x_0. In case the number δ is independent of the particular point x_0 in I, we say that $f(x)$ is *uniformly continuous* in I.

The following important theorem is usually proved† in advanced calculus:

Theorem 1   *If a function $f(x)$ is continuous in a closed interval, it is uniformly continuous in the closed interval.*

Example 1 $f(x) = x^2$ is uniformly continuous in $0 \leq x \leq 1$. Note that

$$|x^2 - x_0^2| = |x - x_0| \cdot |x + x_0| \leq |x - x_0| \cdot (1 + x_0) < \epsilon$$

provided

$$|x - x_0| < \frac{\epsilon}{1 + x_0}$$

If we pick $\delta \leq \epsilon/(1 + x_0)$, we see that $|x - x_0| < \delta$ implies $|x^2 - x_0^2| < \epsilon$. This shows that $f(x) = x^2$ is continuous at x_0. Since δ depends on x_0, it does not show uniform continuity. However, since x_0 can be no greater than 1, we see that $\frac{1}{2}\epsilon \leq \epsilon/(1 + x_0)$. Therefore if we pick $\delta \leq \frac{1}{2}\epsilon$, which is independent of x_0, it is still true that $|x - x_0| < \delta$ implies $|x^2 - x_0^2| < \epsilon$.

Example 2 $f(x) = 1/x$ is continuous at every point in $0 < x \leq 1$ but not uniformly continuous in this interval.

Let $|x - x_0| < \delta$ where $\delta > 0$ and x and x_0 lie in $0 < x \leq 1$. Then $x_0 - \delta < x < x_0 + \delta$ and

$$\left| \frac{1}{x} - \frac{1}{x_0} \right| = \frac{|x - x_0|}{xx_0} < \frac{\delta}{(x_0 - \delta)x_0}$$

provided that we pick δ small enough so that $x_0 - \delta > 0$. Now if we can find δ such that

$$\frac{\delta}{(x_0 - \delta)x_0} < \epsilon$$

for any given $\epsilon > 0$, we will have shown that $1/x$ is continuous at x_0. However, the above inequality is easily solved to yield

$$\delta < \frac{\epsilon x_0^2}{1 + \epsilon x_0}$$

Any such value of δ will ensure that $|1/x - 1/x_0| < \epsilon$. Thus we have shown continuity. We note, however, that as x_0 is taken closer to zero, δ must be taken closer to zero so that the above approach does not yield a value of $\delta > 0$, which will hold for all x_0 in $0 < x_0 \leq 1$. In fact, no such δ exists. For if $|x - x_0| < \delta$, then $x_0 - \delta < x < x_0 + \delta$.

$$\left| \frac{1}{x} - \frac{1}{x_0} \right| = \frac{|x - x_0|}{xx_0} > \frac{|x - x_0|}{(x_0 + \delta)x_0}$$

† See Courant [12], vol. 1, p. 65.

Now pick x so that $|x - x_0| = \frac{1}{2}\delta$; then

$$\left| \frac{1}{x} - \frac{1}{x_0} \right| > \frac{\frac{1}{2}\delta}{(x_0 + \delta)x_0}$$

It is clear that no matter how small δ is chosen, we can find an x_0 small enough so that the right-hand side of the above is greater than 1. Therefore $f(x) = 1/x$ is not uniformly continuous in $0 < x \leq 1$. This example shows that the assumption of a closed interval in Theorem 1 is essential.†

The foregoing remarks are readily extended to a function of two independent variables. If $f(x,y)$ is defined in the closed rectangle $R: a \leq x \leq b, c \leq y \leq d$, we have the following definitions.

Definition 2 $f(x,y)$ *is continuous at a point (x_0,y_0) in R if for every $\epsilon > 0$ there exists a $\delta > 0$ so that*

$$|f(x,y) - f(x_0,y_0)| < \epsilon$$

whenever $|x - x_0| < \delta$ and $|y - y_0| < \delta$.

The function $f(x,y)$ is said to be *continuous in R* if it is continuous at every point of R. If the number δ in the above definition is independent of x_0 and y_0, the function is said to be *uniformly continuous in R*. Finally we have the following theorem.

Theorem 2‡ *If $f(x,y)$ is continuous in a closed rectangle R, it is uniformly continuous in R.*

The mean-value theorem Another important theorem of calculus is the *mean-value theorem.*§ *If $f(x)$ is continuous in $a \leq x \leq b$ and if $f'(x)$ exists in $a < x < b$, then there exists at least one point $\bar{x}, a < \bar{x} < b$, such that*

$$\frac{f(b) - f(a)}{b - a} = f'(\bar{x})$$

This theorem is easy to interpret geometrically. It states (see Fig. 4) that if $f(x)$ satisfies the conditions of the theorem, there exists some point $\bar{x}$ where the slope of the curve is equal to the slope of the chord joining the end points.

We are now prepared to prove Theorems 1 and 2 of Sec. 1-7.

Theorem *If $f(x,t)$ is continuous in a rectangle $R: \alpha \leq x \leq \beta, a \leq t \leq b$, then*

$$F(x) = \int_a^b f(x,t)\, dt$$

is continuous for $\alpha \leq x \leq \beta$.

† Of course, particular functions that are continuous in an open (or half-open) interval may be uniformly continuous, but the example shows this is not generally true.

‡ Courant [12], vol. 2, p. 97.

§ Courant [12], vol. 1, p. 102.

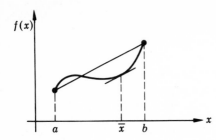

figure 4.   *Mean-value theorem.*

Proof   Let x_0 be any point in $\alpha \le x \le \beta$; we have

$$|F(x) - F(x_0)| = \left| \int_a^b [f(x,t) - f(x_0,t)]\,dt \right| \le \int_a^b |f(x,t) - f(x_0,t)|\,dt\dagger$$

However since $f(x,t)$ is continuous in the closed rectangle R, it is uniformly continuous in R.   Therefore we can make

$$|f(x,t) - f(x_0,t)| < \epsilon$$

by taking x close enough to x_0, and this estimate holds for all t in R.   We then have

$$|F(x) - F(x_0)| \le \epsilon(b - a)$$

Thus $F(x)$ is continuous and the proof is complete.

Theorem   *If $f(x,t)$ and $f_x(x,t)$ are continuous in R, then*

$$F'(x) = \frac{d}{dx} \int_a^b f(x,t)\,dt = \int_a^b f_x(x,t)\,dt$$

Proof   The derivative of any function $F(x)$ at $x = x_0$ is defined by

$$F'(x_0) = \lim_{x \to x_0} \frac{F(x) - F(x_0)}{x - x_0}$$

or equivalently

$$\left| F'(x_0) - \frac{F(x) - F(x_0)}{x - x_0} \right| < \epsilon$$

(for $0 < |x - x_0| < \delta$).

Since $F(x) = \int_a^b f(x,t)\,dt$, we have

$$\frac{F(x) - F(x_0)}{x - x_0} = \int_a^b \frac{f(x,t) - f(x_0,t)}{x - x_0}\,dt$$

$\dagger$ If $g(t)$ is integrable, then $\left| \int_a^b g(t)\,dt \right| \le \int_a^b |g(t)|\,dt$; see Buck [5], p. 105.

Applying the mean-value theorem to the integrand on the right-hand side, we have

$$\frac{F(x) - F(x_0)}{x - x_0} = \int_a^b f_x(\bar{x},t)\,dt$$

where $\bar{x}$ is between x and x_0.   Now we would like to prove that

$$F'(x_0) = \int_a^b f_x(x_0,t)$$

Therefore we must show that

$$\left| F'(x_0) - \frac{F(x) - F(x_0)}{x - x_0} \right| = \left| \int_a^b [f_x(x_0,t) - f_x(\bar{x},t)]\,dt \right|$$

can be made arbitrarily small.   We have

$$\left| \int_a^b [f_x(x_0,t) - f_x(\bar{x},t)]\,dt \right| \le \int_a^b |f_x(x_0,t) - f_x(\bar{x},t)|\,dt$$

Since $f_x(x,t)$ is *uniformly* continuous, the argument used in the proof of Theorem 1 can be applied to the right-hand side of the last expression to complete the proof of Theorem 2.

Special Methods for First-Order Equations

2

2-1 Introduction

Many first-order differential equations can be solved by integration, provided that the integrations are strategically performed. Equations of this type are the subject of the present chapter. We shall be concerned primarily with equations of the first order and first degree, that is, equations of the form

$$M(x,y) + N(x,y)y' = 0 \tag{1}$$

It is often convenient to write (1) in the symmetric differential form

$$M(x,y)\, dx + N(x,y)\, dy = 0 \tag{2}$$

We shall agree in the sequel that (2) is just another way of writing (1).

2-2 Separation of variables

A differential equation is said to have its *variables separated* if it is in the form

$$A(x) + B(y)y' = 0 \tag{3}$$

or in the equivalent form

$$A(x)\, dx + B(y)\, dy = 0 \tag{4}$$

We shall assume that $A(x)$ and $B(y)$ are continuous functions to ensure that certain integrations can be performed. The reader is undoubtedly familiar with the process of solving Eq. (4) by simply integrating both sides to obtain Eq. (7) below. However, in order that the process be thoroughly understood, we shall proceed more carefully. We are looking for a solution $y(x)$ which makes the differential equation (3) an identity in x:

$$A(x) + B(y(x))y'(x) = 0 \tag{5}$$

Assuming a solution exists, both sides of the identity (5) may be integrated with respect to x to give

$$\int A(x)\, dx + \int B(y(x))y'(x)\, dx = c \tag{6}$$

where c is an arbitrary constant. The second integral in the above equation can be rewritten using the usual method of substitution† in an integral. We let $y = y(x)$ and $dy = y'(x)\, dx$ and obtain

$$\int A(x)\, dx + \int B(y)\, dy = c \tag{7}$$

Equation (7) can be thought of as the result of integrating the first term of Eq. (4) with respect to x and the second term with respect to y (this process might be called "implicit integration"). We have shown that *every* solution of Eq. (3) must satisfy (7). Conversely, we must show that if the implicit equation (7) determines a differentiable function $y(x)$, then this $y(x)$ satisfies the differential equation (3). By implicit differentiation we have

$$\frac{d}{dx}\int A(x)\, dx + \frac{d}{dx}\int B(y)\, dy = \frac{dc}{dx}$$

or since

$$\frac{d}{dx}\int A(x)\, dx = A(x)$$

and

$$\frac{d}{dx}\int B(y)\, dy = \frac{d}{dy}\left[\int B(y)\, dy\right]\frac{dy}{dx} = B(y)\frac{dy}{dx}$$

the above reduces to

$$A(x) + B(y)y' = 0$$

which is Eq. (3). Therefore Eq. (7) represents the general solution of the differential equation (3).

We now consider the initial-value problem

$$\begin{aligned} &\text{DE: } A(x)\, dx + B(y)\, dy = 0 \\ &\text{IC: } y(x_0) = y_0 \end{aligned} \tag{8}$$

One way of solving this problem is to find the general solution (7) and then evaluate the constant so that the initial condition is satisfied. It is just as

---

† See Courant [12], vol. 1, p. 207.

easy to integrate the differential equation between the initial point (x_0, y_0) and the general point (x,y) to obtain the solution†

$$\int_{x_0}^{x} A(t)\, dt + \int_{y_0}^{y} B(t)\, dt = 0 \qquad (9)$$

Example 1   Solve: DE: $\sin x\, dx + y\, dy = 0$

$$\text{IC: } y(0) = 1$$

The solution is given by

$$\int_{0}^{x} \sin x\, dx + \int_{1}^{y} y\, dy = 0$$

or

$$-\cos x \Big|_0^x + \tfrac{1}{2} y^2 \Big|_1^y = 0$$

Therefore we obtain the solution as the implicit function

$$y^2 = 2 \cos x - 1$$

To find the x interval for which the solution is defined, we solve the above equation for y to obtain

$$y = +\sqrt{2 \cos x - 1} \qquad -\tfrac{1}{3}\pi < x < \tfrac{1}{3}\pi$$

The restriction on x is necessary for the square root to have meaning; the points $x = \pm \tfrac{1}{3}\pi$ are not included since the derivative of y does not exist at these points. Note also that the positive square root is taken to be consistent with $y(0) = 1$.

Example 2   Solve: $x\, dy + y\, dx = 0$

We note by inspection that $y \equiv 0$ is a solution. To find other solutions, we separate variables by dividing by the product xy to obtain

$$\frac{dy}{y} + \frac{dx}{x} = 0$$

We must of course be careful not to divide by zero. We therefore restrict ourselves to an interval in which x is not zero, say $x > 0$. We must also *assume* that the solution $y(x)$ is not zero for $x > 0$. With these assumptions we can integrate the differential equation to obtain

$$\ln |y| + \ln |x| = \ln c‡$$

---

† The solution may not be unique since Eq. (9) is an implicit equation for $y(x)$ and may determine more than one function $y(x)$.

‡ The formula $\int dy/y = \ln |y|$ covers both the case $y > 0$ and $y < 0$; for if $y > 0$, $\int dy/y = \ln y = \ln |y|$, and if $y < 0$, $\int dy/y = \int -dy/(-y) = \ln (-y) = \ln |y|$.

The constant of integration could have been written as c_1 instead of $\ln c$. However, for every real number c_1, there exists a positive number c such that $c_1 = \ln c$.

where c is a positive constant. This result can be simplified:

$$\ln |xy| = \ln c$$

and therefore

$$|xy| = c \quad \text{or} \quad xy = \pm c$$

Setting $k = \pm c$, we obtain the solutions

$$y = \frac{k}{x} \qquad \begin{matrix} x > 0 \\ k \neq 0 \end{matrix}$$

If $k = 0$, the above derivation does not hold; however, in this case $y \equiv 0$, and this is already known to be a solution. Therefore $y = k/x$ is a solution for all values of k.

In our derivation it was necessary to assume that $y(x)$ was never zero, and it was shown independently that $y \equiv 0$ was a solution; the possibility still remains that solutions exist which are zero for some, but not all, values of $x > 0$. In other words, we cannot be sure that the general solution has been found. For this simple example this question can be settled by a different approach which avoids the process of division. The expression $x\,dy + y\,dx$ is in the familiar form of the differential of a product:

$$d(xy) = x\,dy + y\,dx$$

This is an identity which holds if y is any differentiable function of x. Therefore the differential equation can be written

$$d(xy) = 0$$

If $y(x)$ is any solution of this equation, we obtain

$$xy = k \qquad \text{or} \qquad y = \frac{k}{x}$$

if $x > 0$. Thus *every* solution of the differential equation in the interval $x > 0$ must be given by $y = k/x$.

Problems 2-2

1. Solve the following differential equations. The solutions should be put into as simple a form as possible and checked by substitution into the differential equations. It is not necessary (and it is not forbidden) to find the interval for which the solution is defined or to prove that the general solution has been obtained.

a. $x^2\,dx + y\,dy = 0$

b. $x(1 + y^2)\,dx + y(1 + x^2)\,dy = 0$

c. $\dfrac{du}{dv} = \dfrac{4uv}{v^2 + 1}$

d. $\sin 2x\,dx + \cos y\,dy = 0$

e. $x\,dx + (1 + x)\,dy = 0$

f. $\dfrac{ds}{dt} = \dfrac{3t + ts^2}{s + t^2 s}$

2. Solve: DE: $x^2(1 + y^2)\,dx + 2y\,dy = 0$

   IC: $y(0) = 1$

Determine the x interval for which the solution is defined.

3. Solve: $x\,dy + y\,dx = y\,dy$
4. Does the equation $x^2\,dx + y\,dy = 3$ make sense?   Explain.
5. Explain why the general first-order equation

$$M(x,y)\,dx + N(x,y)\,dy = 0$$

cannot be solved by separation of variables.   Give an example of such an equation.

6. Solve $y' = f(y/x)$ by making the substitution $y = vx$ and showing that the differential equation in the variables v and x can be solved by separation of variables.

2-3 The first-order linear differential equation

A particularly important type of first-order equation is the linear equation

$$a(x)\frac{dy}{dx} + b(x)y = c(x) \tag{10}$$

If $a(x)$ is not zero in some interval, we may divide by $a(x)$ to obtain

$$\frac{dy}{dx} + P(x)y = Q(x) \tag{11}$$

where we have written

$$P(x) = \frac{b(x)}{a(x)} \qquad Q(x) = \frac{c(x)}{a(x)}$$

We shall restrict ourselves to differential equations in the form (11) and, in order to be able to perform certain integrations, we shall assume that $P(x)$ and $Q(x)$ are continuous functions† in some common interval.

We shall first treat the associated homogeneous equation that is obtained from (11) by putting $Q(x) \equiv 0$:

$$\frac{dy}{dx} + P(x)y = 0 \tag{12}$$

---

† Since P and Q are continuous, and the solution y is continuous, we see from (11) that dy/dx is continuous.   Therefore the solutions of (11) possess not only derivatives but also *continuous* derivatives.

One obvious solution of this equation is $y \equiv 0$; this is called the *trivial solution*. Nontrivial solutions can be found by separating the variables:

$$\frac{dy}{y} + P(x)\, dx = 0$$

We must, of course, *assume* that the solution $y(x)$ is never zero (this restriction will be removed later). Proceeding in the usual manner, we obtain successively

$$\ln |y| + \int P(x)\, dx = c$$
$$\ln |y| = c - \int P(x)\, dx$$
$$|y| = e^{c - \int P(x)\, dx} = e^c e^{-\int P(x)\, dx}$$
$$y = \pm e^c e^{-\int P(x)\, dx}$$

Letting $k = \pm e^c$, we obtain the solutions

$$y = k e^{-\int P(x)\, dx} \tag{13}$$

In this derivation the constant k must be different from zero; however, if $k = 0$ in Eq. (13), we obtain $y \equiv 0$ which is already known to be a solution. Therefore (13) provides solutions for all values of k; we shall see below that this is actually the general solution.

To solve the nonhomogeneous equation

$$\frac{dy}{dx} + P(x)y = Q(x) \tag{14}$$

we shall use a method called *variation of parameters*. This is a general method which enables the solution of a nonhomogeneous linear equation (of any order) to be obtained from the general solution of the corresponding homogeneous equation. We have seen that Eq. (13) provides a solution of the *homogeneous* differential equation (12) for any *constant* value of k; we now try to solve the *nonhomogeneous* equation (14) by allowing k to *vary* with x. That is, we set $k = v(x)$ and seek to determine $v(x)$ so that

$$y = v(x) e^{-\int P(x)\, dx} \tag{15}$$

is a solution of the nonhomogeneous equation (14). Substituting (15) into the differential equation (14), we find that $v(x)$ must satisfy

$$v(x) \frac{d}{dx} e^{-\int P(x)\, dx} + e^{-\int P(x)\, dx} \frac{dv}{dx} + P(x)v(x)e^{-\int P(x)\, dx} = Q(x) \tag{16}$$

However,

$$\frac{d}{dx} e^{-\int P(x)\, dx} = e^{-\int P(x)\, dx}(-P(x))$$

and hence (16) may be simplified to obtain

$$\frac{dv}{dx} = Q(x)e^{\int P(x)\,dx} \tag{17}$$

Since the right-hand side is a function of x alone, we can integrate to obtain

$$v = \int Q(x)e^{\int P(x)\,dx}\,dx + c \tag{18}$$

Substituting for v from (15), we obtain

$$y = ce^{-\int P(x)\,dx} + e^{-\int P(x)\,dx}\int Q(x)e^{\int P(x)\,dx}\,dx \tag{19}$$

If we examine our derivation carefully, we find that *every* solution of Eq. (14) must satisfy (19); furthermore the steps are clearly reversible so that (19) satisfies the differential equation (14). Thus we have found the *general* solution of the differential equation.† We also note that the solution is defined in the entire interval in which $P(x)$ and $Q(x)$ are continuous.

The reader should not memorize the complicated formula (19) but should use the method of variation of parameters in each case.

Example 1   Solve: $y' - \dfrac{1}{x}y = x^3$

We restrict ourselves to an interval in which both $-1/x$ and x^3 are continuous, say the interval $x > 0$. The homogeneous equation $y' - y/x = 0$ can be solved by separation of variables to yield $y = cx$. To solve the nonhomogeneous equation, we assume $y = v(x) \cdot x$. Substituting into the differential equation and simplifying, we obtain

$$v' = x^2$$

Therefore $v = c + \frac{1}{3}x^3$ and the general solution is

$$y = \tfrac{1}{3}x^4 + cx$$

Method of the integrating factor   It is instructive to consider the solution of the linear equation from a somewhat different point of view. The method of variation of parameters can be considered as a change of dependent variables from y to v as given by Eq. (15). This equation can be rewritten

$$v = ye^{\int P(x)\,dx} \tag{20}$$

The method is successful because, as was seen above, v satisfies the simple differential equation

$$\frac{dv}{dx} = Q(x)e^{\int P(x)\,dx} \tag{21}$$

---

† If we set $Q(x) \equiv 0$ in Eq. (19), we see that $y = ce^{-\int P(x)\,dx}$ actually is the *general* solution of the homogeneous equation $y' + P(x)y = 0$.

Substituting (20) into (21), we have

$$\frac{d}{dx}\left(ye^{\int P(x)\,dx}\right) = Q(x)e^{\int P(x)\,dx} \tag{22}$$

or, if we compute the derivative on the left-hand side, we obtain

$$e^{\int P(x)\,dx}\left[\frac{dy}{dx} + P(x)y\right] = e^{\int P(x)\,dx}Q(x) \tag{23}$$

This equation is simply the original differential equation

$$\frac{dy}{dx} + P(x)y = Q(x) \tag{24}$$

multiplied through by the nonzero factor

$$I(x) = e^{\int P(x)\,dx} \tag{25}$$

which is called an *integrating factor*. Following these steps in reverse order, we obtain a simple method for solving the linear equation (24).

Method   *To solve*

$$y' + P(x)y = Q(x)$$

multiply both sides by the integrating factor $I = e^{\int P\,dx}$. *The left-hand side can then be written as the derivative of the product of y times the integrating factor:*

$$\frac{d}{dx}\left(ye^{\int P\,dx}\right) = e^{\int P\,dx}Q$$

and the solution can be obtained by integrating both sides.

Example 2   Solve: $\dfrac{dy}{dx} - \dfrac{2y}{x} = x^5$ $x > 0$

The integrating factor is

$$I(x) = e^{-\int 2/x\,dx} = e^{-2\ln x} = e^{\ln x^{-2}} = x^{-2}\dagger$$

Multiplying the differential equation by this factor, we obtain

$$x^{-2}\left(\frac{dy}{dx} - \frac{2y}{x}\right) = x^3 \tag{i}$$

or

$$\frac{d}{dx}(yx^{-2}) = x^3 \tag{ii}$$

† The identity $e^{\ln x^{-2}} = x^{-2}$ comes from the basic property of a logarithm: $\ln u$ is the exponent to which e must be raised to obtain u; therefore $e^{\ln u} = u$.

It is always worthwhile to check Eq. (ii) by performing the differentiation on the left to see if Eq. (i) is obtained.   Integrating (ii), we obtain the solution

$$yx^{-2} = \tfrac{1}{4}x^4 + c$$

or

$$y = \tfrac{1}{4}x^6 + cx^2$$

The method of integrating factors provides a convenient way of solving the initial-value problem

DE: $y' + P(x)y = Q(x)$

IC: $y(x_0) = y_0$ (26)

For this purpose we use the integrating factor

$$I(x) = \exp\left[\int_{x_0}^{x} P(t)\, dt\right]$$

Proceeding as before, we multiply the differential equation by this factor to obtain

$$\frac{d}{dx}[y(x)I(x)] = Q(x)I(x)$$

Integrating from x_0 to x, we obtain

$$y(x)I(x) - y_0 = \int_{x_0}^{x} Q(s)I(s)\, ds$$

since $I(x_0) = 1$.   Substituting for $I(x)$ and solving for y, we obtain

$$y = y_0 \exp\left[-\int_{x_0}^{x} P(t)\, dt\right] + \exp\left[-\int_{x_0}^{x} P(t)\, dt\right]$$
$$\times \int_{x_0}^{x} Q(s) \exp\left[\int_{x_0}^{s} P(t)\, dt\right] ds \qquad (27)$$

This is the *unique* solution of the initial-value problem.   There is of course no reason to memorize this formula; it is a simple matter to work through the method in each particular case.

Example 3   DE: $y' - 2xy = 1$

IC: $y(0) = 1$

We have

$$I(x) = \exp\left(\int_0^x -2x\, dx\right) = e^{-x^2}$$

Multiplying the differential equation by this factor, we obtain

$$\frac{d}{dx}(ye^{-x^2}) = e^{-x^2}$$

Integrating from $x = 0$ to x, we obtain

$$ye^{-x^2} - 1 = \int_0^x e^{-t^2}\, dt$$

If we recall the definition of the non-elementary† function

$$\text{Erf}\,(x) = \frac{2}{\sqrt{\pi}} \int_0^x e^{-t^2}\, dt$$

we see that the solution can be written

$$y = e^{x^2} + \tfrac{1}{2} e^{x^2} \sqrt{\pi}\; \text{Erf}\,(x)$$

Properties of solutions   The following properties of solutions of homogeneous and nonhomogeneous linear equations will be important for later work with higher-order equations.   The reader should have no trouble in verifying these for the first-order case considered above.

1. If $z(x)$ is any nontrivial solution of the homogeneous equation $y' + Py = 0$, then the general solution is $y_h = cz(x)$, where c is an arbitrary constant.

2. If y_p is a particular solution of the nonhomogeneous equation $y' + Py = Q$ and y_h is the general solution of the corresponding homogeneous equation, then the general solution of the nonhomogeneous equation is $y = y_h + y_p$.

We see from these properties that if a particular solution of the non-homogeneous equation can be found by inspection, then the problem is reduced to the solution of the less complicated homogeneous equation. Furthermore, if a nonzero solution of the homogeneous equation can also be found, the general solution of the differential equation can be written down at once.

Example 4   Solve: $y' - 3y = 15$

It is reasonable to look for a particular solution in the form $y_p = k$ (a constant). Substituting in the differential equation, we find

$$y_p = -5$$

The general solution of the homogeneous equation is easily found to be

$$y_h = ce^{3x}$$

Therefore the general solution is

$$y = ce^{3x} - 5$$

† See Sec. 1-8, page 27.

Problems 2-3

1. Find the general solution of:

a. $y' + y = x$

b. $y' + \dfrac{y}{x+1} = x^2$

c. $\dfrac{ds}{dt} - \dfrac{3s}{t} = 5$

d. $y' + y \tan x = \sin 2x$

e. $2y \dfrac{dx}{dy} = x - y + y^3$

f. $y' + y \sin x = kx$

2. Solve:

a. DE: $y' + y = e^{-x}$

b. DE: $y' + y \sin x = kx$

   IC: $y(-1) = 3$

   IC: $y(0) = 0$

3. A certain first-order linear differential equation has a solution $y = x^2$. The associated homogeneous equation has a solution $y = x + 3$. What is the solution to the differential equation with initial conditions $y(2) = 1$?

4. Show that if y_1 and y_2 are particular solutions of the linear equation

$$y' + P(x)y = Q(x)$$

then $y_1 - y_2$ is a solution of the corresponding homogeneous equation. Also show that the general solution of the nonhomogeneous equation is

$$y = y_1 + c(y_1 - y_2)$$

provided $y_1 - y_2 \neq 0$.

5. Prove properties 1 and 2, page 45.

6. Find a particular solution by inspection and then find the general solution:

a. $y' - 2y = 10$

b. $y' - 2y = x + 5$

c. $y' - 2y = e^{3x} + 1$

d. $y' - 2y = e^{2x}$

7. Find the continuous function $p(r)$ which satisfies the integral equation

$$p(r) = a + \int_0^r t^2 p(t)\, dt$$

Hint: Change to a differential equation. Note that the integral equation determines the initial condition.

8. Find the general solution of $y'' - 2y' = 5$. Hint: Let $y' = p$.

9. Find the general solution of:

$$\frac{d^{n+1}y}{dx^{n+1}} - 2 \frac{d^n y}{dx^n} = 5$$

10. If $f(x)$ is a differentiable function satisfying the functional equation

$$f(x + y) = f(x)f(y)$$

show that either $f(x) \equiv 0$ or $f(x) = e^{ax}$.   Hint: Keep x fixed and differentiate with respect to y; then set $y = 0$.

2-4 Exact differential equations

The differential expression

$$M(x,y)\,dx + N(x,y)\,dy \tag{28}$$

is called an exact differential if and only if there exists a function $f(x,y)$ that is differentiable† in some region R of the xy plane and such that

$$\frac{\partial f}{\partial x}(x,y) \equiv M(x,y) \qquad \frac{\partial f}{\partial y}(x,y) \equiv N(x,y) \tag{29}$$

for all (x,y) in R.   We recall that the differential of $f(x,y)$ is

$$df(x,y) = \frac{\partial f}{\partial x}\,dx + \frac{\partial f}{\partial y}\,dy‡ \tag{30}$$

Therefore (28) is exact if and only if there exists a differentiable function $f(x,y)$ such that

$$df(x,y) = M(x,y)\,dx + N(x,y)\,dy \tag{31}$$

For example, the differential expression $x\,dy + y\,dx$ is exact since $d(xy) = y\,dx + x\,dy$.

The differential equation

$$M(x,y)\,dx + N(x,y)\,dy = 0 \tag{32}$$

is called an *exact differential equation* if the left-hand side is an exact differential.   If (32) is an exact differential equation, it can be rewritten as

$$df(x,y) = 0 \tag{33}$$

It is now a simple matter to solve the differential equation.   If $y(x)$ is a solution, then Eq. (33) becomes an identity in x which may be integrated to yield

$$f(x,y) = c \tag{34}$$

Conversely, if (34) defines y as a differentiable function of x, this function is a solution of the differential equation (32), for (34) may be differentiated by

---

† Recall that a function $f(x,y)$ is differentiable if $\partial f/\partial x$, $\partial f/\partial y$ exist and are *continuous*.   See Courant [12], vol. 2, pp. 60–62.

‡ It is important to note that (30) holds also if x and y are differentiable functions of other variables.

the chain rule to give

$$\frac{df}{dx} = \frac{\partial f}{\partial x} + \frac{\partial f}{\partial y}\frac{dy}{dx} = 0 \tag{35}$$

Since $f_x = M$ and $f_y = N$, this reduces to

$$M + N\frac{dy}{dx} = 0$$

which is equivalent to the differential equation (32).

The following theorem provides a simple test to determine whether or not a differential expression is exact.   In the process of proving the theorem we also obtain a method of solving an exact differential equation.

Theorem 1   *If $M(x,y)$ and $N(x,y)$ are continuous functions and have continuous partial derivatives in some rectangle R[†] of the xy plane, then the expression $M(x,y)\,dx + N(x,y)\,dy$ is an exact differential if and only if $M_y = N_x$ throughout R.*

Proof   If $M\,dx + N\,dy$ is exact, there exists a function f for which

$$\frac{\partial f}{\partial x} = M \qquad \frac{\partial f}{\partial y} = N$$

By differentiating the first of these equations with respect to y and the second equation with respect to x, we obtain

$$\frac{\partial M}{\partial y} = \frac{\partial^2 f}{\partial y\,\partial x} \qquad \frac{\partial N}{\partial x} = \frac{\partial^2 f}{\partial x\,\partial y}$$

However, since the first partial derivatives of M and N are continuous, the mixed second partial derivatives of f must be continuous and therefore equal.[‡]   It follows that

$$\frac{\partial M}{\partial y} = \frac{\partial N}{\partial x}$$

This proves the first part of the theorem.

To complete the proof, we assume that $M_y = N_x$ and prove that the differential equation is exact.   That is, a function f must be found satisfying

$$\frac{\partial f}{\partial x} = M \qquad \frac{\partial f}{\partial y} = N$$

---

† We take a rectangular region for simplicity.   The region could be more general.   See Courant [12], vol. 2, pp. 353 et seq.

‡ See Sec. 1-7, Prob. 6, page 26.

A function f will satisfy the first of these relations if and only if

$$f(x,y) = \int M(x,y)\, dx + g(y)$$

where $g(y)$ is an arbitrary function. We now determine the function $g(y)$ so that $f_y = N$. We have

$$\frac{\partial f}{\partial y} = \frac{\partial}{\partial y}[\int M(x,y)\, dx] + g'(y) = N(x,y)$$

and therefore

$$g'(y) = N(x,y) - \frac{\partial}{\partial y}\int M(x,y)\, dx \qquad\qquad\qquad \text{(i)}$$

The function $g(y)$ can be determined from this equation by integration provided that the right-hand side is independent of x. This will be the case if the partial derivative of the right-hand side with respect to x is identically zero. Computing this derivative, we find

$$\frac{\partial}{\partial x}\left[N - \frac{\partial}{\partial y}\int M(x,y)\, dx\right] = \frac{\partial N}{\partial x} - \frac{\partial}{\partial x}\frac{\partial}{\partial y}\int M(x,y)\, dx$$

but

$$\frac{\partial}{\partial x}\left[\frac{\partial}{\partial y}\int M(x,y)\, dx\right] = \frac{\partial}{\partial y}\left[\frac{\partial}{\partial x}\int M(x,y)\, dx\right] = \frac{\partial M}{\partial y}$$

and therefore, remembering $N_x \equiv M_y$, we have

$$\frac{\partial}{\partial x}\left[N - \frac{\partial}{\partial y}\int M(x,y)\, dx\right] = \frac{\partial N}{\partial x} - \frac{\partial M}{\partial y} = 0$$

This shows that $g(y)$ is determined by Eq. (i) and the proof is complete.

    Example 1   Solve: $(3x^2y^2 + x^2)\, dx + (2x^3y + y^2)\, dy = 0$

Since $M = 3x^2y^2 + x^2$ and $N = 2x^3y + y^2$, we find

$$M_y = 6x^2y \qquad N_x = 6x^2y$$

and therefore the equation is exact. To find the function f, we set

$$f(x,y) = \int (3x^2y^2 + x^2)\, dx + g(y) = x^3y^2 + \tfrac{1}{3}x^3 + g(y)$$

This already satisfies $f_x = M$. In order to satisfy $f_y = N$, we set

$$\frac{\partial f}{\partial y} = 2x^3y + g'(y) = 2x^3y + y^2$$

This yields

$$g'(y) = y^2$$
$$g(y) = \tfrac{1}{3}y^3 + C$$

and the function f is

$$f(x,y) = x^3y^2 + \tfrac{1}{3}x^3 + \tfrac{1}{3}y^3 + C$$

The solution of the differential equation is finally

$$x^3y^2 + \tfrac{1}{3}x^3 + \tfrac{1}{3}y^3 + C = 0$$

Example 2   Solve $y \cos x \, dx + \sin x \, dy = 0$. By inspection we see that

$$d(y \sin x) = y \cos x \, dx + \sin x \, dy$$

therefore the solution is

$$y \sin x = C$$

Line integrals   It is well to examine exact differentials from another point of view. Let $M(x,y)$ and $N(x,y)$ be continuous functions in some rectangle R, and let C be the curve $y = h(x)$ with initial point (x_0,y_0) and final point (x,y) lying entirely within R; then by the *line integral*

$$\int_C M(x,y) \, dx + N(x,y) \, dy \tag{36}$$

we mean the ordinary integral obtained by replacing y by $h(x)$ in (36), namely,†

$$\int_C M(x,y) \, dx + N(x,y) \, dy = \int_{x_0}^x [M(x,h(x)) + N(x,h(x))h'(x)] \, dx \tag{37}$$

In general this integral will depend for its value not only on the end points of the curve but also on the path or curve C connecting the end points. In the special case when $M \, dx + N \, dy$ is an exact differential, the line integral is *independent* of the particular curve and depends only on the end points. This is easily seen, for in this case there exists a function f with the property that $df = M \, dx + N \, dy$. Therefore

$$\int_C M \, dx + N \, dy = \int_C df(x,y) = \int_{x_0}^x \frac{df}{dx}(x,y(x)) \, dx = f(x,y) - f(x_0,y_0) \tag{38}$$

and the line integral depends only on the end points (x,y) and (x_0,y_0).

Since the line integral is independent of path, we can evaluate the integral by taking *any* path connecting the initial point (x_0,y_0) with the terminal point (x,y). A convenient choice is the rectangular path shown in Fig. 1,

† If the curve C is given in the parametric form $x = g(t)$, $y = h(t)$, or in the form $x = g(y)$, the line integral can be evaluated in a similar manner. This is a consequence of the rule for changing variables in an integral.

We should mention that a line integral can also be defined directly as a limit of a sum. For details see Courant [12], vol. 2, p. 344.

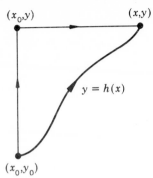

(x_0,y) (x,y)

$y = h(x)$

(x_0,y_0) **figure 1**

consisting of the vertical line connecting (x_0,y_0) and (x_0,y) followed by the horizontal line from (x_0,y) to (x,y). Since $dx = 0$ along the vertical portion and $dy = 0$ along the horizontal portion, we obtain

$$\int_C M\, dx + N\, dy = \int_{y_0}^{y} N(x_0,y)\, dy + \int_{x_0}^{x} M(x,y)\, dx \tag{39}$$

We now apply these considerations to the solution of the initial-value problem for an *exact* differential equation:

DE: $M\, dx + N\, dy = 0$

IC: $y(x_0) = y_0$ \hfill (40)

Integrating both sides of the differential equation along a *solution curve C* from the initial point (x_0,y_0) to (x,y), we obtain

$$\int_C M\, dx + N\, dy = 0 \tag{41}$$

However, since the differential equation is exact, the line integral is independent of path and can be evaluated using Eq. (39) to yield

$$\int_{x_0}^{x} M(x,y)\, dx + \int_{y_0}^{y} N(x_0,y)\, dy = 0 \tag{42}$$

This solves the initial-value problem.

We can also write the general solution of the exact differential equation in (40) in a form similar to (42). According to (38), the line integral is itself a function f such that $df = M\, dx + N\, dy$, and therefore the general solution is

$$\int_{x_0}^{x} M(x,y)\, dx + \int_{y_0}^{y} N(x_0,y)\, dy = C \tag{43}$$

where x_0 and y_0 can be picked arbitrarily. In particular it is often possible to pick x_0 so that the second integral in (43) is simplified.

Example 3   Solve: DE: $(3x^2 + 2xy^2)\, dx + (2x^2y)\, dy = 0$

IC: $y(2) = -3$

The equation is exact since $M_y = N_x$.   The solution is

$$\int_2^x (3x^2 + 2xy^2)\, dx + \int_{-3}^y 2(2^2)y\, dy = 0$$

or

$$x^3 + x^2y^2 = 44$$

Example 4   Solve: $(2x + ye^{xy})\, dx + (xe^{xy})\, dy = 0$
The differential equation is exact since $M_y = N_x$.   The solution is

$$\int_{x_0}^x (2x + ye^{xy})\, dx + \int_{y_0}^y x_0 e^{x_0 y}\, dy = C$$

It is convenient to pick $x_0 = 0$ to simplify the second integral.   The solution is therefore

$$x^2 + e^{xy} = C_1$$

Problems 2-4

1. Test the following for exactness and solve:

    a. $(x + y^2)\, dy + (y - x^2)\, dx = 0$ b. $e^x \cos y\, dx = e^x \sin y\, dy$

    c. $(ax + by)\, dx + (bx + ay)\, dy = 0$ d. $\dfrac{4y^2 - 2x^2}{4xy^2 - x^3}\, dx + \dfrac{8y^2 - x^2}{4y^3 - x^2y}\, dy = 0$

2. Solve: DE: $2(x - 1)y^3\, dx + [3(x - 1)^2 y^2 + 2y]\, dy = 0$

    IC: $y(1) = 3$

3. Derive the following solution of the initial-value problem (40):

$$\int_{x_0}^x M(x,y_0)\, dx + \int_{y_0}^y N(x,y)\, dy = 0$$

4. Explain why a differential equation that is not exact cannot be solved by integrating both sides as in Eqs. (41) and (42).

5. Assume the equation $y^2 \sin x\, dx + yf(x)\, dy = 0$ is exact.   Find all possibilities for $f(x)$.

2-5   Integrating factors

Unfortunately, not all differential equations are exact.   If the equation

$$M\, dx + N\, dy = 0 \tag{44}$$

is not exact, we try to make it exact by multiplying the equation by an appropriate factor $u(x,y)$. A function $u(x,y)$ having the property that

$$uM\ dx + uN\ dy = 0 \tag{45}$$

is an exact differential equation is called an *integrating factor*. Of course, it is important that $u(x,y)$ is not zero. For although every solution of (44) is a solution of (45), the reverse is not necessarily true unless $u(x,y)$ is always different from zero. We have already seen one example of a nonzero integrating factor in the solution of the first-order linear differential equation.

Integrating factors by inspection Sometimes integrating factors can be found by inspection. Consider the simple differential equation

$$x\ dy - y\ dx = 0 \tag{46}$$

This equation is not exact. However, if we recall that

$$d\left(\frac{y}{x}\right) = \frac{x\ dy - y\ dx}{x^2} \qquad x \neq 0 \tag{47}$$

we see that $1/x^2$ is an integrating factor of (46). Using this factor we obtain the solution

$$\frac{y}{x} = c \qquad x \neq 0 \tag{48}$$

The differential equation (46) also has many other integrating factors. From the identities

$$d\left(-\frac{x}{y}\right) = \frac{x\ dy - y\ dx}{y^2} \qquad y \neq 0 \tag{49}$$

$$d\left(\ln\left|\frac{y}{x}\right|\right) = \frac{x\ dy - y\ dx}{xy} \qquad xy \neq 0 \tag{50}$$

$$d\left(\arctan\frac{y}{x}\right) = \frac{x\ dy - y\ dx}{x^2 + y^2} \qquad x^2 + y^2 \neq 0 \tag{51}$$

we see that $1/y^2$, $1/xy$, $1/(x^2 + y^2)$ are also integrating factors of (46). Each of these factors could be used to obtain solutions of (46).

This multiplicity of integrating factors is not an accident. In fact, the following theorem is true.

Theorem 1 *If $u(x,y)$ is an integrating factor of $M\ dx + N\ dy = 0$ and if $dv(x,y) = uM\ dx + uN\ dy$, then $u(x,y)F(v(x,y))$ is also an integrating factor where $F(v)$ is any continuous function of v.*

Proof   Let $G(v)$ be a function whose derivative is $F(v)$.   We have

$$dG(v) = G'(v)\,dv = F(v)\,dv$$

but

$$dv = uM\,dx + uN\,dy$$

Therefore

$$dG(v) = uF(v)(M\,dx + N\,dy)$$

which shows that $uF(v)$ is an integrating factor.

In the case of the differential equation (46) we know that one integrating factor is $u = 1/x^2$.   Also from (47) we have $v = y/x$; therefore integrating factors of (46) are given by $(1/x^2)F(y/x)$.   By specializing the function F we can easily obtain the integrating factors $1/y^2$, $1/xy$, $1/(x^2 + y^2)$ mentioned above.

We now consider a slight generalization of Eq. (46):

$$ay\,dx + bx\,dy = 0 \tag{52}$$

where a and b are nonzero constants.   The identity

$$d(x^ay^b) = ax^{a-1}y^b\,dx + bx^ay^{b-1}\,dy$$
$$= x^{a-1}y^{b-1}(ay\,dx + bx\,dy) \tag{53}$$

shows that $x^{a-1}y^{b-1}$ is an integrating factor of (52).   Use of this integrating factor yields the solution

$$x^ay^b = c \tag{54}$$

Many differential equations can be solved by strategic use of the above identities and results.   We illustrate the technique in the following examples. In these examples the solutions will usually be obtained in implicit form; in order to concentrate on the methods involved, we will not give complete discussions to determine the interval for which the solution is defined or to determine whether or not all solutions have been obtained.

Example 1 $x\,dy - (y + y^2)\,dx = 0$

We group the terms in a promising manner:

$$(x\,dy - y\,dx) - y^2\,dx = 0$$

The term in parentheses has many integrating factors.   We pick an integrating factor which also makes the second term exact, namely, $1/y^2$.   Multiplying by this factor, we obtain

$$-d\left(\frac{x}{y}\right) - dx = 0$$

Integrating and simplifying, we obtain the solution

$$y = \frac{x}{c - x}$$

Example 2 $y\,dx - x\,dy - (x^2 + y^2)\,dx = 0$

An appropriate integrating factor is $1/(x^2 + y^2)$. Multiplying by this factor, we obtain

$$d\left(\arctan \frac{x}{y}\right) - dx = 0$$

and the solution

$$\arctan \frac{x}{y} - x = c \qquad \text{or} \qquad y = \frac{x}{\tan\,(x + c)}$$

Example 3 $(3x^5y^5 - 2y)\,dx + (5x^6y^4 + x)\,dy = 0$

We rearrange the differential equation to exhibit expressions in the form $ay\,dx + bx\,dy$:

$$x^5y^4(3y\,dx + 5x\,dy) + (-2y\,dx + x\,dy) = 0$$

An integrating factor of $(3y\,dx + 5x\,dy)$ is x^2y^4, and an integrating factor of $(-2y\,dx + x\,dy)$ is x^{-3}. Because of the factor x^5y^4 in the first term, both terms will be exact if we multiply by x^{-3}:

$$x^2y^4(3y\,dx + 5x\,dy) + x^{-3}(-2y\,dx + x\,dy) = 0$$

or

$$d(x^3y^5) + d(x^{-2}y) = 0$$

The solution is therefore

$$x^3y^5 + x^{-2}y = c$$

Example 4 $(x - yx^2)\,dy + y\,dx = 0$

This can be rewritten

$$d(xy) - yx^2\,dy = 0$$

The first term is exact and will remain so if it is multiplied by any function of the product xy. Divided by $(xy)^2$, the second term becomes exact:

$$\frac{d(xy)}{(xy)^2} - \frac{dy}{y} = 0$$

Therefore solutions are given by

$$-\frac{1}{xy} - \ln y = c$$

Differential equation for the integrating factor   If $u(x,y)$ is an integrating factor of $M\,dx + N\,dy = 0$, then the equation

$$(uM)\,dx + (uN)\,dy = 0 \tag{55}$$

is exact.   This implies that

$$(uM)_y = (uN)_x \tag{56}$$

or

$$uM_y + u_yM = uN_x + u_xN \tag{57}$$

Thus the integrating factor satisfies a first-order partial differential equation. Generally it is more difficult to solve this partial differential equation than it is to solve the original ordinary differential equation.   Nevertheless, we can obtain some useful results by finding conditions for the integrating factor to be a function of one variable only.

Suppose that the integrating factor is a function of x alone.   Then $u_y = 0$ and Eq. (57) reduces to

$$\frac{u_x}{u} = \frac{M_y - N_x}{N} \tag{58}$$

The left-hand side of (58) is a function of x alone.   In order for this equation to determine $u(x)$, the right-hand side must also be a function of x alone, say

$$\frac{M_y - N_x}{N} \equiv g(x) \tag{59}$$

In this case Eq. (58) can be solved to obtain the integrating factor

$$u = e^{\int g(x)\,dx} \tag{60}$$

In a similar way we find that if

$$\frac{N_x - M_y}{M} \equiv h(y) \tag{61}$$

then an integrating factor is

$$u = e^{\int h(y)\,dy} \tag{62}$$

Example 5   Consider the linear equation

$$y' + P(x)y = Q(x)$$

or

$$(Py - Q)\,dx + dy = 0$$

We have

$$\frac{M_y - N_x}{N} = P(x)$$

Therefore an integrating factor is $u \equiv e^{\int P\,dx}$, the same factor found previously.

Problems 2-5

Solve by finding an integrating factor:
1. $y\,dx + (y^2 - x)\,dy = 0$

2. $x\,dx + (y + x^2 + y^2)\,dy = 0$

3. $(2x - 3x^3y)\,dy - (y + x^2y^2)\,dx = 0$

4. $(3x - y)\,dx + 2x\,dy = 0$

5. The equation $(x^2 + y)\,dx + f(x)\,dy = 0$ is known to have an integrating factor $I(x) = x$. Find all possibilities for f.

6. Using Eqs. (59) through (62), find integrating factors and solve:

a. $x\,dy - y(1 + xy)\,dx = 0$

b. $(xe^{y/x} + 2x^2y)\,dy - ye^{y/x}\,dx = 0$

7. Show that the equation $x^p(\alpha y\,dx + \beta x\,dy) + y^q(\gamma y\,dx + \delta x\,dy) = 0$ has an integrating factor of the form $x^m y^n$ provided that $\alpha\delta - \beta\gamma \neq 0$.

8. Show that the differential equation $(2y + 3x)\,dx + x^2\,dy = 0$ does *not* have an integrating factor of the form $x^m y^n$.

9. The *Bernoulli* differential equation is

$$y' + P(x)y = Q(x)y^n$$

If $I = e^{\int P\,dx}$, this equation can be written

$$\frac{d}{dx}(yI) = IQy^n$$

Solve this equation by finding a suitable integrating factor.

2-6   Use of substitutions

In Sec. 2-5 we studied the solution of nonexact differential equations by finding an integrating factor. An alternative approach is to substitute appropriate new variables so that the differential equation in terms of these new variables is one of the known types. Unfortunately there are very few guides to help us determine the appropriate substitution. Sometimes the form of the differential equation suggests an "obvious" substitution. The following examples illustrate certain cases where substitutions can be used successfully.

Example 1 $y' = f(ax + by)$ (i)

The appearance of $ax + by$ on the right-hand side suggests the substitution

$$u = ax + by$$

We now eliminate the variable y and obtain a new differential equation in the variables u and x; we have

$$u' = a + by'$$

or

$$y' = \frac{u' - a}{b} \qquad \text{assuming } b \neq 0$$

The differential equation becomes

$$\frac{du}{dx} = bf(u) + a \tag{ii}$$

Separating variables, we have the solutions

$$\int \frac{du}{bf(u) + a} = x + c \tag{iii}$$

where we must put $u = ax + by$ after performing the integration.   We note that we must assume that

$$bf(u) + a \neq 0$$

If $bf(u) + a = 0$ for some value $u = \alpha$, we see from the differential equation (ii) that $u = \alpha$ is a solution of (ii) and therefore

$$ax + by = \alpha$$

is a solution of (i).

Example 2 $xy' = yF(xy)$

We try the substitution

$$u = xy \quad \text{or} \quad y = \frac{u}{x}$$

We have

$$\frac{dy}{dx} = \frac{x(du/dx) - u}{x^2}$$

and we obtain the differential equation

$$x\frac{du}{dx} - u = uF(u)$$

or

$$x\,du = u[1 + F(u)]\,dx$$

which can be solved by separation of variables.

Bernoulli differential equation   This equation is

$$\frac{dy}{dx} + P(x)y = Q(x)y^n \tag{63}$$

where n is any real number.   If n is 0 or 1, the equation is linear and can be solved.   To solve the equation for other values of n, we rewrite the equation in the form

$$y^{-n}\frac{dy}{dx} + P(x)y^{-n+1} = Q(x) \tag{64}$$

The appearance of both y^{-n+1} and $y^{-n}\,dy/dx$ suggests the substitution $z = y^{-n+1}$. Performing this substitution, we obtain

$$\frac{1}{1-n}\frac{dz}{dx} + P(x)z = Q(x) \tag{65}$$

which is a linear differential equation.

Example 3 $y' + xy = xy^3$

Multiplying by y^{-3}, we obtain

$$y^{-3}y' + xy^{-2} = x$$

Letting $z = y^{-2}$, we obtain the linear equation

$$z' - 2xz = -2x$$

whose solution is

$$z = 1 + ce^{x^2} \qquad \text{or} \qquad y^2 = \frac{1}{1 + ce^{x^2}}$$

Example 4 $(\cos y)y' - 2x \sin y = -2x$

This is not a Bernoulli differential equation but can be solved by a similar method. We make the substitution $z = \sin y$ and obtain the linear equation

$$z' - 2xz = -2x$$

Using the results of the previous example, we obtain

$$z = \sin y = 1 + ce^{x^2}$$

Homogeneous differential equation   A function $F(x,y)$ is called *homogeneous of degree n* if

$$F(tx,ty) \equiv t^n F(x,y) \tag{66}$$

This equality is to be an *identity* in x, y, and t. For instance,

$$3x^2y^4 - 2xy^5 \qquad \text{and} \qquad 2x - 3y$$

are homogeneous polynomials of degrees 6 and 1, respectively, since

$$3(tx)^2(ty)^4 - 2(tx)(ty)^5 \equiv t^6(3x^2y^4 - 2xy^5)$$

and

$$2(tx) - 3(ty) \equiv t(2x - 3y)$$

Other examples of homogeneous functions are

$$\sin \frac{y}{x} \qquad \text{degree } 0$$

$$\frac{1}{\sqrt{x^2 + y^2}} \qquad \text{degree } -1$$

whereas the functions

$$6x - y^2 \quad \text{and} \quad \frac{1}{x-3}$$

are not homogeneous.

The differential equation

$$M(x,y)\,dx + N(x,y)\,dy = 0 \tag{67}$$

is called *homogeneous*† if $M(x,y)$ and $N(x,y)$ are homogeneous of the *same degree*. A homogeneous equation can always be written in the form

$$\frac{dy}{dx} = f\left(\frac{y}{x}\right) \tag{68}$$

To see this, we solve Eq. (67) for dy/dx:

$$\frac{dy}{dx} = -\frac{M(x,y)}{N(x,y)}$$

Using the definition of homogeneous function, we obtain

$$\frac{dy}{dx} = -\frac{M(x,y)}{N(x,y)} = -\frac{t^{-n}M(tx,ty)}{t^{-n}N(tx,ty)}$$

Since t is an arbitrary number, we can set $t = 1/x$ to obtain

$$\frac{dy}{dx} = -\frac{M(1,y/x)}{N(1,y/x)}$$

The right-hand side of this equation is a function only of the ratio y/x and is therefore in the form (68).

The form of Eq. (68) suggests the substitution $y/x = v$ or $y = vx$. Substituting this in (68), we obtain

$$v + x\frac{dv}{dx} = f(v) \tag{69}$$

This equation can be solved by separating the variables. It is important to note that the differential equation (68) could also be reduced to one in which the variables are separable by means of substitution $x = vy$, which is sometimes more convenient.

† This word *homogeneous* is used here in a sense different from that in connection with linear differential equations.

Example 5 $xy^2 \, dy = (x^3 + y^3) \, dx$

Substituting $y = vx$ and $dy = v \, dx + x \, dv$, we obtain

$$v^2(x \, dv + v \, dx) = (1 + v^3) \, dx \qquad \text{or} \qquad v^2 \, dv = \frac{dx}{x}$$

Therefore

$$\tfrac{1}{3}v^3 = \ln cx \qquad \text{or} \quad y^3 = 3x^3 \ln cx$$

Example 6 $(x^2 + y^2) \, dy = 2xy \, dx$

Since dy is multiplied by a binomial and dx by a monomial, it is more convenient to make the substitution $x = vy$, $dx = v \, dy + y \, dv$:

$$y^2(1 + v^2) \, dy = 2vy^2(v \, dy + y \, dv)$$

Simplifying, we obtain

$$\frac{dy}{y} = \frac{2v \, dv}{1 - v^2}$$

After integration and use of the substitution $v = x/y$, we obtain, after some simplifications,

$$x^2 = y^2 + cy$$

Further examples of homogeneous equations occur in Chap. 10, page 320.

Problems 2-6

Solve the following differential equations:

1. $(x^2 - 2y^2) \, dx + 2xy \, dy = 0$
2. $(x^2 + y^2) \, dx = 2xy \, dy$
3. $(x + y) \, dy + (x - y) \, dx = 0$
4. $\dfrac{dy}{dx} + \dfrac{1}{x} = e^{-y}$

5. $x \dfrac{dy}{dx} + y = xy^{-4}$

6. $\dfrac{dy}{dx} = \dfrac{y}{x - \sqrt{xy}}$

7. By means of an appropriate substitution, reduce the following differential equations to equations in which the variables are separable:

a. $xy' = yF\left(\dfrac{y}{x^n}\right)$

b. $x + yy' = F(x^2 + y^2)$

c. $xy' - ay + y^2 = x^{2a}$

8. Solve: $y(1 + e^{x/y})\, dx + e^{x/y}(y - x)\, dy = 0$

9. Consider the equation

$$\frac{dy}{dx} = f\!\left(\frac{a_1 x + b_1 y + c_1}{a_2 x + b_2 y + c_2}\right) \qquad a_1 b_2 - a_2 b_1 \neq 0$$

Introduce new variables

$$u = x - h$$
$$v = y - k$$

and show that h and k can be found so that the equation becomes

$$\frac{dv}{du} = f\!\left(\frac{a_1 u + b_1 v}{a_2 u + b_2 v}\right)$$

which is homogeneous.

2-7 *Second-order equations reducible to first-order*

Certain second-order equations are nothing but disguised first-order equations.   Consider the second-order equation

$$\frac{d^2 y}{dx^2} = F\!\left(x, y, \frac{dy}{dx}\right) \tag{70}$$

There are two important cases when this equation can be reduced to a first-order equation.   If the right-hand side is *independent of y*, namely,

$$\frac{d^2 y}{dx^2} = F\!\left(x, \frac{dy}{dx}\right) \tag{71}$$

the simple substitution $p = dy/dx$ yields

$$\frac{dp}{dx} = F(x, p) \tag{72}$$

This is a first-order equation for p which may be accessible to the preceding solution methods.   After p has been obtained, y can be found by an integration.

Example 1 $y'' - y' = x$

Let $p = y'$; then $p' - p = x$. This is a linear equation for p. Solving, we obtain

$$p = c e^x - (1 + x)$$

and since $dy/dx = p$, we have

$$y = \int p\, dx = c e^x - \tfrac{1}{2}(1 + x)^2 + c_1$$

If the right-hand side of (70) is *independent of x*, namely, if

$$\frac{d^2y}{dx^2} = F\left(y, \frac{dy}{dx}\right) \tag{73}$$

the equation can also be reduced to a first-order equation by means of the substitution $p = dy/dx$. We consider y as the new independent variable and express y'' in terms of p and y as follows:

$$\frac{d^2y}{dx^2} = \frac{dp}{dx} = \frac{dp}{dy}\frac{dy}{dx} = \frac{dp}{dy}p$$

With these substitutions (73) becomes

$$p\frac{dp}{dy} = F(y,p) \tag{74}$$

which is a first-order equation. After solving for p as a function of y, we can obtain x as a function of y by solving

$$\frac{dy}{dx} = p(y) \tag{75}$$

by separation of variables.

Example 2   The equation of the harmonic oscillator is

$$m\ddot{x} + kx = 0$$

where m and k are positive constants. Letting $\omega^2 = k/m$, we obtain

$$\ddot{x} + \omega^2 x = 0$$

We shall solve this equation by the method of this section; however, a simpler method will be given in Chap. 5. We let

$$\dot{x} = v$$

Then

$$\ddot{x} = v\frac{dv}{dx}$$

Therefore the differential equation becomes

$$v\frac{dv}{dx} + \omega^2 x = 0$$

Separating variables and integrating, we obtain

$$\tfrac{1}{2}v^2 + \tfrac{1}{2}\omega^2 x^2 = \tfrac{1}{2}\omega^2 c_1^2$$

where we have written the constant of integration in the convenient form $\frac{1}{2}\omega^2 c_1$. Solving for v, we obtain

$$v = \omega\sqrt{c_1^2 - x^2}$$

Since $v = dx/dt$, we obtain

$$\frac{dx}{\sqrt{c_1^2 - x^2}} = \omega\,dt$$

Therefore

$$\arcsin\frac{x}{c_1} = \omega t + c_2$$

or

$$x = c_1 \sin\left(\omega t + c_2\right)$$

The solution is a simple sinusoid of angular frequency ω.

Problems 2-7

Solve the differential equations in Probs. 1 through 5.

1. $y'' + 3y' = 2$
2. DE: $y'' = -4y^{-2}$
   IC: $y(0) = 2 \qquad y'(0) = 2$
3. $y'' - a^2 y = 0$
4. $\dfrac{d^{n+1}y}{dx^{n+1}} - \dfrac{2}{x}\dfrac{d^n y}{dx^n} = 5x^2 \qquad n$ a positive integer
5. DE: $m\ddot{x} + kx = mg \qquad \begin{matrix} m > 0 \\ k > 0 \\ g > 0 \end{matrix}$
   IC: $x(0) = x_0$
   $\quad\; v(0) = v_0$
6. Given a general second-order differential equation

$$\frac{d^2 y}{dx^2} = F\left(x, y, \frac{dy}{dx}\right)$$

explain why the substitution $p = dy/dx$ does not help in solving the equation.

2-8 Summary

We have studied three basic types of first-order differential equations:

1. The first-order linear equation
2. Exact differential equations
3. Equations in which the variables are separable

In attempting to solve a first-order equation, the reader should first determine whether or not the equation is of one of these types. In determining whether or not an equation is linear, it should be kept in mind that the equation may be linear in either variable.   If the equation does not fall into any of these three categories, the next step is to try to reduce the equation to one of these types by:

1. Finding an integrating factor
2. Finding a suitable substitution

There are no infallible rules to guide us in finding integrating factors or in finding a suitable substitution.

In attempting to find an integrating factor, one should first of all try to find one by inspection, remembering the rules for differentiation of products, quotients, etc.   Often, grouping of terms will suggest an appropriate integrating factor.

In seeking suitable substitutions, one should recall the standard types such as the Bernoulli equation and the homogeneous equation for which substitutions are known.   Otherwise the form of the differential equation may suggest a suitable substitution.

It should be kept in mind that not all first-order equations can be solved by the methods of this chapter.   It may be necessary to resort to the more general, but less simple, approaches of Chap. 4.

Supplementary problems for chapter 2

1. $(4x^3 + 6xy^2) \, dx + (6x^2y + 3y^2) \, dy = 0$
2. $(1 + u^2) \, dv = (1 + v^2) \, du$
3. $(1 - x^2y) \, dx + (x^2y - x^3) \, dy = 0$
4. $dy = (y \tan x + 2x) \, dx$
5. $2x^3 \, dy - (1 - 3x^2y) \, dx = 0$
6. $y' + 1 = e^x e^y$
7. $x^2y' + xy + 2y^2 = 0$
8. DE: $\dfrac{dN}{dt} = kN(c - N)$

   IC: $N(0) = N_0$

9. DE: $\dfrac{dx}{dt} = (a - x)(b - x)$ $a < b$

   IC: $x(0) = 0$

10. $x + yy' = \sqrt{x^2 + y^2}$
11. $x - yy' = x(x^2 - y^2)$
12. Find an integrating factor for the equation

$$xy' - y = \varphi(x) F\left(\frac{y}{x}\right)$$

13. Prove that if $y = u(x)$ is a particular solution of $[y^2 + Q(x)] \, dx + dy = 0$, then $e^{-2\int u \, dx}/(y - u)^2$ is an integrating factor.

14. Use the result of Prob. 13 to solve

$$y' + y^2 = 1 + x^2$$

which has the obvious solution $y = x$.

15. The equation

$$y' + P(x)y + Q(x)y^2 = f(x)$$

is called the *Ricatti equation*. The solutions of this nonlinear equation cannot generally be expressed in terms of a finite number of integrations.

a. If a particular solution $y_1(x)$ of the Ricatti equation is known, show that the substitution $u = y - y_1$ reduces the equation to the Bernoulli equation

$$u' + (P + 2y_1Q)u + Qu^2 = 0$$

b. The above Bernoulli equation can be written

$$\frac{u'}{u} + Qu + (P + 2y_1Q) = 0$$

A similar equation can be written for $v = y - y_2$ if $y_2(x)$ is a second solution of the Ricatti equation.   By subtracting the two equations for u and v, show that solutions of the Ricatti equation are given by

$$y - y_1 = c(y - y_2)e^{\int Q(y_1 - y_2)dx}$$

c. If a third solution $y_3(x)$ of the Ricatti equation is known, show that the solution of the Ricatti equation is given by

$$\frac{y - y_1}{y_3 - y_1} \div \frac{y - y_2}{y_3 - y_2} = c$$

where c is an arbitrary constant.

16. Solve the Ricatti equation

$$y' + x^3y - x^2y^2 = 1$$

which has a particular solution $y = x$.

Applications of First-order Equations

3

3-1 Introduction

The origin and development of differential equations has always been closely associated with physical applications. The laws of nature in many different fields of science are formulated in terms of differential equations. It is important to note that the same type of differential equation often describes seemingly unrelated physical phenomena. Thus mathematics serves as a unifying thread connecting the various parts of science. This is one of the reasons why the study of differential equations is so important for a student of science or engineering. In this chapter we present a variety of applications of first-order differential equations. The applications will be simple but significant illustrations of the role of mathematics in studying physical phenomena.

3-2 Falling bodies with air resistance

Consider a body of mass m that is falling vertically under the influence of gravity. If there is no air resistance, the only force on the body is its weight. A simple application of Newton's second law yields the differential equation

$$m\ddot{x} = mg† \quad \text{or} \quad \ddot{x} = g \tag{1}$$

where x is the displacement measured positively *downwards* and g is the acceleration due to gravity. In addition to (1) we assume that the initial

† We assume we are dealing with motions near the surface of the earth. Thus we may assume g is a constant.

position and velocity of the body are prescribed:

$$x(0) = x_0$$
$$\dot{x}(0) = v_0 \tag{2}$$

Two successive integrations of (1) using (2) yield the familiar results

$$v = \dot{x} = v_0 + gt \tag{3}$$

$$x = x_0 + v_0 t + \tfrac{1}{2}gt^2 \tag{4}$$

for the velocity and displacement of the body at any time. We note that the velocity increases linearly with time and the displacement increases quadratically with time.

Now let us consider the effect of air resistance. The force due to the air resistance is a complicated function of the shape and velocity of the body and of the characteristics of air. It is found experimentally that under certain conditions this force can be closely approximated by a constant times a power of the velocity. With this assumption the differential equation of motion becomes

$$m\ddot{x} = mg - k(\dot{x})^n \tag{5}$$

where k is a positive constant and n is an integer. Introducing the velocity $v = \dot{x}$, we obtain the first-order equation

$$m\dot{v} = mg - kv^n \tag{6}$$

For "small" velocities n can be taken as unity and the differential equation (6) becomes a linear equation:

$$\frac{dv}{dt} + \frac{k}{m}v = g \tag{7}$$

The solution of this equation satisfying the initial condition $v(0) = v_0$ can easily be obtained by multiplying by the integrating factor $e^{kt/m}$ and integrating from $t = 0$ to t. The final result is

$$v = \frac{mg}{k} + \left(v_0 - \frac{mg}{k}\right)e^{-kt/m} \tag{8}$$

The displacement is found by integrating Eq. (8):

$$x - x_0 = \int_0^t v\,dt$$

$$x = x_0 + \frac{mg}{k}t + \frac{m}{k}\left(v_0 - \frac{mg}{k}\right)(1 - e^{-kt/m}) \tag{9}$$

where $x(0) = x_0$ is the initial displacement.

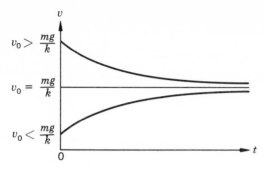

figure 1

It is interesting to note that the velocity as given by Eq. (8) does not increase indefinitely as t increases but approaches a *limiting or terminal velocity* v_l given by

$$v_l = \lim_{t \to \infty} v = \frac{mg}{k} \tag{10}$$

If $v_0 < mg/k$, the velocity increases toward the limiting velocity, whereas if $v_0 > mg/k$, the velocity decreases toward the limiting velocity. If $v_0 = mg/k$, the velocity remains constant. These results are shown in Fig. 1.

The behavior of the displacement can be seen by examining Eq. (9). Assuming v_0 is positive, it is not difficult to show that x always increases with t. When t is very large, the exponential term in (9) is almost zero, and x increases almost linearly with t. This is consistent with the fact that the velocity is almost constant for large t.

It is an interesting exercise to compute the limit of the velocity and displacement given in Eqs. (8) and (9) as the parameter k approaches zero, that is, as the air resistance approaches zero. Rewriting Eq. (8), we get

$$v = v_0 e^{-kt/m} + mg\left(\frac{1 - e^{-kt/m}}{k}\right)$$

The limit of the first term as k approaches zero is simply v_0. The limit of the second term can be found by L'Hospital's rule. The final result is

$$\lim_{k \to 0} v = v_0 + gt$$

which is simply Eq. (3) for the case of no air resistance. In a similar manner it can be shown that Eq. (9) approaches Eq. (4) as k approaches zero. These results are of course to be expected since Eq. (7) approaches Eq. (1) as k approaches zero. However, it should be remarked that taking the limit of a parameter in a differential equation and then solving the differential equation is *not* always the same as solving the differential equation and then taking the limit of the parameter.

We return briefly to the general case of Eq. (6):

$$m\dot{v} = mg - kv^n$$

If $n > 1$, this equation is not linear.   However, the variables can be separated as follows:

$$\frac{m\,dv}{mg - kv^n} = dt$$

The integral of the left-hand side can be evaluated for various values of n, but the results are not very simple.   The case $n = 2$ is left as an exercise (see Prob. 3).   Without solving the differential equation (6), we can, in a sense, still find the limiting velocity.   For if the body were moving with the constant limiting velocity v_l, the acceleration would be zero.   Equation (6) then becomes

$$mg - kv_l^n = 0$$

or

$$v_l = \sqrt[n]{\frac{mg}{k}} \tag{11}$$

It must be remembered that this result is based on the *assumption* that a limiting velocity exists and therefore cannot claim to be a proof of this fact.

Problems 3-2

1. A body of weight 32 lb is dropped from rest from a height of 100 ft in a medium offering an air resistance proportional to the velocity.   If the limiting velocity is 400 ft/sec, find the velocity and displacement at any time.   Find the time at which the velocity is 200 ft/sec.

2. A freely falling body of weight $\frac{5}{4}$ lb encounters an air resistance equal in pounds to $2v + v^2$ where v is the speed.   Set up the differential equation for the velocity.   Without solving the equation, find the limiting velocity.

3. Solve: DE: $m\dot{v} = mg - kv^2$
   IC: $v(0) = 0$
and show that the limiting velocity is $\sqrt{mg/k}$.

4. Repeat Prob. 1 with air resistance proportional to the square of the velocity.

5. A body of mass m is thrown vertically into the air with a speed of v_0. The body encounters an air resistance proportional to the velocity.   Set up and solve the differential equation for the velocity of the body *while it is rising*.   At what time does the body reach its maximum height?

6. A particle of mass m moves along the x axis subject to a resisting force proportional to the cube of the velocity.   Find the velocity and the displacement in terms of the time t, the initial velocity, and the initial displacement.

3-3   Motion on a given curve

A particle of mass m moving along a given curve satisfies the equation

$$m\ddot{s} = F \tag{12}$$

where s is the arc length of the curve measured from some fixed point and F is the force on the particle in the direction of the tangent to the curve. We shall assume that the force F depends only on the position s of the particle. In this case the differential equation is

$$m\ddot{s} = F(s) \tag{13}$$

and does not depend explicitly on the time t; therefore we can reduce (13) to a first-order equation by letting

$$\dot{s} = v \tag{14}$$

and

$$\ddot{s} = v\frac{dv}{ds}$$

With these substitutions we obtain

$$mv\frac{dv}{ds} = F(s) \tag{15}$$

in which the variables are separable. Integrating (15) yields

$$\tfrac{1}{2}mv^2 - \int_{s_0}^{s} F(u)\,du = c_1 \tag{16}$$

where s_0 is an arbitrary point on the curve and c_1 is the constant of integration.

The terms in Eq. (16) have a physical significance. The term

$$T(v) = \tfrac{1}{2}mv^2 \tag{17}$$

is called the *kinetic energy* of the particle. The kinetic energy is due to the motion of the particle. The term

$$V(s) = -\int_{s_0}^{s} F(u)\,du \tag{18}$$

is called the *potential energy*. This energy does not depend on the motion of the particle but only on its position. From (18) we see that the potential energy is the negative of the work done on the particle by the force F in moving the particle from s_0 to s. We note also that the potential energy is determined only up to an additive constant because of the arbitrariness of s_0.

In terms of the kinetic energy and potential energy, Eq. (16) becomes

$$T + V = c_1 \tag{19}$$

That is, the *sum of the potential energy and the kinetic energy is constant* during the motion. Thus we have obtained the important *law of conservation of energy*. The constant c_1 is called the *total energy*.

To obtain the position s of the particle, we solve (16) for v:

$$v = \pm \sqrt{\frac{2}{m} [c_1 - V(s)]} \tag{20}$$

Regarding the sign of the square root, the plus sign should be taken if the particle is moving in the direction of increasing arc length and the negative sign in the opposite case. Recalling that $v = ds/dt$, we find from (20) that

$$t = \pm \int \frac{ds}{\sqrt{(2/m)[c_1 - V(s)]}} + c_2 \tag{21}$$

where c_2 is an arbitrary constant. This equation gives t as a function of s; it can also be considered as an implicit equation for s as a function of t.

A particle sliding down a curve   An interesting example of the above is the problem of a particle sliding down a given curve, without friction, under the influence of gravity.

Let the equation of the curve be given in the parametric form

$$\begin{aligned} x &= x(s) \\ y &= y(s) \end{aligned} \tag{22}$$

where s is the arc length (see Fig. 2). The vertical force on the particle is $-mg$ and the tangential force is

$$F = -mg \cos \alpha \tag{23}$$

where α is the angle the tangent line makes with the vertical. This can also be written

$$F = -mg \frac{dy}{ds} \tag{24}$$

The equation of motion is therefore

$$m\ddot{s} = -mg \frac{dy}{ds} \tag{25}$$

Using the substitutions (14), we can easily obtain the first integral of (25):

$$\tfrac{1}{2}v^2 = g(y_0 - y) \tag{26}$$

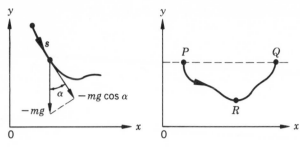

figure 2   figure 3

where we have assumed that $v = 0$ at $y = y_0$. Thus the speed of the particle is given by

$$v = \pm\sqrt{2g(y_0 - y)} \tag{27}$$

and the time of fall is given by

$$t = \pm\int \frac{ds}{\sqrt{2g[y_0 - y(s)]}} + c \tag{28}$$

We can deduce the general nature of the motion without performing the integration explicitly. Suppose the curve has the shape shown in Fig. 3 with arc length measured from point P toward point Q. We assume that P and Q are at equal distances above the x axis. We note that dy/ds is negative on the arc PR and positive on the arc RQ. Therefore, according to (25), the acceleration $\ddot{s}$ is positive on PR and negative on RQ. If the particle starts at P with zero velocity, the velocity will increase (the acceleration is positive), reaching a maximum at point R. At this point the acceleration changes sign and the velocity decreases until point Q is reached. At Q the velocity must be zero according to Eq. (27). Since the acceleration is negative at Q, the particle will reverse itself and start the reverse swing toward P. It is clear that the time to move from P to Q is the same as the time for the reverse swing. The particle therefore oscillates periodically between P and Q.

Escape velocity   Another interesting example is the problem of finding the least velocity with which to fire a projectile so that it will leave the earth and never return. It is convenient to consider the problem in reverse. We shall find the velocity with which a body strikes the earth when falling from rest from some initial distance and then find the limit of this velocity as the initial distance approaches infinity.

We neglect all forces except the force of attraction between the body and the earth as given by Newton's universal law of gravitation. This force is

$$\frac{-kmM}{x^2}$$

where m and M are the masses of the body and the earth respectively, x is the distance of the body from the center of the earth, and k is the gravitational constant. The equation of motion is

$$m\ddot{x} = \frac{-kmM}{x^2} \tag{29}$$

The constant k can be evaluated by noting that $\ddot{x} = -g$ when $x = R$, the radius of the earth. Therefore $k = R^2g/M$ and the differential equation (29) becomes

$$\ddot{x} = \frac{-R^2g}{x^2} \tag{30}$$

Letting $v = \dot{x}$, we obtain

$$v\frac{dv}{dx} = \frac{-R^2g}{x^2}$$

The solution satisfying $v(x_0) = 0$ is easily obtained:

$$\frac{v^2}{2} = R^2g\left(\frac{1}{x} - \frac{1}{x_0}\right) \tag{31}$$

Setting $x = R$ in this equation, we obtain the velocity with which the body strikes the earth:

$$v = -\sqrt{2R^2g\left(\frac{1}{R} - \frac{1}{x_0}\right)} \tag{32}$$

where the minus sign is chosen since the direction of the velocity is downward. Letting x_0 approach infinity, we find

$$\lim_{x_0 \to \infty} v = -\sqrt{2Rg} \tag{33}$$

The escape velocity is therefore $\sqrt{2Rg}$ or approximately 6.9 miles/sec.

Problems 3-3

1. A particle of unit mass moves along the x axis and is acted on by a force $F(x) = -1/x^3$. If the particle starts from rest at $x = 1$, find its position at any time. Also find the time for the particle to reach the origin.

2. A particle of unit mass moves along the x axis and is acted on by a force $F(x) = -\sin(x/4)$. If the particle starts at $x = 0$ with a velocity $v = 4$, find the position of the particle at any time and the limiting position as time increases.

3. If a pendulum hanging at rest is given an initial velocity v_0, how high will it rise? (See Sec. 1-3, Prob. 3.)

4. A particle starts from rest at $x = a$ and moves along the x axis under the force $F(x) = -k/x$, $k > 0$. Show that the time to reach the origin is

$$t = a\sqrt{\frac{m}{2k}} \int_0^1 \left[\ln\left(\frac{1}{u}\right)\right]^{-\frac{1}{2}} du$$

Evaluate this integral in terms of the gamma function (see Prob. 6, page 31) to obtain $t = a(m\pi/2k)^{\frac{1}{2}}$.

3-4   Linear motion with variable mass

If a body of constant mass m is moving along the x axis under the action of a force F, the equation of motion is given by Newton's law

$$m\ddot{x} = F \tag{34}$$

Since m is a constant this is equivalent to

$$\frac{d}{dt}(mv) = F \tag{35}$$

where v is the velocity of the body.   The quantity mv is called the *momentum* of the body.   Thus Newton's law can be stated as *the rate of change of momentum equals the applied force.*

Denoting the momentum by p and integrating (35) between the times t_1 and t_2 we obtain

$$p(t_2) - p(t_1) = \int_{t_1}^{t_2} F\, dt \tag{36}$$

The integral $\int_{t_1}^{t_2} F\, dt$ therefore measures the *change in momentum* in the time interval from t_1 to t_2.

We now consider the case when the mass of the body *varies* with time. Such is the case, for example, in rocket motion or in a falling evaporating raindrop.   Let mv be the momentum of the body at time t.   Suppose at time $t + \Delta t$ a mass Δm is added to the body.   Before joining the body we assume the mass Δm has velocity $(v + u)$, that is a velocity u relative to the body.   We assume that at the instant when the joining occurs, the velocity of the mass Δm changes from v to $v + \Delta v$ and Δm takes on the velocity of m. Thus the momentum of the bodies is as follows:

at time t: $mv + \Delta m(v + u)$

at time $t + \Delta t$: $(m + \Delta m)(v + \Delta v)$

According to (36), the change in momentum in the time interval Δt must satisfy

$$(m + \Delta m)(v + \Delta v) - (mv + \Delta m(v + u)) = \int_{t}^{t+\Delta t} F\, d\tau$$

or simplifying

$$m\,\Delta v - u\,\Delta m + \Delta v\,\Delta m = \int_{t}^{t+\Delta t} F\, d\tau \tag{37}$$

Dividing by Δt we have

$$m\frac{\Delta v}{\Delta t} - u\frac{\Delta m}{\Delta t} + \frac{\Delta v}{\Delta t}\Delta m = \frac{1}{\Delta t}\int_{t}^{t+\Delta t} F\, d\tau \tag{38}$$

Assuming that as Δt approaches zero, Δm and Δv and $\int_{t}^{t+\Delta t} F \, d\tau$ all approach zero in such a manner that the indicated limits exist, we obtain

$$m \frac{dv}{dt} - u \frac{dm}{dt} = F \tag{39}$$

This is the equation of motion for a time varying mass.

Example 1   A raindrop of initial mass m_0 falls from rest through a stationary cloud.   Assume that the resistance to motion is proportional to the velocity and that the mass of the raindrop increases at a constant rate a.   Find the velocity of the raindrop.

Since the mass increases at a constant rate we have

$$m = m_0 + at$$

The equation of motion is

$$m \frac{dv}{dt} - u \frac{dm}{dt} = mg - kv$$

Since the added mass has zero velocity $v + u = 0$ or $u = -v$.   Therefore, the equation of motion is

$$\frac{dv}{dt} + \frac{(a + k)}{m_0 + at} v = g$$

This is a linear equation whose solution is

$$v = \frac{g(m_0 + at)}{2a + k} + \frac{c}{(m_0 + at)^{1+k/a}}$$

where the constant c can be found from the condition $v(0) = 0$.

Example 2   A rocket of initial mass m_0 travels vertically upwards with initial velocity v_0.   The rocket loses mass at a constant rate so that the mass at any time is

$$m = m_0 - at$$

where a is a positive constant.   It is assumed that the lost mass travels backwards at a constant speed b relative to the rocket.   Neglecting all external forces on the rocket except its weight mg, we find the equation of motion (39) is (assume the positive direction for x is upwards):

$$(m_0 - at) \frac{dv}{dt} + (-a)b = -(m_0 - at)g$$

or

$$\frac{dv}{dt} = \frac{ab}{m_0 - at} - g$$

Integrating and using $v(0) = v_0$, we find

$$v = v_0 - gt - a \ln \left(1 - \frac{bt}{m_0} \right)$$

A second integration, using $x(0) = 0$, yields

$$x = (v_0 - a)t - \tfrac{1}{2}gt^2 - \frac{am_0}{b} \left(1 - \frac{bt}{m_0} \right) \ln \left(1 - \frac{bt}{m_0} \right)$$

This gives the height of the rocket at any time t up to the time when the entire mass of the rocket is expended.

Problems 3-4

1. One end of a 3-ft chain is held so that the other end just touches the floor. It is then released. Find the force with which the high end hits the ground.

2. A chain is coiled up on the ground. One end is lifted with a constant force. Find the velocity.

3. A raindrop falls through dry air. The raindrop is assumed to lose mass at a constant rate a by evaporation. Assume that the evaporating mass has the same velocity as the raindrop. If the air resistance is proportional to the velocity, find the velocity of the raindrop.

4. A spherical raindrop falls from rest through dry air. The raindrop loses mass at a rate proportional to its surface area. Find the velocity of the raindrop at any time, ignoring air resistance. Assume lost mass has zero velocity.

5. A rocket is moving in a straight line along horizontal rails. Assume the rocket loses mass at a constant rate a and the lost mass is thrown backwards at a constant speed b relative to the rocket. If the initial velocity is zero, find the velocity at any time if:

   a. There is no air resistance.
   b. The air resistance is proportional to the velocity.

3-5   Newton's law of cooling

This law states that the time rate of change of the temperature of a cooling body is proportional to the temperature difference between the body and its surroundings. Therefore the differential equation for the temperature T of the cooling body is

$$\frac{dT}{dt} = -k(T - T_s) \tag{40}$$

where T_s is the temperature of the surroundings and k is a positive constant.
In certain situations the temperature of the surroundings does not

change appreciably and T_s can be considered to be constant.   The differential equation (40) is then the linear differential equation

$$\frac{dT}{dt} + kT = kT_s \tag{41}$$

The solution of this equation satisfying $T(0) = T_0$ is

$$T = T_s + (T_0 - T_s)e^{-kt} \tag{42}$$

The temperature T therefore approaches the limiting temperature T_s as t increases.

Example 1   A body at a temperature of $100°$ is placed in a room of temperature $50°$.   The room temperature does not change appreciably.   If after 10 min the temperature of the body has decreased to $90°$, when will the body be at a temperature of $60°$?

Solution   We have $T_0 = 100$ and $T_s = 50$.   Therefore the temperature is given by (42):

$$T = 50 + 50e^{-kt}$$

In order to evaluate k, we use the fact that $T = 90$ when $t = 10$:

$$90 = 50 + 50e^{-10k}$$

Solving for k, we obtain

$$e^{-10k} = \tfrac{4}{5}$$
$$-10k = \ln \tfrac{4}{5}$$

or

$$k = -\tfrac{1}{10} \ln \tfrac{4}{5} = \tfrac{1}{10} \ln \tfrac{5}{4} = 0.014$$

Therefore the temperature at any time is

$$T = 50 + 50e^{-0.014t}$$

To find the time for a given temperature, we solve this equation for t to obtain

$$t = -\tfrac{1}{0.014} \ln \frac{T - 50}{50}$$

From this equation we find that $T = 60$ when

$$t = -\tfrac{1}{0.014} \ln \frac{60 - 50}{50} = \tfrac{1}{0.014} \ln 5$$

or

$$t = 115 \text{ min}$$

We now consider a case where the temperature of the surroundings is changing.   We shall assume that the heat lost by the cooling body is completely absorbed by the surroundings.   Let the weights (in pounds) of the body and the surroundings be denoted by w and w_s, respectively, and their specific heats† (in calories per pound per degree) by c and c_s, respectively. Equating the heat lost by the body to the heat gained by the surroundings, we obtain

$$wc(T_0 - T) = w_s c_s(T_s - T_{s0})$$

where T_0 and T_{s0} are initial temperatures and all temperatures are in degrees Fahrenheit. Solving this equation for T_s, we obtain

$$T_s = T_{s0} + A(T_0 - T) \tag{43}$$

where we have set

$$A = \frac{wc}{w_s c_s}$$

Using (43), the differential equation (40) becomes

$$\frac{dT}{dt} + k(1 + A)T = k(T_{s0} + AT_0) \tag{44}$$

This is again a linear differential equation in the same form as (40). The solution is

$$T = T_1 + (T_0 - T_1)e^{-k(1+A)t} \tag{45}$$

where T_1 is the limiting temperature

$$T_1 = \frac{T_{s0} + AT_0}{1 + A} \tag{46}$$

Therefore the temperature T decreases exponentially to the limiting temperature T_1. At the same time the temperature T_s increases exponentially to T_1, as can be seen from Eq. (43). The temperature T_1 is the equilibrium temperature; if the body and the surroundings were initially at temperature T_1, no heat transfer would occur, and the body and the surroundings would remain at this temperature.

Problems 3-5

1. A body at a temperature of 100° is placed in a room of unknown temperature. The room temperature does not change appreciably. If after 10 min the body has cooled to 90° and after 20 min to 85°, find the temperature of the surroundings.

† The specific heat of a body is the amount of heat (in calories) necessary to raise the temperature of one pound of the body by one degree Fahrenheit.

2. A body of temperature 100° is placed in water of temperature 50°. After 10 min the temperature of the body is 80° and the temperature of the water is 60°. Assuming all the heat lost by the body is absorbed by the water, find the temperature of the body and of the water at any time. Find the equilibrium temperature.

3. Assuming all the heat lost by the cooling body is absorbed by the surroundings, find the differential equation satisfied by the temperature of the surroundings.

3-6 Dilution problems

A tank initially contains V gal of fresh water. At $t = 0$ a brine solution, containing c lb of salt per gallon, is poured into the tank at a rate of a gal/min. The contents of the tank are stirred to maintain homogeneity, and the dilute solution is pumped out at a rate of a gal/min. The problem is to find the amount of salt in the tank at any time.

Let x (in pounds) be the amount of salt in the tank at any time. The volume of the solution in the tank is always V; therefore the concentration of salt is x/V (in pounds per gallon). Salt leaves the tank at the rate of $a \cdot x/V$ lb/min and salt enters the tank at ca lb/min. The rate of change of the amount of salt in the tank is therefore

$$\frac{dx}{dt} = ca - \frac{ax}{V} \tag{47}$$

This is the same equation encountered on several previous occasions. The solution satisfying $x(0) = 0$ is

$$x = cV(1 - e^{-at/V}) \tag{48}$$

The concentration y of salt in the tank at any time is given by

$$y = \frac{x}{V} = c(1 - e^{-at/V}) \tag{49}$$

The concentration of the salt in the tank therefore approaches c, the concentration of the brine, as time increases.

Problems 3-6

1. A tank initially contains 100 gal of brine solution containing 10 lb of salt. At $t = 0$ fresh water is poured into the tank at 3 gal/min and the uniform dilute solution leaves at the same rate. Find the concentration of salt in the tank at any time. How long will it take for the dilute solution to reach a concentration equal to one-half the initial concentration?

2. Two tanks initially contain 100 gal of pure water. Brine, with a concentration of 2 lb/gal, is poured into the first tank at a rate of 3 gal/min. The

uniform solution in the first tank is pumped into the second tank at 3 gal/min and the solution in the second tank is pumped out at the same rate. Find the concentration in each tank at any time.

3. A leaky tank loses liquid at the rate of 5 gal/min. Brine, containing 3 lb of salt per gallon, is pumped in at the rate of 4 gal/min, the solution being constantly stirred. If, initially, the tank contains 100 gal of pure water, find the number of pounds of salt in the tank after t min.

4. A 100-gal tank initially contains 50 gal of brine with a concentration of 1 lb of salt per gallon. Pure water is poured into the tank at a rate of 4 gal/min and the solution, which is kept uniform, is pumped out at the rate of 2 gal/min. What is the concentration of salt in the container at the instant of overflow?

5. Tank A contains 100 gal of fresh water. Tank B contains 100 gal of a brine solution with a salt concentration of 2 lb/gal. Both tanks are stirred to keep their contents uniform. The contents of tank A are pumped into tank B at 3 gal/min and the contents of B are pumped into A at the same rate. Find the concentration of salt in each tank at any time. Find the equilibrium concentration.

6. Generalize Prob. 2 to n tanks.

3-7   *Chemical reactions*

Chemical reactions do not occur instantaneously but take a definite time for their completion. The rate at which a reaction proceeds is a very complicated matter which cannot be discussed here in any detail. We present only a few simple considerations to show how differential equations enter into the study of reaction rates.

First-order reactions   If each molecule of a substance has a tendency to decompose spontaneously into smaller molecules, and this tendency is unaffected by the presence of other molecules, the reaction is called a *first-order* or *unimolecular* reaction. The radioactive decomposition of radium is an example of such a reaction. It is reasonable to expect that the number of molecules that decompose by a first-order process in unit time will be proportional to the number present. Since the number of molecules is proportional to the mass, the mass will decrease according to the law

$$\frac{dm}{dt} = -km \tag{50}$$

where k is a positive constant called the *rate constant*. The solution of this equation is

$$m = m_0 e^{-kt} \tag{51}$$

where m_0 is the initial mass. We see that the mass decreases exponentially to zero. The time required for the mass to decrease to one-half the initial mass is called the *half-life* of the substance.   Setting $m = m_0/2$ in Eq. (51) and solving for t, we obtain

$$t = \frac{1}{k} \ln 2 \tag{52}$$

for the half-life.   For radium the half-life is approximately 1,733 years.

Second-order reactions   If a reaction occurs by collision and interaction of one molecule of substance A with one molecule of substance B, the reaction is called *second-order* or *bimolecular*.   It is reasonable to expect that the time rate of reaction will be proportional to the number of collisions in unit time.   Furthermore the number of collisions in unit time is proportional to the product of the concentrations of substances A and B. Hence, if the concentrations (in grams per liter) of substances A and B are denoted by C_1 and C_2, respectively, we may write the differential equation

$$\text{Time rate of reaction} = \frac{-dC_1}{dt} = \frac{-dC_2}{dt} = kC_1C_2 \tag{53}$$

Now, if A and B have initial concentrations of a and b, respectively, and if x is the concentration of A (or B) that has already reacted, we have

$$C_1 = a - x \quad \text{and} \quad C_2 = b - x$$

The differential equation for the reaction is therefore

$$\frac{dx}{dt} = k(a - x)(b - x) \tag{54}$$

This equation can be solved by separation of variables and the use of partial fractions.   If a is less than b, we obtain, using the condition $x(0) = 0$,

$$kt = \frac{1}{a - b} \ln \frac{1 - x/a}{1 - x/b} \tag{55}$$

or, solving for x,

$$x = \frac{ab(1 - e^{-(b-a)kt})}{b - ae^{-(b-a)kt}} \tag{56}$$

Since $a < b$, we see that, as time increases, x tends to a, the initial concentration of A.   This means that all of substance A has reacted and there remains an excess $b - a$ of substance B.

Problems 3-7

1. If the half-life of a radioactive substance is 1,000 years, how much of it is left after 100 years?

2. If $m(t)$ is the amount of mass left at time t in a first-order reaction, show that the rate constant is given by

$$k = \frac{1}{t_2 - t_1} \ln\left(\frac{m_1}{m_2}\right)$$

where $m(t_1) = m_1$ and $m(t_2) = m_2$.

3. Complete the derivation of Eq. (56) in the text.

4. Solve the differential equation (54) for the case $a = b$. Find the half-life. Find $\lim_{t \to \infty} x(t)$.

5. The differential equation for a certain reaction is

$$\frac{dx}{dt} = k_1 a - k_2 x^2 \qquad k_1 > 0 \qquad k_2 > 0$$

where a is the initial concentration and $x(0) = 0$. Solve this equation and find the limiting concentration.

3-8   Population growth

Some interesting mathematical problems arise in biology.† One example is the problem of predicting the growth of a population. This is, of course, an extremely difficult problem depending on many complicated parameters. We shall present here a highly simplified approach to the problem.

Let $N(t)$ be the number of people in an isolated population at time t. The function $N(t)$ is clearly an integer-valued function; it is constant in intervals of time when the population does not change, and it is discontinuous when a birth or death changes the population. Such discrete-valued functions are difficult to analyze. Our first simplifying approximation is to replace $N(t)$ by a continuous and differentiable function $N(t)$. This allows us to use the powerful tools of calculus. $N(t)$ can be thought of as a smoothed-out version of the actual population.

Our next simplifying assumption is that the time rate of change of N is a function only of N; that is,

$$\frac{dN}{dt} = F(N) \tag{57}$$

The function $F(N)$ is called the *growth function*. We shall further assume that:

1. $F(0) = 0$
2. $F(C) = 0$ where $C > 0$

† E.g., see Bailey [2] or Lotka [25].

Assumption (1) implies that $N \equiv 0$ is a solution of the differential equation (57); that is, if no people are initially present, then no people will ever be present. Assumption (2) implies that $N \equiv C$, a constant, is a solution of (57). This can be thought of as an upper limit to the population which could occur, for example, because of a limited food supply. The simplest function $F(N)$ satisfying assumptions 1 and 2 is

$$F(N) = KN(C - N) \tag{58}$$

where $K > 0$ for a growing population.

Thus the simplest differential equation for population growth is

$$\frac{dN}{dt} = KN(C - N) \tag{59}$$

This can be written as

$$\frac{dN}{dt} = KCN - KN^2 \tag{60}$$

The first term on the right-hand side can be considered the *birth rate* and the second term *death rate*.

Equation (60) can be solved either by separation of variables or as a Bernoulli equation. The solution satisfying the initial condition $N(t_0) = N_0$ is

$$N = \frac{C}{1 + (C/N_0 - 1)e^{-KC(t-t_0)}} \tag{61}$$

We see from Eq. (61) that N approaches the limiting value C as t increases and that N approaches zero as t approaches negative infinity. A graph of the solution is shown in Fig. 4. For our purposes only the middle portion of the graph is significant; this exhibits an S shape which is characteristic of many population-growth problems.

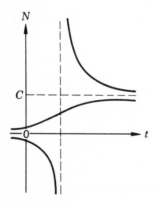

figure 4

Year	Population in millions
1790	4
1850	23
1910	92

Formula (61) has been applied by Pearl and Reed[†] to the population growth of the United States. The data in the accompanying table were used to evaluate the three constants N_0, C, and K. Using these data (see Prob. 2 below), formula (61) becomes

$$N = \frac{210}{1 + 51.5e^{-0.03t}} \tag{62}$$

The Pearl and Reed formula provides remarkably accurate predictions of the population of the United States. For instance, the population for 1950 as predicted from (62) is 151 million, which is the same as the population figure. We note that this formula predicts a limiting population of 210 million, which would be closely approximated by about the year 2075. However, such a forecast as this, based on such simplified assumptions and made without regard to a multiplicity of important physical factors, must, of course, be accepted with reservation.

Problems 3-8

1. In formula (61) set $x = 1/N$ and show that

$$\ln\left(\frac{Cx - 1}{Cx_0 - 1}\right) = -KC(t - t_0)$$

Show also that if t_0, t_1, t_2 are three *equally* spaced instants of time and x_0, x_1, x_2 are the corresponding values of x, then the constants C and K are given by

$$C = \frac{2x_1 - (x_0 + x_2)}{x_1^2 - x_0 x_2}$$

$$KC = \frac{1}{t_1 - t_0} \ln\left(\frac{Cx_1 - 1}{Cx_0 - 1}\right)$$

2. Using the data in the text and the results of Prob. 1, obtain the Pearl and Reed formula (62).

3. Derive Eq. (61).

3-9 A simple electrical circuit

The determination of the current in an electrical circuit furnishes an interesting and useful application of differential equations. We consider here one

[†] R. Pearl and L. J. Reed, *Proc. Nat. Acad. Sci.*, **6**: 275 (1920).

simple example. The reader who is unfamiliar with the basic laws of electrical circuits should consult Sec. 6-4, where these matters are discussed more fully.

The circuit shown in Fig. 5 contains a resistance R, an inductance L, and a source of voltage $E(t)$. According to Kirchhoff's law, the current $I(t)$ satisfies the differential equation

$$L\frac{dI}{dt} + RI = E(t) \tag{63}$$

This is a first-order linear equation. The solution satisfying the initial condition $I(0) = I_0$ can be readily obtained:

$$I = I_0 e^{-(R/L)t} + e^{-(R/L)t}\int_0^t e^{(R/L)s}E(s)\,ds \tag{64}$$

Since R and L are positive constants, we see that the term

$$I_0 e^{-(R/L)t} \tag{65}$$

approaches zero as t approaches infinity. This term is therefore called a *transient*. In some applications the ratio R/L is so large that the transient current is practically zero after a very short time and is often neglected by electrical engineers. In other applications dealing with high-speed systems such as modern electronic computers, the transient current cannot be neglected.

The simplest voltage source is a battery which provides a constant voltage. If we set $E \equiv E_b$, a constant, in (64), the integral can be evaluated and the current is

$$I(t) = \frac{E_b}{R}(1 - e^{-(R/L)t}) \tag{66}$$

where we have set $I(0) = 0$. We see that the current builds up from its initial value of zero and approaches the *steady-state* value I_s given by

$$I_s = \frac{E_b}{R} \tag{67}$$

as time increases. Equation (67) is just *Ohm's law* for the circuit with no inductance.

As a final illustration we consider the voltage $E(t)$ to be sinusoidal of

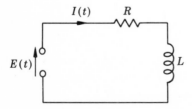

figure 5

circular frequency ω, that is,

$$E = E_0 \cos \omega t \qquad \text{or} \qquad E = E_0 \sin \omega t$$

Instead of substituting these expressions into (64) and evaluating the integral, we shall use a different approach which will introduce the notion of *complex impedance*. First we recall that the general solution of (63) can be written

$$I = I_h + I_p \tag{68}$$

where I_h is the general solution of the corresponding homogeneous equation

$$L \frac{dI}{dt} + RI = 0 \tag{69}$$

and I_p is a particular solution of (63). We easily find that

$$I_h = ce^{-(R/L)t} \tag{70}$$

This solution is clearly a transient.

To find a particular solution, we consider the two differential equations

$$L \frac{dI_1}{dt} + RI_1 = E_0 \cos \omega t \tag{71}$$

and

$$L \frac{dI_2}{dt} + RI_2 = E_0 \sin \omega t \tag{72}$$

We now combine these two equations into one equation for the complex function†

$$I = I_1 + iI_2 \tag{73}$$

where $i = \sqrt{-1}$. Multiplying (72) by i and adding it to (71), we obtain

$$L \frac{dI}{dt} + RI = E_0 \left(\cos \omega t + i \sin \omega t \right) \tag{74}$$

or, recalling Euler's formula,

$$e^{i\omega t} = \cos \omega t + i \sin \omega t \tag{75}$$

we obtain the differential equation

$$L \frac{dI}{dt} + RI = E_0 e^{i\omega t} \tag{76}$$

---

† See Appendix B for a review of complex numbers. See also Sec. 5-3 for a discussion of complex-valued solutions of a differential equation.

This equation for the *complex current I* is equivalent to the two equations (71) and (72) for I_1 and I_2.   The advantage of (76) is that a particular solution I_p can easily be found, and the solution is in a very simple and convenient form.   Once a particular solution I_p has been found, we can find particular solutions of (71) and (72) by simply taking the real and imaginary parts of I_p.   For, according to (73), we have

$$I_1 = \mathrm{Re}\,(I) \tag{77}$$

$$I_2 = \mathrm{Im}\,(I) \tag{78}$$

To find a particular solution of (76), we assume†

$$I_p = Ae^{i\omega t} \tag{79}$$

where A must be determined.   Substituting in (76), we obtain

$$(R + i\omega L)Ae^{i\omega t} = E_0 e^{i\omega t} \tag{80}$$

Therefore

$$A = \frac{E_0}{R + i\omega L}$$

and

$$I_p = \frac{E_0 e^{i\omega t}}{R + i\omega L} \tag{81}$$

This equation has a simple interpretation.   The numerator

$$E = E_0 e^{i\omega t} \tag{82}$$

is called the *complex voltage,* and the denominator

$$Z = R + i\omega L$$

is called the *complex impedance.*   With these notations (81) becomes

$$I_p = \frac{E}{Z} \tag{83}$$

Thus the complex current can be obtained by dividing the complex voltage by the complex impedance.   Equation (83) has exactly the same form as *Ohm's law* for direct-current circuits.   It turns out that the concept of impedance can be used for any alternating-current circuit.   Further, complicated circuits can be handled by combining impedances in the same manner as resistances in direct-current circuits.   Therefore alternating-current circuits can be handled by *complex-number arithmetic* instead of solving the differential equations directly.

† See Example 6, page 125.

To obtain the particular solutions I_{1p} and I_{2p}, we must find the real and imaginary parts of (81). To do this, we write the complex impedance in the *polar form* (see Fig. 6).

$$Z = R + i\omega L = |Z| e^{i\delta} \tag{84}$$

where

$$|Z| = \sqrt{R^2 + \omega^2 L^2} \tag{85}$$

and δ is the *positive acute* angle given by

$$\delta = \arctan \frac{\omega L}{R} \tag{86}$$

Using (84) in Eq. (83), we obtain

$$I_p = \frac{E_0 e^{i\omega t}}{|Z| e^{i\delta}} = \frac{E_0}{|Z|} e^{i(\omega t - \delta)} = \frac{E_0}{|Z|} [\cos (\omega t - \delta) + i \sin (\omega t - \delta)] \tag{87}$$

Therefore

$$I_{1p} = \frac{E_0}{|Z|} \cos (\omega t - \delta) \tag{88}$$

$$I_{2p} = \frac{E_0}{|Z|} \sin (\omega t - \delta) \tag{89}$$

These solutions are called the *steady-state* currents for the alternating-current circuits. The complete current can be found by adding the transient current (70) to these steady-state currents. We note that the steady-state current is a sinusoid of the same circular frequency ω as the applied voltage. However, there is a *phase difference* δ/ω† between the current and the voltage. Further, the amplitude of the steady-state current is $E_0/|Z|$ where E_0 is the amplitude of the applied voltage and $|Z| = \sqrt{R^2 + \omega^2 L^2}$ is the magnitude of the impedance.

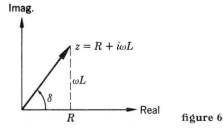

figure 6

† Note that $\cos (\omega t - \delta) = \cos \omega(t - \delta/\omega)$ so that the graph of this function (plotted against t) is obtained from the graph of $\cos \omega t$ by a translation of δ/ω units to the right along the t axis.

Circuits containing resistance, inductance, and capacitance will be discussed in Chap. 6.

Problems 3-9

1. Find a particular solution of

$$L\frac{dI}{dt} + RI = E_0 e^{\alpha t}$$

assuming $\alpha \neq -R/L$. Could this solution have been obtained by the "impedance" approach described in the text?

2. An RL circuit has no voltage source but has an initial current $I(0) = I_0$. Find the current at any time and draw a graph of $I(t)$.

3. An RL circuit contains a battery and a switch in series as shown in the figure. The switch is closed at $t = 0$ and opened at $t = t_1 > 0$. Assume that $I(0) = 0$ and that $I(t)$ is continuous for $t \geq 0$ even though the voltage is discontinuous at $t = t_1$.

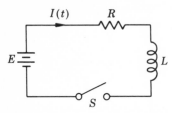

a. Without solving the differential equation, sketch a graph of $I(t)$.
b. Solve analytically for $I(t)$ for the two intervals $0 \leq t \leq t_1$ and $t \geq t_1$.

4. In Prob. 3 assume the switch is opened and closed periodically. Without solving the differential equation, sketch the graph of the current.

5. The current in a circuit containing a resistance, a capacitance, and a source of voltage in series satisfies the differential equation

$$R\frac{dI}{dt} + \frac{1}{C}I = \frac{dE}{dt}$$

where C is the capacitance.

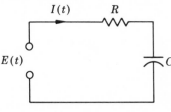

a. If $E = E_0 e^{\alpha t}$, show that

$$I = \frac{E_0 e^{\alpha t}}{R + 1/(\alpha C)} + ke^{-t/RC}$$

b. Use the impedance method to find the steady-state current if the voltage $E(t)$ is sinusoidal, i.e., $E = E_0 \cos \omega t$ or $E = E_0 \sin \omega t$. Find the complex impedance of the circuit.

3-10   Families of curves and orthogonal trajectories

The set of curves determined by an equation

$$f(x,y,c) = 0 \tag{90}$$

by allowing the parameter c to vary over some interval of real numbers is called a one-parameter family of curves.   For instance

$$x^2 + y^2 = c \qquad c > 0 \tag{91}$$

represents the one-parameter family of circles with center at the origin and radius $\sqrt{c}$.   Likewise,

$$y = mx \qquad -\infty < m < \infty \tag{92}$$

represents the one-parameter family of straight lines through the origin (except the line $x = 0$).   These two families of curves are shown in Fig. 7.

From simple geometry we know that *each* of the circles of the family (91) intersects *all* of the straight lines (92) at right angles and conversely.   Whenever this situation prevails between two families of curves, we say that each family is the *orthogonal trajectory* of the other.

Orthogonal families of curves occur often in physics and engineering problems.   For instance, the curves along which heat flows, *the flow lines*, are orthogonal to *the isothermal lines*.   Similarly, the *lines of force* are orthogonal to the *equipotential lines*.

We now consider the problem of finding the orthogonal trajectory of a given family of curves.   In order to do this, we shall first find a differential equation satisfied by all members of the given family.   Consider, for

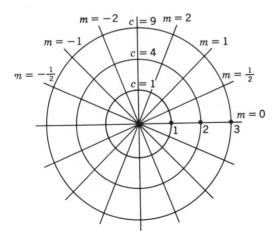

figure 7

example, the family of lines

$$y = mx \tag{92}$$

By differentiation we find

$$\frac{dy}{dx} = m \tag{93}$$

A *particular* member of the family corresponding to the parameter value m satisfies the differential equation (93). In order to find a differential equation satisfied by *all* members of the family (92), we eliminate the parameter m from the two equations (92) and (93) to obtain

$$\frac{dy}{dx} = \frac{y}{x} \tag{94}$$

This is called the *differential equation of the family* (92). It is important to note that *every* member of the family (92) satisfies this *one* differential equation. Obviously *the differential equation of a family must be independent of the parameter defining the family.*

After finding the differential equation of the family, it is an easy matter to find the differential equation of the orthogonal trajectories. In the case of the family of lines (92), the differential equation of the orthogonal trajectory is simply

$$\frac{dy}{dx} = -\frac{x}{y} \tag{95}$$

For, according to (94), the slope of the curve passing through the point (x,y) is y/x, and therefore the slope of the orthogonal curve through the same point is the negative reciprocal $(-x/y)$. Solving Eq. (95) by separation of variables, we find the orthogonal trajectories are the curves

$$x^2 + y^2 = c$$

in agreement with our previous discussion. The following method is a generalization of this procedure.

Method   *To obtain the orthogonal trajectories of a given family of curves $f(x,y,c) = 0$, obtain the differential equation of the family of curves by eliminating the parameter c between the given equation and the equation obtained by differentiating implicitly with respect to x. If the resulting differential equation of the family is*

$$\frac{dy}{dx} = g(x,y)$$

then the differential equation of the orthogonal trajectory is

$$\frac{dy}{dx} = -\frac{1}{g(x,y)}$$

and the orthogonal trajectories can be obtained by solving this differential equation.

Problem 3-10

1. For each of the following, describe and sketch the family of curves, find the differential equation of the family, and find the orthogonal trajectories:

a. $x^2 + y^2 = c$
c. $y = cx^2$
e. $x^2 + y^2 = cx$

b. $x^2 - y^2 = c$
d. $y = ce^x$

Supplementary problems for chapter 3

1. The velocity with which water flows out of an orifice is the same as it would have in falling freely from the water surface to the orifice. The stream of water contracts on leaving the orifice so that the effective area of the orifice is about six-tenths of the actual area. If x is the height of water above the orifice and $A(x)$ is the cross-sectional area of the vessel at this height, then the height x satisfies the equation

$$A(x)\frac{dx}{dt} = -a\sqrt{2gx}$$

where a is the effective area of the orifice.

a. Derive the differential equation.
b. Find the time to empty a cylindrical tank 3 ft high and 1 ft in radius through an orifice in the bottom with an effective area of 4 sq in.

2. Show that if a vessel filled with a liquid is rotated at a uniform velocity about a vertical axis, the surface assumes the shape of a paraboloid of revolution. Hint: The resultant force on a particle of the liquid at the surface is normal to the surface.

3. A particle of mass m is projected with an initial speed of v_0 at an elevation angle of α. If the particle encounters an air resistance proportional to its velocity, find the parametric equations of the trajectory.

4. Two points A and B are directly opposite each other on the banks of a river of width w. A man starts at A and rows across the river always heading directly toward B. Assume that the river flows with a uniform speed of v_1 and the man's rate of rowing in still water is v_2. Find the path followed by the man:

a. If $v_2 = v_1$.
b. If $v_2 = 2v_1$.

5. *Curve of pursuit.* Find the path of a dog which runs to overtake its master, both moving with uniform speed and the latter in a straight line.

6. A destroyer spots an enemy submarine 4 miles away. The submarine immediately descends and departs at full speed in a straight course of unknown direction. The speed of the destroyer is three times the speed of the submarine. Discover a course the destroyer should take to be sure to overtake the submarine.

Existence and Uniqueness and Methods of Approximation

4

4-1 Introduction

Up to this point we have attempted to find solutions of differential equations and to represent them in a simple analytic form. Often this is not possible. In this chapter we put aside the matter of finding explicit expressions for the solutions and take a more general point of view. Our aim is to find conditions that guarantee the existence and the uniqueness of solutions and to consider general methods of approximating these solutions.

We shall begin with a simple but useful geometric interpretation of solutions of differential equations.

4-2 The direction field

Consider

$$\frac{dy}{dx} = f(x,y) \tag{1}$$

The geometric interpretation of this equation is that the integral curve passing through (x,y) must have a slope given by $f(x,y)$.

Example 1 $y' = \dfrac{-x}{y}$ $y > 0$

The slope of the integral curve through (x,y) is $-x/y$. Therefore, at each of its points, the integral curve is perpendicular to the radius vector connecting $(0,0)$ to (x,y). We deduce from elementary geometry that the integral curves are circular arcs about the origin. This can readily be verified analytically by showing that $y = \sqrt{a^2 - x^2}$ satisfies the differential equation.

At each point (x,y) where the function $f(x,y)$ in Eq. (1) is defined, we may imagine a short straight line of slope $f(x,y)$ to be drawn. Such a line is called a *lineal element*. The collection of all lineal elements is called a direction field. Thus, geometrically, the differential equation (1) prescribes a direction field. The problem of solving a differential equation is to find curves which "fit" the direction field, that is, curves that are at each point (x,y) tangent to the lineal element through (x,y).

By drawing lineal elements through various points, we may often obtain a good enough idea of the direction field to be able to "see" the integral curves as in Fig. 1.

Example 2 $y' = x + y + \frac{1}{4}$

The direction field for this equation is shown in Fig. 1. For example, the slope of the lineal element through $(1,0)$ is $\frac{5}{4}$. The integral curve through the origin, shown in Fig. 1, is obtained by drawing a curve through the origin which is everywhere tangent to a lineal element. An inspection of Fig. 1 discloses that there is an integral curve through the point $(0,-\frac{5}{4})$ which is a straight line having a slope of -1; we can easily check this by showing that $y = -x - \frac{5}{4}$ satisfies the differential equation. Integral curves could be drawn through every point y_0 on the y axis. This suggests that the general solution is a one-parameter family of curves where y_0 can be taken as parameter.

By differentiating both sides of the differential equation, we can obtain the second derivative of the solution

$$y'' = 1 + y' = \frac{5}{4} + x + y$$

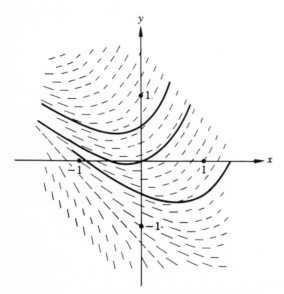

figure 1.   *Direction field for* $y' = x + y + \frac{1}{4}$.

From this equation we see that $y'' > 0$ where $\frac{5}{4} + x + y > 0$, $y'' = 0$ where $\frac{5}{4} + x + y = 0$, and $y'' < 0$ where $\frac{5}{4} + x + y < 0$. Thus above the line $\frac{5}{4} + x + y = 0$ the integral curves are concave up, and below this line they are concave down.

Useful aids to drawing the direction field are the *isoclines*. An isocline is a curve, such that the lineal elements through the points on this curve all have the same slope. The isoclines for $y' = f(x,y)$ are simply the curves $f(x,y) = c$, where c represents the slope of the lineal element through each point of the isocline. If the isoclines are simple curves, this provides a method for rapidly sketching the direction field.

Example 3 $y' = 1 - x^2 - y^2$

The isoclines are the circles $x^2 + y^2 = 1 - c$. These are sketched in Fig. 2 together with lineal elements of slope c, for various values of c. By means of the direction field thus obtained, the integral curve through the origin is sketched.

A polygonal approximation to the integral curve through (x_0,y_0) may be obtained by following the lineal element through (x_0,y_0) for a "short" distance to a point (x_1,y_1) and then following the lineal element through (x_1,y_1) a "short" distance to a point (x_2,y_2) and so on.

One might expect that, as the polygonal segments are made shorter, the approximations improve and in the limit the integral curve through (x_0,y_0) is

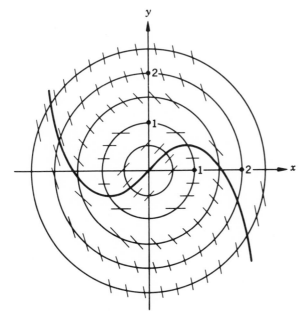

figure 2.   *Isoclines for* $y' = 1 - x^2 - y^2$.

obtained. With proper restrictions on $f(x,y)$ this is in fact true and furnishes the basis of a method of approximating solutions and of proving the existence of a solution. (See Secs. 4-3 and 4-5.)

Problems 4-2

1. Deduce the solutions of $y' = y/x$ for $x \neq 0$ solely from the geometric interpretation of the equation.

2. Without solving the equations, determine whether or not any straight-line solutions exist for:

a. $y' = 2x - y$
b. $y' = 2x^2 - y$

3. Sketch the direction field for the following equations and sketch various integral curves. Are there any points through which there would appear to be more than one integral curve?

a. $yy' = x$ b. $xy' = -y$
c. $y' = 3x$ d. $y' = 2x - y$
e. $y' = x^2 + y^2$ f. $yy' = 1 - y$
g. $y' = \sqrt{|y|}$ h. $y' + e^{y/x} = 0$
i. $y' = \dfrac{x^2 + y}{x} \qquad x \neq 0$ j. $y' = xy + y^2$
k. $y' = \max{(x,y)}$ l. $y' = \sin\sqrt{x^2 + y^2}$

4. *Liénard's construction.*† To find the direction field for an equation of the form

$$\frac{dy}{dx} = -\frac{x + f(y)}{y}$$

we first draw the *characteristic curve* $x = -f(y)$. To find the lineal element through $P(x,y)$, draw a horizontal line from P to a point Q_1 on the characteristic and then drop a perpendicular line to a point Q_2 on the x axis. The linear element at P is then perpendicular to PQ_2. (See Fig. 3.) This construction can easily be performed using a straight edge and a right triangle.

a. Show that this construction gives the lineal element at P.

Sketch the various integral curves for:

b. $y' = \dfrac{x - y}{y}$

c. $y' = -\dfrac{x + y^2}{y}$

† See also Sec. 10-4.

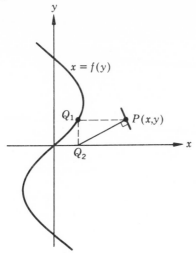

figure 3

5. *Phase plane.* Many second-order equations important in applications have the form

$$F(\ddot{x}, \dot{x}, x) = 0 \tag{i}$$

i.e., the equations do not depend explicitly on t. Letting $v = \dot{x}$, Eq. (i) is equivalent to a first-order equation for v as a function of x:

$$f\left(v \frac{dv}{dx}, v, x\right) = 0 \tag{ii}$$

The xv plane is called the *phase* plane and the integral curves of (ii) in the xv plane are called the *phase trajectories* of (i). The phase trajectories are oriented by the time t. As t increases, since $v = dx/dt$, the phase trajectories are traversed so as to increase x if $v > 0$ and to decrease x if $v < 0$. Also periodic solutions of (i) correspond to closed curves in the phase plane.

Find analytical representations of the phase trajectories for:

a. $\ddot{x} + x = 0$ b. $\ddot{x} + \dot{x} = 0$

c. $\ddot{x} = 1$ d. $\ddot{x} - x = 0$

e. Find the phase trajectories for the *Van der Pol*† equation using Liénard's construction (Prob. 4).

$$\ddot{x} - \left(1 - \frac{\dot{x}^2}{3}\right)\dot{x} + x = 0$$

6. Sketch some of the phase trajectories for the equation

$$\ddot{x} + x = -1 \qquad \dot{x} > 0$$
$$\ddot{x} + x = +1 \qquad \dot{x} < 0$$

† See Chap. 10, page 324.

assuming the phase trajectories are continuous curves. (This is an example of what is called *coulomb friction*.)† If we define sgn $z = +1$, $z > 0$, and sgn $z = -1$, $z < 0$, the above equation can be written $\ddot{x} + x = -\text{sgn } \dot{x}$.

7. Sketch some of the phase trajectories of:

a. $\ddot{x} = -\text{sgn } x$ b. $\ddot{x} = -\text{sgn } (x + y)$

c. $\ddot{x} = -\text{sgn } (x - y)$

These equations are examples of dynamical equations for simple *control systems* where x represents the error or deviation of the system from some desired state and the right-hand sides are various *control functions* or correcting forces. For example, in part a of this problem the control function $-\text{sgn } x$ has a sign opposite to the sign of x and thus applies a force to the system which tends to reduce the error.

8. Sketch the phase trajectories of $\ddot{x} + x = -\text{sgn } x$.

4-3   Existence and uniqueness of solutions

Let R be a rectangle in the xy plane

$$R: a \leq x \leq b \qquad c \leq y \leq d \tag{2}$$

and consider the differential equation

$$\frac{dy}{dx} = f(x,y) \tag{3}$$

From the viewpoint of the direction field, the following existence theorem may appear plausible (see Fig. 4).

Theorem 1 *If $f(x,y)$ is continuous in R, then there exists at least one solution of $y' = f(x,y)$ through every interior point‡ (x_0,y_0) in R; furthermore the solution(s)§ through (x_0,y_0) exists up to the boundary of R.*

We shall not prove this theorem here¶; we shall prove Theorem 2 below, which is less general than Theorem 1, but which also guarantees the uniqueness of the solution. The following example shows that the conditions in Theorem 1 are not strong enough to ensure uniqueness.

† See Sec. 10-5.

‡ An interior point of R is a point of R that is not on the boundary, i.e., a point (x,y) such that $a < x < b$, $c < y < d$.

§ Note that since $y' = f(x,y)$ with $f(x,y)$ continuous, a solution y has not only a derivative but a *continuous* derivative.

¶ See Hurewicz [18], chap. 1, pt. A.

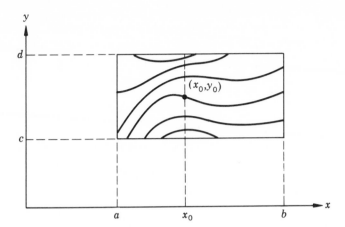

figure 4

Example 1   DE: $y' = |y|^{\frac{1}{2}}$

IC: $y(0) = 0$

We note that $|y|^{\frac{1}{2}}$ is continuous in the entire xy plane.   It is clear that one solution is $y \equiv 0$.   A second solution can be obtained by separating variables.   Assuming $y(x) > 0$, we find $y = \frac{1}{4}x^2$ (for $x \geq 0$).   Similarly assuming $y(x) < 0$, we find $y = -\frac{1}{4}x^2$ (for $x \leq 0$).   We may piece these two parts together as follows:

$$y = \begin{cases} \dfrac{x^2}{4} & x \geq 0 \\[2mm] \dfrac{-x^2}{4} & x \leq 0 \end{cases}$$

and we can easily check that the above is a second solution of the differential equation in the interval $-\infty < x < \infty$.

A sufficient condition for uniqueness is that both $f(x,y)$ and $\partial f(x,y)/\partial y$ be continuous in R.   We see that this condition is violated in Example 1, for in this case $f(x,y) = |y|^{\frac{1}{2}}$ and $f_y = 1/(2\sqrt{y})$ if $y > 0$, $f_y = -1/(2\sqrt{-y})$ if $y < 0$; thus f_y is discontinuous at $y = 0$.

If $\partial f(x,y)/\partial y$ is continuous in R and (x,y_1), (x,y_2) are two points in R, then by the mean-value theorem we can write

$$f(x,y_1) - f(x,y_2) = f_y(x,\bar{y}_1)(y_1 - y_2) \tag{4}$$

where $\bar{y}_1$ is between y_1 and y_2.   Now since f_y is continuous in R, it is bounded,† say $|f_y(x,y)| \leq A$ for all (x,y) in R.   From Eq. (4) we find

$$|f(x,y_1) - f(x,y_2)| \leq A \, |y_1 - y_2| \tag{5}$$

† See Courant [12], vol. 2, p. 97.

for all points in R.   If Eq. (5) holds, f is said to satisfy a *Lipschitz* condition in R with respect to the variable y.   We have just shown that if f_y is continuous in R, then f satisfies a Lipschitz condition in R with respect to y. The converse is not true (see Prob. 4).   Therefore the Lipschitz condition is weaker than the condition that f_y is continuous.   However, it turns out that the Lipschitz condition is sufficient to ensure existence and uniqueness of the solution of the differential equation and the following theorem holds.

Theorem 2   *If $f(x,y)$ is continuous in R and satisfies a Lipschitz condition in R with respect to y, then there exists a unique solution of $y' = f(x,y)$ through each point (x_0,y_0) in the interior of R; the solution has a continuous derivative and is defined for some interval $|x - x_0| \leq \alpha$.*

We shall prove this theorem in the appendix to this chapter.   We note (see Fig. 5) that the solution through a given point may leave the rectangle R through either the vertical side or the horizontal side.   Therefore the x interval over which a solution is defined is in general only a subinterval of $a \leq x \leq b$.   It is not in general easy to predict the largest interval in which the solution is defined.†   The following example illustrates this (see also Probs. 1 and 3).

Example 2   DE: $y' = 1 + y^2$

IC: $y(0) = 0$

Both $f(x,y) = 1 + y^2$ and $f_y = 2y$ are continuous in the entire xy plane.   Therefore the rectangle R can be taken as large as desired.   The unique solution of

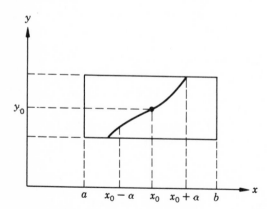

figure 5

† Linear equations are an exception.   See Sec. 4-8.

the problem is easily found to be $y = \tan x$. This solution is defined for $-\frac{1}{2}\pi < x < \frac{1}{2}\pi$. If a solution is desired for a larger interval, say $-\pi < x < \pi$, no such solution exists.

If we keep the point x_0 fixed (see Fig. 5), Theorem 2 guarantees a unique solution for each value of y_0 within the rectangle. These solutions can be represented by the equation

$$y = g(x,y_0) \tag{6}$$

with $y(x_0) = y_0$. This equation represents a one-parameter family of curves with parameter y_0. If we restrict ourselves to a small enough interval around x_0, Eq. (6) represents all solutions of the differential equation for which $c < y < d$. In this case we can say that the *general solution* is the one-parameter family of functions given by (6).

Problems 4-3

1. By modifying Example 2, construct a differential equation such that a solution exists in the interval $-a < x < a$, where a is a positive number (possibly very small) less than $\pi/2$ but does not exist in $-b < x < b$ for $b > a$.

2. Construct a nonlinear differential equation whose solution exists for all x.

3. Suppose $f(x,y)$ is continuous in R: $|x| \le a$, $|y| \le b$, and $|f(x,y)| < M$ in R. For what x interval would you expect the solution of $y' = f(x,y)$, $y(0) = 0$, to exist?

4. Show that the function $f(x) = |x|$ satisfies a Lipschitz condition for x in the interval $-1 \le x \le 1$. Show also that $f(x)$ does not have a derivative with respect to x at $x = 0$.

5. Does Example 1 have any other solutions in $-\infty < x < \infty$ besides the solutions given?

6. Show that the problem $y' = 3y^{\frac{2}{3}}$, $y(0) = 0$, does not have a unique solution.

7. For what values of $\alpha > 0$ does $y' = y^{\alpha}$, $y(0) = 0$, have a unique solution?

8. For what regions R in the xy plane do the following satisfy the conditions of Theorem 2 and therefore possess a unique solution through a given point?

a. $y' = \dfrac{y}{x}$ b. $y' = -\dfrac{x}{y}$

c. $y' = y^{\frac{1}{3}}$ d. $y' = \dfrac{xy}{x^2 + y^2}$

e. $y' = e^{x/y}$ f. $y' = \sin(xy)$

g. $y' + P(x)y = Q(x)$, $P(x)$ and $Q(x)$ continuous for $a \le x \le b$

4-4   The Picard method

One of the methods of proving the existence theorem is an iteration method called the Picard method in honor of Emile Picard. This method consists of constructing an appropriate sequence of approximate solutions to the differential equation which converges to the solution (most existence proofs use this general approach). In this section we shall describe the method and illustrate it with two examples. The proof of Theorem 2 using the Picard method is given in the appendix to this chapter.

Consider the initial-value problem

$$\text{DE: } \frac{dy}{dx} = f(x,y)$$
$$\text{IC: } y(x_0) = y_0 \tag{7}$$

where we assume that $f(x,y)$ is continuous in a rectangle R containing (x_0,y_0). We convert (7) to the integral equation

$$y(x) = y_0 + \int_{x_0}^{x} f(t,y(t))\, dt \tag{8}$$

If $y(x)$ is any solution of (7), then we may obtain (8) by integrating the differential equation in (7) from x_0 to x. Conversely, if $y(x)$ is any continuous function satisfying (8), we see that $y(x_0) = y_0$ and by differentiation we find that $y(x)$ also satisfies the differential equation. Therefore finding a solution of the initial-value problem (7) is equivalent to finding a continuous solution of the integral equation (8).

We now turn to solving the integral equation (8) by the method of iteration. We are looking for a function which when placed in the right-hand side of (8) will reproduce itself on the left-hand side. We start by taking as an initial approximation any continuous function $y_0(x)$ having the property that $y_0(x_0) = y_0$. Often we pick $y_0(x) \equiv y_0$, but it will be shown that the choice of $y_0(x)$ does not matter. Substituting $y_0(x)$ for $y(x)$ in the right-hand side of (8), we obtain

$$y_1(x) = y_0 + \int_{x_0}^{x} f(t,y_0(t))\, dt \tag{9}$$

We then take $y_1(x)$ as our next approximation. We note that whereas $y_0(x)$ has the same value at x_0 as the true solution $y(x)$, the function $y_1(x)$ has not only the same value at x_0 but also the same derivative at x_0 as the true solution $y(x)$. We may therefore expect that $y_1(x)$ is a better approximation to the solution than $y_0(x)$. To obtain the next approximation, we repeat the process to obtain

$$y_2(x) = y_0 + \int_{x_0}^{x} f(t,y_1(t))\, dt \tag{10}$$

In general, we obtain the $(n + 1)$st approximation from the nth approximation by means of the relation

$$y_{n+1}(x) = y_0 + \int_{x_0}^{x} f(t,y_n(t))\, dt \qquad n = 0, 1, 2, \ldots \tag{11}$$

If the sequence of functions $y_n(x)$ converges uniformly† to a continuous limit function $y(x)$ in some interval containing x_0, then we may pass to the limit in both sides of (11) to obtain

$$y(x) = y_0 + \int_{x_0}^{x} f(t,y(t))\, dt \tag{12}$$

so that $y(x)$ is our desired solution.

In the appendix to this chapter we shall show under the conditions of Theorem 2, Sec. 4-3, that the sequence $y_n(x)$ does indeed converge to the unique solution of the initial-value problem regardless of the choice of the initial function $y_0(x)$. We shall illustrate this convergence in Example 1 below for a simple differential equation.

Although the Picard method is extremely important in the theoretical study of differential equations, it is not always a convenient method for obtaining an explicit approximate solution to a given differential equation because of the many integrations that must be performed. In Example 2 below we give an example in which the integrations can be explicitly carried out.

Example 1   DE: $y' = -y$

IC: $y(0) = 1$

Rewriting the differential equation as an integral equation, we obtain

$$y(x) = 1 - \int_{0}^{x} y(t)\, dt$$

Letting $y_0(x) \equiv 1$, we obtain

$$y_1(x) = 1 - \int_{0}^{x} dt = 1 - x$$

$$y_2(x) = 1 - \int_{0}^{x} (1 - t)\, dt = 1 - x + \frac{x^2}{2!}$$

$$y_3(x) = 1 - \int_{x_0}^{x} \left(1 - t + \frac{t^2}{2!}\right) dt = 1 - x + \frac{x^2}{2!} - \frac{x^3}{3!}$$

It is easy to prove by induction that

$$y_n(x) = 1 - x + \frac{x^2}{2!} + \cdots + (-1)^n \frac{x^n}{n!}$$

† See Appendix A.

Recalling the Taylor-series expansion of e^{-x}, we see that

$$\lim_{n \to \infty} y_n(x) = e^{-x}$$

Therefore the approximate solution approaches the true solution.

Example 2   DE: $y' = x + y^2$
IC: $y(1) = 1$

Rewriting the differential equation as an integral equation, we obtain

$$y(x) = 1 + \int_1^x [t + y^2(t)] \, dt$$

The first two Picard approximations are

$$y_0 = 1$$
$$y_1 = 1 + \int_1^x (t + 1) \, dt$$

It is convenient to have the answer in powers of $x - 1$ rather than in powers of x. We therefore rewrite the integrand:

$$y_1 = 1 + \int_1^x [(t - 1) + 2] \, dt = 1 + 2(x - 1) + \tfrac{1}{2}(x - 1)^2$$

The next approximation is

$$y_2 = 1 + \int_1^x \{(t - 1) + 1 + [1 + 2(t - 1) + \tfrac{1}{2}(t - 1)^2]^2\} \, dt$$
$$= 1 + 2(x - 1) + \tfrac{5}{2}(x - 1)^2 + \tfrac{5}{3}(x - 1)^3 + \tfrac{1}{4}(x - 1)^4 + \tfrac{1}{20}(x - 1)^5$$

From this approximation we can calculate approximate values of y near $x = 1$ and expect some degree of accuracy. We obtain, for example, $y(1.1) \approx 1.2$, $y(1.2) \approx 1.5$.

Problems 4-4

1. In the following problems show that the sequence of Picard approximations converges to the true solution:

a. $y' = x + y$ $y(0) = 0$ b. $y' = -y$ $y(1) = 1$

2. Find the Picard approximations for:

a. $y' = 1 + y^2$ $y(0) = 0$ (first four approximations)
b. $y' = x^2 + y^2$ $y(0) = 1$ (first three approximations)

3. If $f(x,y)$ has continuous partial derivatives of all orders, show that the nth Picard approximation $y_n(x)$ has the same value at x_0 and the same derivatives up to order n at x_0 as the true solution.

4-5   The Cauchy-Euler method

One of the basic concepts in calculus is that the tangent line to a curve furnishes a "good" approximation to the curve "near" the point of tangency. The Cauchy-Euler method uses this simple idea to construct an approximate solution to a differential equation (see also the last paragraph of Sec. 4-2). By refining the approximation, it is possible to construct a solution in terms of a limiting process.

We consider the initial-value problem

$$\text{DE:} \frac{dy}{dx} = f(x,y)$$
$$\text{IC:} \; y(x_0) = y_0 \tag{13}$$

We shall find approximate values of the solution at the equidistant points spaced a distance h apart:

$$x_n = x_0 + nh \qquad n = 0, 1, 2, \ldots$$

Let us denote the actual solution of (13) by $y(x)$ and an approximate solution by $Y(x)$.   Also let

$$y(x_k) = y_k \qquad k = 0, 1, \ldots \tag{14}$$
$$Y(x_k) = Y_k \qquad k = 0, 1, \ldots \tag{15}$$

Since the initial condition is given, we set $Y_0 = y_0$.   To find Y_1, we replace the actual solution in the interval x_0, $x_0 + h$ by the tangent line having slope $f(x_0, Y_0)$.   Therefore (see Fig. 6)

$$Y_1 = Y_0 + hf(x_0, Y_0) \tag{16}$$

We take (x_1, Y_1) as an approximation to the point (x_1, y_1) on the integral curve. At (x_1, Y_1) we again approximate the solution by a tangent line, this time of

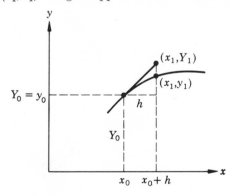

figure 6

slope $f(x_1,Y_1)$, to obtain

$$Y_2 = Y_1 + hf(x_1,Y_1) \tag{17}$$

Proceeding in this manner, we find

$$Y_{n+1} = Y_n + hf(x_n,Y_n) \qquad n = 0, 1, 2, \ldots \tag{18}$$

Equation (18) is called a *recurrence relation* or a *difference*† *equation* which enables the approximations Y_1, Y_2, ... to be determined one after another.

From this process we obtain what is geometrically a polygonal approximation to the integral curve through (x_0,y_0). Under the conditions on f stated in Theorem 1, Sec. 4-3, it can be proved‡ that this polygonal approximation approaches the true solution as the step size h approaches zero. The first example below illustrates this for a simple differential equation.

For a specific value of h we may use (18) to obtain an approximate numerical solution§ to any first-order equation of the form (13). This is illustrated in Example 2.

Example 1   DE: $\dfrac{dy}{dx} = -y$

IC: $y(0) = 1$

Let $x_0 = 0$, $x_1 = h, \ldots, x_n = nh$. We have

$$Y_0 = 1$$
$$Y_{n+1} = Y_n - hY_n$$

For this simple example it is possible to solve this difference equation for a general Y_n. We see that

$$Y_1 = 1 - h$$
$$Y_2 = (1 - h)Y_1 = (1 - h)^2$$

It is easy to guess, and prove by mathematical induction, that

$$Y_n = (1 - h)^n$$

Letting $nh = x$, we may write the solution in the form

$$Y_n = \left(1 - \frac{x}{n}\right)^n$$

Now passing to the limit as $h \to 0$ and $n \to \infty$ in such a manner that $nh = x$

† See Chap. 11 for a further discussion of difference equations.
‡ See Hurewicz [18], Chap. 1, pt. A.
§ See Chap. 12 for more refined numerical methods.

remains fixed, we obtain

$$\lim_{n \to \infty} Y_n = \lim_{n \to \infty} \left(1 - \frac{x}{n}\right)^n = e^{-x}\dagger$$

It is easily verified that e^{-x} is the true solution of our problem. Therefore, in this special case, we have demonstrated that the approximate solution approaches the true solution as the increment approaches zero.

Example 2   Obtain an approximate solution of

DE: $\dfrac{dy}{dx} = x + y^2$

IC: $y(1) = 1$

at $x_0 = 1$, $x_1 = 1.1, \ldots, x_5 = 1.5$. We have $x_0 = 1$, $h = 0.1$, $y_0 = Y_0 = 1$. Therefore

$$Y_1 = Y_0 + hf(x_0, Y_0)$$
$$= 1 + 0.1(1 + 1) = 1.2$$

$$Y_2 = Y_1 + hf(x_1, Y_1)$$
$$= 1.2 + 0.1(1.1 + 1.2^2)$$
$$= 1.5 \qquad \text{approximately}$$

Proceeding in a similar manner, we obtain $Y(1.3) \approx 1.8$, $Y(1.4) \approx 2.2$, $Y(1.5) \approx 2.9$. These results agree with Example 2 in Section 4-4.

Problems 4-5

1. Consider $y' + y/x = 1$, $y(1) = \frac{1}{2}$. Use an increment h and show that the approximate solution approaches the actual solution as $h \to 0$.

2. Using $h = 0.1$, find approximate values of the solution for five increments to the right of the initial point for the following equations:

a. $y' = x - y^2$ $y(0) = 1$ (Use two decimal places.)
b. $y' = x^2 + y^2$ $y(0) = 0$ (Use three decimal places.)

4-6   Taylor series

Another very useful method of approximating solutions is through the use of Taylor series. We recall that the Taylor series for an infinitely differentiable function $f(x)$ about a point $x = a$ is

$$\sum_{k=0}^{\infty} \frac{f^{(k)}(a)}{k!}(x - a)^k = f(a) + f'(a)(x - a) + \frac{f''(a)}{2!}(x - a)^2 + \cdots \tag{19}$$

$\dagger$ See Courant [12], vol. 1, p. 175.

It is true[†] for many (but not all) infinitely differentiable functions that the Taylor series for $f(x)$ converges to $f(x)$ in some interval about $x = a$. We note that the Taylor series for a function about $x = a$ can be written down as soon as *all* the derivatives are known at the one point $x = a$.

Consider now the problem

DE: $y' = f(x,y)$

IC: $y(x_0) = y_0$ (20)

where we assume for simplicity that $f(x,y)$ is a *polynomial* in x and y. If we assume that the solution $y(x)$ possesses a Taylor series in the vicinity of $x = x_0$, then it is easy in principle to find this series. We have from the initial condition that $y(x_0) = y_0$ and from the differential equation $y'(x_0) = f(x_0,y_0)$. To find the second derivative, we differentiate the differential equation to obtain

$$y''(x) = \frac{\partial f}{\partial x} + \frac{\partial f}{\partial y} y'(x) \tag{21}$$

Thus $y''(x_0)$ can be calculated. By repeating this process, we may obtain as many of the higher derivatives of y evaluated at x_0 as desired. Consequently the tentative solution can be expressed as the infinite series

$$y(x) = y(x_0) + y'(x_0)(x - x_0) + \frac{y''(x_0)}{2!}(x - x_0)^2 + \cdots \tag{22}$$

If this series converges in some interval about x_0, the properties of power series[‡] ensure that it is indeed a solution of the differential equation. We illustrate this convergence in Example 1 below. It can be shown[§] that a Taylor-series solution exists when $f(x,y)$ is an analytic function[¶] of x and y about (x_0,y_0).

A few terms of the Taylor series often provide a close approximation to the value of $y(x)$ for values of x near x_0. (See Example 2, page 111.)

Example 1 $\dfrac{dy}{dx} = -y$ $y(0) = 1$

We have

$$y'(0) = -y(0) = -1 \qquad y''(0) = -y'(0) = +1 \qquad y'''(0) = -1$$

By induction we may prove that $y^{(n)}(0) = (-1)^n$. Therefore

$$y = 1 - x + \frac{x^2}{2!} - \frac{x^3}{3!} + \cdots \tag{i}$$

---

† See Courant [12], vol. 1, p. 336.

‡ See Appendix A, page 497, and Sec. 7-2.

§ See Martin and Reissner [26], pp. 59–66, or Ince [19], pp. 282–284.

¶ I.e., $f(x,y) = \sum_0^\infty \sum_0^\infty a_{mn}(x - x_0)^m (y - y_0)^n$.

The series on the right is the series for e^{-x}. Therefore the series converges for all x to the solution $y = e^{-x}$.

Alternatively, we could assume that

$$y(x) = \sum_{k=0}^{\infty} a_k x^k \tag{ii}$$

where the coefficients a_k must be determined so as to satisfy the differential equation.

Substituting Eq. (ii) into the differential equation, we find

$$\sum_{0}^{\infty} k a_k x^{k-1} = -\sum_{0}^{\infty} a_k x^k \tag{iii}$$

or since $\sum_{0}^{\infty} k a_k x^{k-1} = \sum_{k=0}^{\infty} (k + 1) a_{k+1} x^k$, we find that (iii) yields

$$\sum_{k=0}^{\infty} [(k + 1)a_{k+1} + a_k] x^k \equiv 0$$

Since a power series that is identically zero must have every coefficient equal to zero,† we obtain the *recurrence relation*

$$(k + 1)a_{k+1} + a_k = 0 \qquad k = 0, 1, \ldots \tag{iv}$$

From the initial condition we know that $a_0 = 1$; from (iv) we find $a_1 = -1$, $a_2 = -\frac{1}{2}a_1 = 1/2!, \ldots, a_n = (-1)^n/n!$. Thus we again obtain the solution (i).

Example 2 $\dfrac{dy}{dx} = x + y^2 \qquad y(1) = 1$

We have

$$y'(1) = 2 \qquad y''(1) = 1 + 2y(1)y'(1) = 5 \qquad y'''(1) = 18$$

The first four terms of the solution are therefore

$$1 + 2(x - 1) + \tfrac{5}{2}(x - 1)^2 + \frac{18}{3!}(x - 1)^3$$

The first three terms agree with the terms obtained by the Picard method.

Problems 4-6

1. Find the Taylor-series solution about $x = 0$ for the following by both of the methods used in Example 1. In each case show that the series converges to the actual solution.

  a. $y' = 2y \qquad y(0) = 2$ b. $y' = 2y \qquad y(0) = 0$
  c. $y' = x + y \qquad y(0) = 0$ d. $y' = x + y \qquad y(1) = 4$
  e. $y' = x^2 + y$ (general solution)
  f. $y'' + y = 0 \qquad y(0) = 0$
                        $y'(0) = 1$

† See Appendix A.

2. Find the first few terms of the Taylor-series solution for:

a. $y' = 1 + y^2$ $y(0) = 0$ b. $y' = x^2 + y^2$ $y(1) = 3$

c. $y'' + xy = 0$ $y(0) = 1$ d. $y'' = x^2 + y^2$ $y(0) = 1$

                      $y'(0) = -1$ $y'(0) = 0$

4-7   Existence and uniqueness theorems for systems of equations and higher-order equations

Consider a system of two equations

$$\text{DE: } \frac{dy}{dx} = f(x,y,z)$$

$$\frac{dz}{dx} = g(x,y,z) \tag{23}$$

$$\text{IC: } y(x_0) = a \qquad z(x_0) = b$$

Geometrically the solution of this system represents a curve in xyz space, the tangent to this curve having direction numbers $1, f, g$. As with a single equation, the continuity of f and g in some region R is not sufficient to ensure a unique solution. A sufficient condition for the existence of a unique solution is that f and g not only are continuous but satisfy a Lipschitz condition in each of the variables y and z.

A similar situation prevails for the initial-value problem for a system of n first-order equations

$$\text{DE: } \frac{dy_i}{dx} = f_i(x, y_1, y_2, \ldots, y_n)$$
$$\text{IC: } y_i(x_0) = c_i \qquad i = 1, 2, \ldots, n \tag{24}$$

Let R stand for the set of points $(x, y_1, \ldots, y_n)$† which satisfy

$$a_0 \leq x \leq b_0 \qquad a_i \leq y_i \leq b_i \qquad i = 1, 2, \ldots, n \tag{25}$$

Then the following theorem holds.

     Theorem 1   *If $f_i(x, y_1, \ldots, y_n)$ are continuous in R and satisfy Lipschitz conditions in R with respect to each of the y_j,‡ then there exists a unique solution of*

$$\frac{dy_i}{dx} = f_i(x, y_1, \ldots, y_n)$$

---

† The ordered set of $n + 1$ numbers $(x, y_1, \ldots, y_n)$ may be considered to be an element or point in an $(n + 1)$-dimensional vector space. See Chap. 11, page 355.

‡ I.e., $|f_i(x, y_1, \ldots, \bar{y}_j, \ldots, y_n) - f_i(x, y_1, \ldots, y_j, \ldots, y_n)| \leq k_i |\bar{y}_j - y_j|$.

through each interior point in R.   The $y_i(x)$ have continuous derivatives and are defined in some x interval containing x_0.

This theorem is proved in the appendix to this chapter.

We now consider an nth order equation which is solved for the nth derivative:

$$\text{DE: } y^{(n)} = f(x\, y, y^{(1)}, \ldots, y^{(n-1)}), \tag{26}$$

with initial conditions

$$\text{IC: } y(x_0) = c_1,\, y'(x_0) = c_2, \ldots,\, y^{(n-1)}(x_0) = c_n \tag{27}$$

By making a simple substitution, we may reduce (26) to a system of first-order equations.   We let $y = y_1$ and we introduce new functions $y_2, \ldots, y_n$ by

$$y_1' = y_2$$
$$y_2' = y_3$$

$$\cdot$$
$$\cdot \tag{28}$$
$$\cdot$$

$$y_{n-1}' = y_n$$
$$y_n' = f(x, y_1, \ldots, y_n)$$

The last equation comes from (26) since $y_n' = y_1^{(n)}$.   We note also that the initial conditions (27) become

$$y_i(x_0) = c_i \qquad i = 1, 2, \ldots, n \tag{29}$$

Thus the solution of the nth-order equation is equivalent to the solution of a system of n first-order equations.   Rephrasing Theorem 1, we have the following:

Theorem 2   *If $f(x, y_1, \ldots, y_n)$ is continuous in the region R given by (25) and satisfies a Lipschitz condition in R with respect to $y_1, \ldots, y_n$ and if $(x_0, c_1, \ldots, c_n)$ is in the interior of R, then Eq. (26) with initial conditions (27) possesses a unique solution.   The solution has a continuous nth derivative and is defined in some x interval around x_0.*

Problems 4-7

1. Let R be the region described by Eq. (25). If $f(x, y_1, \ldots, y_n)$ is continuous in R and satisfies a Lipschitz condition in R with respect to each of the y_j, show that

$$|f(x, y_1, \ldots, y_n) - f(x, z_1, \ldots, z_n)| \le A(|y_1 - z_1| + \cdots + |y_n - z_n|)$$

for all points in R.

2. Describe the regions R for which the following equations are assured to possess unique solutions:

a. $y'' = x^2 + y^2$ b. $y' = x + z$ $z' = \dfrac{y + x}{z}$

c. $x^2 y'' + xy' + (x^2 - 4)y = 0$

4-8   Existence and uniqueness theorems for linear equations

For linear equations the existence and uniqueness theorem can be somewhat strengthened. Let us recall the results for the initial-value problem for a first-order linear equation

$$\text{DE: } a(x)y' + b(x)y = f(x)$$
$$\text{IC: } y(x_0) = y_0$$

In Sec. 2-3 [Eq. (27)] we demonstrated that if $a(x)$, $b(x)$, $f(x)$ were continuous in a common interval I containing the point x_0, and $a(x)$ was never zero in I, then the initial-value problem possessed a unique solution *defined throughout I*. We note that the solution is guaranteed for the whole interval I, not just for a small neighborhood of x_0.

For an nth-order linear equation we have the following theorem.

Theorem 1 *Let $a_i(x)$ and $f(x)$ be continuous and $a_0(x)$ never zero in an interval I containing x_0; then the initial-value problem*

$$\text{DE: } a_0(x)y^{(n)} + a_1(x)y^{(n-1)} + \cdots + a_n(x)y = f(x)$$
$$\text{IC: } y(x_0) = b_0 \qquad y'(x_0) = b_1, \ldots, y^{(n-1)}(x_0) = b_{n-1}$$

possesses a unique solution. The solution has a continuous nth derivative and is defined throughout I.

This theorem is a special case of the following theorem for a system of first-order linear equations which will be proved in the appendix to this chapter.

Theorem 2 *Let $a_{ij}(x)$ and $f_i(x)$ be continuous in an interval I containing x_0; then the problem*

$$\text{DE: } \frac{dy_i}{dx} = \sum_{j=1}^{n} a_{ij}(x)y_j + f_i(x) \qquad i = 1, \ldots, n$$
$$\text{IC: } y_i(0) = b_i$$

possesses a unique solution. The $y_i(x)$ have continuous derivatives and are defined throughout I.

Problems 4-8

1. What is the unique solution of the initial-value problem in Theorem 1 when $f(x) \equiv 0$ and $b_0 = b_1 = \cdots = b_{n-1} = 0$?

2. What is the interval I in Theorem 1 when the $a_i(x)$ are constants?

3. Suppose the $a_i(x)$ and $f(x)$ in Theorem 1 are continuous for all x; can the initial-value problem have a solution $y = \tan x$? Why?

4. Show that Theorem 1 follows from Theorem 2.

5. A linear differential equation is called homogeneous if $f(x) \equiv 0$. Suppose that such an equation possesses a solution $y(x)$ such that $y(0) = y'(0) = y''(0) = 0$ but $y(x)$ is not identically zero. What can be said about the order of the differential equation?

6. Write the linear differential equation in Theorem 1 in the form $y^{(n)}(x) = F(x, y, \ldots, y^{(n-1)})$. In what region does F satisfy a Lipschitz condition in the variables $y, y', \ldots, y^{(n-1)}$?

Appendix: Proof of existence and uniqueness theorems

In the proof of these theorems, essential use is made of the concept of uniform convergence. This is discussed in Appendix A, page 496.

Uniqueness and existence for a single first-order equation. We now prove the basic theorem, Theorem 2, Sec. 4-3:

If $f(x,y)$ is continuous in a rectangle R and satisfies a Lipschitz condition in R with respect to y, then there exists a unique solution of $y' = f(x,y)$ through each interior point (x_0, y_0) in R; the solution has a continuous derivative and is defined for some interval about x_0.

Proof Let a, b be such that the rectangle R': $|x - x_0| \le a$, $|y - y_0| \le b$ lies entirely within R. If M is the maximum of $|f(x,y)|$ in R, we shall prove that a unique solution of $y' = f(x,y)$ with $y(x_0) = y_0$ exists for $|x - x_0| \le \alpha$, where α is the smaller of the numbers a, b/M.

This choice of α ensures that any solution $y(x)$ satisfying $y(x_0) = y_0$ must remain in R' (and therefore in R) for $|x - x_0| \le \alpha$. For, since $|y'(x)| \le M$, we must have $|y(x) - y_0| \le M|x - x_0|$, and since $|x - x_0| \le b/M$, this ensures that $|y(x) - y_0| \le b$ for $|x - x_0| \le \alpha$.

Existence. We have seen that solving the initial-value problem is equivalent to finding a continuous solution of the integral equation

$$y(x) = y_0 + \int_{x_0}^{x} f(t, y(t))\, dt \qquad |x - x_0| \le \alpha \tag{i}$$

It is convenient to assume $x \ge x_0$; a similar proof will hold for $x \le x_0$. Proceeding by the Picard method, we let

$$y_0(x) \equiv y_0$$

$$y_n(x) = y_0 + \int_{x_0}^{x} f(t, y_{n-1}(t))\, dt \qquad n = 1, 2, \ldots \tag{ii}$$

We first show that $|y_n(x) - y_0| \leq b$ for $x_0 \leq x \leq x_0 + \alpha$. Clearly the inequality holds for $n = 0$. Assuming that $|y_{n-1}(x) - y_0| \leq b$, we have from (ii)

$$|y_n(x) - y_0| \leq \int_{x_0}^x |f(t,y_{n-1}(t))| \, dt \leq (x - x_0)M \leq \alpha M \leq b$$

Therefore, by induction, $|y_n(x) - y_0| \leq b$ for all n. Thus it is ensured that each of the $y_n(x)$ lies within R for $x_0 \leq x \leq x_0 + \alpha$. This is needed below in order that the Lipschitz condition be applicable to $f\, x,y_n(x))$.

We now show that the sequence $y_n(x)$ converges uniformly to a continuous function $y(x)$ which satisfies (i). Consider the infinite series

$$y_0(x) + [y_1(x) - y_0(x)] + [y_2(x) - y_1(x)] + \cdots + [y_n(x) - y_{n-1}(x)] + \cdots \tag{iii}$$

The nth partial sum of this series is just $y_n(x)$. We shall show that the infinite series [and therefore $y_n(x)$] converges uniformly. Estimating the terms of the series, we find

$$|y_1(x) - y_0(x)| \leq M\,|x - x_0|$$

$$|y_2(x) - y_1(x)| \leq \int_{x_0}^x |f(t,y_1(t)) - f(t,y_0(t))| \, dt$$

Since all $y_n(x)$ lie in R for $x_0 \leq x \leq x_0 + \alpha$, we may apply the Lipschitz condition for f to obtain

$$|y_2(x) - y_1(x)| \leq A \int_{x_0}^x |y_1(t) - y_0(t)| \, dt \leq AM \int_{x_0}^x (t - x_0) \, dt$$

$$\leq \frac{MA(x - x_0)^2}{2!} \leq \frac{MA\alpha^2}{2!}$$

It is easy to prove by induction (Prob. 1)

$$|y_n(x) - y_{n-1}(x)| \leq \frac{MA^{n-1}\alpha^n}{n!} \qquad n = 1, 2, \ldots \tag{iv}$$

Now it is easily seen that the series $M \sum_{n=1}^\infty A^{n-1}\alpha^n/n!$ (see Prob. 2) converges for every α; therefore by the Weierstrass M test (see Appendix A, page 497) the series (iii) converges uniformly to a continuous function $y(x)$. Passing to the limit on both sides of (ii), we find

$$y(x) = \lim_{n \to \infty} y_n(x) = y_0 + \lim_{n \to \infty} \int_{x_0}^x f(t,y_{n-1}(t)) \, dt$$

$$= y_0 + \int_{x_0}^x f(t, \lim_{n \to \infty} y_{n-1}(t)) \, dt$$

$$= y_0 + \int_{x_0}^x f(t,y(x)) \, dt$$

These steps are justified because of the uniform convergence of $y_n(x)$ and the continuity of $f(x,y)$ (see Appendix A, page 497). Thus we have shown the existence of a solution. Since $y' = f(x,y)$, and $f(x,y)$ is continuous, the solution has a continuous derivative.

Uniqueness. If $z(x)$ is any solution of (i) for $x_0 \leq x \leq x_0 + \alpha$, then

$$z(x) = y_0 + \int_{x_0}^{x} f(t, z(t)) \, dt$$

and

$$|y_n(x) - z(x)| \leq \int_{x_0}^{x} |f(t, y_{n-1}(t)) - f(t, z(t))| \, dt$$

As above, we may show (see Prob. 3) that

$$|y_n(x) - z(x)| \leq \frac{MA^n \alpha^{n+1}}{(n+1)!} \qquad x_0 \leq x \leq x_0 + \alpha \tag{v}$$

Letting $n \to \infty$, we have that $|y(x) - z(x)| \leq 0$. Therefore $y(x) \equiv z(x)$ for $x_0 \leq x \leq x_0 + \alpha$. Thus the solution is unique.

Problems

1. Derive (iv).
2. Prove $\sum_{n=1}^{\infty} A^{n-1} \alpha^n / n!$ converges for every α. To what value does it converge?
3. Derive (v).

Existence and uniqueness for a system of first-order equations We now prove Theorem 1, Sec. 4-7.

If $f_i(x, y_1, \ldots, y_n)$ are continuous in R and satisfy a Lipschitz condition in R with respect to each of the y_j, then there exists a unique solution of

$$\frac{dy_i}{dx} = f_i(x, y_1, \ldots, y_n) \qquad i = 1, 2, \ldots, n \tag{i}$$

satisfying the initial conditions $y_i(x_0) = \beta_i$ where $(x_0, \beta_1, \ldots, \beta_n)$ is an interior point of R; the $y_i(x)$ have continuous derivatives and are defined in some interval $|x - x_0| \leq \alpha$.

Proof We take R to be the set of points $(x, y_1, \ldots, y_n)$ satisfying $|x - x_0| \leq a$, $|y_1 - \beta_1| \leq b_1, \ldots, |y_n - \beta_n| \leq b_n$. If $M_i = \max |f_i(x, y_1, \ldots, y_n)|$ for points in R, we shall prove that a unique solution exists for $|x - x_0| \leq \alpha$, where $\alpha = \min \{a, b_1/M_1, b_2/M_2, \ldots, b_n/M_n\}$. This choice of α ensures that any set of functions $y_i(x)$ satisfying Eq. (i), with $y_i(x_0) = \beta_i$, must remain in R for $|x - x_0| \leq \alpha$. For, since $|y_i'(x)| \leq M_i$, we have $|y_i(x) - \beta_i| \leq M_i |x - x_0|$, and, since $|x - x_0| \leq b_i/M_i$, we have $|y_i(x) - \beta_i| \leq b_i$ for $|x - x_0| \leq \alpha$.

As in the case of a single equation, we replace (i) by a system of integral equations

$$y_i(x) = \beta_i + \int_{x_0}^{x} f_i(t, y_1(t), y_2(t), \ldots, y_n(t)) \, dt \tag{ii}$$

We solve (ii) by the Picard method, where we use an initial approximation $y_i^{(0)}(x) = \beta_i$ and define $y_i^{(k)}(x)$ for $k \geq 1$ by

$$y_i^{(k)}(x) = \beta_i + \int_{x_0}^{x} f_i(t, y_1^{(k-1)}(t), \ldots, y_n^{(k-1)}(t)) \, dt \tag{iii}$$

We shall show that each of the sequences $y_i^{(k)}(x)$ for $i = 1, \ldots, n$ converge uniformly as $k \to \infty$ to continuous functions $y_i(x)$ satisfying (ii). Again it is convenient to assume that $x \geq x_0$. In Prob. 1 we show that $|y_i^{(k)}(x) - \beta_i| \leq b_i$ for $x_0 \leq x \leq x_0 + \alpha$ for every k so that all of the Picard approximations lie within R. We again consider $y_i^{(k)}(x)$ to be the kth partial sum of the infinite series

$$y_i^{(0)}(x) + [y_i^{(1)}(x) - y_i^{(0)}(x)] + \cdots + [y_i^{(k)}(x) - y_i^{(k-1)}(x)] + \cdots \qquad \text{(iv)}$$

Estimating the terms of the series, we find

$$|y_i^{(1)}(x) - y_i^{(0)}(x)| \leq M_i(x - x_0)$$

$$|y_i^{(2)}(x) - y_i^{(1)}(x)| \leq \int_{x_0}^x |f_i(t, y_1^{(1)}(t), \ldots, y_n^{(1)}(t))$$
$$- f_i(t, y_1^{(0)}(t), \ldots, y_n^{(0)}(t))| \, dt \qquad \text{(v)}$$

From the Lipshitz condition and Prob. 1, Sec. 4-7, the integrand on the right-hand side satisfies

$$|f_i(t, y_1^{(1)}(t), \ldots, y_n^{(1)}(t)) - f_i(t, y_1^{(0)}(t), \ldots, y_n^{(0)}(t))|$$
$$\leq A\,[|y_1^{(1)}(t) - y_1^{(0)}(t)| + |y_2^{(1)}(t) - y_2^{(0)}(t)| + \cdots + |y_n^{(1)}(t) - y_n^{(0)}(t)|]$$
$$\leq A[M_1\,|t - x_0| + \cdots + M_n\,|t - x_0|]$$

Therefore from (v) we derive

$$|y_i^{(2)}(x) - y_i^{(1)}(x)| \leq A[\tfrac{1}{2}(x - x_0)^2(M_1 + \cdots + M_n)] \leq AM\tfrac{1}{2}(x - x_0)^2$$

where $M = M_1 + M_2 + \cdots + M_n$. It is easy to prove by induction (see Prob. 2) that

$$|y_i^{(k)}(x) - y_i^{(k-1)}(x)| \leq Mn^{k-2}\,A^{k-1}\,\frac{(x - x_0)^k}{k!} \leq Mn^{k-2}\,\frac{A^{k-1}\alpha^k}{k!}$$
$$= M(n^2A)^{-1}\,\frac{(nA\alpha)^k}{k!} \qquad \text{(vi)}$$

Since the series $\sum_{k=1}^{\infty} (nA\alpha)^k/k!$ converges for every α, the series (iv), for each i, converges uniformly to a continuous function $y_i(x)$. Passing to the limit on both sides of (iii), we see that the $y_i(x)$ satisfy the integral equation. Thus we have proved the existence of a solution for $x_0 \leq x \leq x_0 + \alpha$. Since the right-hand sides of (i) are continuous, the $y_i(x)$ have continuous derivatives. Uniqueness is proved in Prob. 3.

Problems

1. Prove that $|y_i^{(k)}(x) - \beta_i| \leq b_i$ for $x_0 \leq x \leq x_0 + \alpha$.
2. Derive (vi).
3. Prove the uniqueness of the solution.
4. *Existence of a solution for a system of linear differential equations.* Prove Theorem 2, Sec. 4-8. Hint: We have already proved the existence of a solution in an interval $x - x_0 \leq \alpha$. It is only necessary to show that the solution exists in the entire interval I in which the coefficients $a_{ij}(x)$ are continuous.

Linear Differential Equations

5

5-1 Introduction

It would be difficult to overestimate the importance of linear differential equations. These equations find extensive use in all the fields of science and engineering. In this chapter we shall study the second-order linear differential equation

$$a(x)y'' + b(x)y' + c(x)y = f(x) \tag{1}$$

and the corresponding equation of nth order. Certain general properties of the solutions of (1) will be derived. However only the simplest special cases will be explicitly solved, e.g., the case when the coefficients a, b, c are all constants. The actual solutions in more general cases will be considered in Chap. 7.

5-2 Fundamental theory of second-order linear equations

We begin by recalling from Chap. 4 the fundamental existence theorem for Eq. (1).

Theorem 1 *Consider the initial-value problem*

$$\text{DE: } a(x)y'' + b(x)y' + c(x)y = f(x)$$
$$\text{IC: } y(x_0) = y_0 \qquad y'(x_0) = y_0' \tag{2}$$

If $a(x)$, $b(x)$, $c(x)$, $f(x)$ are all continuous functions in a common interval I containing the point x_0 and $a(x)$ is never zero in I, then the initial-value problem (2) possesses a unique solution. The solution has a continuous second derivative and is defined throughout I.

We note in particular that the *unique* solution of the homogeneous problem

DE: $a(x)y'' + b(x)y' + c(x)y = 0$

IC: $y(x_0) = 0 \qquad y'(x_0) = 0$

is the trivial solution $y(x) \equiv 0$.

In order to guarantee the applicability of this theorem, we shall assume throughout this chapter, without further mention, that the functions a, b, c, f of Eq. (1) are all continuous in a common interval I and that $a(x)$ is never zero in I.

It is convenient to introduce an abbreviation for the left-hand side of (1), namely,

$$L(y) \equiv a(x)y'' + b(x)y' + c(x)y \tag{3}$$

With this notation Eq. (1) becomes simply

$$L(y) = f(x) \tag{4}$$

and if $f(x) \equiv 0$, (4) becomes

$$L(y) = 0 \tag{5}$$

Equation (4) with $f(x) \not\equiv 0$ is called *nonhomogeneous*, whereas Eq. (5) is called *homogeneous*.

L can be thought of as an *operator* which acts on a twice-differentiable function $y(x)$ to produce the function $L(y)$ as given by (3). The linearity property of the operator L, as expressed in the following theorem, will be used extensively.

Theorem 2   *The operator L defined by (3) has the "linearity property"; that is,*

$$L(c_1 y_1 + c_2 y_2) = c_1 L(y_1) + c_2 L(y_2) \tag{6}$$

where y_1 and y_2 are any two twice-differentiable functions and c_1 and c_2 are any two constants.

Proof   Using the definition of L, we have

$$L(c_1 y_1 + c_2 y_2) = a(x)(c_1 y_1 + c_2 y_2)'' + b(x)(c_1 y_1 + c_2 y_2)'$$
$$+ c(x)(c_1 y_1 + c_2 y_2)$$
$$= c_1[a(x)y_1'' + b(x)y_1' + c(x)y_1]$$
$$+ c_2[a(x)y_2'' + b(x)y_2' + c(x)y_2]$$
$$= c_1 L(y_1) + c_2 L(y_2)$$

We now state several important theorems concerning the solutions of homogeneous and nonhomogeneous linear equations.

Theorem 3   *If $y_1(x)$ and $y_2(x)$ are solutions of the homogeneous equation $L(y) = 0$, then $y = c_1 y_1(x) + c_2 y_2(x)$ is also a solution.*

Proof   From (6) we have

$$L(c_1 y_1 + c_2 y_2) = c_1 L(y_1) + c_2 L(y_2)$$

However, since y_1 and y_2 are solutions of $L(y) = 0$, we have $L(y_1) = 0$ and $L(y_2) = 0$. Therefore $L(c_1 y_1 + c_2 y_2) = 0$ and $y = c_1 y_1 + c_2 y_2$ is a solution of $L(y) = 0$.

Example 1   By substitution we find that e^{2x} and e^{-2x} are solutions of

$$y'' - 4y = 0$$

Therefore

$$y = c_1 e^{2x} + c_2 e^{-2x}$$

is also a solution.   (We shall see later that this is in fact the general solution.)

Theorem 3 tells us that the linear combination $c_1 y_1 + c_2 y_2$ is a solution of $Ly = 0$ if y_1 and y_2 are solutions.   However, it does not guarantee that the general solution is obtained in this manner.   To clear up this point, we need the concept of *linear independence* of two functions.   We shall prove below that if y_1 and y_2 are linearly independent solutions of $L(y) = 0$, then the combination $c_1 y_1 + c_2 y_2$ is indeed the general solution (see Theorem 6).

Definition   *Two functions $y_1(x)$ and $y_2(x)$ are called linearly dependent (LD) in an interval I if it is possible to find two constants c_1 and c_2, not both zero, so that $c_1 y_1(x) + c_2 y_2(x) \equiv 0$ in I.   Two functions are called linearly independent (LI) if they are not linearly dependent, i.e., if $c_1 y_1(x) + c_2 y_2(x) \equiv 0$ in I only when $c_1 = c_2 = 0$.*

If two functions are linearly dependent in I, then one of the functions is identically equal to a constant times the other in I; if two functions are linearly independent, it is impossible to express either function as a constant times the other.   The concept of linear independence allows us to distinguish when two functions are "essentially" different.

Example 2   The functions x^2 and x are linearly independent in every interval I.   For $c_1 x^2 + c_2 x \equiv 0$ in I implies $c_1 = c_2 = 0$.   If either c_1 or c_2 were not zero, the quadratic equation $c_1 x^2 + c_2 x = 0$ could hold for at most two values of x, whereas it must hold for *all* x in I.

Example 3   Two functions are linearly dependent if one of the functions is the *zero function*.   For if $y_1 \equiv 0$, we have $1 \cdot y_1 + 0 \cdot y_2 \equiv 0$.

Example 4   If $\lambda_1 \neq \lambda_2$, $e^{\lambda_1 x}$ and $e^{\lambda_2 x}$ are linearly independent in every interval I. For if $c_1 e^{\lambda_1 x} + c_2 e^{\lambda_2 x} \equiv 0$, we would have $c_1 \equiv -c_2 e^{(\lambda_2 - \lambda_1)x}$. But this can hold only if $\lambda_2 = \lambda_1$.

In Theorems 4, 5, and 6 we need the following two facts from algebra :†

1. *Two linear algebraic equations in two unknowns have a unique solution if and only if the determinant of the coefficients is not zero.*
2. *Two homogeneous linear equations in two unknowns have a nontrivial solution if and only if the determinant of the coefficients is zero.*

A useful test for linear independence of two *differentiable* functions can be given in terms of the Wronskian, defined by

$$W[y_1(x), y_2(x)] \equiv \begin{vmatrix} y_1(x) & y_2(x) \\ y_1'(x) & y_2'(x) \end{vmatrix} \equiv y_1(x)y_2'(x) - y_2(x)y_1'(x)$$

Theorem 4   *If the Wronskian $W[y_1(x), y_2(x)]$ of two differentiable functions is different from zero for at least one point in an interval I, then $y_1(x)$ and $y_2(x)$ are linearly independent in I.*

Proof   Assume that y_1 and y_2 are linearly dependent. Then there exist constants c_1 and c_2, not both zero, so that

$$c_1 y_1(x) + c_2 y_2(x) \equiv 0 \tag{i}$$

By differentiation of this identity we have

$$c_1 y_1'(x) + c_2 y_2'(x) \equiv 0 \tag{ii}$$

For each fixed x Eqs. (i) and (ii) are two homogeneous linear equations for c_1 and c_2. By assumption these equations have a nontrivial solution at each x. Therefore the determinant of the coefficients must be zero for each x. But the determinant is $W[y_1(x), y_2(x)]$ which is assumed to be different from zero for at least one point. Therefore y_1 and y_2 cannot be linearly dependent.

Corollary   *If y_1 and y_2 are linearly dependent in I, then $W[y_1, y_2] \equiv 0$ in I.*

The converse of Theorem 4 is not true in general. For example, the differentiable functions x^3 and $|x|^3$ can be easily shown to be linearly independent in the interval $[-1, 1]$, and yet $W[x^3, |x|^3] \equiv 0$. (See Prob. 4.)

However, if we consider two *solutions* of $L(y) = 0$ instead of any two functions, a strengthened converse holds.

† See Thomas [35], pp. 420 and 426.

Theorem 5 *If $y_1(x)$ and $y_2(x)$ are linearly independent solutions of $Ly = 0$ in I, the Wronskian $W[y_1(x), y_2(x)]$ is never zero in I.*

Proof If $W[y_1(x), y_2(x)]$ is zero at a point x_0 in I, then the equations

$$c_1 y_1(x_0) + c_2 y_2(x_0) = 0$$

$$c_1 y_1'(x_0) + c_2 y_2'(x_0) = 0$$

have a nontrivial solution. For these values c_1 and c_2 the function

$$y(x) = c_1 y_1(x) + c_2 y_2(x)$$

satisfies the differential equation $L(y) = 0$ and the initial conditions $y(x_0) = y'(x_0) = 0$. However, from Theorem 1 the function $y(x) \equiv 0$ is a solution of this problem and the *only* solution. Therefore

$$c_1 y_1(x) + c_2 y_2(x) \equiv 0$$

This means that y_1 and y_2 are linearly dependent contrary to hypothesis. Therefore $W[y_1(x), y_2(x)]$ can never vanish in I.

Corollary *The Wronskian of two solutions of $L(y)=0$ is either identically zero (if the solutions are linearly dependent) or never zero (if the solutions are linearly independent).*

We are now in a position to prove the following theorem.

Theorem 6 *If $y_1(x)$ and $y_2(x)$ are linearly independent solutions of $Ly = 0$, then $y = c_1 y_1 + c_2 y_2$ is the general solution.*

Proof Let $y(x)$ be *any* solution of $Ly = 0$, let x_0 be any point in the interval under consideration, and let $y(x)$ take on the initial values $y(x_0) = \alpha$, $y'(x_0) = \beta$. We determine two constants c_1 and c_2 by solving the equations

$$\alpha = c_1 y_1(x_0) + c_2 y_2(x_0)$$

$$\beta = c_1 y_1'(x_0) + c_2 y_2'(x_0)$$

This can always be done since the determinant of the coefficients is the Wronskian of y_1 and y_2 evaluated at x_0 and is different from zero, since y_1 and y_2 are linearly independent. Using these values of c_1 and c_2, consider the function $w(x)$ defined by

$$w(x) = c_1 y_1(x) + c_2 y_2(x)$$

We see that $w(x)$ satisfies the differential equation $L(y) = 0$, and $w(x)$ takes on the initial values $w(x_0) = \alpha$ and $w'(x_0) = \beta$. By the uniqueness theorem there is only one function having these properties; therefore $w(x) \equiv y(x)$.

Example 5   Since e^{2x} and e^{-2x} are linearly independent solutions of $y'' - 4y = 0$, Theorem 6 guarantees that $y = c_1 e^{2x} + c_2 e^{-2x}$ is the general solution.

We can in particular find the solution satisfying arbitrary initial conditions by properly picking the constants c_1 and c_2. For example, if the initial conditions are $y(0) = 1$, $y'(0) = 0$, we solve the equations

$$1 = c_1 + c_2$$
$$0 = 2c_1 - 2c_2$$

to obtain $c_1 = \frac{1}{2}$, $c_2 = \frac{1}{2}$. The solution is therefore $y = \frac{1}{2}(e^{2x} + e^{-2x})$.

From Theorem 6 we see that the problem of finding the general solution of $Ly = 0$ is reduced to finding any two linearly independent solutions. The important question of whether or not two linearly independent solutions of $L(y) = 0$ actually exist in general is answered by the following theorem.

Theorem 7   *There exist two linearly independent solutions of $Ly = 0$.*

Proof   Let $y_1(x)$ be the *unique* solution of $L(y) = 0$ satisfying $y_1(x_0) = 1$, $y_1'(x_0) = 0$, and let $y_2(x)$ be the unique solution satisfying $y_2(x_0) = 0$, $y_2'(x_0) = 1$. The *existence* of y_1 and y_2 is ensured by Theorem 1. To prove that y_1 and y_2 are linearly independent, we merely calculate the Wronskian at x_0

$$W[y_1(x_0), y_2(x_0)] = \begin{vmatrix} 1 & 0 \\ 0 & 1 \end{vmatrix} = 1$$

From Theorem 5 we see that y_1 and y_2 are linearly independent.

We now consider the nonhomogeneous equation.

Theorem 8   *If y_p is any particular solution of the nonhomogeneous equation $L(y) = f(x)$ and y_h is the general solution of the homogeneous equation $L(y) = 0$, then the general solution of $L(y) = f(x)$ is $y = y_h + y_p$.*

Proof   Every function of the form $y_h + y_p$ is a solution of $L(y) = f$ since

$$L(y_h + y_p) = Ly_h + Ly_p = 0 + f = f$$

We now show that every solution $y(x)$ of $Ly = f$ can be written in the form $y = y_h + y_p$. If $y(x)$ is any solution, then $y - y_p$ is a solution of the homogeneous equation since $L(y - y_p) = L(y) - L(y_p) = f - f = 0$. Therefore, if $y - y_p = z$, we have $y = z + y_p$, where z is a solution of the homogeneous equation, as was to be proved.

Thus the nonhomogeneous equation can be solved by first solving the homogeneous equation and then finding a particular solution of the nonhomogeneous equation.

Example 6   Solve: $y'' - 4y = e^x$

From Example 5 the general solution of the homogeneous equation is

$$y_h = c_1 e^{2x} + c_2 e^{-2x}$$

To find a particular solution of the nonhomogeneous equation, we assume (see Sec. 5-5)

$$y_p = A e^x$$

Substituting in the differential equation, we get

$$A e^x - 4 A e^x = e^x$$

Therefore

$$-3A = 1 \qquad \text{or} \qquad A = -\tfrac{1}{3}$$

and the particular solution is $y_p = -e^x/3$. The general solution is

$$y = c_1 e^{2x} + c_2 e^{-2x} - \tfrac{1}{3} e^x$$

A simple, but extremely important theorem for linear equations is the following.

Theorem 9   *Principle of superposition.   If y_1 is a solution of $L(y) = f_1$ and y_2 is a solution of $L(y) = f_2$, then $y = y_1 + y_2$ is a solution of $L(y) = f_1 + f_2$.*

Proof $L(y_1 + y_2) = L(y_1) + L(y_2) = f_1 + f_2$

This theorem allows us to replace the problem of solving $L(y) = f_1 + f_2$ by the usually simpler problems of solving $L(y) = f_1$ and $L(y) = f_2$.

Example 7   Solve:

$$y'' - 4y = e^x + 5 \tag{i}$$

In Example 6 we found that $y = -\tfrac{1}{3} e^x$ is a particular solution of $y'' - 4y = e^x$. To find a particular solution of $y'' - 4y = 5$, we assume that $y_p \equiv k$. Substituting into the differential equation, we find $k = -\tfrac{5}{4}$. The general solution of (i) is therefore

$$y = c_1 e^{2x} + c_2 e^{-2x} - \tfrac{1}{3} e^x - \tfrac{5}{4}$$

Clearly the principle of superposition can be extended to any finite number of terms: *If, for $i = 1, 2, \ldots, n$, y_i is a solution of $L(y) = f_i$, then $y = \sum_{i=1}^{n} y_i$ is a solution of $L(y) = \sum_{i=1}^{n} f_i$.*

Example 8   Solve:

$$y'' + y = \sum_{k=1}^{n} a_k e^{kx} \tag{i}$$

Consider the differential equation with only a single term on the right:

$$y'' + y = a_k e^{kx} \tag{ii}$$

To find a particular solution of (ii), assume $y_k = Ae^{kx}$. Substituting in (ii), we find $A = a_k/(1 + k^2)$ and $y_k = a_k e^{kx}/(1 + k^2)$. Thus a particular solution of (i) is

$$y = \sum_{k=1}^{n} y_k = \sum_{k=1}^{n} \frac{a_k e^{kx}}{1 + k^2}$$

From a physical point of view, we can interpret the solution y of $L(y) = f$ as the *response* of a physical system to an *excitation* or *forcing function f*. The principle of superposition then has the interpretation that *the response to a superposition (sum) of excitations is the superposition of the responses to the individual excitations*. This statement holds if the physical system is linear, i.e., governed by a linear differential equation; it does *not* hold if the system is nonlinear.

Problems 5-2

1. Which of the following pairs of functions are linearly independent?

   a. $\sin ax$, $\cos ax$, $a \neq 0$, in $(-\infty, \infty)$
   b. $e^{ax} \sin bx$, $e^{ax} \cos bx$, $b \neq 0$, in $(-\infty, \infty)$
   c. $|x|$, x in $(-1,1)$
   d. $|x|$, x in $(0,1)$
   e. x^m, x^n in $(0, \infty)$
   f. $\sin (x - 1)$, $\sin x$ in $(-\infty, \infty)$

2. Three particular solutions of a certain second-order linear equation $Ly = f$ are $y = x$, $y = e^x + x$, $y = 2e^x + 1 + x$. What is the general solution?

3. Referring to Theorem 1, prove that two solutions that are identical in a subinterval of I are identical throughout I.

4. Show that x^3 and $|x|^3$ are linearly independent in $-1 \leq x \leq 1$ but that $W[x^3, |x|^3] \equiv 0$.

5-3   *Complex-valued solutions*

The treatment of linear equations with constant coefficients, discussed in the next two sections, is considerably simplified and unified if we use complex-valued functions.

A complex-valued function f of a real variable assigns a complex number to each real number x in the domain of f. Thus for each x

$$f(x) = u(x) + iv(x)$$

where u and v are real functions.

The calculus of complex-valued functions is discussed in Appendix B from a somewhat more general point of view than is needed here. For our present purposes the following definition suffices.

Definition 1 *If $f = u + iv$ where u and v are real functions, f is said to be continuous if u and v are continuous; f is said to be differentiable if u and v are differentiable and $f'(x) = u'(x) + iv'(x)$.*

The sum, product, quotient, and chain rules for derivatives of complex functions are the same as for real functions (see Prob. 1).

Example 1 $\dfrac{d}{dx}(3x + ix^2) = 3 + i2x$

Example 2 $\dfrac{d}{dx}(3x + ix^2)^2 = 2(3x + ix^2)(3 + i2x) = 2(9x - 2x^3) + i18x^2$

Example 3   If $E(x) = e^{ax}(\cos bx + i \sin bx)$, we have by the product rule

$$E'(x) = e^{ax}(-b \sin bx + ib \cos bx) + ae^{ax}(\cos bx + i \sin bx)$$
$$= e^{ax}[(a \cos bx - b \sin bx) + i(a \sin bx + b \cos bx)]$$
$$= (a + ib)e^{ax}(\cos bx + i \sin bx)$$

that is,

$$E'(x) = (a + ib)E(x)$$

The most important complex function for our purpose is the complex exponential *defined* by

$$e^{(a+ib)x} = e^{ax} \cos bx + ie^{ax} \sin bx \tag{7}$$

When $a = 0$, we have Euler's formula

$$e^{ibx} = \cos bx + i \sin bx$$

so that (7) can be written

$$e^{(a+ib)x} = e^{ax}e^{ibx}$$

The derivative of the complex exponential is

$$\frac{d}{dx} e^{(a+ib)x} = (a + ib)e^{(a+ib)x}$$

as derived in Example 3 above.   The justification for this definition is that the complex exponential satisfies the same laws of exponents and differentiation as its real counterpart e^{rx} for real r.   (See Prob. 2.)   An alternative approach, which *defines* the complex exponential by means of a power series and then *derives* Eq. (7), is discussed in Appendix B.

A complex-valued function f is called a solution to an nth-order differential equation if f possesses n derivatives and satisfies the differential equation identically.

Example 4 $y = e^{ix}$ satisfies $y'' + y = 0$, for

$$y' = ie^{ix} \qquad y'' = -e^{ix} \qquad e^{ix} - e^{ix} \equiv 0$$

The following theorem shows the connection between real and complex solutions of a linear differential equation with *real* coefficients.

Theorem 1   *Consider the differential equation*

$$a(x)y'' + b(x)y' + c(x)y = 0$$

where a, b, c are real functions.   The complex function $y = u + iv$ where u, v are real is a solution if and only if u and v are solutions.

Proof   As usual, denote the left-hand side of the differential equation by $L(y)$.   It is easy to prove (see Prob. 3) that $L(y) = L(u) + iL(v)$ where $L(u)$ and $L(v)$ are real.   Therefore y is a solution if and only if $L(y) = L(u) + iL(v) \equiv 0$.   Since a complex number is zero if and only if its real and imaginary parts are zero, we have $L(y) = 0$ if and only if $L(u) = 0$ and $L(v) = 0$.

Example 5   In Example 4 we have seen that $y = e^{ix}$ satisfies $y'' + y = 0$. Since $e^{ix} = \cos x + i \sin x$, the above theorem shows that $\cos x$ and $\sin x$ are real solutions.

Problems 5-3

1. If $f = u + iv$ and $g = r + is$ are differentiable complex functions of a real variable, prove that:

a. $(f + g)' = f' + g'$

b. $(fg)' = fg' + f'g$

c. $\left(\dfrac{f}{g}\right)' = \dfrac{gf' - fg'}{g^2}$

d. $(cf)' = cf'$ c a complex constant

2. Verify that e^{cx} where c is complex and x is real satisfies:

a. $e^{cx_1}e^{cx_2} = e^{c(x_1+x_2)}$

b. $\dfrac{e^{cx_1}}{e^{cx_2}} = e^{c(x_1-x_2)}$

c. Recalling De Moivre's formula (n a positive integer)

$(\cos bx + i \sin bx)^n = \cos nbx + i \sin nbx$, show that $(e^{cx})^n = e^{cxn}$.

3. If $y = u + iv$ and $Ly = a(x)y'' + b(x)y' + c(x)y$, where a, b, c are real functions, show that $L(y) = L(u) + iL(v)$.

4. Show that $y = e^{(1+i)x}$ satisfies $y'' - 2y' + 2y = 0$.   What are real solutions of the equation?

5. Define $x^{(a+ib)} = e^{(a+ib) \ln x}$ for real $x > 0$.

a. Show that $x^{a+ib} = x^a [\cos (b \ln x) + i \sin (b \ln x)]$.

b. Show that

$$\frac{d}{dx} x^{a+ib} = (a + ib) x^{(a-1)+ib}$$

(i.e., the usual rule for differentiation holds).

6. Show that $y = x^{1+i}$ satisfies $x^2 y'' - xy' + 2y = 0$. What are real solutions of the equation?

7. Show that $y = e^{ix}$ satisfies $y' - iy = 0$. Are the real and imaginary parts of y solutions of the equation? Why?

8. Let u and v be real linearly independent solutions of $a(x)y'' + b(x)y' + c(x)y = 0$ where a, b, c are real functions. Show that $y_1 = u + iv$ and $\bar{y}_1 = u - iv$ ($\bar{y}$ is the complex conjugate of y) are complex solutions. Show that the general complex solution is $y = c_1 y_1 + c_2 \bar{y}_1$ where c_1 and c_2 are arbitrary complex numbers.

5-4   Homogeneous linear equations with constant coefficients

A simple, but important, class of linear differential equations is that with constant coefficients. We consider the homogeneous equation

$$ay'' + by' + cy = 0 \tag{8}$$

where a, b, and c are *real* constants and $a \neq 0$. Since this equation does not contain x explicitly, it could be solved by the method of Sec. 2-7. We prefer to use a simpler, but less direct, approach.

Let us consider the types of functions that could possibly satisfy Eq. (8). The solution could not be a function like $y = \ln x$, for the derivatives of this function are $y' = 1/x$ and $y'' = -1/x^2$. Upon substitution into the equation there would be no term to cancel the term $c (\ln x)$. It is clear that the solution must be a function whose derivative does not differ greatly in form from the function itself. The functions $e^{\lambda x}$, $\sin \beta x$, $\cos \beta x$ come to mind immediately. The functions $\sin \beta x$ and $\cos \beta x$ are simply combinations of complex exponentials and need not be considered separately. Therefore it is reasonable to look for solutions in the form

$$y = e^{\lambda x} \tag{9}$$

where λ must be determined so that (9) satisfies the differential equation (8). The derivatives of (9) are $y' = \lambda e^{\lambda x}$ and $y'' = \lambda^2 e^{\lambda x}$. Substituting these into Eq. (8), we find

$$e^{\lambda x}(a\lambda^2 + b\lambda + c) \equiv 0$$

If this equation holds for all x, then $e^{\lambda x}$ is a solution for all x; however, the above can hold for all x only if

$$a\lambda^2 + b\lambda + c = 0 \tag{10}$$

This equation, which determines λ, is called the *auxiliary* or *characteristic* equation. The roots of (10) are

$$\lambda = \frac{-b \pm \sqrt{b^2 - 4ac}}{2a}$$

If the discriminant $\Delta = b^2 - 4ac$ is greater than zero, the roots are real and distinct; if Δ is equal to zero, the roots are real and equal; and if Δ is less than zero, the roots are complex conjugate numbers. We discuss each of these cases.

Case I. *Real, distinct roots,* $\Delta > 0$. If $\Delta > 0$, the roots are

$$\lambda_1 = \frac{-b + \sqrt{b^2 - 4ac}}{2a}$$

$$\lambda_2 = \frac{-b - \sqrt{b^2 - 4ac}}{2a}$$

and are real, distinct numbers. Therefore $e^{\lambda_1 x}$ and $e^{\lambda_2 x}$ are both solutions of the differential equation. Since these functions are obviously linearly independent, the general solution is

$$y = c_1 e^{\lambda_1 x} + c_2 e^{\lambda_2 x} \tag{11}$$

Case II. *Real, equal roots,* $\Delta = 0$. In this case we have only one root given by

$$\lambda_1 = -\frac{b}{2a} \tag{12}$$

Therefore $e^{\lambda_1 x}$ is a solution of the differential equation. Of course, $c_1 e^{\lambda_1 x}$ is also a solution, but we need a second linearly independent solution. To find this, we use the method of variation of parameters.† Assume a solution of the form $y = v(x)e^{\lambda_1 x}$. Substituting this into the differential equation (8), we obtain

$$av'' + (2a\lambda_1 + b)v' + (a\lambda_1^2 + b\lambda_1 + c)v = 0 \tag{13}$$

---

† This method, when used in this manner to find a second solution of a homogeneous linear differential equation, when one solution is known, is often called *reduction of order* (see Sec. 5-6, Prob. 9).

Since λ_1 is a root of the characteristic equation, the coefficient of v vanishes, and since $\lambda_1 = -b/2a$, the coefficient of v' also vanishes. The differential equation (13) therefore reduces to

$$v'' = 0 \tag{14}$$

The general solution of this equation is

$$v = c_1 + c_2 x$$

Since $y = v(x)e^{\lambda_1 x}$, we have that

$$y = c_1 e^{\lambda_1 x} + c_2 x e^{\lambda_1 x} \tag{15}$$

is a solution of the differential equation (8). Since the functions $e^{\lambda_1 x}$ and $x e^{\lambda_1 x}$ are linearly independent, the general solution is given by Eq. (15).

Case III.   Complex conjugate roots, $\Delta < 0$.   In this case the roots are

$$\lambda_1 = \alpha + i\beta = -\frac{b}{2a} + i\sqrt{\frac{c}{a} - \frac{b^2}{4a^2}}$$

$$\lambda_2 = \alpha - i\beta = -\frac{b}{2a} - i\sqrt{\frac{c}{a} - \frac{b^2}{4a^2}}$$

and are conjugate complex numbers (since a, b, c are real). These yield the two complex solutions†

$$e^{(\alpha+i\beta)x} = e^{\alpha x}(\cos \beta x + i \sin \beta x) \tag{16}$$

$$e^{(\alpha-i\beta)x} = e^{\alpha x}(\cos \beta x - i \sin \beta x) \tag{17}$$

Since the differential equation (8) has *real* coefficients, the real and imaginary parts of either of these functions, namely,

$$e^{\alpha x}\cos \beta x \qquad e^{\alpha x}\sin \beta x \tag{18}$$

are real solutions (see Theorem 1, Sec. 5-3). The Wronskian of these functions is easily shown to be nonzero. Therefore the functions (18) are linearly independent solutions and the general solution for this case is

$$y = e^{\alpha x}(A \cos \beta x + B \sin \beta x) \tag{19}$$

where A, B are arbitrary real constants.

We also note that the general complex-valued solution of the differential equation (8) is (see Prob. 8, Sec. 5-3)

$$y = c_1 e^{(\alpha+i\beta)x} + c_2 e^{(\alpha-i\beta)x} \tag{20}$$

† See Sec. 5-3.

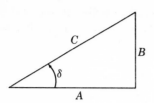

figure 1

where c_1 and c_2 are *arbitrary complex* constants.   Equation (20) can be written, using the right-hand sides of (16) and (17),

$$y = e^{\alpha x}[(c_1 + c_2) \cos \beta x + i(c_1 - c_2) \sin \beta x] \tag{21}$$

If we now take c_1 and c_2 to be *conjugate* complex constants, then it is easily shown that $c_1 + c_2$ and $i(c_1 - c_2)$ are both real numbers.   Calling these numbers A and B respectively, we have again the real general solution (19).

It is often convenient to write the solution for this case in still another form.   Let A and B be thought of as the coordinates of a point in the plane. Introducing polar coordinates C and δ of the point, we have (see Fig. 1)

$$\begin{matrix} A = C \cos \delta & & C = \sqrt{A^2 + B^2} \\ & \text{or} & \\ B = C \sin \delta & & \delta = \arctan \dfrac{B}{A} \;\dagger \end{matrix} \tag{22}$$

Substituting these in Eq. (19), we obtain

$$y = Ce^{\alpha x}(\cos \delta \cos \beta x + \sin \delta \sin \beta x) \tag{23}$$

Recalling the addition law for cosines, we obtain the solution in the compact form

$$y = Ce^{\alpha x} \cos (\beta x - \delta) \tag{24}$$

In a similar way we obtain

$$y = Ce^{\alpha x} \sin (\beta x + \theta) \tag{25}$$

where the angle θ is the complement of δ.

Summary   *The solutions of*

DE: $ay'' + by' + cy = 0$ $a \neq 0$

$\dagger$ By the angle $\delta = \arctan B/A$ we do not mean the principal value of $\arctan B/A$ but rather the angle δ such that

$$\sin \delta = \frac{B}{\sqrt{A^2 + B^2}} \quad \text{and} \quad \cos \delta = \frac{A}{\sqrt{A^2 + B^2}}$$

are obtained by first solving the characteristic equation

$$a\lambda^2 + b\lambda + c = 0$$

and, according to the type of roots, using the results of the accompanying table.

Type of root	Linearly independent solutions	General solution
Real, distinct $\lambda_1 \neq \lambda_2$	$e^{\lambda_1 x},\ e^{\lambda_2 x}$	$c_1 e^{\lambda_1 x} + c_2 e^{\lambda_2 x}$
Real, equal $\lambda_1 = \lambda_2$	$e^{\lambda_1 x},\ x e^{\lambda_1 x}$	$c_1 e^{\lambda_1 x} + c_2 x e^{\lambda_1 x}$
Complex conjugate $\lambda_1 = \alpha + i\beta$ $\lambda_2 = \alpha - i\beta$ $\beta \neq 0$	$\begin{cases} e^{(\alpha+i\beta)x},\ e^{(\alpha-i\beta)x} \\ \text{or} \\ e^{\alpha x} \cos \beta x,\ e^{\alpha x} \sin \beta x \end{cases}$	$\begin{cases} c_1 e^{(\alpha+i\beta)x} + c_2 e^{(\alpha-i\beta)x} \\ \text{or} \\ e^{\alpha x} (A \cos \beta x + B \sin \beta x) \\ \text{or} \\ C e^{\alpha x} \cos (\beta x - \delta) \\ \text{or} \\ C e^{\alpha x} \sin (\beta x + \theta) \end{cases}$

Most scientists and engineers encounter linear equations with constant coefficients so often that the above results, and the methods used to obtain them, are committed to memory.

Example 1 $2y'' - y' - 3y = 0$

The characteristic equation is

$$2\lambda^2 - \lambda - 3 = 0$$

Therefore

$$\lambda_1 = -1 \qquad \lambda_2 = \tfrac{3}{2}$$

and the general solution is

$$y = c_1 e^{-x} + c_2 e^{3x/2}$$

Example 2 $y'' - 4y' + 4y = 0$

The characteristic equation is

$$\lambda^2 - 4\lambda + 4 = 0$$

$$\lambda = 2, 2$$

The general solution is

$$y = (c_1 + c_2 x)e^{2x}$$

Example 3 $32y'' - 40y' + 17y = 0$

The characteristic equation is

$$32\lambda^2 - 40\lambda + 17 = 0$$

$$\lambda_1 = \tfrac{5}{8} + i\tfrac{3}{8} \qquad \lambda_2 = \tfrac{5}{8} - i\tfrac{3}{8}$$

The solution can be written in the form

$$y = c_1 e^{(\frac{5}{8} + i\frac{3}{8})x} + c_2 e^{(\frac{5}{8} - i\frac{3}{8})x}$$

or

$$y = e^{\frac{5}{8}x}(A \cos \tfrac{3}{8}x + B \sin \tfrac{3}{8}x)$$

or

$$y = ce^{\frac{5}{8}x} \cos (\tfrac{3}{8}x - \delta)$$

Problems 5-4

1. Find the general solution of:

a. $y'' = 0$
c. $y'' - a^2y = 0$
e. $y'' + y' = 0$
g. $3y'' + 14y' + 8y = 0$

b. $y'' - 2y' = 0$
d. $y'' + a^2y = 0$
f. $y'' + 2y' + y = 0$
h. $y'' + y' + y = 0$

2. Suppose the characteristic equation for differential equation (8) has distinct, real roots, λ_1 and λ_2. Show that $(e^{\lambda_1 x} - e^{\lambda_2 x})/(\lambda_1 - \lambda_2)$ is a solution of Eq. (8). Show also that

$$\lim_{\lambda_2 \to \lambda_1} \frac{e^{\lambda_1 x} - e^{\lambda_2 x}}{\lambda_1 - \lambda_2} = xe^{\lambda_1 x}$$

This limiting process is another way of finding the second linearly independent solution in the case of equal roots.

3. Suppose the roots of the characteristic equation are real and distinct. The roots can then be written as $\lambda_{1,2} = A \pm B$ where A and B are real. Show that the solution of the differential equation can be written as $y = e^{Ax}(c_1 \cosh Bx + c_2 \sinh Bx)$. Recall the definitions of the hyperbolic functions:

$$\cosh x = \frac{e^x + e^{-x}}{2} \qquad \sinh x = \frac{e^x - e^{-x}}{2}$$

4. Write down a homogeneous second-order linear differential equation with constant real coefficients whose solutions are:

a. $1, x$
c. $3 \cos 4x, 5 \sin 4x$
e. $5 \cosh 2x, 9 \sinh 2x$
g. $e^x \cos (x - 1), e^x \cos (x - 2)$

b. $2e^x, e^{-5x}$
d. $e^{-x} \cos x, e^{-x} \sin x$
f. $4e^{3x}, 7xe^{3x}$
h. x, e^x

5. We note that the solutions of the second-order differential equation with constant coefficients are defined for all x. Could this have been predicted from the existence theorem?

6. Referring to the existence theorem, explain why the function $\ln x$ cannot be the solution of a homogeneous linear equation with constant coefficients.

7. Solve: DE: $y'' + y' - 6y = 0$
          IC: $y(0) = 1$ $y'(0) = -2$

8. Solve the following third-order equations by extending the method in the text:

a. $y''' = 0$ b. $y''' - y = 0$

c. $y''' - 7y'' + 16y' - 12y = 0$ d. $y''' - 3y'' + 3y' - y = 0$

5-5 *Undetermined coefficients*

We have seen that the general solution of the nonhomogeneous equation

$$L(y) = ay'' + by' + cy = f(x) \qquad a \neq 0 \tag{26}$$

is

$$y = y_h + y_p \tag{27}$$

where y_h is the general solution of the homogeneous equation and y_p is a particular solution of Eq. (26). In Sec. 5-4 we solved the homogeneous equation with constant coefficients. We now investigate methods of finding particular solutions of Eq. (26) when $f(x)$ is an exponential, a sinusoid, a polynomial, or a product of such functions. These functions appear often in applications.

I. $f(x) = ke^{\alpha x}$. We seek a particular solution of the differential equation

$$ay'' + by' + cy = ke^{\alpha x} \tag{28}$$

Because of the exponential term on the right-hand side of the equation, we look for a solution in the form

$$y_p = Ae^{\alpha x} \tag{29}$$

where the undetermined coefficient A will be determined so that the differential equation is satisfied. The substitution of (29) into the differential equation (28) results in

$$(a\alpha^2 + b\alpha + c)Ae^{\alpha x} = ke^{\alpha x} \tag{30}$$

or

$$A = \frac{k}{a\alpha^2 + b\alpha + c}$$

Therefore

$$y_p = \frac{ke^{\alpha x}}{a\alpha^2 + b\alpha + c} \tag{31}$$

provided the denominator is not zero. We note that the denominator is the characteristic polynomial

$$p(\lambda) = a\lambda^2 + b\lambda + c \tag{32}$$

evaluated at $\lambda = \alpha$. The particular solution can therefore be written in the convenient form

$$y_p = \frac{ke^{\alpha x}}{p(\alpha)} \qquad p(\alpha) \neq 0 \tag{33}$$

If $p(\alpha) = 0$, Eq. (33) does not determine a particular solution. In this case α is a root of the characteristic equation. Therefore $e^{\alpha x}$ is a solution of the homogeneous equation and cannot possibly also be a solution of the nonhomogeneous equation. To find a particular solution if $p(\alpha) = 0$, we assume

$$y_p = Axe^{\alpha x}\dagger \tag{34}$$

Substituting into the differential equation (28), we obtain

$$(a\alpha^2 + b\alpha + c)Axe^{\alpha x} + (2a\alpha + b)Ae^{\alpha x} = ke^{\alpha x}$$

Since $p(\alpha) = a\alpha^2 + b\alpha + c = 0$, we have $A = k/(2a\alpha + b)$ and

$$y_p = \frac{kxe^{\alpha x}}{2a\alpha + b} \tag{35}$$

provided the denominator is not zero. We note that $2a\alpha + b = p'(\alpha)$; therefore

$$y_p = \frac{kxe^{\alpha x}}{p'(\alpha)} \qquad p(\alpha) = 0, p'(\alpha) \neq 0 \tag{36}$$

If $p(\alpha) = 0$ and $p'(\alpha) \neq 0$, then α is a simple root of the characteristic equation. If both $p(\alpha)$ and $p'(\alpha)$ are zero, α is a double root of the characteristic equation.‡ This means that both $e^{\alpha x}$ and $xe^{\alpha x}$ are solutions of the homogeneous equation. In this case we assume

$$y_p = Ax^2e^{\alpha x} \tag{37}$$

† Alternatively we could assume $y_p = v(x)e^{\alpha x}$ and determine $v(x)$ by substituting into (28).

‡ If α is a root of $p(\lambda) = 0$, then $p(\lambda) = (\lambda - \alpha)g(\lambda)$ where $g(\lambda)$ is a linear polynomial. If $g(\alpha) \neq 0$, α is a simple root and if $g(\alpha) = 0$, α is a double root. We see that $p'(\lambda) = (\lambda - \alpha)g'(\lambda) + g(\lambda)$; therefore $p'(\alpha) = g(\alpha)$. This shows that if α is a simple root, $p'(\alpha) \neq 0$ and if α is a double root, $p(\alpha) = p'(\alpha) = 0$.

Proceeding as above, we obtain

$$y_p = \frac{kx^2 e^{\alpha x}}{2a} = \frac{kx^2 e^{\alpha x}}{p''(\alpha)} \qquad p(\alpha) = p'(\alpha) = 0 \qquad (38)$$

The denominator cannot be zero since, by assumption, a is different from zero. Summarizing the above we have the following convenient rule: *A particular solution of $L(y) = ke^{\alpha x}$ is given by*

$$y_p = \begin{cases} \dfrac{ke^{\alpha x}}{p(\alpha)} & p(\alpha) \neq 0 \\[2.5ex] \dfrac{kxe^{\alpha x}}{p'(\alpha)} & p(\alpha) = 0, \quad p'(\alpha) \neq 0 \\[2.5ex] \dfrac{kx^2 e^{\alpha x}}{p''(\alpha)} & p(\alpha) = p'(\alpha) = 0 \end{cases}$$

Example 1   Solve

$$y'' - 5y' + 4y = 3 + 2e^x$$

$$p(\lambda) = \lambda^2 - 5\lambda + 4 = 0 \qquad \lambda = 4, 1$$

$$y_h = c_1 e^x + c_2 e^{4x}$$

$$y_p = \frac{3}{p(0)} + \frac{2xe^x}{p'(1)} = \tfrac{3}{4} - \tfrac{2}{3}xe^x$$

$$y = c_1 e^x + c_2 e^{4x} + \tfrac{3}{4} - \tfrac{2}{3}xe^x$$

II. $f(x) = k \cos \beta x$ *or* $f(x) = k \sin \beta x$. We shall handle this case by exploiting the relationship between sinusoids and complex exponentials. At first sight the method may appear artificial. However, the method is much used and is the basis of the impedance method used in solving alternating-current circuits.† Consider the differential equation

$$L(y) = ay'' + by' + cy = k \cos \beta x \qquad (39)$$

Also consider the companion equation

$$L(v) = av'' + bv' + cv = k \sin \beta x \qquad (40)$$

We now combine these two real equations into one equation for the complex function

$$w = y + iv \qquad (41)$$

† See Secs. 3-9 and 6-4.

Since the operator L is linear, we have

$$L(w) = L(y) + iL(v)$$
$$= k(\cos \beta x + i \sin \beta x)$$

Therefore the complex function w satisfies

$$L(w) = ke^{i\beta x} \tag{42}$$

This equation has an exponential on the right-hand side and can be solved as in Case I.   After w is found, y can be found by taking the real part of w, and v by taking the imaginary part.   We summarize this method below.

Method   *To find a particular solution of $L(y) = k \cos \beta x$ [or $L(y) = k \sin \beta x$], find a particular solution of $L(w) = ke^{i\beta x}$ and take the real part [or imaginary part] of the result.*

Example 2   Find a particular solution of

$$y'' + 7y' + 12y = 3 \cos 2x \tag{43}$$

We have

$$p(\lambda) = \lambda^2 + 7\lambda + 12 \tag{44}$$

Consider $y = \text{Re } w = $ real part of w, where w satisfies the equation

$$w'' + 7w' + 12w = 3e^{i2x} \tag{45}$$

A particular solution of Eq. (45) is

$$w_p = \frac{3e^{i2x}}{p(2i)} = \frac{3e^{i2x}}{8 + 14i} \tag{46}$$

In order to find $\text{Re } w_p = y_p$, we must put w_p in an appropriate form:

$$w_p = \frac{3e^{i2x}}{8 + 14i} \frac{8 - 14i}{8 - 14i}$$

$$= \tfrac{3}{260}(\cos 2x + i \sin 2x)(8 - 14i)$$

$$= \tfrac{3}{260}(8 \cos 2x + 14 \sin 2x) + \tfrac{3}{260}i(8 \sin 2x - 14 \cos 2x)$$

$$y_p = \text{Re } w_p = \tfrac{6}{65} \cos 2x + \tfrac{21}{130} \sin 2x \tag{47}$$

Alternatively, we could have written

$$8 + 14i = \sqrt{260}e^{i\theta} \qquad \theta = \arctan \tfrac{14}{8}$$

and

$$w_p = \frac{3e^{i2x}}{\sqrt{260}e^{i\theta}} = \frac{3}{\sqrt{260}} e^{i(2x-\theta)} \tag{48}$$

Therefore

$$y_p = \frac{3}{\sqrt{260}} \cos (2x - \theta) \tag{49}$$

The two expressions (47) and (49) for y_p are exactly equivalent.

Example 3 $y'' + 4y = 3 \sin 2x$

$$p(\lambda) = \lambda^2 + 4 = 0 \qquad \lambda = \pm 2i$$

$$y = \operatorname{Im} w$$

$$w'' + 4w = 3e^{i2x}$$

$$w_p = \frac{3xe^{i2x}}{p'(2i)} = \frac{3xe^{i2x}}{4i} = -\tfrac{3}{4}ixe^{i2x}$$

$$y_p = \operatorname{Im} w_p = -\tfrac{3}{4}x \cos 2x$$

An alternative way of solving $L(y) = k \cos \beta x$ or $L(y) = k \sin \beta x$ is to assume a solution of the form

$$y_p = A \cos \beta x + B \sin \beta x \tag{50}$$

and to determine the coefficients A and B by substituting into the differential equation and equating the coefficients of $\cos \beta x$ and $\sin \beta x$ on both sides of the equation. If $p(i\beta) = 0$, that is, if $\cos \beta x$ (and $\sin \beta x$) is a solution of the homogeneous equation, the form

$$y_p = x(A \cos \beta x + B \sin \beta x) \tag{51}$$

must be used in place of (50).

Example 4 $y'' + 7y' + 12y = 3 \cos 2x$

$$y_p = A \cos 2x + B \sin 2x$$

Substituting into the differential equation, we obtain

$$(8A + 14B) \cos 2x + (8B - 14A) \sin 2x = 3 \cos 2x$$

Therefore (why?)

$$(8A + 14B) = 3 \qquad 8B - 14A = 0$$

or

$$A = \tfrac{6}{65} \qquad B = \tfrac{21}{130}$$

and

$$y_p = \tfrac{6}{65} \cos 2x + \tfrac{21}{130} \sin 2x$$

III. $f(x) = B_0 + B_1 x + \cdots + B_n x^n$ The differential equation

$$ay'' + by' + cy = B_0 + B_1 x + \cdots + B_n x^n \tag{52}$$

clearly has a polynomial for a particular solution. If $p(0) = c \neq 0$, the substitution of

$$y_p = Q_n(x) = A_0 + A_1 x + \cdots + A_n x^n \tag{53}$$

into the differential equation will yield a polynomial of degree n on the left-hand side of the equation. The coefficients A_k can then be obtained by equating the coefficients of like powers of x on both sides of the equation. If $p(0) = c = 0$ but $p'(0) = b \neq 0$, we must assume $y_p = x Q_n(x)$ in order to obtain a polynomial of degree n on the left side of the equation. Similarly, if $p(0) = p'(0) = 0$, we must assume $y_p = x^2 Q_n(x)$.

Example 5 $y'' + 3y' = 2x^2 + 3x$

Assume

$$y_p = x(Ax^2 + Bx + C)$$

Substituting into the equation, we obtain after simplification

$$9Ax^2 + (6A + 6B)x + 2B + 3C = 2x^2 + 3x$$

and equating coefficients of different powers of x, we obtain

$$9A = 2 \qquad 6A + 6B = 3 \qquad 2B + 3C = 0$$

Therefore

$$A = \tfrac{2}{9} \qquad B = \tfrac{5}{18} \qquad C = -\tfrac{5}{27}$$

and the particular solution is

$$y_p = \tfrac{2}{9} x^3 + \tfrac{5}{18} x^2 - \tfrac{5}{27} x$$

IV. $f(x) = (B_0 + B_1 x + \cdots + B_n x^n)e^{\alpha x}$ For the equation

$$ay'' + by' + cy = (B_0 + B_1 x + \cdots + B_n x^n)e^{\alpha x} \tag{54}$$

we must assume

$$y_p = Q_n(x)e^{\alpha x} \qquad p(\alpha) \neq 0 \tag{55}$$

or

$$y_p = xQ_n(x)e^{\alpha x} \qquad p(\alpha) = 0, p'(\alpha) \neq 0 \tag{56}$$

or

$$y_p = x^2 Q_n(x)e^{\alpha x} \qquad p(\alpha) = p'(\alpha) = 0 \tag{57}$$

By allowing α in these equations to be a complex number, we can also solve the equation

$$ay'' + by' + cy = (B_0 + B_1 x + \cdots + B_n x^n)e^{\alpha x} \cos \beta x \tag{58}$$

or the same equation with $\cos \beta x$ replaced by $\sin \beta x$.

Example 6 $y'' + 4y = xe^x \sin x$

Let $y = \text{Im } w$.

$$w'' + 4w = xe^x e^{ix} = xe^{(1+i)x}$$

Assume

$$w_p = (Ax + B)e^{(1+i)x}$$

Substituting, we obtain

$$(2i + 4)Ax + (4 + 2i)B + (2 + 2i)A = x$$

Therefore

$$A = \frac{1}{2i + 4} \quad \text{and} \quad B = -\frac{1 + i}{6 + 8i}$$

and

$$w_p = \frac{xe^x e^{ix}}{4 + 2i} - \frac{1 + i}{6 + 8i} e^x e^{ix}$$

$$y_p = \text{Im } w_p = \tfrac{1}{10} xe^x(-\cos x + 2 \sin x) - \tfrac{1}{50} e^x(-\cos x + 7 \sin x)$$

Problems 5-5

Find particular solutions of the following equations:

1. $y'' + 3y' - 5y = 4e^{2x} + 6e^{-3x}$
2. $y'' + 3y' + 5y = 2 \sin 3x$
3. $y'' + 9y = 4 \cos 3x$
4. $y'' + 3y' - 4y = 2e^{-4x} + 5$
5. $y'' + 4y' + 4y = 3e^{-2x} + e^{-x}$
6. $y'' - 3y' + 4y = x^3 + 3x$
7. $y'' - 3y' + y = 3e^x \sin x$
8. $y'' - 3y' = 2x^2 + 3e^x$
9. $y'' + 2y' + 2y = 2e^x \cos x$
10. $y'' - 4y = 3xe^{2x}$

5-6   *Variation of parameters*

The method of variation of parameters enables us to find a particular solution of a nonhomogeneous equation whenever two linearly independent solutions of the homogeneous equation are known. This method works for any linear equation *even when the coefficients are not constant.* We consider the equation

$$a(x)y'' + b(x)y' + c(x)y = f(x) \tag{59}$$

where all the functions are continuous in some interval I and $a(x) \neq 0$ in I. We assume that $y_1(x)$ and $y_2(x)$ are linearly independent solutions of the homogeneous equation. The general solution of the homogeneous equation is therefore

$$y_h = c_1 y_1(x) + c_2 y_2(x) \tag{60}$$

We now try to find a solution of the nonhomogeneous equation (59) by replacing the constants c_1 and c_2 by functions of x. Therefore we look for a solution in the form

$$y = v_1(x)y_1(x) + v_2(x)y_2(x) \tag{61}$$

By substituting this expression into the differential equation (59), we will get only one condition that the two functions $v_1(x)$ and $v_2(x)$ must satisfy. We therefore can impose another condition. This can be rather arbitrarily imposed but should be such as to simplify the determination of $v_1(x)$ and $v_2(x)$.

Let us calculate the derivatives of (61). We have

$$y' = v_1 y_1' + v_2 y_2' + (v_1' y_1 + v_2' y_2) \tag{62}$$

Before proceeding to calculate y'', we note that if we require

$$v_1' y_1 + v_2' y_2 \equiv 0 \tag{63}$$

then no second derivatives of v_1 and v_2 will appear in y''. We therefore take (63) as one condition on v_1 and v_2. Calculating y'', with the condition (63), we obtain

$$y'' = v_1 y_1'' + v_1' y_1' + v_2 y_2'' + v_2' y_2' \tag{64}$$

Substituting (61), (62), and (64) into the differential equation, we obtain

$$(ay_1'' + by_1' + cy_1)v_1 + (ay_2'' + by_2' + cy_2)v_2 + v_1' ay_1' + v_2' ay_2' = f(x) \tag{65}$$

The coefficients of v_1 and v_2 are zero because y_1 and y_2 are solutions of the homogeneous equation. Therefore we have the two equations

$$v_1' y_1 + v_2' y_2 = 0$$
$$v_1' y_1' + v_2' y_2' = \frac{f(x)}{a(x)} \tag{66}$$

to determine v_1 and v_2. These equations can be solved for v_1' and v_2' provided

$$\begin{vmatrix} y_1 & y_2 \\ y_1' & y_2' \end{vmatrix} = y_1 y_2' - y_2 y_1' \neq 0 \tag{67}$$

This determinant is the Wronskian of y_1 and y_2 and is always different from zero if y_1 and y_2 are linearly independent. Solving Eqs. (66), we obtain

$$v_1' = -\frac{y_2(x)f(x)}{a(x)W[y_1,y_2]} \qquad v_2' = \frac{y_1(x)f(x)}{a(x)W[y_1,y_2]}$$

Therefore

$$v_1 = -\int \frac{y_2(x)f(x)\,dx}{a(x)W[y_1,y_2]} \qquad v_2 = \int \frac{y_1(x)f(x)\,dx}{a(x)W[y_1,y_2]} \tag{68}$$

and a solution of the nonhomogeneous equation is

$$y = v_1(x)y_1(x) + v_2(x)y_2(x) \tag{69}$$

Example 1   Solve: $y'' + y = \sec x$

Linearly independent solutions of the homogeneous equation are

$$y_1 = \cos x \qquad y_2 = \sin x$$

and the general solution of the homogeneous equation is

$$y_h = c_1 \cos x + c_2 \sin x$$

We have

$$W[y_1,y_2] = y_1 y_2' - y_2 y_1' = \cos x\,(\cos x) - (\sin x)(-\sin x) = 1$$

Therefore

$$v_1 = -\int \sin x \sec x\,dx = -\int \frac{\sin x\,dx}{\cos x} = \ln|\cos x|$$

$$v_2 = \int \cos x \sec x\,dx = \int dx = x$$

$$y_p = \cos x \ln|\cos x| + x \sin x$$

$$y = y_h + y_p = c_1 \cos x + c_2 \sin x + \cos x \ln|\cos x| + x \sin x$$

Problems 5-6

Solve by variation of parameters:

1. $y'' + y = \tan x$
2. $y'' - 5y' + 6y = 2e^{4x}$
3. $y'' - y = \sin^2 x$
4. $y'' - y = x - 2e^{-x}$
5. $y'' + 4y' + 4y = \dfrac{e^{-2x}}{x^2}$

6. By variation of parameters show that the solution of

DE: $y'' + \lambda^2 y = f(x) \qquad \lambda > 0$

IC: $y(0) = 0$

$\qquad y'(0) = 0$

is

$$y = \frac{1}{\lambda}\int_0^x f(t)\sin\lambda(x - t)\,dt$$

7. Show that the solution of the initial-value problem

DE: $a(x)y'' + b(x)y' + c(x)y = f(x)$

IC: $y(x_0) = y'(x_0) = 0$

is

$$y = \int_{x_0}^{x} \frac{y_1(t)y_2(x) - y_1(x)y_2(t)}{a(t)\,W[y_1(t),y_2(t)]}\, f(t)\, dt$$

where $y_1(x)$ and $y_2(x)$ are any two linearly independent solutions of the corresponding homogeneous equation.

8. In the interval $x > 0$, find two linearly independent solutions of $x^2y'' - 2y = 0$ by assuming a solution of the form $y = x^m$. Then solve the equation

$$x^2y'' - 2y = 2x^5$$

9. Assume that one nontrivial solution $y_1(x)$ of the homogeneous equation $a(x)y'' + b(x)y' + c(x)y = 0$ is known. Show by the method of variation of parameters that a second solution is given by

$$y = y_1(x) \int^{} \frac{\exp\left\{-\int[b(x)/a(x)]\, dx\right\}}{y_1{}^2(x)}\, dx$$

in any interval where $y_1(x)$ does not vanish.

10. Extend the method of Prob. 9 to show that the general solution of $a(x)y'' + b(x)y' + c(x)y = f(x)$ can be determined if one nontrivial solution of the corresponding homogeneous equation is known.

5-7 Euler's equation

The differential equation

$$x^2y'' + pxy' + qy = f(x) \tag{70}$$

where p and q are constants is called *Euler's* equation. This equation is encountered in some important problems of potential theory (see Sec. 14-4) and also will serve as a useful example in the study of series solutions in Chap. 7.

We consider the homogeneous equation corresponding to (70), namely,

$$x^2y'' + pxy' + qy = 0 \tag{71}$$

Once the general solution of (71) has been found, (70) can be solved by variation of parameters. We shall restrict ourselves to the interval $x > 0$†
so that the coefficient of y'' does not vanish and the existence theorem will apply. Therefore for the interval $x > 0$ we need only find two linearly independent solutions in order to obtain the general solution.

---

† The interval $x < 0$ could also be used.

The simplest method of solving (71) is to guess the form of the solution. We note that the power of x in each term of (71) is the same as the order of the derivative. This suggests that a solution exists in the form

$$y = x^m$$

where m is a constant, for, upon substitution into the differential equation, each term becomes simply a constant times x^m. Hence we shall determine the *index* m so that x^m is a solution. (For another method see Prob. 2.)

Substituting in Eq. (71), we have

$$x^2 \cdot m(m - 1)x^{m-2} + px \cdot mx^{m-1} + q \cdot x^m = 0$$

or

$$x^m[m^2 + (p - 1)m + q] = 0$$

Since $x^m \neq 0$ for $x > 0$, we must have

$$m^2 + (p - 1)m + q = 0 \tag{72}$$

This quadratic equation for the index m is called the *indicial equation*. Three different cases can arise, depending on the value of the discriminant

$$\Delta = (p - 1)^2 - 4q \tag{73}$$

Case I. $\Delta > 0$. In this case Eq. (72) has two distinct, real roots, $m_1 \neq m_2$, and hence x^{m_1} and x^{m_2} are linearly independent solutions. The general solution of (71) is therefore

$$y = c_1 x^{m_1} + c_2 x^{m_2} \tag{74}$$

Case II. $\Delta = 0$. In this case the indicial equation (72) has the double root

$$m_1 = \frac{1 - p}{2} \quad .$$

and our method yields only one solution, $y = x^{m_1}$. A second solution can be obtained by variation of parameters. Substituting $y = ux^{m_1}$ into Eq. (71), we get, after some simplification,

$$xu'' + u' = 0$$

Letting $v = u'$, we have

$$xv' + v = 0$$

Hence

$$v = \frac{c_1}{x}$$

and, on integrating,

$$u = c_1 \ln x + c_2$$

Therefore a second linearly independent solution is $x^{m_1} \ln x$ and the general solution is

$$y = x^{m_1}(c_1 \ln x + c_2) \tag{75}$$

Case III. $\Delta < 0$.   The roots of the indicial equation (72) are conjugate complex numbers

$$m_1 = a + ib \qquad m_2 = a - ib$$

Since x^{m_1} and x^{m_2} are linearly independent, the general solution of (71) is

$$y = c_1 x^{m_1} + c_2 x^{m_2}$$

or

$$y = x^a(c_1 x^{ib} + c_2 x^{-ib}) \tag{76}$$

This solution can be expressed in terms of real functions by noting that†

$$x^{ib} = e^{ib \ln x} = \cos (b \ln x) + i \sin (b \ln x)$$

Hence the solution (76) can be written in the form

$$y = x^a[A \cos (b \ln x) + B \sin (b \ln x)] \tag{77}$$

Example 1   Solve the equation

$$x^2 y'' - xy' - 8y = 0$$

The indicial equation is

$$m^2 - 2m - 8 = 0$$

with solutions $m = -2$ and $m = 4$.   Therefore the general solution is

$$y = c_1 x^{-2} + c_2 x^4$$

Problems 5-7

1. Find the general solution of each of the following equations for $x > 0$:

a. $2x^2 y'' + xy' - y = 3x - 5x^2$ b. $x^2 y'' + 7xy' + 5y = 10 - \dfrac{4}{x}$

2. a. Show that Euler's equation (70) can be transformed into a linear equation with *constant* coefficients by making the substitution $x = e^z$ or $z = \ln x$.

   b. Use the method of part a to solve the equation

$$x^2 y'' - 2xy' + 2y = x^2 + 4$$

3. Sketch the curve $y = x \sin (\ln x)$.

† See Sec. 5-3, Prob. 5.

5-8   *Formulas of Lagrange and Abel*

In this section we develop some useful formulas concerning the solution of the homogeneous and nonhomogeneous equations.

Let

$$L(y) = a(x)y'' + b(x)y' + c(x)y \tag{78}$$

where as usual a, b, c are continuous and $a \neq 0$ in some interval I.   If u and v are any two twice-differentiable functions, we have

$$uL(v) - vL(u) = a(x)(uv'' - vu'') + b(x)(uv' - vu') \tag{79}$$

We note that the coefficient of $b(x)$ is the Wronskian

$$W[u,v] = uv' - vu'$$

A brief computation shows that the coefficient of $a(x)$ is $dW[u,v]/dx$.   Therefore (79) can be written in the compact form

$$uL(v) - vL(u) = a(x)\frac{dW[u,v]}{dx} + b(x)W[u,v] \tag{80}$$

This is *Lagrange's formula*, which holds for any twice-differentiable functions u and v.

If we now let u and v be solutions of the homogeneous equation $L(y) = 0$, we have

$$a(x)\frac{dW}{dx} + b(x)W = 0 \tag{81}$$

Solving this first-order linear equation, we obtain *Abel's formula* for the Wronskian of any two solutions of $L(y) = 0$:

$$W[u,v] = k \exp\left[-\int \frac{b(x)}{a(x)}\,dx\right] \tag{82}$$

From this formula we obtain another proof of the fact that the Wronskian of two solutions is either identically zero (if $k = 0$) or never zero (if $k \neq 0$).

We now make use of the above results to derive the solution of the nonhomogeneous problem

DE: $L(y) = f(x)$

IC: $y(x_0) = 0$

$y'(x_0) = 0$

assuming that two linearly independent solutions y_1 and y_2 of the homogeneous equation are known.

Letting $u = y_1$ and $v = y$ in Lagrange's formula (80), we obtain

$$\frac{dW[y_1,y]}{dx} + \frac{b(x)}{a(x)}\, W[y_1,y] = \frac{y_1 f(x)}{a(x)} \tag{83}$$

Solving this first-order equation using the integrating factor

$$I(x) = \exp\left[\int_{x_0}^{x} \frac{b(t)}{a(t)}\, dt\right]$$

we obtain

$$W[y_1,y] = \exp\left[-\int_{x_0}^{x} \frac{b(t)}{a(t)}\, dt\right]\int_{x_0}^{x} \frac{\exp\left\{\int_{x_0}^{s} [b(t)/a(t)]\, dt\right\} y_1(s) f(s)}{a(s)}\, ds \tag{84}$$

where we have used the initial conditions $y(x_0) = y'(x_0) = 0$. Using Abel's formula, Eq. (84) reduces to

$$y_1 y' - y y_1' = W[y_1,y_2]\int_{x_0}^{x} \frac{y_1(s) f(s)}{a(s)\, W[y_1(s),y_2(s)]}\, ds \tag{85}$$

where $W[y_1,y_2]$ is never zero since y_1 and y_2 are linearly independent solutions of $L(y) = 0$. In the same manner we may obtain a similar formula replacing y_1 by y_2:

$$y_2 y' - y y_2' = W[y_1,y_2]\int_{x_0}^{x} \frac{y_2(s) f(s)}{a(s)\, W[y_1(s),y_2(s)]}\, ds \tag{86}$$

Solving the two simultaneous equations (85) and (86) for y, we obtain

$$y(x) = y_2(x)\int_{x_0}^{x} \frac{y_1(s) f(s)\, ds}{a(s)\, W[y_1(s),y_2(s)]} - y_1(x)\int_{x_0}^{x} \frac{y_2(s) f(s)\, ds}{a(s)\, W[y_1(s),y_2(s)]} \tag{87}$$

or equivalently

$$y(x) = \int_{x_0}^{x} \frac{[y_1(s)y_2(x) - y_2(s)y_1(x)] f(s)}{a(s)\, W[y_1(s),y_2(s)]}\, ds \tag{88}$$

a result previously obtained by variation of parameters.   (See Sec. 5-6, Prob. 7.)

Example 1   Solve: DE: $y'' + y = f(x)$

$$\text{IC: } y(0) = y'(0) = 0$$

The solutions of the homogeneous equation are

$$y_1(x) = \sin x \qquad y_2(x) = \cos x$$

and

$$W[y_1,y_2] = -1$$

Substituting into (88), we obtain

$$y(x) = \int_{x_0}^{x} \frac{\sin s \cos x - \cos s \sin x}{-1}\, f(s)\, ds = \int_{x_0}^{x} \sin(x - s) f(s)\, ds$$

Problems 5-8

1. Use Abel's formula to prove that if y_1 and y_2 are linearly independent solutions of $L(y) = 0$, then $y = c_1 y_1 + c_2 y_2$ is the general solution.

2. Assume that $y_1(x)$ is a solution of $L(y) = 0$. Use Abel's formula to derive a second solution

$$y_2(x) = y_1(x) \int \frac{\exp\left\{-\int [b(x)/a(x)]\, dx\right\}}{y_1^2(x)}\, dx$$

3. Solve: DE: $y'' - y = f(x)$

$$y(0) = y'(0) = 0$$

4. The operator L is called *self-adjoint* if it can be written as

$$L(y) = [a(x)y']' + c(x)y$$

that is, if in Eq. (78) we have $b(x) = a'(x)$. If L is self-adjoint, show that

$$\int_{x_0}^{x_1} (uLv - vLu)\, dx = a(x)W[u(x),v(x)] \Big|_{x_0}^{x_1}$$

This is called *Green's formula*. Show also that if u and v are solutions of $L(y) = 0$, then

$$a(x)W[u(x),v(x)] \equiv \text{const}$$

5. Let y_1 and y_2 be solutions of Bessel's differential equation

$$x^2 y'' + xy' + (x^2 - n^2)y = 0$$

Find $W[y_1,y_2]$.

5-9 Linear equations of the nth order

In this section we extend the fundamental theory for second-order equations to nth-order equations. The proofs of Theorems 2 through 9 are in principle the same as for the second-order case already given in Sec. 5-2 and will be left to the reader.

Let L denote the operator defined by

$$L(y) = a_0(x)y^{(n)} + a_1(x)y^{(n-1)} + \cdots + a_n(x)y \tag{89}$$

The fundamental existence and uniqueness theorem states (see Chap. 4):

Theorem 1 *Let $a_i(x)$ and $f(x)$ be continuous and $a_0(x)$ never zero in an interval I and let x_0 be in I; then the initial-value problem*

DE: $L(y) = f(x)$

IC: $y(x_0) = b_0,\ y'(x_0) = b_1,\ \ldots,\ y^{(n-1)}(x_0) = b_{n-1}$

possesses a unique solution. The solution has a continuous nth derivative and is defined throughout I.

Theorem 2   *If* $y_1, \ldots, y_k$ *are functions possessing n derivatives through-out I and* $c_1, \ldots, c_k$ *are constants, then*

$$L[c_1 y_1(x) + c_2 y_2(x) + \cdots + c_k y_k(x)] = c_1 L(y_1) + c_2 L(y_2) + \cdots + c_k L(y_k)$$

Theorem 3   *If* $y_1, \ldots, y_k$ *are solutions of* $L(y) = 0$, *so is* $y = c_1 y_1 + c_2 y_2 + \cdots + c_k y_k$.

Definition 1   *The set of functions* $y_1(x)$, $y_2(x)$, $\ldots$, $y_k(x)$ *is called linearly dependent in an interval I if there exist constants* $c_1, c_2, \ldots, c_k$, *not all zero, such that* $c_1 y_1(x) + \cdots + c_k y_k(x) \equiv 0$. *The set of functions is called linearly independent if it is not linearly dependent, that is, if* $c_1 y_1(x) + \cdots + c_k y_k(x) \equiv 0$ *implies* $c_1 = c_2 = \cdots = c_k = 0$.

Example 1   The set of functions $1, x, x^2, \ldots, x^{k-1}$ is linearly independent in every interval, for if

$$c_0 + c_1 x + \cdots + c_{k-1} x^{k-1} \equiv 0$$

then $c_0 = c_1 = \cdots = c_{k-1} = 0$. (Otherwise the polynomial of degree $k - 1$ would have more than $k - 1$ roots.)

Example 2   The functions $e^{\lambda_1 x}, \ldots, e^{\lambda_k x}$ are linearly independent in every interval if the λ_i's are distinct. The proof is by induction. The statement is obviously true for $k = 1$. We assume that it has been proved for $k - 1$ functions. If the functions were linearly dependent, there would exist constants $c_1, \ldots, c_k$, not all zero, such that

$$c_1 e^{\lambda_1 x} + \cdots + c_k e^{\lambda_k x} \equiv 0$$

Divide by $e^{\lambda_k x}$ and put $\lambda_i - \lambda_k = \alpha_i$ to obtain

$$c_1 e^{\alpha_1 x} + \cdots + c_{k-1} e^{\alpha_{k-1} x} + c_k \equiv 0$$

Differentiating, we obtain

$$\alpha_1 c_1 e^{\alpha_1 x} + \cdots + \alpha_{k-1} c_{k-1} e^{\alpha_{k-1} x} \equiv 0$$

which means that $e^{\alpha_1 x}, \ldots, e^{\alpha_{k-1} x}$ are linearly dependent, contrary to our inductive assumption.

Definition 2   *The Wronskian of k functions which are* $(k - 1)$ *times differentiable is*

$$W[y_1(x), \ldots, y_k(x)] = \begin{vmatrix} y_1(x) & y_2(x) & \ldots & y_k(x) \\ y_1'(x) & y_2'(x) & \ldots & y_k'(x) \\ \cdots\cdots\cdots\cdots\cdots\cdots\cdots\cdots \\ y_1^{(k-1)}(x) & y_2^{(k-1)}(x) & \ldots & y_k^{(k-1)}(x) \end{vmatrix} \tag{90}$$

In the proofs of the following three theorems the following facts from algebra† are needed: (1) *A set of n linear algebraic equations in n unknowns has a unique solution if and only if the determinant of the coefficients is not zero.* (2) *A set of n homogeneous linear equations in n unknowns has a non-trivial solution if and only if the determinant of the coefficients is zero.* With these facts Theorems 4, 5, and 6 are proved in exactly the same manner as in Sec. 5-2.

Theorem 4 *If $W[y_1(x), \ldots, y_k(x)]$ is different from zero for at least one point in an interval I, then $y_1, \ldots, y_k$ are linearly independent in I.*

Theorem 5 *If $y_1, \ldots, y_k$ are linearly independent solutions of $L(y) = 0$ in an interval I, then $W[y_1, \ldots, y_k]$ is never zero in I.*

Theorem 6 *If $y_1, \ldots, y_n$ are n linearly independent solutions of the nth-order equation $L(y) = 0$, then $y = c_1 y_1 + \cdots + c_n y_n$ is the general solution.*

Theorem 7 *There exist n linearly independent solutions of $Ly = 0$.*

Theorem 8 *The general solution of $L(y) = f(x)$ is $y = y_h + y_p$ where y_p is a particular solution and y_h is the general solution of $L(y) = 0$.*

Theorem 9 *If y_1 is a solution of $Ly = f_1$ and y_2 is a solution of $Ly = f_2$, then $y = y_1 + y_2$ is a solution of $Ly = f_1 + f_2$.*

Homogeneous equation with constant coefficients Consider

$$L(y) = a_0 y^{(n)} + a_1 y^{(n-1)} + \cdots + a_n y = 0 \tag{91}$$

where the a_i are *real* constants and $a_0 \neq 0$. As in the second-order case we assume a solution of the form $y = e^{\lambda x}$. Substituting into (91), we find that λ must satisfy the characteristic equation

$$p(\lambda) = a_0 \lambda^n + a_1 \lambda^{n-1} + \cdots + a_n = 0 \tag{92}$$

If λ_1 is a real root of this equation, then $e^{\lambda_1 x}$ is a solution of (91). If $\lambda = \alpha + i\beta$ is a complex root, then (since the equation has real coefficients) $\bar{\lambda} = \alpha - i\beta$ is also a root and $e^{(\alpha+i\beta)x}$ and $e^{(\alpha-i\beta)x}$ are complex-valued solutions of (91); from these we obtain the real solutions $e^{\alpha x} \cos \beta x$ and $e^{\alpha x} \sin \beta x$.

If λ_1 is a root of multiplicity k, then it can be shown (see Prob. 2 below) that

$$e^{\lambda_1 x}, \, xe^{\lambda_1 x}, \, x^2 e^{\lambda_1 x}, \, \ldots, \, x^{k-1} e^{\lambda_1 x}$$

† See Thomas [35], pp. 420 and 426.

are k linearly independent solutions.   By finding all the roots of (92), we can find a total of n solutions which are linearly independent (see Prob. 4); therefore the general solution can be found.

Example 3 $y^{(4)} - y = 0$

$$p(\lambda) = \lambda^4 - 1 = 0$$

$$\lambda = 1, -1, +i, -i$$

$$y = c_1 e^x + c_2 e^{-x} + c_3 \cos x + c_4 \sin x$$

Example 4   Solve

$$y^{(5)} - y^{(3)} - 4y^{(2)} - 3y^{(1)} - 2y = 0$$

$$p(\lambda) = \lambda^5 - \lambda^3 - 4\lambda^2 - 3\lambda - 2 = 0 = (\lambda - 2)(\lambda^2 + \lambda + 1)^2 = 0$$

$$\lambda = 2, \ -\tfrac{1}{2} \pm i \frac{\sqrt{3}}{2}, \ -\tfrac{1}{2} \pm i \frac{\sqrt{3}}{2}$$

$$y = c_1 e^{2x} + (c_2 + c_3 x)e^{-\frac{1}{2}x} \cos \frac{\sqrt{3}}{2} + (c_4 + c_5 x) e^{-\frac{1}{2}x} \sin \frac{\sqrt{3}}{2} x$$

In each of the above examples the roots of the characteristic equation were easily obtained.   However, in most cases, finding the roots of the characteristic equation is the most difficult part of the problem.   In practice this would usually be done by some method of numerical approximation.

Undetermined coefficients   If the right-hand side of the constant coefficient equation

$$L(y) = a_0 y^{(n)} + \cdots + a_n y = f(x) \tag{93}$$

is an exponential, a sinusoid, or a polynomial, a particular solution may be found by undetermined coefficients in exactly the same manner as already discussed in Sec. 5-5.

If $f(x) = ce^{\alpha x}$, we assume $y_p = Ae^{\alpha x}$ and obtain upon substitution into (93)

$$L(Ae^{\alpha x}) = Ap(\alpha)e^{\alpha x} = ce^{\alpha x} \tag{94}$$

where $p(\alpha)$ is the characteristic polynomial evaluated at α.   Therefore if $p(\alpha) \neq 0$,

$$y_p = \frac{ce^{\alpha x}}{p(\alpha)} \tag{95}$$

In case $p(\alpha) = 0$ but $p'(\alpha) \neq 0$, we obtain (see Prob. 6)

$$y_p = \frac{cxe^{\alpha x}}{p'(\alpha)} \tag{96}$$

and so on.

If $f(x) = A \cos \alpha x$ [or $f(x) = A \sin \alpha x$], we let $y = \operatorname{Re} w$ [or $y = \operatorname{Im} w$] and solve $L(w) = Ae^{i\alpha x}$ by the preceding method.

If $f(x)$ is a polynomial of degree k, we assume y_p is a polynomial with undetermined coefficients of degree k (or higher if necessary) so that upon substitution into (93) we obtain on the left a polynomial of degree k. Equating coefficients of like powers of x on each side will allow y_p to be determined.

Example 5 $y^{(4)} - y = 2e^{2x} + 3e^x$

$$p(\lambda) = \lambda^4 - 1$$

$$y_p = \frac{2e^{2x}}{p(2)} + \frac{3xe^x}{p'(1)} = \frac{2e^{2x}}{15} + \frac{3xe^x}{4}$$

Example 6 $y^{(4)} - y = 3 \cos 2x$

Let $y = \operatorname{Re} w$; then

$$w^{(4)} - w = 3e^{i2x}$$

$$w_p = \frac{3e^{i2x}}{p(2i)} = \frac{3e^{i2x}}{15}$$

$$y_p = \tfrac{3}{15} \cos 2x$$

Example 7 $y^{(4)} - y' = 2x + 1$

$$y_p = x(A + Bx) = Ax + Bx^2$$

$$- A - 2Bx = 2x + 1$$

Therefore

$$A = -1 \qquad B = -1$$

$$y_p = -x - x^2$$

Variation of parameters   The method of variation of parameters extends to the nth-order equation

$$L(y) = a_0(x)y^{(n)} + a_1(x)y^{(n-1)} + \cdots + a_n(x)y = f(x) \tag{97}$$

provided n linearly independent solutions $y_1, \ldots, y_n$ of $L(y) = 0$ are known. We assume

$$y_p = \sum_{i=1}^{n} v_i(x)y_i(x) \tag{98}$$

where the $v_i(x)$ must be determined so that (97) is satisfied. We also set the following $(n - 1)$ conditions on $v_i(x)$ in such a way that when the successive derivatives of y_p are computed [see Eq. (100)], only first derivatives of

$v_i(x)$ appear:

$$\Sigma v_i' y_i = 0$$
$$\Sigma v_i' y_i' = 0$$

.
.   (99)
.

$$\Sigma v_i' y_i^{(n-2)} = 0$$

With (99) the derivatives of y_p are

$$y_p' = \Sigma v_i y_i'$$
$$y_p'' = \Sigma v_i y_i''$$

.
.   (100)
.

$$y_p^{(n-1)} = \Sigma v_i y_i^{(n-1)}$$
$$y_p^{(n)} = \Sigma v_i' y_i^{(n-1)} + \Sigma v_i y_i^{(n)}$$

Substituting into (97), we obtain

$$a_0(x) \Sigma v_i' y_i^{(n-1)} = f(x) \tag{101}$$

This equation together with Eqs. (99) form n equations for the functions $v_i'(x)$ with determinant $W[y_1, \ldots, y_n]$. Since the Wronskian is not zero, these equations have a unique solution for $v_i'(x)$, from which $v_i(x)$ can be found by integration.

Example 8   DE: $y''' = f(x)$
IC: $y(0) = y'(0) = y''(0) = 0$

where $f(x)$ is assumed to be continuous. We find a particular solution by variation of parameters.

$$y_1 = 1 \qquad y_2 = x \qquad y_3 = x^2 \qquad y_p = v_1(x) + v_2(x)x + v_3(x)x^2$$

where $v_i(x)$ satisfy

$$v_1' + v_2' x + v_3' x^2 = 0$$
$$v_2' + 2v_3' x = 0$$
$$2v_3' = f$$

Solving, we get

$$v_3' = \tfrac{1}{2}f \qquad v_2' = -xf \qquad v_1' = \tfrac{1}{2}x^2 f$$

Therefore

$$v_1 = \int_0^x \tfrac{1}{2}t^2 f(t)\, dt \qquad v_2 = -\int_0^x tf(t)\, dt \qquad v_3 = \int_0^x \tfrac{1}{2}f(t)\, dt$$

and

$$y_p = \int_0^x \tfrac{1}{2}t^2 f(t)\, dt - x\int_0^x t f(x)\, dt + x^2\int_0^x \tfrac{1}{2}f(t)\, dt = \int_0^x \tfrac{1}{2}(x-t)^2 f(t)\, dt$$

We see that the particular solution found also satisfies the initial conditions. We note that the original problem could also be solved by integrating directly to obtain

$$y = \int_0^x \int_0^t \int_0^s f(r)\, dr\, ds\, dt$$

Since the solution is unique, we obtain the identity

$$\int_0^x \int_0^t \int_0^s f(r)\, dr\, ds\, dt = \int_0^x \frac{(x-t)^2}{2} f(t)\, dt$$

Problems 5-9

1. Prove Theorems 4 through 9.
2. If λ_1 is a root of multiplicity k of the characteristic equation (92):

   a. Show that $e^{\lambda_1 x}, xe^{\lambda_1 x}, \ldots, x^{k-1}e^{\lambda_1 x}$ are solutions of (91).
   b. Show that these solutions are linearly independent.

3. Prove that if $\lambda_1, \ldots, \lambda_k$ are distinct numbers and $P_1(x), \ldots, P_k(x)$ are arbitrary polynomials of degree $n_1, n_2, \ldots, n_k$ respectively, then the functions $f_1(x) = P_1(x)e^{\lambda_1 x}$, $f_2(x) = P_2(x)e^{\lambda_2 x}, \ldots, f_k(x) = P_k(x)e^{\lambda_k x}$ are linearly independent.

4. If $\lambda_1, \ldots, \lambda_k$ are the distinct roots of multiplicity $n_1, \ldots, n_k$, respectively, of the characteristic polynomial (92), then $P(\lambda) = (\lambda - \lambda_1)^{n_1}(\lambda - \lambda_2)^{n_2} \cdots (\lambda - \lambda_k)^{n_k}$ where $n_1 + n_2 + \cdots + n_k = n$. From Prob. 2 we see that the functions

$$e^{\lambda_i x}, xe^{\lambda_i x}, \ldots, x^{n_i - 1}e^{\lambda_i x} \qquad i = 1, 2, \ldots, k$$

are solutions of (91). Prove these n solutions are linearly independent.

5. Find general solutions:

   a. $y''' - 2y'' - y' + 2y = 0$ b. $y''' + y = 0$
   c. $y^{(4)} + y = 0$ d. $y^{(4)} - y = 0$
   e. $y^{(4)} + y' = 0$ f. $y^{(4)} + 2y'' + y = 0$
   g. $y''' - 3y'' + 3y' - y = 0$ h. $y^{(4)} - 3y'' + 2y = 0$
   i. $y^{(4)} + 2y''' + 3y'' + 2y' + y = 0$
      Note: $p(\lambda) = (\lambda^2 + \lambda + 1)^2$.

6. Prove Eq. (96).
7. Find particular solutions:

   a. $y''' - 3y'' + 3y' - y = 1 + e^{2x} + e^x$
   b. $y^{(4)} - y = \sin x + \cos x$ c. $y^{(4)} + 2y'' + y = \sin x$
   d. $y''' + 2y'' + y' = 2x^2 + 1$ e. $y''' + y = e^x \sin x$

Applications of Second-order Linear Differential Equations

6

6-1 Introduction

Many of the problems that arise in the studies of mechanical vibrations, electrical circuits, planetary motions, etc., can be formulated as second-order linear differential equations with constant coefficients. In this chapter we shall derive some of these equations, solve them in specific cases, and incidentally, discover a remarkable similarity between the behavior of mechanical and electrical systems.

In the case of mechanical vibrations the simplest system that illustrates the general properties is a mass suspended by a spring. In Sec. 1-3 we derived the general equation for such a system:

$$m\ddot{x} + r\dot{x} + kx = F(t) \qquad \dot{x} \equiv \frac{dx}{dt} \tag{1}$$

In this equation x (feet) is the displacement of the mass below its equilibrium position, t (seconds) is the time, m (slugs) is the mass, r is the damping constant, k is the spring constant, and $F(t)$ represents an external force, in pounds, applied to the mass.

Although Eq. (1) was derived for the simple case of the motion of a mass on a spring, the reader should be aware of the fact that this equation also governs the motion of many other mechanical systems. See, for example, the problems at the end of Sec. 6-2.

The differential equation representing a simple electrical circuit and the equations of planetary motion will be derived and solved below (see Secs. 6-4 and 6-5).

6-2   Free vibrations

If the mechanical system governed by Eq. (1) is vibrating *freely*, that is, with no external forces acting, then $F(t) \equiv 0$ and the differential equation for the system becomes

$$m\ddot{x} + r\dot{x} + kx = 0 \tag{2}$$

This is a *homogeneous* linear differential equation with constant coefficients and can be solved explicitly by the method of Chap. 5.   For simplicity, however, we shall first consider a special case.

Free undamped vibrations   If there are no damping (or friction) forces present in the system, then $r = 0$, and the equation becomes

$$m\ddot{x} + kx = 0 \tag{3}$$

The general solution of Eq. (3) can be written in the form

$$x = c_1 \cos \omega_n t + c_2 \sin \omega_n t \tag{4}$$

where the constant ω_n is given by

$$\omega_n = \sqrt{\frac{k}{m}} \tag{5}$$

Thus the mass executes periodic vibrations. Any motion which is described by an equation of the form (4) is called *simple harmonic motion*.

The quantity $\omega_n = \sqrt{k/m}$ is called the *circular frequency* of the system. The *frequency*, that is, the number of complete oscillations (cycles) per second, is given by

$$f_n = \frac{\omega_n}{2\pi} = \frac{1}{2\pi}\sqrt{\frac{k}{m}} \tag{6}$$

The quantity f_n is called the *natural frequency* of the system.   The *period* T of one complete oscillation is given by

$$T = \frac{1}{f_n} = \frac{2\pi}{\omega_n} \tag{7}$$

The solution (4) can be written in another form which is sometimes more convenient:†

$$x = A \cos (\omega_n t - \delta) \tag{8}$$

† See Sec. 5-4, page 132.

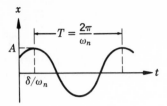

figure 1

The new constants A and δ are related to c_1 and c_2 by the equations

$$A = \sqrt{c_1^2 + c_2^2} \qquad \cos \delta = \frac{c_1}{A}$$

$$\sin \delta = \frac{c_2}{A} \tag{9}$$

From Eq. (8) we see that A can be interpreted as the *amplitude* of the vibration, and δ as the *phase*. (See Fig. 1.)

Example 1   A steel ball weighing 64 lb is suspended by a spring which is stretched 4 ft by the weight. If, at time $t = 0$, the ball is displaced 6 in. below its equilibrium position and released, what is the differential equation describing the motion? What will the position of the ball be at time t sec? Assume that friction may be neglected and take $g = 32$ ft/sec^2.

Solution   Since the weight w is 64 lb, the mass m is given by

$$m = \frac{w}{g} = 2 \text{ slugs}$$

In order to determine the spring constant k, Hooke's law gives

$$w = ks$$

and so

$$k = \frac{w}{s} = \tfrac{64}{4} = 16 \text{ lb/ft}$$

Substituting for m and k in Eq. (3), we have

$$2\ddot{x} + 16x = 0$$

or

$$\ddot{x} + 8x = 0 \tag{10}$$

This is the differential equation of motion. The general solution is

$$x = c_1 \cos 2\sqrt{2}\,t + c_2 \sin 2\sqrt{2}\,t \tag{11}$$

In order to determine the particular solution, we apply the initial conditions

$$\text{IC: } x(0) = \tfrac{1}{2} \qquad \dot{x}(0) = 0 \tag{12}$$

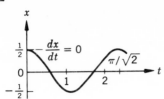

figure 2

Applying the first initial condition to Eq. (11), we have

$$\tfrac{1}{2} = c_1 \tag{13}$$

Differentiating (11) and applying the second initial condition, we have

$$0 = 2\sqrt{2}c_2$$

Thus

$$c_1 = \tfrac{1}{2} \qquad c_2 = 0 \tag{14}$$

and the solution of our problem is

$$x = \tfrac{1}{2} \cos 2\sqrt{2}t \tag{15}$$

Therefore the ball will execute simple harmonic motion with amplitude $\tfrac{1}{2}$ ft and frequency $\sqrt{2}/\pi$ cycles per second. The period of the oscillations is $\pi/\sqrt{2}$, i.e., every $\pi/\sqrt{2}$ sec the ball will reach the same position with the same velocity. See Fig. 2.

Free damped vibrations   If damping or friction forces are included, then $r \neq 0$ and the differential equation of motion is

$$m\ddot{x} + r\dot{x} + kx = 0 \tag{16}$$

With friction present the motion will no longer be simple harmonic.   In fact, physical intuition indicates that the motion will be *damped*; that is, the oscillations will tend to die out as time increases.   We shall prove that this is true and also obtain insight into some other physical phenomena associated with such motions.

To solve the differential equation (16), assume a solution of the form $e^{\lambda t}$ to obtain

$$m\lambda^2 + r\lambda + k = 0 \tag{17}$$

Solving for λ,

$$\lambda = \frac{-r \pm \sqrt{r^2 - 4mk}}{2m} \tag{18}$$

or

$$\lambda = -\frac{r}{2m} \pm \sqrt{\left(\frac{r}{2m}\right)^2 - \frac{k}{m}} \tag{19}$$

Three different cases arise, depending on the sign of the discriminant Δ, where

$$\Delta = \left(\frac{r}{2m}\right)^2 - \frac{k}{m} = \frac{1}{4m^2}(r^2 - 4km) \tag{20}$$

Case I. $\Delta > 0$ *Overdamped case.* For this case the values of λ are real and distinct,

$$\lambda = -a \pm b$$

where

$$a = \frac{r}{2m} \qquad b = \sqrt{\left(\frac{r}{2m}\right)^2 - \frac{k}{m}}$$

The general solution of Eq. (16) is then

$$x = c_1 e^{-(a-b)t} + c_2 e^{-(a+b)t} \tag{21}$$

Since this solution is composed of real exponential terms, there will be no oscillations. Also, since $a > b$ (why?), x will approach the limit zero as $t \to \infty$. See Fig. 3.

Physically, this case arises when the resistance r is large, since $\Delta > 0$ if and only if $r > 2\sqrt{mk}$ [see Eq. (20)]. For example, if the weight is immersed in a heavy motor oil, or if there is a shock absorber attached to the weight, then, if r is large enough, the weight will not oscillate. It will slowly return to its equilibrium position.

Case II. $\Delta = 0$ *Critically damped case.* In this case Eq. (17) has a double root

$$\lambda = -\frac{r}{2m}$$

and the solution of (16) is

$$x = e^{(-r/2m)t}(c_1 + c_2 t) \tag{22}$$

The behavior of the system in this case will be similar to that of Case I (see Fig. 3). Note that we again have $x \to 0$ as $t \to \infty$.

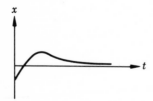

figure 3

Case III. $\Delta < 0$ *Oscillatory case.* In this case the values of λ are conjugate complex numbers

$$\lambda = -\alpha \pm i\beta$$

where

$$\alpha = \frac{r}{2m} \qquad \beta = \sqrt{\frac{k}{m} - \left(\frac{r}{2m}\right)^2} = \sqrt{\omega_n^2 - \alpha^2} \tag{23}$$

The general solution is

$$x = e^{-\alpha t}(c_1 \cos \beta t + c_2 \sin \beta t) \tag{24}$$

or, equivalently,

$$x = Ae^{-\alpha t} \cos (\beta t - \delta) \tag{25}$$

Since the cosine term in (25) is multiplied by the steadily decreasing factor $Ae^{-\alpha t}$, the motion will be oscillatory, with decreasing amplitude. The graph of the motion will be similar to that shown in Fig. 4, page 162.

From Eq. (20) it is easily seen that this case ($\Delta < 0$) will occur if $r < 2\sqrt{mk}$. For example, if r is very small, as in the case of air resistance, then the system will oscillate with slowly decreasing amplitude and the body will approach its equilibrium position as a limit. If r is very small, then the *frequency* β of the oscillations will be very nearly equal to ω_n. See Eq. (23).

General remarks   In all three of the above cases the solution contains a factor of the form $e^{-(r/2m)t}$ which ensures that the motion will tend to die out as t becomes large.   But these three cases represent all possible motions of vibrating systems which are governed by the differential equation (16). Therefore all such motions tend to die out and are consequently referred to as *transient motions*. These remarks depend, of course, on the assumption that $r \neq 0$, but since there is always some friction present in a mechanical system, this assumption is always valid.

Example 2   Suppose that a steel ball on a spring is set in motion as in Example 1 (see page 158). However, in the present case assume further that the medium through which the ball travels offers a resisting force which is numerically equal to four times the velocity. Describe the motion.

Solution   We have $r = 4$ and, since $m = 2$ and $k = 16$, the differential equation of motion is obtained by substituting in Eq. (16)

$$2\ddot{x} + 4\dot{x} + 16x = 0$$

or

$$\text{DE: } \ddot{x} + 2\dot{x} + 8x = 0$$
$$\text{IC: } x(0) = \tfrac{1}{2} \qquad \dot{x}(0) = 0 \tag{26}$$

The auxiliary equation is

$$\lambda^2 + 2\lambda + 8 = 0$$

and hence

$$\lambda = -1 \pm i\sqrt{7} \tag{27}$$

This is the oscillatory case and the general solution is

$$x = e^{-t}(c_1 \cos \sqrt{7}t + c_2 \sin \sqrt{7}t) \tag{28}$$

Applying the initial conditions, we find

$$c_1 = \tfrac{1}{2} \qquad c_2 = \frac{\sqrt{7}}{14}$$

and hence the solution of our problem is

$$x = e^{-t}(\tfrac{1}{2} \cos \sqrt{7}t + \tfrac{1}{14}\sqrt{7} \sin \sqrt{7}t) \tag{29}$$

This solution can also be written in the form

$$x = Ae^{-t} \cos (\sqrt{7}t - \delta)\dagger \tag{30}$$

where

$$A = \sqrt{\tfrac{2}{7}} \qquad \text{and} \qquad \delta = \arctan (\tfrac{1}{7}\sqrt{7})$$

The form of the solution in Eq. (30) is especially useful when drawing a sketch of the solution. Since the cosine oscillates between $+1$ and -1, x will oscillate between $+\sqrt{\tfrac{2}{7}}e^{-t}$ and $-\sqrt{\tfrac{2}{7}}e^{-t}$. See Fig. 4.

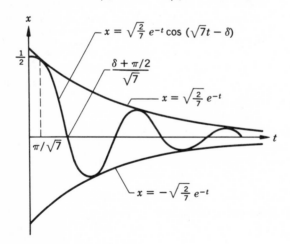

figure 4

† See Eq. (9) for the definitions of A and δ.

Special properties of the oscillatory solution   We have shown that
for the oscillatory case ($\Delta < 0$) the general solution of

$$m\ddot{x} + r\dot{x} + kx = 0 \tag{31}$$

is

$$x = Ae^{-\alpha t} \cos (\beta t - \delta) \tag{32}$$

where

$$\alpha = \frac{r}{2m} \qquad \beta = \sqrt{\frac{k}{m} - \left(\frac{r}{2m}\right)^2} = \sqrt{\omega_n^2 - \alpha^2} \tag{33}$$

It is clear that the oscillating mass passes through equilibrium, that is, $x = 0$,
whenever the cosine term vanishes.   This occurs when

$$\beta t - \delta = \tfrac{1}{2}(2n + 1)\pi \qquad n = 0, 1, 2, \ldots$$

or when

$$t = \frac{1}{\beta}[\delta + \tfrac{1}{2}(2n + 1)\pi] \qquad n = 0, 1, 2, \ldots \tag{34}$$

See Fig. 4, where $\beta = \sqrt{7}$.

By differentiating Eq. (32), it is easily seen that x has a maximum or
minimum value when

$$-\alpha \cos (\beta t - \delta) - \beta \sin (\beta t - \delta) = 0\dagger$$

or

$$\tan (\beta t - \delta) = -\frac{\alpha}{\beta}$$

Hence

$$\beta t - \delta = \arctan \left(-\frac{\alpha}{\beta}\right) + n\pi$$

or

$$t = \frac{1}{\beta}\left[\delta + \arctan \left(-\frac{\alpha}{\beta}\right)\right] + \frac{n\pi}{\beta}$$

Let

$$H = \frac{1}{\beta}\left[\delta + \arctan \left(-\frac{\alpha}{\beta}\right)\right] \tag{35}$$

and then

$$t = H + \frac{n\pi}{\beta}$$

Therefore the displacement x has a relative maximum (or minimum) whenever

$$t = H + \frac{n\pi}{\beta} \qquad n = 0, 1, 2, \ldots \tag{36}$$

† See Prob. 9.

The displacement x_n at this time is obtained by substituting into Eq. (32). Thus

$$x_n = Ae^{-\alpha(H + n\pi/\beta)} \cos\left[\beta\left(H + \frac{n\pi}{\beta}\right) - \delta\right] \tag{37}$$

It is interesting to note that if x_n is a maximum, then the displacement x_{n+2} at the next maximum can be expressed in terms of x_n by a strikingly simple formula.   We have

$$x_{n+2} = Ae^{-\alpha[H + (n+2)\pi/\beta]} \cos\left[\beta\left(H + \frac{n+2}{\beta}\pi\right) - \delta\right]$$

$$= Ae^{-2\pi\alpha/\beta}e^{-\alpha(H + n\pi/\beta)} \cos\left[\beta\left(H + \frac{n\pi}{\beta}\right) - \delta + 2\pi\right]$$

$$= e^{-2\pi\alpha/\beta}Ae^{-\alpha(H + n\pi/\beta)} \cos\left[\beta\left(H + \frac{n\pi}{\beta}\right) - \delta\right]$$

or, on comparing with Eq. (37),

$$x_{n+2} = e^{-2\pi\alpha/\beta}x_n \tag{38}$$

Equation (38) is equivalent to the following equation, which gives the ratio of two successive maximums (or minimums):

$$\frac{x_n}{x_{n+2}} = e^{2\pi\alpha/\beta} \tag{39}$$

If we now set

$$L = \frac{2\pi\alpha}{\beta} \tag{40}$$

then Eq. (39) yields

$$L = \ln\frac{x_n}{x_{n+2}} \tag{41}$$

This constant L is known as the *logarithmic decrement*.   It is somewhat surprising to see that L is independent of n.   In fact, on substituting for α and β from Eq. (33), we have

$$L = \frac{2\pi\alpha}{\beta} = \frac{2\pi r}{\sqrt{4mk - r^2}} \tag{42}$$

Problems 6-2

1. A ball weighing 32 lb is hanging from a spring which is stretched 2 ft by the weight.   The ball is set in motion at its equilibrium position with an initial velocity of 6 ft/sec.   Set up and solve the differential equation of motion.   What is the natural frequency of the motion?   When will the ball pass through its equilibrium position?   Sketch the graph of the displacement versus time. Neglect friction and take $g = 32$ ft/sec².

2. A mass on a spring is set in motion at $x = -3$ ft with initial velocity $\dot{x}(0) = 3$ ft/sec. If the oscillating mass reaches a maximum displacement of $2\sqrt{3}$ ft, at what speed does it pass through its equilibrium position? Neglect friction.

3. A simple pendulum consists of a mass m supported by a weightless rod of length L pivoted at the top, as shown. The tangential component of the

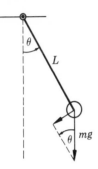

acceleration is $L\ddot{\theta}$ where θ (in radians) is the angular displacement. The tangential component of the weight force is $-mg \sin \theta$. If friction is neglected, the differential equation of motion is

$$L\ddot{\theta} + g \sin \theta = 0$$

a. Solve for the angular velocity $\dot{\theta}$.

b. If the angle θ is small, $\sin \theta$ may be replaced by θ in the differential equation. Solve the resulting equation and prove that the period of the motion is $2\pi\sqrt{L/g}$.

4. A 360-lb block of wood in the form of a 2-ft cube is floating in liquid which has a density of 60 lb/ft³. If the cube is depressed, so that its upper face is level with the surface of the liquid, and then released, find the differential equation of motion. Solve the equation and find the period of the motion. Neglect friction.

5. A bead of mass m is fastened to the midpoint of a tightly stretched horizontal string of length $2L$. If the bead is displaced a small vertical distance s_0 and released, it will oscillate about its equilibrium position. Neglect friction and gravity and derive the differential equation of the motion assuming a constant tension T lb on the string. Find the frequency of the vibration.

6. A 96-lb weight is hanging from a spring whose constant is 5 lb/ft. The weight is set in motion at $x = 3$ ft with an initial velocity of -6 ft/sec. Discuss the resulting motion if the surrounding medium offers a resistance numerically equal to four times the velocity. Assume $g = 32$ ft/sec². Sketch the graph of displacement versus time.

7. A 256-lb weight is hanging from a spring which is stretched 6 ft by the weight. Suppose that the weight is set in motion and it is observed that it crosses its equilibrium position every π sec. Assume a resistance proportional to the velocity and find the constant of proportionality.

8. A uniform bar of length L and weight W rests on two horizontal rollers which are rotating inwards in opposition to each other with constant angular velocity ω. Let d, $d < L$, be the distance between the rollers. Let μ be the coefficient of sliding friction between the bar and the rollers. Set up the differential equation which governs the motions of the bar. Determine the period of the motion.

9. For the case of damped oscillations ($\Delta < 0$) the displacement curve is given by Eq. (32)

$$x = Ae^{-\alpha t} \cos (\beta t - \delta)$$

Show that the relative maxima of x occur shortly before the curve becomes tangent to the bounding curve $x = Ae^{-\alpha t}$.

6-3 Forced vibrations

If an exterior force $F(t)$ is applied to the oscillating mass, the differential equation of motion becomes

$$m\ddot{x} + r\dot{x} + kx = F(t) \tag{43}$$

In order to obtain the general solution of this equation, we first solve the homogeneous equation as in the previous section. Then we must find a particular solution of the nonhomogeneous equation (43). Although it is possible to write a general expression for the particular solution, we shall find it more instructive to examine an important special case.

The most important types of forcing function that occur in practice are the form†

$$F(t) = F_0 \cos \omega t \tag{44}$$

or

$$F(t) = F_0 \sin \omega t \tag{45}$$

where the amplitude F_0 and the frequency ω are constants. Instead of solving Eq. (43) for each of these cases separately, we can solve both simultaneously by using the technique of complex exponentials (see Chap. 5, page 137). That is, we assume

$$F(t) = F_0 e^{i\omega t} = F_0 \cos \omega t + i F_0 \sin \omega t \tag{46}$$

and solve the equation

$$m\ddot{x} + r\dot{x} + kx = F_0 e^{i\omega t} \tag{47}$$

† Since the differential equation (43) is linear, the principle of superposition holds, and hence the results obtained here for a force in the form (44) or (45) can easily be extended to linear combinations of such terms. In fact, these results can be extended to very general classes of forcing functions by the use of Fourier series. (See Chap. 13.)

The real and imaginary parts of the solution will then be the solutions for $F = F_0 \cos \omega t$ and $F = F_0 \sin \omega t$, respectively.

The characteristic polynomial $p(\lambda)$ for Eq. (47) is

$$p(\lambda) = m\lambda^2 + r\lambda + k$$

and hence the particular integral x_p of (47) is

$$x_p = \frac{F_0 e^{i\omega t}}{p(i\omega)} = \frac{F_0 e^{i\omega t}}{-m\omega^2 + ir\omega + k} \dagger \qquad \text{for } p(i\omega) \neq 0 \tag{48}$$

On computing the real and imaginary parts of x_p, we obtain the desired particular solutions of (43).

For $F(t) = F_0 \cos \omega t$ we have

$$x_p = \frac{F_0[(k - m\omega^2) \cos \omega t + r\omega \sin \omega t]}{(k - m\omega^2)^2 + (r\omega)^2} \tag{49}$$

and for $F(t) = F_0 \sin \omega t$ we have

$$x_p = \frac{F_0[(k - m\omega^2) \sin \omega t - r\omega \cos \omega t]}{(k - m\omega^2)^2 + (r\omega)^2} \tag{50}$$

We now define the amplitude A and phase δ of the particular solutions by the equations

$$A = \frac{F_0}{\sqrt{(k - m\omega^2)^2 + (r\omega)^2}} \tag{51}$$

and

$$\cos \delta = \frac{k - m\omega^2}{\sqrt{(k - m\omega^2)^2 + (r\omega)^2}} \qquad \sin \delta = \frac{r\omega}{\sqrt{(k - m\omega^2)^2 + (r\omega)^2}} \tag{52}$$

Hence we can now write our results in the form

$$x_p = A \cos (\omega t - \delta) \qquad \text{for} \quad F = F_0 \cos \omega t \tag{53}$$

and

$$x_p = A \sin (\omega t - \delta) \qquad \text{for} \quad F = F_0 \sin \omega t \tag{54}$$

It is important to note that, in contrast to the *transient* solution of the homogeneous equation, the *particular solutions* as given by (53) and (54) *do not die out* as t increases. In fact, they represent simple harmonic vibrations with *constant amplitude* A and the same frequency ω as the forcing function.

Finally, the general solution of Eq. (43) is given by the sum of the transient solution x_h of the homogeneous equation and the particular

† The quantity $p(i\omega) = -m\omega^2 + ir\omega + k$ is called the *complex impedance* of the system. This concept plays an important role in the theory of electrical circuits. See Sec. 6-4, page 178.

solution x_p. That is,

$$x = x_h + x_p \tag{55}$$

Since x_h approaches zero as t increases, the motion for large values of t approaches that given by x_p. For this reason the particular solution x_p of the nonhomogeneous equation (43) is called the *steady-state solution*.

Example 1 Consider the weight on a spring as described in Sec. 6-2, Examples 1 and 2. That is, $m = 2, r = 4, k = 16, x(0) = \frac{1}{2}, \dot{x}(0) = 0$. Suppose now that the top of the spring is given the displacement y (positive downwards) where

$$y = \cos 2t$$

In order to compute the resulting force on the suspended weight, we note that if y is positive, the spring is being *compressed* a distance of y ft. Since a downward displacement x of the weight *stretches* the spring, it is clear that the resultant *stretching* of the spring is given by $x - y$. The corresponding force, by Hooke's law, is then $F = -k(x - y)$. Thus Newton's law gives

$$m\ddot{x} = -r\dot{x} - k(x - y)$$

or

$$m\ddot{x} + r\dot{x} + kx = ky \tag{56}$$

Hence the differential equation of motion is

$$2\ddot{x} + 4\dot{x} + 16x = 16 \cos 2t \tag{57}$$

The solution x_h of the homogeneous equation from Example 2, page 161, is

$$x_h = e^{-t}(c_1 \cos \sqrt{7}t + c_2 \sin \sqrt{7}t) \tag{58}$$

The particular solution x_p of (57) can be obtained directly or by substituting into Eq. (49). The result is

$$x_p = \cos 2t + \sin 2t \tag{59}$$

Thus the general solution is

$$x = e^{-t}(c_1 \cos \sqrt{7}t + c_2 \sin \sqrt{7}t) + (\cos 2t + \sin 2t) \tag{60}$$

On applying the initial conditions to the *general solution* (60), we have

$$x = e^{-t}\left(-\tfrac{1}{2} \cos \sqrt{7}t - \frac{5}{2\sqrt{7}} \sin \sqrt{7}t\right) + (\cos 2t + \sin 2t) \tag{61}$$

It should be noted that the values of c_1 and c_2 obtained here are different from the ones in the solution (29) of the *free* vibration problem in Example 2. The first term in the last equation is the *transient* term and the second term is the *steady-state* term. The graph of Eq. (61) appears in Fig. 5.

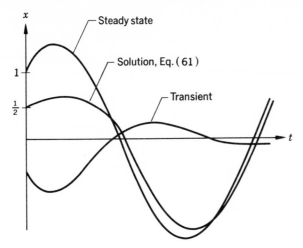

x

Steady state

Solution, Eq. (61)

Transient

1

$\frac{1}{2}$

t

figure 5

Resonance In many applications to problems in vibrations, the frequency ω of the applied force can be varied. If the other parameters are assumed to remain constant, then the amplitude A of the steady-state vibrations can be treated as a function of ω.

That is, from Eq. (51) we have

$$A(\omega) = \frac{F_0}{\sqrt{(k - m\omega^2)^2 + (r\omega)^2}} \tag{62}$$

It is easily shown (see Prob. 4) that if r is small, this function has a maximum when

$$\omega = \omega_r = \sqrt{\frac{k}{m} - \frac{r^2}{2m^2}} = \sqrt{\omega_n^2 - 2\alpha^2} \qquad \alpha = \frac{r}{2m} \tag{63}$$

This frequency ω_r, which yields maximum amplitude for the forced vibrations, is called the *resonant frequency*.† Substituting ω_r in Eq. (62), we obtain the amplitude $A(\omega_r)$, at resonance

$$A(\omega_r) = \frac{F_0}{r\sqrt{k/m - (r/2m)^2}} \qquad 0 < r \le \sqrt{2mk} \tag{64}$$

It is evident from Eq. (64) that $A(\omega_r)$ becomes large when the damping constant r is very small. The variation of A with ω is shown in Fig. 6 for various values of r. The values of the resonant amplitude $A(\omega_r)$ are also shown for two small values of r.

The phenomenon of resonance plays an important role in the design of any mechanical system in which vibratory forces are present. If the frequency of an

† Since we have assumed that r is small, ω_r is approximately equal to ω_n. Hence engineers frequently use ω_n as an approximation to the resonant frequency.

$A(\omega)$

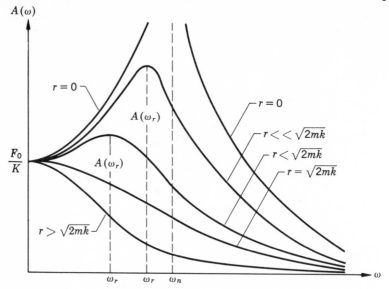

figure 6

applied force is at or near the resonant frequency, then the resulting vibrations may have an amplitude great enough to cause a breakdown in the system. Although resonance is usually to be avoided in a mechanical system, we shall see in Sec. 6-4 that the occurrence of resonance in an electrical circuit is often desirable and quite useful.

It should be noted that the particular integral in Eq. (48) was obtained on the assumption that the right-hand side of Eq. (46) was not a solution of the homogeneous equation. However, if $r = 0$ and $\omega = \omega_n$, then the particular solution for $F = F_0 \cos \omega t$ is

$$x_p = \frac{F_0}{2\sqrt{km}} t \sin \omega_n t \tag{65}$$

This is the case of *pure resonance* in which the amplitude increases linearly with t.

Beats   Consider the case when $r = 0$ and ω is close to, but not equal to, ω_n, and $F = F_0 \cos \omega t$.   The differential equation of motion is

$$m\ddot{x} + kx = F_0 \cos \omega t \tag{66}$$

and the solution satisfying initial conditions $x(0) = 0$ and $\dot{x}(0) = 0$ is

$$x = \frac{F_0}{m(\omega_n^2 - \omega^2)} (\cos \omega t - \cos \omega_n t) \tag{67}$$

or, using a trigonometric identity,

$$x = \frac{-2F_0}{m(\omega_n^2 - \omega^2)} \sin \left[\tfrac{1}{2}(\omega - \omega_n)\right]t \sin \left[\tfrac{1}{2}(\omega + \omega_n)\right]t \tag{68}$$

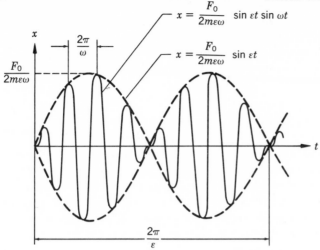

figure 7

Now, since $\omega_n \sim \omega$, we have $\frac{1}{2}(\omega + \omega_n) \sim \omega$, and if we set $\epsilon = \frac{1}{2}(\omega - \omega_n)$, then Eq. (68) can be written in the approximate form

$$x = \frac{F_0}{2m\epsilon\omega} \sin \epsilon t \sin \omega t \tag{69}$$

But since ϵ is small, the term $\sin \epsilon t$ can be considered as a slowly varying amplitude for the principle vibration $\sin \omega t$. Such a motion illustrates the phenomenon of *beats*, as shown in Fig. 7.

Problems 6-3

1. A 32-lb weight stretches a spring 0.32 ft. The weight is pulled $\frac{1}{2}$ ft below its equilibrium position and then released. The upper end of the spring is made to vibrate so that its displacement y from its mean position at any time t is $2 \sin 10t$. The resistance of the medium is 10 times the velocity of the weight. Assuming $g = 32$ ft/sec², set up and solve the differential equation of the weight. How does the weight vibrate for large values of t?

2. A 64-lb weight is hanging at rest on a spring whose constant is 4 lb/ft. At time $t = 0$, the top of the spring is suddenly raised a distance of 2 ft and then held fixed. Discuss the motion of the weight if the resistance of the medium is six times the velocity. Use $g = 32$ ft/sec².

3. Given a mechanical system in which $m = 1$, $r = \sqrt{2}$, and $k = 8$, suppose a force $F_0 \cos \omega t$ is applied to the system. Find the amplitude c of the steady-state vibrations when $\omega = 5$ and when $\omega = \sqrt{7}$. Why is the amplitude so much greater for $\omega = \sqrt{7}$?

4. The amplitude $A(\omega)$ of a forced vibration is given by Eq. (62). Find the value of ω which makes $A(\omega)$ a maximum. Assume $0 < r \le \sqrt{2mk}$.

5. At time $t = 0$, a 64-lb weight is hanging at rest on a spring whose constant is 18 lb/ft. The top of the spring is then given the displacement $y = \cos 3.2t$. Find the exact solution for the resulting displacement. (Assume $r = 0$ and $g = 32$ ft/sec^2.) Compare with the approximate solution as given by Eq. (69). Plot the solution as given by Eq. (69).

6-4   *Electrical circuits*

In order to use the theory of differential equations to investigate the behavior of electrical circuits, it is necessary that we understand the basic properties of circuits. In this subsection we shall give a brief introduction to the physical theory and derive the differential equation which governs the behavior of a simple series circuit.† The circuit that we shall study is shown schematically in Fig. 8, where E represents the source of electromotive force (emf). This may be a battery or a generator which produces a *potential difference* (or *voltage*) which causes the electric *current* I to flow through the circuit when the switch S is closed. The symbol R represents a *resistance* to the flow of current such as that produced by a light bulb, a toaster, etc. When current flows through a coil of wire (L), a magnetic field is produced which opposes any change in the current through the coil. The change in voltage produced by the coil is proportional to the rate of change of the current, and the constant of proportionality is called the *inductance* L of the coil. A *capacitance* (or condenser), indicated by C, usually consists of two metal plates separated by a material through which very little current can flow. A capacitor has the effect of reversing the flow of current as one plate or the other becomes charged.

Let $Q(t)$ be the *charge* on the capacitor at time t. Then, if $Q(t)$ is measured in *coulombs*, the *current* $I(t)$ which flows in the circuit at time t is, *by definition*,

$$I(t) = \frac{dQ(t)}{dt} \tag{70}$$

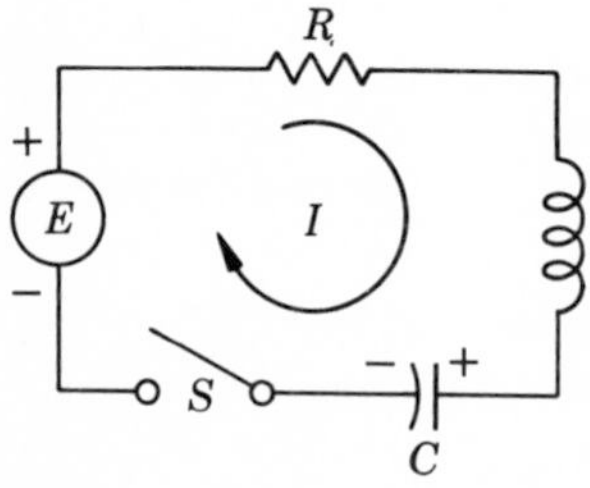

figure 8

† For a more complete discussion of the fundamentals of electricity see, for example, Clement and Johnson [10].

$I(t)$ is measured in *amperes* and is the same in all parts of a series circuit at any instant. The charge Q can be expressed in terms of the current I by integrating (70) to get

$$Q(t) = Q_0 + \int_{t_0}^{t} I(s)\,ds†$$ (71)

Here Q_0 is a constant which represents the initial charge on the capacitor at time $t = t_0$.

The main problem we shall consider is that of finding the current $I(t)$ when we are given the resistance R (in *ohms*), the inductance L (in *henrys*), and the capacitance C (in *farads*). R, L, and C are assumed to be constants. It will also be necessary to know the impressed voltage $E(t)$ (in *volts*) and, as initial conditions, the charge Q_0 and current I_0 at some instant $t = t_0$.

Each of the components of the circuit produces a drop in voltage which is governed by the following physical laws:

1. The voltage drop across a resistance of R ohms equals RI.
2. The voltage drop across an inductance of L henrys equals $L\,dI/dt$.
3. The voltage drop across a capacitance of C farads equals Q/C.

The voltage drops in a circuit satisfy one of the most important of all the laws of electricity, *Kirchhoff's law: The sum of the voltage drops around a simple closed series circuit equals zero.*

For the circuit under consideration, Kirchhoff's law yields the equation

$$L\frac{dI}{dt} + RI + \frac{Q}{C} - E(t) = 0$$ (72)

However, since this equation involves the two unknowns, I and Q, it is desirable to transform it into an equation involving only one unknown. This can be done by substituting $I = dQ/dt$ to obtain

$$L\frac{d^2Q}{dt^2} + R\frac{dQ}{dt} + \frac{1}{C}Q = E(t)$$ (73)

This is the differential equation for Q which will be solved in detail in the next section. The general solution of Eq. (73) involves two arbitrary constants which can be evaluated if one knows the charge Q_0 and current $I_0 = Q'(t_0)$ at some instant $t = t_0$. When $Q(t)$ has been found, $I(t)$ is then obtained by differentiating, since $I = dQ/dt$.

† In some books Eq. (71) is replaced by the statement $Q = \int I\,dt$, which ignores the constant of integration. This constant Q_0 can be quite important, however. For example, it is critical in determining the response of an electrical network to a television signal.

Before proceeding to the solution of Eq. (73), it is instructive to examine two other equations which can be derived from the basic equation (72). First, we can obtain an equation for I by replacing Q in Eq. (72) by the expression in Eq. (71). We have then

$$L\frac{dI}{dt} + RI + \frac{1}{C}\left[Q_0 + \int_{t_0}^{t} I(s)\,ds\right] = E(t) \tag{74}$$

Since this equation involves the unknown I in a derivative and in an integral, it is called an *integro-differential equation*. Such equations are solved most conveniently by operational methods (see Chap. 9). Since one is usually interested in finding the current rather than the charge, Eq. (74) is often used in practice because it can be solved directly for I. Also the initial charge Q_0 is displayed explicitly. Electrical engineers sometimes prefer to replace the term Q_0/C by the initial *voltage* v_0, where

$$v_0 = -\frac{Q_0}{C}$$

is the voltage *rise* across the capacitor when $t = t_0$. Equation (74) could then be written

$$E(t) + v_0 - L\frac{dI}{dt} - RI - \frac{1}{C}\int_{t_0}^{t} I(s)\,ds = 0 \tag{75}$$

Another equation for the current I can be obtained by differentiating (73) or (75) to get

$$L\frac{d^2I}{dt^2} + R\frac{dI}{dt} + \frac{1}{C}I = \frac{dE}{dt} \tag{76}$$

This equation will be valid if the impressed voltage $E(t)$ is a differentiable function. The current I is completely determined by Eq. (76) if we know the initial values of I and dI/dt. Since Q_0 and I_0 are assumed to be known, we could find dI/dt at $t = t_0$ by substituting $t = t_0$ in Eq. (74). This gives

$$\left.\frac{dI}{dt}\right|_{t=t_0} = \frac{E(0)}{L} - \frac{R}{L}I_0 - \frac{1}{LC}Q_0$$

Solution of the differential equation   We have derived the differential equation

$$L\frac{d^2Q}{dt^2} + R\frac{dQ}{dt} + \frac{1}{C}Q = E(t) \tag{77}$$

for the charge Q. This is a linear second-order differential equation with constant coefficients which could be solved directly by applying the methods of Chap. 5. We shall, however, take a different approach.

In the earlier sections of this chapter we have presented a fairly complete solution of the equation which governs mechanical vibrations. This equation is

$$m\frac{d^2x}{dt^2} + r\frac{dx}{dt} + kx = F(t) \tag{78}$$

If one compares these two equations, (77) and (78), one discovers a surprising and important fact. The differential equations for electrical and mechanical vibrations are *identical in form*! This is one of the most striking examples of the way that mathematics helps to show a fundamental relationship between apparently unrelated branches of science.

The similarity between Eqs. (77) and (78) has many important ramifications. The most obvious is the fact that Eq. (77) need not be solved in detail. Its solution can be inferred from the solution of (78). In addition to this pleasant circumstance, the similarity of these equations enables us to interpret electrical phenomena in terms of their more concrete mechanical analogs. Finally, engineers sometimes utilize this similarity by determining the properties of a proposed mechanical system from experiments with a simple electrical analog.

The correspondence between Eqs. (77) and (78) depends on the following correspondence between the quantities in the two systems.

Mechanical quantities		Electrical quantities
displacement x	$\longleftrightarrow$	charge Q
mass m	$\longleftrightarrow$	inductance L
friction r	$\longleftrightarrow$	resistance R
spring constant k	$\longleftrightarrow$	reciprocal of capacitance $1/C$
external force $F(t)$	$\longleftrightarrow$	impressed voltage $E(t)$
velocity $v = \dot{x}$	$\longleftrightarrow$	current $I = dQ/dt$

Since the mathematical solutions of both problems are the same, we can solve electrical problems simply by referring to the solution of the corresponding mechanical problem in Secs. 6-2 and 6-3.

Free electrical vibrations   In the circuit shown in Fig. 8, suppose that the voltage source is removed at time $t = 0$ while current is flowing in the closed circuit. We shall now determine the charge on the capacitor and the current in the circuit at any later time t.

The differential equation for the charge Q is

$$L\frac{d^2Q}{dt^2} + R\frac{dQ}{dt} + \frac{1}{C}Q = 0 \tag{79}$$

This corresponds to Eq. (16) for free mechanical vibrations. In the mechanical case the behavior of the solution depends upon the value of

$$\Delta = \left(\frac{r}{2m}\right)^2 - \frac{k}{m} \tag{80}$$

In terms of electrical quantities we have

$$\Delta = \left(\frac{R}{2L}\right)^2 - \frac{1}{LC} \tag{81}$$

We have seen (Sec. 6-3) that the oscillatory case ($\Delta < 0$) occurs if $r < 2\sqrt{km}$. In terms of electrical quantities this corresponds to the case

$$R < 2\sqrt{\frac{L}{C}} \tag{82}$$

Since the capacitance C (in farads) is usually very small in an electrical circuit, this inequality (82) will usually be satisfied and hence the charge Q will be given by a formula analogous to Eq. (25):

$$Q = Ae^{-at} \cos(bt - d) \tag{83}$$

Here we have

$$a = \frac{R}{2L} \qquad b = \sqrt{\frac{1}{LC} - \left(\frac{R}{2L}\right)^2} \tag{84}$$

and A and d are arbitrary constants which would be determined by the initial values of the charge and current.

The current I is obtained by differentiating Eq. (83). The result can be written in the form

$$I = \frac{dQ}{dt} = Be^{-at} \cos(bt - p) \tag{85}$$

Hence the current will keep changing direction and will tend to die out as t increases.

Regardless of the value of Δ, both the charge and current will contain the exponential damping factor (since R is never zero in a real electrical circuit) and hence will tend to die out. They are called *transients* in this case.

Circuit with voltage source   When there is a voltage source present in the circuit, the differential equation for Q is

$$L\frac{d^2Q}{dt^2} + R\frac{dQ}{dt} + \frac{1}{C}Q = E(t) \tag{86}$$

Two types of impressed voltage are especially important. If the source is a battery, then $E(t) = E_0$, a constant. This case is considered in Prob. 2. If the voltage source is an ordinary *alternating-current generator*, then $E(t)$ can be written in the form

$$E(t) = E_0 \cos (\omega t - d) \tag{87}$$

where $\omega/2\pi$ is the frequency and d the phase of the impressed voltage.
Since Eq. (87) can be written in the form

$$E(t) = A \cos \omega t + B \sin \omega t$$

we have to find particular integrals of Eq. (86) with right-hand sides $\cos \omega t$ and $\sin \omega t$. Both problems can be solved simultaneously if we use the complex exponential form $e^{i\omega t}$ for the right-hand side. Hence we consider the equation

$$L \frac{d^2 Q}{dt^2} + R \frac{dQ}{dt} + \frac{1}{C} Q = E_0 e^{i\omega t} \tag{88}$$

By using the method of Sec. 6-3, page 166, the particular solution† may be written down immediately:

$$Q_p = \frac{E_0}{p(i\omega)} = \frac{E_0}{-L\omega^2 + iR\omega + 1/C} e^{i\omega t} \tag{89}$$

By differentiating, we have the *complex current*

$$I_p = \frac{E_0 i\omega}{-L\omega^2 + iR\omega + 1/C} e^{i\omega t} \tag{90}$$

This equation may be written in the form

$$I_p = \frac{E_0 e^{i\omega t}}{i\omega L + R + 1/(i\omega C)} \tag{91}$$

Since I_p represents the steady-state current which flows in the circuit when a voltage $E = E_0 e^{i\omega t}$ is applied, we have

$$I_p = \frac{E}{Z} \tag{92}$$

where

$$Z = i\omega L + R + \frac{1}{i\omega C} \tag{93}$$

† The solution Q_p of Eq. (88) represents the *complex charge*. The actual charge Q which satisfies Eq. (86) is easily obtained from the real and imaginary parts of Q_p.

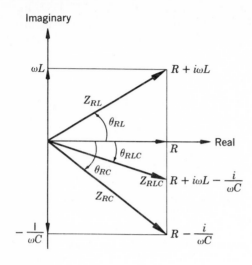

figure 9

or

$$Z = R + i\left(\omega L - \frac{1}{\omega C}\right) \tag{94}$$

This quantity Z is called the *complex impedance*† (see Fig. 9).

Electrical engineers find it convenient to express the complex impedance Z in polar form $Z = |Z|\, e^{i\theta}$:

$$|Z| = \sqrt{R^2 + \left(\omega L - \frac{1}{\omega C}\right)^2} \tag{95}$$

$$\theta = \tan^{-1}\frac{\omega L - 1/(\omega C)}{R} \tag{96}$$

Our result can now be written in the following form:

$$I_p = \frac{E_0}{|Z|}\, e^{i(\omega t - \theta)} \tag{97}$$

when

$$E(t) = E_0 e^{i\omega t} \tag{98}$$

Thus the current I_p will oscillate with the same frequency ω as the impressed voltage. The amplitude of the current is given by $E_0/|Z|$, and θ represents the phase difference.

It is interesting to separate the effects of the three types R, L, and C of electrical components. In a pure resistive circuit $|Z| = R$, $\theta = 0$. If

† See Sec. 6-3, page 167, and Sec. 3-9, page 88.

there is no capacitor in the circuit, then $|Z| = \sqrt{R^2 + (\omega L)^2}$ and $\theta = $ arctan $(\omega L/R)$, i.e., $\theta > 0$ and represents the angle by which the current *leads* the voltage. If there is no inductance in the circuit, then

$$|Z| = \sqrt{R^2 + 1/(\omega C)^2}$$

and $\theta = $ arctan $(-1/\omega CR)$, and hence $\theta < 0$ and the current *lags* the voltage. (See Fig. 9.)

Among other similarities with mechanical vibrations, electrical circuits have the property of *resonance*, and this is put to work in many ways. For example, the tuning knob of a radio is used to vary the capacitance in the tuning circuit. Thus the resonant frequency is changed until it agrees with the frequency of one of the incoming radio signals. The amplitude of the current produced by this signal will then be much greater than that of all other signals. In this way the tuning circuit picks out the desired station.

Problems 6-4

1. Consider an electrical circuit as in Fig. 8 with $L = 1$ henry, $R = 200$ ohms, and $C = 2 \times 10^{-4}$ farads and the applied voltage $E(t) = E_0 \cos 100t$. Assume that there are no charges present and no current flowing at time $t = 0$ when the voltage is applied. Find the steady-state charge and current.

2. Suppose that the RLC circuit illustrated in Fig. 8 is powered by a battery which supplies a constant voltage E_0. Find the current in the circuit t seconds after the switch is closed if there is no initial charge on the capacitor. Identify the transient and steady-state currents.

3. Consider a series circuit with $L = 0.2$ henry, $C = 1 \times 10^{-4}$ farad, and $R = 40$ ohms. Find the transient charge on the capacitor and the transient current in the circuit if the charge is zero and the current is A_0 amp at the instant that the voltage source is removed from the circuit.

4. Prove that a series circuit without inductance cannot have an oscillatory transient current.

5. Explain why the circuit in Prob. 1 does not have a resonant frequency.

6. The instantaneous power delivered to an RLC circuit is, by definition, $p = EI$. Find the average steady-state power delivered during 1 cycle of the voltage $E = E_0 \cos \omega t$. Hint: The average of $f(t)$ over the interval from $t = 0$ to $t = T$ is given by $1/T \int_0^T f(t)\, dt$.

7. When a voltage $E = E_0 \cos \omega t$ is applied to a simple series circuit, the amplitude of the resulting steady-state current is $E_0/|Z|$ where $|Z|$, the magnitude of the impedance, is given by Eq. (95). Show that maximum amplitude occurs when $\omega = 1/\sqrt{LC}$, that is, when the frequency of the applied voltage equals the natural frequency of the undamped circuit. Show that in this case the current satisfies the law $I = E/R$.

6-5   The equations of planetary motion

The motion of a planet revolving about the sun, the motion of a satellite about its mother planet, and the motion of an electron in its orbit about the nucleus of an atom are all described by similar mathematical equations. In this section we shall set up and solve the differential equations which govern these motions.   We shall treat the problem for the particular case of the motion of a planet, but it will then be an easy task to extend these results to satellite and electron motion.

Suppose that a planet P with mass m is moving at a distance r from the sun S (see Fig. 10).   It is convenient to adopt a coordinate system which is determined by the vectors $\mathbf{u}_r$ and $\mathbf{u}_\theta$, as shown. $\mathbf{u}_r$ is a unit vector directed away from the sun along the radial line from S to P, and $\mathbf{u}_\theta$ is the unit vector which is obtained by turning $\mathbf{u}_r$ counterclockwise through an angle of 90°.

Now, according to Newton's law of gravitation, the gravitational force $\mathbf{F}$ on the planet is directed toward the sun† and has a magnitude Km/r^2, where K is a constant.   In vector notation we have

$$\mathbf{F} = -\frac{Km}{r^2}\,\mathbf{u}_r \tag{99}$$

It is important to note that there is no component of force in the direction of $\mathbf{u}_\theta$.‡

Newton's second law states that

$$\mathbf{F} = m\mathbf{a} \tag{100}$$

where $\mathbf{a}$ is the acceleration vector.   In terms of the vectors $\mathbf{u}_r$ and $\mathbf{u}_\theta$ we may

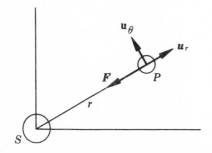

figure 10

† In this simplified treatment we are neglecting the force on P which is due to the attraction of other planets in the solar system.   Also we assume that the motion takes place in a plane and that the sun remains stationary.

‡ When all the force acting on a body is directed toward or away from a fixed point, then the body is said to be moving in a *central force field*.

express $m\mathbf{a}$ in the form

$$m\mathbf{a} = ma_r\mathbf{u}_r + ma_\theta\mathbf{u}_\theta \tag{101}$$

where a_r and a_θ are the components of $\mathbf{a}$ in the directions of $\mathbf{u}_r$ and $\mathbf{u}_\theta$. Equating the components of $\mathbf{F}$ and $m\mathbf{a}$ from Eqs. (99) and (101), we have the pair of equations

$$ma_r = -\frac{Km}{r^2} \tag{102}$$

$$ma_\theta = 0$$

The components a_r and a_θ of the acceleration are expressed most conveniently in terms of the polar coordinates (r,θ) of the point P. We have†

$$a_r = \ddot{r} - r(\dot\theta)^2$$
$$a_\theta = r\ddot\theta + 2\dot\theta\dot r \tag{103}$$

Hence Eqs. (102) may be written in the form

$$\ddot{r} - r(\dot\theta)^2 = -\frac{K}{r^2} \tag{104a}$$

$$r\ddot\theta + 2\dot\theta\dot r = 0 \tag{104b}$$

These are the differential equations which govern the motion of a planet. Actually, these equations are a system of two simultaneous differential equations in the two unknowns r and θ. Such systems are discussed in general in Chap. 8. However, the system (104) is simple enough for us to solve with the techniques that have already been developed.

Equation (104b) is easily integrated to give an important result. If (104b) is multiplied by r, we have

$$r^2\ddot\theta + 2r\dot r\dot\theta \equiv \frac{d}{dt}(r^2\dot\theta) = 0 \tag{105}$$

and hence

$$r^2\dot\theta = h = \text{const} \tag{106}$$

This last equation‡ is equivalent to Kepler's second law of planetary motion, which states that the area swept out by the radius vector is proportional to time. This area A satisfies

$$\dot A = \tfrac{1}{2}r^2\dot\theta = \tfrac{1}{2}h \tag{107}$$

† For a derivation of these formulas see, for example, Thomas [35], Chap. 11.

‡ The fact that $r^2\dot\theta$ is constant could have been obtained immediately from the principle of *conservation of angular momentum* since the angular momentum of the planet is $mr(r\dot\theta) = mh$.

and hence

$$A = \tfrac{1}{2}ht \qquad \text{Kepler's second law†} \tag{108}$$

Returning now to Eq. (104a), one might be tempted to substitute Eq. (106) and thus eliminate θ. This would yield a differential equation for r as a function of t. (See Prob. 9.) However, since we are looking for the *path* of the planet, we want to have r as a function of θ. Hence we shall eliminate t from Eq. (104a) by using the chain rule. We have

$$\dot{r} = \frac{dr}{d\theta}\,\dot{\theta} = hr^{-2}\frac{dr}{d\theta} \tag{109}$$

$$\ddot{r} = h\left(-2r^{-3}\,\dot{r}\,\frac{dr}{d\theta} + r^{-2}\frac{d^2r}{d\theta^2}\,\dot{\theta}\right)$$

or

$$\ddot{r} = -h^2r^{-2}\left[2r^{-3}\left(\frac{dr}{d\theta}\right)^2 - r^{-2}\frac{d^2r}{d\theta^2}\right] \tag{110}$$

The quantity in brackets in this last equation is exactly equal to the second derivative of $1/r$ with respect to θ. Hence, if we introduce

$$u = \frac{1}{r} \tag{111}$$

then Eq. (110) becomes

$$\ddot{r} = -h^2u^2\frac{d^2u}{d\theta^2} \tag{112}$$

Finally, on substituting for r, $\ddot{r}$, and $\dot{\theta}$ in Eq. (104a), we have

$$\frac{d^2u}{d\theta^2} + u = \frac{K}{h^2} \tag{113}$$

This is a linear equation with constant coefficients and the solution can be written immediately

$$u = \frac{K}{h^2} + B\cos(\theta - \delta) \tag{114}$$

where B and δ are constants which would be determined by the position and velocity of the planet at some instant. Substituting $u = 1/r$ in (114), we have

$$r = \left[\frac{K}{h^2} + B\cos(\theta - \delta)\right]^{-1} \tag{115}$$

† It is interesting to note that this law [Eq. (108)] holds for motion in any central force field, regardless of whether or not the inverse-square law holds. See Prob. 1.

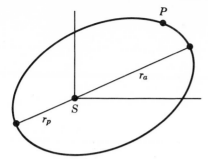

figure 11

which can be written in a more familiar form

$$r = \frac{ep}{1 + e \cos (\theta - \delta)} \tag{116}$$

if we set

$$e = \frac{Bh^2}{K} \qquad p = \frac{1}{B} \tag{117}$$

Equation (116) represents the orbit of a planet, and it should be recognized as the polar equation of a conic section with eccentricity e. If $e < 1$ (as it is for a planet,) then the orbit is an ellipse with one focus at the origin (i.e., at the sun). See Fig. 11. This is Kepler's first law. Kepler's third law is left as an exercise (see Prob. 7).

We have derived Kepler's laws by solving the differential equations which resulted from applying Newton's laws to the motion of a planet. This seems natural now, but historically the process was reversed. Kepler obtained his results by a laborious analysis of observational data, and then Newton used Kepler's laws as a starting point in proving that the motion of the planets must be produced by a central force inversely proportional to the square of the distance from the sun. See Prob. 2.

The results of this section have been obtained for the motion of a planet around the sun. However, the motion of a satellite or an electron is also produced by a central force of the form of Eq. (99). Hence the same results obtain and Eq. (116) represents the orbit.

Problems 6-5

1. Derive the equations of motion for a particle of mass m moving in a central force field given by $\mathbf{F} = P(r,\theta)\mathbf{u}_r$. Prove that Kepler's second law, Eq. (108), holds for any such motion.

2. Show that if a planet moves along the elliptical path given by

$$r = \frac{ep}{1 + e \cos \theta}$$

then the only *central force field* which could produce this motion is given by

$$\mathbf{F} = -\frac{Km}{r^2}\mathbf{u}_r$$

That is, derive Newton's law of gravitation from Kepler's first two laws.

3. Suppose that a body of mass m moves in a central force field given by $\mathbf{F} = -(Km/r^3)\mathbf{u}_r$. Set up and solve the differential equations of motion. Discuss the motion of the body in the special case when $K = h^2$ where h is given by Eq. (106).

4. Suppose a body moves on the circle $r = a\cos\theta$ under the influence of a central force field directed toward the origin. Show that this force is given by $\mathbf{F} = -(K/r^5)\mathbf{u}_r$.

5. The positions in the orbit when a planet is nearest to the sun and farthest from the sun are called *perigee*, r_p, and *apogee*, r_a, respectively (see Fig. 11). Using Eq. (116), find the distance from sun to planet at perigee and at apogee.

6. Using the result of Prob. 5, show that the length $2a$ of the major axis of the elliptical orbit is

$$2a = \frac{2ep}{1 - e^2}$$

Show that the length of the minor axis is

$$2b = 2\sqrt{epa}$$

7. When a planet makes one complete revolution about the sun, the area swept out by a radius vector equals the area πab of the elliptical orbit. Using the result of Prob. 6 and Eq. (108), obtain the following formula for the period T:

$$T^2 = \frac{4\pi^2}{K} a^3$$

This is Kepler's third law.

8. An artificial satellite, in its orbit around the earth, reaches a maximum height of 1,510 miles and a minimum of 218 miles. Using the results of Probs. 5, 6, and 7, determine the period of its motion. Assume the earth is a sphere with radius $R = 4,000$ miles.

9. Obtain a differential equation for r as a function of t by substituting (106) into (104a). Solve this equation for $\dot{r}$ in terms of r.

Linear Differential Equations with Variable Coefficients

7

7-1 Introduction

In the previous chapters we have developed a general method for solving second-order linear differential equations with constant coefficients. For homogeneous equations we obtained explicit solutions in terms of elementary functions. Even for nonhomogeneous equations it was possible to express the solution in terms of a finite number of integrations.

The situation is quite different when one considers equations with variable coefficients, that is, equations of the type

$$a_n(x) \frac{d^n y}{dx^n} + a_{n-1}(x) \frac{d^{n-1} y}{dx^{n-1}} + \cdots + a_1(x) \frac{dy}{dx} + a_0(x)y = f(x) \tag{1}$$

where $a_0, a_1, a_2, \ldots, a_n$ and f are prescribed functions of x. Except in certain special cases,† equations of this type simply do not have solutions which are expressible in closed form. In general it is necessary to express the solution, if any, in the form of an infinite series or to devise a method for obtaining an acceptable numerical approximation.

Equations of type (1) arise very frequently in physical and engineering problems, and their study has proved to be of tremendous importance in the development of modern mathematical analysis.

In Secs. 5-2 and 5-9 many important general properties of linear equations were developed. It should be emphasized at this time that these results, namely, all theorems in Secs. 5-2 and 5-9, still hold, even if the coefficients of the equations are variable. It is only in the *methods* of *solution* that we must develop new techniques.

† E.g., Euler's equation. See Sec. 5-7.

In Chap. 2 the general linear equation of *first order* was solved completely, even for the case of variable coefficients.   In this chapter we shall restrict our study to equations of the second order, that is, equations of the type

$$a_2(x)\frac{d^2y}{dx^2} + a_1(x)\frac{dy}{dx} + a_0(x)y = f(x) \tag{2}$$

In any interval in which $a_2(x) \neq 0$ we can divide both sides of this equation by $a_2(x)$, and hence we can write the equation in the form

$$y'' + P(x)y' + Q(x)y = R(x) \tag{3}$$

The peculiar difficulties associated with variable coefficients first appear in this second-order case.   Higher-order equations present no new difficulties.   Also, in physical applications, many of the important equations with variable coefficients are of the second order.

7-2   Solution by power series†

One of the most powerful methods of solving the homogeneous equation

$$y'' + P(x)y' + Q(x)y = 0 \tag{4}$$

is the method of *power series*.   We shall see that, in many cases, a solution of this equation which is valid near a point $x = a$ can be expressed in a power series about $x = a$.   That is, we shall seek a solution in the form

$$y = a_0 + a_1(x - a) + a_2(x - a)^2 + \cdots + a_n(x - a)^n + \cdots \tag{5}$$

where the coefficients $a_0, a_1, a_2, \ldots$ are constants which have to be determined.   This series will be the Taylor-series expansion of the solution y and may or may not represent an elementary function.   In order to determine the coefficients, we shall assume the existence of a solution of this form and substitute in Eq. (4).   We shall see that it will then be possible to find the constants a_n if P and Q are themselves expressible in Taylor-series form.   Finally, we must determine the interval of convergence of the resulting series and prove that it is actually a solution of the differential equation.

Before stating the conditions under which Eq. (4) has a power-series solution, we need two definitions.

† The fundamental properties of power series are reviewed in Appendix A. See also Sec. 4-6.

Definition 1   *A function $f(x)$ is said to be analytic at $x = a$ if it can be expanded in a power series, in powers of $x - a$, which converges to $f(x)$ in an open interval containing $x = a$.†*   *This series is the Taylor series for $f(x)$.*

A *necessary* condition for $f(x)$ to be analytic is that the function and its derivatives of all orders must exist at $x = a$.   If $f(x)$ is *not* analytic at $x = a$, it is said to be *singular* or to have a *singularity* at $x = a$.

Example 1   The function $(1 - x)^{-1}$ is analytic at $x = 0$ since

$$(1 - x)^{-1} = 1 + x + x^2 + \cdots + x^n + \cdots \qquad -1 < x < 1$$

But $(1 - x)^{-1}$ is *not* analytic at $x = 1$ since it approaches infinity as x approaches 1 and hence cannot be represented in a power series in powers of $(x - 1)$.

Example 2   The function $f(x) = x^{\frac{1}{3}}$ is *not* analytic at $x = 0$ even though it is continuous there.   The reader should try to expand $f(x)$ in a Maclaurin series in order to see why the series does not exist.

Example 3   All rational functions, that is, functions of the form $f(x) = P_1(x)/P_2(x)$, where $P_1(x)$ and $P_2(x)$ are polynomials in x, are analytic everywhere except for those values of x for which $P_2(x)$ vanishes.‡

Example 4   The function $f(x) = (x^2 + 1)^{-1}$ is analytic for all *real* values of x.   However, it is sometimes necessary (see Theorem 2, page 192) to consider also the behavior for *complex* values of x.   We see then that $f(x)$ has singularities at $x = \pm i$.

Definition 2   *The point $x = a$ is called an ordinary point of the differential equation*

$$y'' + P(x)y' + Q(x)y = 0$$

if both $P(x)$ and $Q(x)$ are analytic at $x = a$.   If either $P(x)$ or $Q(x)$ is not analytic at $x = a$, then the point is called a singular point or a singularity.

† In order to simplify the subsequent discussion of series solutions, we shall call a function $f(x)$ analytic at $x = a$ even if the series expansion is not valid at $x = a$ itself, *provided* that it is possible to redefine the function *at $x = a$* so that the series represents the *new function* in the full interval.   For example, $f(x) = (\sin x)/x$ is analytic at $x = 0$ since, if we define $f(0) = 1$, then the series $f(x) = 1 - x^2/3! + x^4/5! - \cdots$ represents $f(x)$ for all x.   In a case such as this, the function, e.g., $(\sin x)/x$, is said to have a *removable singularity*.

‡ If P_1 also vanishes at a point $x = a$ where P_2 vanishes, then $f(x)$ *may* still be analytic at $x = a$ according to our usage of the term "analytic."   See Prob. 1.

Example 5   Euler's differential equation

$$x^2y'' + pxy' + qy = 0 \qquad p, q \text{ const}$$

has a singularity at $x = 0$ since $P(x) = p/x$, $Q(x) = q/x^2$. All other points are ordinary points for this equation.

We shall now state an important theorem which gives the conditions which ensure the existence of a power-series solution.†

Theorem 1 *If $x = a$ is an ordinary point of the differential equation*

$$y'' + P(x)y' + Q(x)y = 0$$

then there exist two linearly independent power-series solutions of the form

$$y = a_0 + a_1(x - a) + a_2(x - a)^2 + \cdots + a_n(x - a)^n + \cdots$$

These solutions will be valid in some interval containing $x = a$.‡

Method of solution   The method of solution in series form in the neighborhood of an *ordinary point* is best explained by the use of an example. Consider the equation

$$y'' + xy' + 2y = 0 \tag{6}$$

Here $P(x) = x$ and $Q(x) = 2$, both of which are analytic for all x, and in particular at $x = 0$. Hence by Theorem 1 there exist solutions of the form

$$y = a_0 + a_1x + a_2x^2 + a_3x^3 + \cdots + a_nx^n + \cdots \tag{7}$$

The coefficients a_n can be determined by substituting the series (7) into the differential equation (6). We first compute the derivatives y' and y'' (assuming, for the present, that termwise differentiation is permissible):

$$y' = a_1 + 2a_2x + 3a_3x^2 + \cdots + na_nx^{n-1} + \cdots \tag{8}$$

$$y'' = 2a_2 + 3 \cdot 2a_3x + 4 \cdot 3a_4x^2 + \cdots + n(n - 1)a_nx^{n-2} + \cdots$$

Substituting into Eq. (6) and collecting terms, we have

$$(2a_2 + 2a_0) + (3 \cdot 2a_3 + a_1 + 2a_1)x + (4 \cdot 3a_4 + 2a_2 + 2a_2)x^2$$
$$+ \cdots + [n(n - 1)a_n + (n - 2)a_{n-2} + 2a_{n-2}]x^{n-2} + \cdots = 0 \tag{9}$$

Now, in order for a power series to vanish for every value of x in some interval, it is necessary that the coefficient of each power of x must vanish.§ Hence

---

† For a proof see, for example, Coddington and Levinson [11].

‡ The reader should note that the theorem does not exclude the possibility of a series solution at a singular point. This matter is discussed further in Sec. 7-3.

§ This follows from the theorem on the uniqueness of power series. See Appendix A.

we have

$$2(a_2 + a_0) = 0$$
$$3(2a_3 + a_1) = 0 \tag{10}$$
$$4(3a_4 + a_2) = 0$$

.
.
.

and, in general,

$$n[(n-1)a_n + a_{n-2}] = 0 \quad \text{if} \quad n \geq 2 \quad \text{(recurrence formula)}$$

Therefore

$$a_2 = -a_0$$
$$a_3 = -\tfrac{1}{2}a_1$$
$$a_4 = -\tfrac{1}{3}a_2 \tag{11}$$

.
.
.

and, in general,

$$a_n = -\frac{1}{n-1} a_{n-2} \quad \text{if } n \geq 2 \quad \text{(recurrence formula)}$$

Upon examining these last equations, we see that any coefficient a_n can be expressed ultimately in terms of either a_0 or a_1. For example,

$$a_4 = -\frac{1}{3} a_2 = +\frac{1}{3} a_0 \qquad a_5 = -\frac{1}{4} a_3 = +\frac{1}{4 \cdot 2} a_1 \qquad \text{etc.}$$

If we now substitute these coefficients back into the trial series (7), we have†

$$y = a_0 \left[1 - x^2 + \frac{1}{3} x^4 - \frac{1}{3 \cdot 5} x^6 + \cdots \right.$$
$$\left. + (-1)^n \frac{1}{3 \cdot 5 \cdots (2n-1)} x^{2n} + \cdots \right]$$
$$+ a_1 \left[x - \frac{1}{2} x^3 + \frac{1}{2 \cdot 4} x^5 - \cdots \right.$$
$$\left. + (-1)^{n-1} \frac{1}{2 \cdot 4 \cdots (2n-2)} x^{2n-1} + \cdots \right] \tag{12}$$

† The general terms in these series can be inferred from the first few terms and then established by mathematical induction. The reader should be warned, however, that it is not always possible to write a simple expression for the general term.

This is the general solution of Eq. (6) with a_0 and a_1 as arbitrary constants. In order to determine the interval of validity of the solution (12), we must determine the interval of convergence of the series. Using the recurrence formula from (11), we can apply the ratio test† as follows:

$$\lim_{n \to \infty} \left| \frac{a_n x^n}{a_{n-2} x^{n-2}} \right| = \lim_{n \to \infty} \frac{1}{n-1} x^2 = 0 \qquad \text{for all } x$$

Hence the series in (12) converge for all x. The properties of power series ensure that these series are differentiable for all x, and therefore Eq. (12) represents the general solution of Eq. (6) for all x.

If we are given initial conditions, e.g.,

$$y(0) = 2 \qquad y'(0) = 3$$

then substitution into Eq. (12) immediately yields

$$a_0 = 2 \qquad a_1 = 3$$

Several features of this example bear further examination. In the first place, the fact that the coefficients of the series depended on either a_0 or a_1 but not both was accidental. In general, a_n can be expected to depend on both a_0 and a_1. Second, the fact that both P and Q were analytic at $x = 0$ was crucial in determining the a_n. The reader should try to solve the equation

$$x^3 y'' + x^4 y' + 2y = 0$$

by the above method in order to see why it breaks down. Finally, in order to determine the interval of convergence of the series in the solution (12), we did not need to find the nth term in the series. All that was needed was the recurrence formula from (11). The convergence can usually be established in this way although the process can become complicated if the recurrence formula involves three or more coefficients instead of just two as in this example. (See Prob. 8.)

Series solution using summation notation An alternative method of carrying out the computations in a series solution involves the use of the summation notation for power series.‡ In order to solve the differential equation (6) by this method, we assume

$$y = \sum_{n=0}^{\infty} a_n x^n \tag{13}$$

† See Appendix A.

‡ The summation notation will be used frequently throughout the remainder of this chapter, and the reader will find it very helpful to master the techniques developed in this section.

Then

$$y' = \sum_{n=0}^{\infty} n a_n x^{n-1} \qquad y'' = \sum_{n=0}^{\infty} n(n-1)a_n x^{n-2} \;\dagger$$

Substituting in (6), we have

$$\sum_{n=0}^{\infty} n(n-1)a_n x^{n-2} + \sum_{n=0}^{\infty} n a_n x^n + 2\sum_{n=0}^{\infty} a_n x^n = 0 \tag{14}$$

Now, in order to obtain the equations for the coefficients, it is necessary that the three sums in (14) be combined into a single sum. This can be done as follows. Rewrite the first sum with a new index of summation, that is,

$$\sum_{n=0}^{\infty} n(n-1)a_n x^{n-2} = \sum_{k=0}^{\infty} k(k-1)a_k x^{k-2}$$

Next let $k - 2 = n$ (and hence $k = n + 2$) so that the power of x will be the same as in the last two sums in Eq. (14). The first sum then becomes

$$\sum_{n=-2}^{\infty} (n+2)(n+1)a_{n+2}x^n$$

or, since the first two terms in the sum are zero, we can write

$$\sum_{n=0}^{\infty} (n+2)(n+1)a_{n+2}x^n$$

Equation (14) can now be written‡

$$\sum_{n=0}^{\infty} (n+2)(n+1)a_{n+2}x^n + \sum_{n=0}^{\infty} n a_n x^n + 2\sum_{n=0}^{\infty} a_n x^n = 0 \tag{15}$$

or finally, combining the sums,

$$\sum_{n=0}^{\infty} [(n+2)(n+1)a_{n+2} + (n+2)a_n]x^n = 0 \tag{16}$$

Now, since each coefficient must vanish, we have

$$(n+2)(n+1)a_{n+2} + (n+2)a_n = 0 \qquad n \geq 0 \tag{17}$$

or since $(n + 2)(n + 1) \neq 0$ for $n \geq 0$, the general *recurrence formula* for the coefficients becomes

$$a_{n+2} = -\frac{1}{n+1}a_n \qquad n \geq 0 \tag{18}$$

† Note that the first term in y' and the first two terms in y'' vanish, and hence we could just as well have written

$$y' = \sum_{n=1}^{\infty} n a_n x^{n-1} \qquad y'' = \sum_{n=2}^{\infty} n(n-1)a_n x^{n-2}$$

‡ After the reader has acquired a little skill in the manipulation of sums, the changes in indices can be performed mentally.

As above, all the coefficients a_n can now be computed by successive application of (18). It should be noted that the recurrence formula in (18) is equivalent to the one in Eq. (11).

Interval of convergence   In the previous paragraphs the interval of convergence of the series solution has been determined by the use of the ratio test. Another, more general, method of determining the interval of convergence of series solutions involves the use of complex numbers.†

Theorem 2   *If $x = a$ is an ordinary point for the differential equation*

$$y'' + P(x)y' + Q(x)y = 0$$

then there exist two linearly independent series solutions of the form

$$y(x) = a_0 + a_1(x - a) + a_2(x - a)^2 + \cdots + a_n(x - a)^n + \cdots$$

These series will converge at least for all values of x in the interval, $|x - a| < R$, where R is the distance‡ from the point $x = a$ to the nearest singular point of the differential equation in the complex plane.§

Example 6 $(x^2 + 1)y'' + 3xy' + 2y = 0$

A solution near the ordinary point $x = 1$ can be written in the form

$$y = a_0 + a_1(x - 1) + a_2(x - 1)^2 + \cdots + a_n(x - 1)^n + \cdots$$

The singularities of the differential equation are at $x = \pm i$. The distance between $x_1 = 1$ and $x_2 = i$ is $|x_1 - x_2| = [(1 - 0)^2 + (0 - 1)^2]^{\frac{1}{2}} = \sqrt{2}$, and hence the series converges for all values of x which satisfy

$$|x - 1| < \sqrt{2}$$

That is, it converges for all real values of x in the interval

$$1 - \sqrt{2} < x < 1 + \sqrt{2}$$

It should be mentioned here that Theorem 2 does *not necessarily* imply divergence for points outside this interval. However, in this particular example, $\sqrt{2}$ *is* the actual radius of convergence. See Prob. 9.

---

† See Appendix B for a review of the fundamental properties of complex numbers.

‡ The reader should recall that the distance $|z_1 - z_2|$ between two complex numbers $z_1 = a_1 + ib_1$ and $z_2 = a_2 + ib_2$ is given by $|z_1 - z_2| = [(a_1 - a_2)^2 + (b_1 - b_2)^2]^{\frac{1}{2}}$.

§ For a proof see, for example, Ince [19], chap. 12.

Example 7   The equation

$$y'' + xy' + 2y = 0$$

was solved in the preceding paragraph.   Since this equation has no singularities in the entire complex plane, the power-series solution about any point $x = a$ will converge for all x.

Taylor-series method   In solving an initial-value problem, that is, one in which $y(a)$ and $y'(a)$ are given, the following method is often more convenient than the general series method developed above.

The reader will recall that if a function $y(x)$ can be represented by a power series near $x = a$, then this series is the Taylor series:†

$$y(x) = y(a) + y'(a)(x - a)$$
$$+ \frac{y''(a)}{2!}(x - a)^2 + \cdots + \frac{y^{(n)}(a)}{n!}(x - a)^n + \cdots \tag{19}$$

In particular, if $x = 0$ is an ordinary point of the differential equation

$$y'' + P(x)y' + Q(x)y = 0 \tag{20}$$

then the solution near $x = 0$ is

$$y(x) = y(0) + y'(0)x + \frac{y''(0)}{2!}x^2 + \cdots + \frac{y^{(n)}(0)}{n!}x^n + \cdots \tag{21}$$

Example 8   Find the solution of

$$y'' + (x + 3)y' - 2y = 0 \tag{22}$$

which satisfies the initial conditions

$$y(0) = 1 \qquad y'(0) = 2 \tag{23}$$

Since $x = 0$ is an ordinary point of the differential equation (22), the solution will be of the form (21).   The values of $y(0)$ and $y'(0)$ are given.   The quantity $y''(0)$ can be computed from the differential equation (22) itself, since

$$y'' = -(x + 3)y' + 2y$$

and hence for $x = 0$ we have

$$y''(0) = -3y'(0) + 2y(0) = -4$$

In order to obtain $y'''(0)$, we differentiate Eq. (22) to get

$$y''' + (x + 3)y'' + y' - 2y' = 0$$

and hence

$$y'''(0) = -3y''(0) + y'(0) = 14$$

† See Appendix A for a review of the Taylor series.   See also Sec. 4-6.

For y^{iv} we have

$$y^{iv} + (x + 3)y''' + 3y'' - y'' = 0$$

and

$$y^{iv}(0) = -3y'''(0) - 2y''(0) = -34$$

Therefore, after substituting into Eq. (21), the solution of (22) and (23) is

$$y(x) = 1 + 2x - 2x^2 + \tfrac{7}{3}x^3 - \tfrac{17}{12}x^4 + \cdots \tag{24}$$

By Theorem 2 this series converges for all x, since the differential equation (22) has no singularities.

This technique has the advantage of simplicity and is especially useful when one desires only to approximate a solution by the first few terms. On the other hand, the general term of the series is not easily obtained by this method.

Although we have been considering only *linear* equations in this chapter, it should be mentioned at this point that the Taylor-series method can often be used to approximate the solution of *nonlinear* equations.

Problems 7-2

1. Find all singularities of the following functions:

a. $f(x) = \dfrac{x - 2}{x^3 - 2x^2 + x}$

b. $f(x) = \dfrac{x - 1}{x^3 - 2x^2 + x}$

c. $f(x) = \dfrac{(x - 1)^2}{x^3 - 2x^2 + x}$

d. $f(x) = (x^3 + 1)^{-1}$

e. $f(x) = \tan x + \ln (1 + x)$

f. $f(x) = \begin{cases} e^{-1/x^2} & \text{if } x \neq 0 \\ 0 & \text{if } x = 0 \end{cases}$

2. Find all singularities of the following differential equations:

a. $y'' + xy' + 3y = 0$
b. $(x^2 - 3x + 2)y'' + \sqrt{xy'} + x^2y = 0$
c. $(1 - x^2)y'' - 2xy' + n(n + 1)y = 0$
d. $(x^2 - x)y'' + x^2y' - 3xy = 0$
e. $(e^x - 1)y'' + xy = 0$
f. $x(x^2 + 2x + 2)y'' + (x^2 + 1)y' + 3y = 0$

3. Find the general solution near $x = 0$ of the equation $y'' + 4y = 0$ by the power-series method. Check your result by solving the equation directly and expanding the result in power series.

4. Find the general solution near $x = 0$ by the power-series method. Find the recurrence formula and use it to prove the convergence of the series.

a. $y'' - xy' + 2y = 0$

b. $y'' + xy = 0$

5. Solve by the power-series method.   Find the recurrence formula and use it to prove convergence of the series.   Write down the first six nonzero terms in the series.

a. $y'' + x^2 y = 0$
   $y(0) = 12, \; y'(0) = 20$

b. $y'' + xy' + y = 0$
   $y(0) = 1, \; y'(0) = 0$

c. $y'' + xy' + 2y = 0$
   $y(0) = 1, \; y'(0) = 2$

6. Use the power-series method to find the solution of $x^2 y'' - xy' + 2y = 0$ which satisfies $y(2) = y'(2) = 1$.   (Hint: Substitute $t = x - 2$ and find a solution in powers of t.)   Find the recurrence formula and determine the interval of convergence of the series.   Check by using Theorem 2.

7. Use the Taylor-series method to find the first four nonzero terms in the solution of the following initial-value problems.   Use Theorem 2 to determine the interval of convergence, if possible.

a. $y'' + xy' + y = 0$
   $y(0) = 1, \; y'(0) = 0$

b. $y'' + xy' + 2y = 0$
   $y(0) = 1, \; y'(0) = 2$

c. $(x + 2)y'' + 3y = 0$
   $y(0) = 0, \; y'(0) = 1$

d. $y'' + (\sin x)y' + (x - 1)y = 0$
   $y(0) = 1, \; y'(0) = 0$

e. $y'' + y^2 = 0$
   $y(0) = 1, \; y'(0) = 0$

f. $xy'' + y' - 3xy = 0$
   $y(1) = 1, \; y'(1) = 0$

g. $(x^2 + 4)y'' + y = 0$
   $y(0) = 1, \; y'(0) = 2$

8. Solve $(x^2 - 1)y'' + y' - 2y = 0$ in a power series about $x = 0$.   Show that the recurrence formula can be written in the form

$$\frac{a_{n+2}}{a_{n+1}} - \frac{n-2}{n+2}\frac{a_n}{a_{n+1}} = \frac{1}{n+2}$$

Assume that $\lim_{n \to \infty} a_{n+1}/a_n = L$ exists and use this to determine the interval of convergence.   Compare with the result obtained by the use of Theorem 2.

9. Solve the equation in Example 6 in the text, i.e., $(x^2 + 1)y'' + 3xy' + 2y = 0$, in a power series about $x = 1$.   Use the method of Prob. 8 to show that the radius of convergence is $\sqrt{2}$.

7-3   Solution near a singular point

In the previous section we have seen a method of solving a linear differential equation near an ordinary point. Although this is the simplest case, a great deal can be learned about the overall behavior of solutions of the equation

$$y'' + P(x)y' + Q(x)y = 0 \tag{25}$$

by examining the solutions near singular points, that is, points where either P or Q is *not* analytic. For example, the Euler equation

$$x^2 y'' + pxy' + qy = 0$$

when written in the form of Eq. (25) is

$$y'' + \frac{p}{x}y' + \frac{q}{x^2}y = 0 \tag{26}$$

and since $P(x) = p/x$ and $Q(x) = q/x^2$, there is a singular point at $x = 0$. On examining Sec. 5-7, we see that the solutions to such an equation can contain terms like x^{-2} or $\ln x$ which become infinite as x approaches zero and hence cannot be expressed in power series there.

Method of Frobenius Although the general problem of finding solutions near a singular point has never been solved completely, there is one important type of singularity that can be treated by an extension of the power-series technique, known as the *method of Frobenius*.

Definition *A point $x = a$ is said to be a regular singular point†[or a regular singularity of the differential equation (25) if:*

1. *$x = a$ is a singular point, and*
2. *$(x - a)P(x)$ and $(x - a)^2 Q(x)$ are analytic‡ at $x = a$.*

For example, the Euler equation (26) has a regular singularity at $x = 0$ since $xP(x) = p$ and $x^2 Q(x) = q$, both of which are analytic since p and q are constants. On the other hand, the equation

$$y'' + \frac{3}{(x - 1)^2}y' + xy = 0$$

has an *irregular singularity* at $x = 1$ since $(x - 1)P(x) = 3(x - 1)^{-1}$, which is not analytic at $x = 1$.

---

† Some authors use this term for points at which slightly different conditions apply. E.g., see Ince [19].

‡ See Definition 1, page 187.

If the differential equation (25) has a regular singularity at a point $x = a$, it is possible to obtain a solution by an extension of the power-series method of the previous section.† In this case we assume a solution of the form

$$y = (x - a)^\alpha [a_0 + a_1(x - a) + a_2(x - a)^2 + \cdots + a_n(x - a)^n + \cdots]$$
(27)

and hope to determine the index α‡ in addition to the coefficients of the series. The form of the trial solution (27) is suggested by the form of the solution of Euler's equation, which is the simplest equation with a regular singularity. The method of solving a differential equation by assuming a trial solution of the form of (27) will be called the method of Frobenius.§

In order to simplify the following discussion, let us assume that the differential equation (25) has a regular singularity at the point $x = 0$.¶ Hence we shall seek a solution of the form

$$y = x^\alpha (a_0 + a_1 x + a_2 x^2 + \cdots + a_n x^n + \cdots)$$
(28)

or

$$y = a_0 x^\alpha + a_1 x^{\alpha+1} + a_2 x^{\alpha+2} + \cdots + a_n x^{\alpha+n} + \cdots$$
(29)

where, without loss of generality (why?), we can always assume

$$a_0 \neq 0$$

Differentiating, we have

$$y' = a_0 \alpha x^{\alpha-1} + a_1(\alpha + 1)x^\alpha + a_2(\alpha + 2)x^{\alpha+1} + \cdots$$
$$+ a_n(\alpha + n)x^{\alpha+n-1} + \cdots \quad (30)$$

$$y'' = a_0 \alpha(\alpha - 1)x^{\alpha-2} + a_1(\alpha + 1)\alpha x^{\alpha-1} + a_2(\alpha + 2)(\alpha + 1)x^\alpha + \cdots$$
$$+ a_n(\alpha + n)(\alpha + n - 1)x^{\alpha+n-2} + \cdots$$

Now, since we intend to substitute these series into the differential equation (25) and collect like powers of x, it is necessary that we expand P and Q in some kind of series. But since there is a regular singularity at

---

† The solution near an *irregular singularity* is usually a much more difficult problem. See Probs. 6 and 7 for an indication of some of the difficulties that can be encountered. See also Coddington and Levison [11], chap. 5.

‡ The index α may or may not be an integer. If α is a nonnegative integer, then (27) reduces to an ordinary power series.

§ Here again there is no universal agreement on the use of this phrase, many authors reserving it for the method of obtaining a second solution in degenerate cases. See page 212.

¶ A singularity at $x = a$ can be shifted to the origin by making the substitution $t = x - a$.

$x = 0$, we know that $xP(x)$ and $x^2Q(x)$ are analytic and hence have power-series expansions of the form

$$xP(x) = p_0 + p_1x + p_2x^2 + p_3x^3 + \cdots + p_nx^n + \cdots$$
$$x^2Q(x) = q_0 + q_1x + q_2x^2 + q_3x^3 + \cdots + q_nx^n + \cdots \tag{31}$$

Therefore, for $x \neq 0$, P and Q can be written in the form

$$P(x) = p_0x^{-1} + p_1 + p_2x + p_3x^2 + \cdots + p_nx^{n-1} + \cdots$$
$$Q(x) = q_0x^{-2} + q_1x^{-1} + q_2 + q_3x + \cdots + q_nx^{n-2} + \cdots \tag{32}$$

If we now substitute the series for y, y', y'', P, and Q into Eq. (25) and collect terms, we obtain

$$a_0[\alpha(\alpha - 1) + p_0\alpha + q_0]x^{\alpha-2} + \{a_1[(\alpha + 1)\alpha + p_0(\alpha + 1) + q_0]$$
$$+ a_0(p_1\alpha + q_1)\}x^{\alpha-1} + \cdots = 0 \quad (33)$$

By multiplying by $x^{-(\alpha-2)}$ and using the uniqueness theorem for power series, we again have the result that the coefficient of each power of x must vanish. For the first coefficient, since $a_0 \neq 0$, we have

$$\alpha(\alpha - 1) + p_0\alpha + q_0 = 0$$

or

$$\alpha^2 + (p_0 - 1)\alpha + q_0 = 0 \tag{34}$$

This equation is called the *indicial equation* since it is used to determine the index α. Compare with the method of solving Euler's equation (see Sec. 5-7). Once α has been determined, then the coefficients a_n can be computed successively (see Example 2, page 199).

We shall state without proof† the general theorem regarding the existence of solutions of the type (29).

Theorem 1   A. *If the differential equation*

$$y'' + P(x)y' + Q(x)y = 0$$

has a regular singularity at $x = 0$ and if the roots α_1 and α_2 of the indicial equation are distinct and do not differ by an integer, then there are two linearly independent solutions of the form

$$y_1(x) = x^{\alpha_1} \sum_{n=0}^{\infty} a_nx^n \qquad y_2(x) = x^{\alpha_2} \sum_{n=0}^{\infty} b_nx^n$$

These solutions are valid at least in some interval $0 < x < R$.‡

† For a proof of this theorem see Ince [19], chap. 15.

‡ If the indicial roots are complex numbers, then the solutions given here would not be real. However, real solutions can be obtained in the same manner as for Euler's equation (see page 146) by using the identity $x^{a \pm ib} = x^a e^{\pm ib \ln x} = x^a[\cos(b \ln x) \pm i \sin(b \ln x)]$.

B. *If the roots α_1 and α_2 of the indicial equation are equal or differ by an integer, then there are linearly independent solutions of the form*

$$y_1(x) = x^{\alpha_1} \sum_{n=0}^{\infty} a_n x^n \qquad y_2(x) = x^{\alpha_2} \sum_{n=0}^{\infty} b_n x^n + C y_1(x) \ln x$$

where the constant C is not zero when $\alpha_1 = \alpha_2$ and may or may not be zero when α_1 and α_2 differ by an integer.

It should be noted that, in any case, there is at least one solution of the form $y = x^z \Sigma a_n x^n$ if $x = 0$ is a regular singularity. The peculiar difficulties which arise when the indicial roots differ by an integer will be discussed in detail in connection with Bessel's equation (Sec. 7-4).

Example 1   Discuss the singularities and the *form* of the solution of the equation

$$y'' + \left(2 - \frac{7}{3x}\right)y' + \left(\frac{1}{x^2} + x\right)y = 0 \tag{35}$$

The only singularity is at $x = 0$, and this is seen to be a regular singularity since

$$xP(x) = x\left(2 - \frac{7}{3x}\right) = -\tfrac{7}{3} + 2x$$

$$x^2 Q(x) = x^2\left(\frac{1}{x^2} + x\right) = 1 + x^3$$

and hence $xP(x)$ and $x^2Q(x)$ are analytic at $x = 0$. From (34) the indicial equation is

$$\alpha^2 - \tfrac{10}{3}\alpha + 1 = 0$$

with roots $\alpha_1 = 3$, $\alpha_2 = \tfrac{1}{3}$. Therefore, from Theorem 1A, the general solution of (35) near $x = 0$ can be written in the form

$$y = x^3(a_0 + a_1 x + \cdots + a_n x^n + \cdots) + x^{\frac{1}{3}}(b_0 + b_1 x + \cdots + b_n x^n + \cdots)$$

The constants a_n and b_n could be obtained by substitution in Eq. (35). Near any ordinary point $x = a \neq 0$ the general solution could be written in the power-series form

$$y = [A_0 + A_1(x - a) + \cdots + A_n(x - a)^n + \cdots]$$
$$+ [B_0 + B_1(x - a) + \cdots + B_n(x - a)^n + \cdots]$$

Example 2   Solve by the method of Frobenius:

$$4xy'' + 2y' - y = 0 \tag{36}$$

Since this equation has a regular singularity at $x = 0$, we assume a solution of the form

$$y = x^z \sum_{n=0}^{\infty} a_n x^n$$

or

$$y = \sum_{n=0}^{\infty} a_n x^{\alpha+n}$$

Differentiating, we have

$$y' = \sum_{n=0}^{\infty} (\alpha + n)a_n x^{\alpha+n-1} \qquad y'' = \sum_{n=0}^{\infty} (\alpha + n)(\alpha + n - 1)a_n x^{\alpha+n-2} \qquad (37)$$

and on substituting into (36), we have

$$4\sum_{n=0}^{\infty} (\alpha + n)(\alpha + n - 1)a_n x^{\alpha+n-1} + 2\sum_{n=0}^{\infty} (\alpha + n)a_n x^{\alpha+n-1} - \sum_{n=0}^{\infty} a_n x^{\alpha+n} = 0$$

Since the power of x is the same in the first two sums, they can be combined to give

$$\sum_{n=0}^{\infty} [4(\alpha + n)(\alpha + n - 1) + 2(\alpha + n)]a_n x^{\alpha+n-1} - \sum_{n=0}^{\infty} a_n x^{\alpha+n} = 0$$

or

$$\sum_{n=0}^{\infty} 2(\alpha + n)(2\alpha + 2n - 1)a_n x^{\alpha+n-1} - \sum_{n=0}^{\infty} a_n x^{\alpha+n} = 0 \qquad (38)$$

In order to combine these two sums, we can adjust the index of the second sum† so that (38) becomes

$$\sum_{n=0}^{\infty} 2(\alpha + n)(2\alpha + 2n - 1)a_n x^{\alpha+n-1} - \sum_{n=1}^{\infty} a_{n-1} x^{\alpha+n-1} = 0 \qquad (39)$$

Finally, in order to combine these sums, the range of the index n must be the same in both. This can be achieved by separating out the $n = 0$ term from the first sum. Equation (39) then becomes

$$2\alpha(2\alpha - 1)a_0 x^{\alpha-1} + \sum_{n=1}^{\infty} [2(\alpha + n)(2\alpha + 2n - 1)a_n - a_{n-1}]x^{\alpha+n-1} = 0 \qquad (40)$$

Since each coefficient must vanish, we have

$$2\alpha(2\alpha - 1)a_0 = 0 \qquad (41)$$

$$2(\alpha + n)(2\alpha + 2n - 1)a_n - a_{n-1} = 0 \qquad n = 1, 2, 3, \ldots \qquad (42)$$

Now, since $a_0 \neq 0$, Eq. (41) gives us the *indicial equation* ‡

$$\alpha(2\alpha - 1) = 0 \qquad (43)$$

and hence α can have either of the values

$$\alpha = 0 \qquad \text{or} \qquad \alpha = \tfrac{1}{2} \qquad (44)$$

† The process of adjusting the index is described on page 191.

‡ The indicial equation (43) could have been obtained directly from the differential equation (36) by using formula (34). However, the recurrence formula (42) can be obtained only by a process such as the one used in this example.

The coefficients a_n of the series corresponding to the index $\alpha = 0$ can now be computed by substituting $\alpha = 0$ in the recurrence formula (42). We have then

$$2n(2n - 1)a_n - a_{n-1} = 0 \qquad n = 1, 2, \ldots \tag{45}$$

or, since $2n(2n - 1) \neq 0$ for $n = 1, 2, \ldots$,

$$a_n = \frac{1}{2n(2n - 1)} a_{n-1} \tag{46}$$

Hence

$$a_1 = \frac{1}{2 \cdot 1} a_0$$

$$a_2 = \frac{1}{4 \cdot 3} a_1 = \frac{1}{4 \cdot 3 \cdot 2 \cdot 1} a_0$$

$$\cdot$$
$$\cdot$$
$$\cdot$$

$$a_n = \frac{1}{2n(2n - 1)} a_{n-1} = \cdots = \frac{1}{(2n)!} a_0$$

Therefore the solution when $\alpha = 0$ is

$$y = a_0 \left[1 + \frac{x}{2!} + \frac{x^2}{4!} + \cdots + \frac{x^n}{(2n)!} + \cdots \right]$$

The solution for $\alpha = \frac{1}{2}$ is found in the same way,† and the general solution is then

$$y = a_0 \left[1 + \frac{x}{2!} + \frac{x^2}{4!} + \cdots + \frac{x^n}{(2n)!} + \cdots \right]$$
$$+ b_0 x^{\frac{1}{2}} \left[1 + \frac{x}{3!} + \frac{x^2}{5!} + \cdots + \frac{x^n}{(2n + 1)!} + \cdots \right] \tag{47}$$

It is easy to show that these power series converge for all x. However, the presence of the function $x^{\frac{1}{2}}$ limits the domain of validity of the solution (47) to positive values of x.

The alert reader may recognize these series and hence realize that the solution is simply

$$y = a_0 \cosh \sqrt{x} + b_0 \sinh \sqrt{x}$$

It should be emphasized, however, that the method of Frobenius will usually lead to series that *cannot* be expressed in terms of elementary functions.

Solution for large x   In addition to examining the singularities which occur for finite values of x, it is often desirable to investigate the behavior at infinity, that is, for large values of x. The simplest way to do this is to introduce a new independent variable s by the transformation

$$s = \frac{1}{x} \tag{48}$$

† The reader should carry out the computations.

Then as $x \to \infty$, $s \to 0$ and if we examine the new equation near $s = 0$, we can deduce the behavior of the original equation as $x \to \infty$.

We have

$$\frac{dy}{dx} = \frac{dy}{ds}\frac{ds}{dx} = \frac{dy}{ds}\left(-\frac{1}{x^2}\right) = -s^2\frac{dy}{ds} \tag{49}$$

and

$$\frac{d^2y}{dx^2} = \frac{d}{ds}\left(\frac{dy}{dx}\right)\frac{ds}{dx} = \left(-s^2\frac{d^2y}{ds^2} - 2s\frac{dy}{ds}\right)(-s^2) \tag{50}$$

Hence the equation

$$\frac{d^2y}{dx^2} + P(x)\frac{dy}{dx} + Q(x)y = 0 \tag{51}$$

becomes

$$s^4\frac{d^2y}{ds^2} + \left[2s^3 - s^2P\left(\frac{1}{s}\right)\right]\frac{dy}{ds} + Q\left(\frac{1}{s}\right)y = 0 \tag{52}$$

or

$$\frac{d^2y}{ds^2} + \left[2s^{-1} - s^{-2}P\left(\frac{1}{s}\right)\right]\frac{dy}{ds} + s^{-4}Q\left(\frac{1}{s}\right)y = 0 \tag{53}$$

Equation (53) has zero as an ordinary point if $2s^{-1} - s^{-2}P(1/s)$ and $s^{-4}Q(1/s)$ are both analytic at $s = 0$. If either of these functions is not analytic, then the equation has a regular singularity at $s = 0$ if $2 - s^{-1}P(1/s)$ and $s^{-2}Q(1/s)$ are both analytic at $s = 0$.

Example 3   An important equation which will be studied in detail in Sec. 7-6 is *Legendre's equation*:

$$(1 - x^2)y'' - 2xy' + n(n + 1)y = 0 \tag{54}$$

where n is a constant, not necessarily an integer.

This equation has regular singularities at $x = \pm 1$, and in order to determine the behavior at infinity, we let $s = 1/x$ and obtain the equation

$$\frac{d^2y}{ds^2} + \left[\frac{2}{s} - \frac{2}{s(s^2 - 1)}\right]\frac{dy}{ds} + \frac{n(n + 1)}{s^2(s^2 - 1)}y = 0 \tag{55}$$

Since this equation has a regular singularity at $s = 0$, Legendre's equation (54) has a regular singularity at infinity. This means that there is a solution of (54) in the form

$$y = \left(\frac{1}{x}\right)^\alpha \sum_{n=0}^\infty a_n\left(\frac{1}{x}\right)^n \tag{56}$$

which is valid for large x.

Problems 7-3

1. Find all singularities of the following differential equations for finite values of x. For all regular singularities find and solve the indicial equation and state the *form* of the general solution.

a. $y'' + \dfrac{8}{3x} y' - \dfrac{2}{3x^2} y = 0$ b. $x^2 y'' + xy' + (x^2 + 1)y = 0$

c. $(x - 1)^2 y'' + xy' + (x - 1)y = 0$

d. $2x^2 y'' - 5(\sin x)y' + 3y = 0$

e. $e^x \left(y'' + \dfrac{y'}{4x} \right) + \sqrt{xy} = 0$

2. Find the general solution of the following differential equations by using the method of Frobenius:

a. $x^2 y'' - 2xy' + (2 + x^2)y = 0$ b. $xy'' - y' + 2y = 0$

c. $x^2 y'' + xy' + (x^2 - \frac{1}{4})y = 0$

3. Examine the following equations for singularities at infinity:

a. $xy'' + y' + xy = 0$ b. $x^5 y'' + 2x^4 y' + y = 0$

c. $x^3 y'' + 3x^2 y' + (x + 4)y = 0$

4. The third-order linear differential equation

$$y''' + P_1(x)y'' + P_2(x)y' + P_3(x)y = 0$$

has a regular singularity at $x = 0$ if $P_1(x)$, $P_2(x)$, and $P_3(x)$ are *not* all analytic at $x = 0$ but $xP_1(x)$, $x^2 P_2(x)$, and $x^3 P_3(x)$ *are* all analytic at $x = 0$. Find all of the regular singularities of the equation

$$x^2(x - 1)y''' - 3y'' + 4xy' + y = 0$$

5. Generalize the definition of a regular singularity in Prob. 4 and show that the point $x = 0$ is a regular singularity of Euler's equation of order n

$$x^n \frac{d^n y}{dx^n} + p_{n-1} x^{n-1} \frac{d^{n-1} y}{dx^{n-1}} + \cdots + p_1 x \frac{dy}{dx} + p_0 y = 0$$

where $p_0, p_1, \ldots, p_{n-1}$ are constants.

6. Given the differential equation with *constant* coefficients, $ay'' + by' + cy = 0$:

a. Show that this equation has an *irregular* singularity at infinity.

b. Show that no solution exists which can be written in the form $\sum_0^\infty a_n x^{-n}$.

7. Given the following first-order differential equation which has an *irregular* singularity at $x = 0$, $dy/dx - (1/x^2)y = -1/x$:

a. Assume a solution in the form of a power series $y = \sum_1^\infty a_n x^n$ and then show that this series *diverges* for all $x \neq 0$.

b. Show that $y_p = e^{-1/x} \int_{-\infty}^{1/x} \frac{e^t}{t} \, dt$, $x < 0$, satisfies the given differential equation.

c. Use integration by parts repeatedly on the solution y_p in part b to obtain

$$y_p = \sum_{n=1}^N a_n x^n + R_N(x)$$

where the coefficients a_n are the same as the ones obtained in part a and $R_N(x) = N!e^{-1/x} \int_{-\infty}^{1/x} t^{-N-1}e^t \, dt$.

d. Show that $\lim_{N \to \infty} R_N(x) = \infty$ (i.e., the series diverges) but $\lim_{x \to 0} R_N(x) = 0$ and hence a *fixed number of terms* in the series gives a better and better approximation to the solution, as x approaches zero. Such series are called *asymptotic series* and are surprisingly useful even though divergent!

7-4   Bessel's equation

One of the most important differential equations with variable coefficients is *Bessel's equation.*†

$$x^2 y'' + xy' + (x^2 - p^2)y = 0 \tag{57}$$

where p is any constant. (We shall restrict ourselves to real values of p and assume $p \geq 0$.) The solutions of this equation are known as *Bessel functions*. These functions are among the most important non-elementary functions, and they arise in a wide variety of problems in mechanics, electrical engineering, thermodynamics, aerodynamics, etc.

Dividing Eq. (57) by x^2, we have for $x > 0$

$$y'' + \frac{1}{x}y' + \left(1 - \frac{p^2}{x^2}\right)y = 0 \tag{58}$$

This equation has a regular singularity at $x = 0$, and hence we assume a solution of the form

$$y = x^\alpha(a_0 + a_1 x + \cdots + a_n x^n + \cdots)$$

---

† Bessel's equation is encountered in many physical problems which are phrased as boundary-value problems for a partial differential equation. See Chap. 15.

or, using the summation notation,

$$y = \sum_{n=0}^{\infty} a_n x^{\alpha+n} \tag{59}$$

After differentiating and substituting into Bessel's equation (57), we have

$$\sum_{n=0}^{\infty} (\alpha + n)(\alpha + n - 1)a_n x^{\alpha+n} + \sum_{n=0}^{\infty} (\alpha + n)a_n x^{\alpha+n}$$
$$+ \sum_{n=0}^{\infty} (x^2 - p^2)a_n x^{\alpha+n} = 0 \tag{60}$$

or

$$\sum_{n=0}^{\infty} [(\alpha + n)^2 - p^2]a_n x^{\alpha+n} + \sum_{n=0}^{\infty} a_n x^{\alpha+n+2} = 0 \tag{61}$$

Adjusting the index of the second sum so that the power of x is the same as in the first sum, we have

$$\sum_{n=0}^{\infty} [(\alpha + n)^2 - p^2]a_n x^{\alpha+n} + \sum_{n=2}^{\infty} a_{n-2} x^{\alpha+n} = 0 \tag{62}$$

The terms for $n = 0$ and $n = 1$ in the first sum must be written separately, and then we can combine the two sums to get

$$(\alpha^2 - p^2)a_0 x^{\alpha} + [(\alpha + 1)^2 - p^2]a_1 x^{\alpha+1}$$
$$+ \sum_{n=2}^{\infty} \{[(\alpha + n)^2 - p^2]a_n + a_{n-2}\}x^{\alpha+n} = 0 \tag{63}$$

Once again each coefficient must vanish and hence

$$(\alpha^2 - p^2)a_0 = 0 \tag{64}$$

$$[(\alpha + 1)^2 - p^2]a_1 = 0 \tag{65}$$

$$[(\alpha + n)^2 - p^2]a_n + a_{n-2} = 0 \qquad n = 2, 3, 4, \ldots \tag{66}$$

As usual, we assume $a_0 \neq 0$ and hence Eq. (64) yields the indicial equation

$$\alpha^2 - p^2 = 0 \tag{67}$$

and thus the index α is given by

$$\alpha = \pm p \tag{68}$$

We shall first find the solution corresponding to the nonnegative index

$$\alpha = p \geq 0$$

The solution for $\alpha = -p$ will be considered later.

If we substitute $\alpha = p$ in Eqs. (65) and (66), we obtain

$$(2p + 1)a_1 = 0 \tag{68a}$$

$$(2pn + n^2)a_n + a_{n-2} = 0 \qquad n \geq 2 \qquad \text{(recurrence formula)} \tag{68b}$$

Since $p \geq 0$, Eq. (68a) implies $a_1 = 0$. Then from (68b) we see that $a_3 = 0$, $a_5 = 0, \ldots, a_{2n+1} = 0$. Hence only the even-numbered coefficients remain. Since $p \geq 0$, $2pn + n^2 \neq 0$ and we can solve for a_n in (68b) to get

$$a_n = -\frac{1}{2pn + n^2} a_{n-2} \tag{69}$$

By repeated application of this formula we obtain

$$a_2 = -\frac{1}{4p + 4} a_0 = -\frac{1}{4(p + 1)} a_0$$

$$a_4 = -\frac{1}{8p + 16} a_2 = +\frac{1}{2.4(p + 2)} \frac{1}{4(p + 1)} a_0$$

$$= +\frac{1}{2!4^2(p + 2)(p + 1)} a_0$$

$$a_6 = -\frac{1}{12p + 36} a_4 = -\frac{1}{3!4^3(p + 3)(p + 2)(p + 1)} a_0$$

and in general

$$a_{2k} = (-1)^k \frac{1}{k!4^k(p + k)(p + k - 1) \cdots (p + 2)(p + 1)} a_0 \qquad k \geq 1 \tag{70}$$

Hence one solution of Bessel's equation can be written in the form

$$y = a_0 \sum_{k=0}^{\infty} (-1)^k \frac{x^{p+2k}}{k!4^k(p + k)(p + k - 1) \cdots (p + 2)(p + 1)}$$

By a proper choice of a_0, this function can be written in the conventional form of the *Bessel function of the first kind of order* p, usually denoted by $J_p(x)$. We set

$$a_0 = \frac{1}{2^p \Gamma(p + 1)}$$

where $\Gamma(p + 1)$ is the gamma function.† We have then

$$J_p(x) = \sum_{k=0}^{\infty} \frac{(-1)^k x^{p+2k}}{2^{p+2k} k!(p + k)(p + k - 1) \cdots (p + 2)(p + 1)\Gamma(p + 1)}$$

or, since the expression $(p + k) \cdots (p + 1)\Gamma(p + 1)$ is merely the expanded form of $\Gamma(p + k + 1)$, we can write finally

$$J_p(x) = \sum_{k=0}^{\infty} \frac{(-1)^k}{k!\Gamma(p + k + 1)} \left(\frac{x}{2}\right)^{p+2k} \tag{71}$$

This is the Bessel function of the first kind of order p.

† See Sec. 1-9 for a discussion of the gamma function.

In order to prove that $J_p(x)$ is actually a solution of Bessel's equation (57), we must establish the convergence of the infinite series. Applying the ratio test and the recurrence formula (69), we have

$$\lim_{n \to \infty} \left| \frac{a_n x^{n+p}}{a_{n-2} x^{n+p-2}} \right| = \lim_{n \to \infty} \frac{1}{2pn + n^2} x^2 = 0$$

and hence the series converges for all x and all $p \geq 0$. Therefore $J_p(x)$ is a solution to Bessel's equation at least for $x \geq 0$ if $p \geq 0$. If p is an integer, then the solution (71) is valid for all x.

The reader may possibly be disturbed by the fact that the Bessel functions are defined in terms of infinite series. But this situation is neither new nor unusual. The familiar function $\sin x$ can be properly defined only in terms of an infinite series (or in terms of a definite integral). The tables of values of $\sin x$ are obtained from computations with these infinite series, and the same is true of the Bessel functions. They have been tabulated[†] (for special values of p), and hence as much information is available about them as there is about $\sin x$. The similarity does not end here. We shall see below that the Bessel functions have many properties which are analogous to those of the trigonometric functions. Nevertheless, it should be emphasized that in general (for p not equal to half an odd integer)[‡], the Bessel functions are non-elementary functions. That is, they cannot be expressed as a finite algebraic combination of elementary functions.

If p is not equal to zero or a positive integer, we can obtain a second, linearly independent, solution to Bessel's equation by replacing p by $-p$ in formula (71). In this case the general solution of (57) for $x > 0$ is given by

$$y = c_1 J_p(x) + c_2 J_{-p}(x)$$

where

$$J_{-p}(x) = \sum_{k=0}^{\infty} \frac{(-1)^k}{k! \Gamma(-p + k + 1)} \left(\frac{x}{2} \right)^{-p+2k} \tag{72}$$

If $p = 0$, (72) would be identical to (71) and we would have to look elsewhere for a second solution. (See page 210.) If p is a positive integer, then with a proper interpretation of (72)[§] it can be shown that J_{-p} is proportional to J_p and, again, a second solution must be obtained by a different method. (See page 210 and Prob. 11.) Both these exceptional cases

---

† For tables and graphs of the Bessel functions see, for example, Jahnke, Emde, and Lösch [20].

‡ See Prob. 6.

§ See Prob. 5.

exhibit the difficulties resulting from the fact that the roots of the indicial equation are equal or differ by an integer (see Theorem 1B, page 199).†

Some properties of Bessel functions   In this subsection we shall use the series representation (71) in order to develop some interesting and useful properties of the Bessel functions of various orders.

We have shown that

$$J_p(x) = \sum_{k=0}^{\infty} \frac{(-1)^k}{k!\Gamma(p+k+1)} \left(\frac{x}{2}\right)^{p+2k} \tag{73}$$

Since this series converges for all x, we can differentiate termwise to get

$$J_p'(x) = \sum_{k=0}^{\infty} \frac{(-1)^k(p+2k)}{k!\Gamma(p+k+1)} \left(\frac{x}{2}\right)^{p+2k-1} \cdot \frac{1}{2}$$

and hence

$$xJ_p'(x) = \sum_{k=0}^{\infty} (p+2k) \frac{(-1)^k}{k!\Gamma(p+k+1)} \left(\frac{x}{2}\right)^{p+2k}$$

$$= p\sum_{k=0}^{\infty} \frac{(-1)^k}{k!\Gamma(p+k+1)} \left(\frac{x}{2}\right)^{p+2k} + \sum_{k=1}^{\infty} 2k \frac{(-1)^k}{k!\Gamma(p+k+1)} \left(\frac{x}{2}\right)^{p+2k}$$

$$= pJ_p(x) + x\sum_{k=1}^{\infty} \frac{(-1)^k}{(k-1)!\Gamma(p+k+1)} \left(\frac{x}{2}\right)^{p+2k-1}$$

$$= pJ_p(x) + x\sum_{k=0}^{\infty} \frac{(-1)^{k+1}}{k!\Gamma(p+k+2)} \left(\frac{x}{2}\right)^{p+2k+1}$$

or finally

$$xJ_p'(x) = pJ_p(x) - xJ_{p+1}(x) \tag{74}$$

In a similar manner it can be shown that

$$xJ_p'(x) = -pJ_p(x) + xJ_{p-1}(x) \tag{75}$$

By subtracting formula (74) from (75), we obtain the recurrence relation connecting Bessel functions of successive orders

$$\frac{2p}{x} J_p(x) = J_{p-1}(x) + J_{p+1}(x) \tag{76}$$

or, rearranging terms,

$$J_{p+1}(x) = \frac{2p}{x} J_p(x) - J_{p-1}(x) \tag{77}$$

† When p is equal to half an odd integer, the indicial roots again differ by an integer. In this case, however, it can be shown that J_p and J_{-p} are linearly independent; e.g., see Prob. 6.

As an example of the usefulness of these formulas, we set $p = 1$ in (77) and obtain

$$J_2(x) = \frac{2}{x} J_1(x) - J_0(x) \tag{78}$$

Hence, if we know the values of J_0 and J_1, we can easily compute J_2. The value of J_n for any integer can be computed in a similar manner. Thus, since J_0 and J_1 have been extensively tabulated (e.g., see, Jahnke, Emde, and Lösch [20]), the values of J_n are available for all positive integers n.

Bessel functions of integral order If the constant p in Bessel's equation is equal to zero or a positive integer n, then the general expression (71) for the Bessel function can be simplified somewhat. If n is an integer, then

$$\Gamma(n + k + 1) = (n + k)!$$

and (71) becomes

$$J_n(x) = \sum_{k=0}^{\infty} \frac{(-1)^k}{k!(n+k)!} \left(\frac{x}{2}\right)^{n+2k} \tag{79}$$

Two cases of special interest in applications arise for $n = 0$ and $n = 1$. We have

$$J_0(x) = \sum_{k=0}^{\infty} \frac{(-1)^k}{(k!)^2} \left(\frac{x}{2}\right)^{2k} = 1 - \frac{x^2}{2^2}$$

$$+ \frac{x^4}{2^4(2!)^2} + \cdots + (-1)^k \frac{x^{2k}}{2^{2k}(k!)^2} + \cdots \tag{80}$$

$$J_1(x) = \sum_{k=0}^{\infty} \frac{(-1)^k}{k!(k+1)!} \left(\frac{x}{2}\right)^{2k+1} = \frac{x}{2} - \frac{x^3}{2^3 2!}$$

$$+ \frac{x^5}{2^5 2!3!} + \cdots + (-1)^k \frac{x^{2k+1}}{2^{2k+1}k!(k+1)!} + \cdots \tag{81}$$

The graphs of these functions appear in Fig. 1. Note that they have the general appearance of damped sinusoids.†

Since the series for $J_0(x)$ converges for all x, we can differentiate termwise to get

$$J_0'(x) = \sum_{k=1}^{\infty} \frac{(-1)^k x^{2k-1}}{(k!)(k-1)!2^{2k-1}} = -\sum_{k=0}^{\infty} \frac{(-1)^k}{(k+1)!k!} \left(\frac{x}{2}\right)^{2k+1}$$

and hence, since this last expression is just $-J_1(x)$, we have

$$J_0'(x) = -J_1(x) \tag{82}$$

† See Prob. 9.

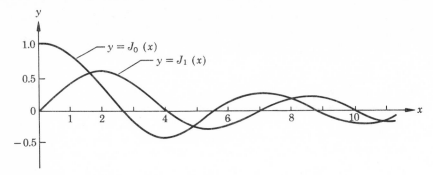

figure 1. *The Bessel functions $J_0(x)$ and $J_1(x)$.*

Similarly we can show that

$$\frac{d}{dx}[xJ_1(x)] = xJ_0(x) \tag{83}$$

These relations are special cases of the general properties of Bessel functions which were derived in the previous subsection.

Second solution for Bessel's equation of integral order If n is zero or a positive integer, then Bessel's equation

$$x^2y'' + xy' + (x^2 - n^2)y = 0 \tag{84}$$

has an indicial equation with roots that are either equal (when $n = 0$) or differ by an integer (when $n = 1, 2, \ldots$). In this case Theorem 1 (page 198) assures us of only one solution of Frobenius type. This is the solution

$$J_n(x) = \sum_{k=0}^{\infty} \frac{(-1)^k}{k!(n + k)!}\left(\frac{x}{2}\right)^{n+2k} \tag{85}$$

The form of the second solution, according to Theorem 1B, is

$$y_2(x) = x^{-n}\sum_{k=0}^{\infty} b_k x^k + CJ_n(x)\ln x \tag{86}$$

The constants b_k and C may be obtained by substituting this expression into Bessel's equation (84). However, there are several other methods for obtaining the second solution directly, that is, without appealing to Theorem 1. We shall illustrate two of these methods in the following paragraphs.†

Since we already know one solution, $J_n(x)$, we can apply the method of *reduction* of *order* (see Sec. 5-6, Prob. 9). That is, we seek a solution of the form

$$y = u(x)J_n(x) \tag{87}$$

† See Prob. 15 for still another method of obtaining a second solution.

On substituting into Eq. (84), we see that $u(x)$ must satisfy the equation

$$xJ_n(x)u''(x) + [2xJ'_n(x) + J_n(x)]u'(x) = 0 \tag{88}$$

If we now let

$$v(x) = u'(x) \tag{89}$$

then $v(x)$ must satisfy the first-order linear differential equation

$$xJ_n(x)v'(x) + [2xJ'_n(x) + J_n(x)]v(x) = 0 \tag{90}$$

which is easily solved to obtain

$$v(x) = \frac{C}{xJ_n^2(x)} \tag{91}$$

This solution is valid as long as $xJ_n^2(x)$ does not vanish. However, from the continuity of $J_n(x)$ it is easy to show that there is some positive number x_0 such that $xJ_n^2(x) \neq 0$ for $0 < x \leq x_0$. (See Prob. 8.) Integrating $v(x)$, we have

$$u(x) = C_2 \int_{x_0}^{x} \frac{dt}{tJ_n^2(t)} + C_1 \tag{92}$$

and hence the general solution of Bessel's equation (84) when n is an integer can be written in the form

$$y = C_1 J_n(x) + C_2 J_n(x) \int_{x_0}^{x} \frac{dt}{tJ_n^2(t)} \qquad 0 < x \leq x_0 \tag{93}$$

Since the integral in this equation cannot be evaluated directly, this form of the solution is difficult to apply. The integral can be evaluated by the use of infinite series, but the expansion can be obtained by another method which has more general interest. We first consider the situation for $n = 0$.

Using the notation of linear operators, Bessel's equation of order zero can be written in the form

$$L(y) \equiv x^2 y'' + xy' + x^2 y = 0 \tag{94}$$

If we applied the method of Frobenius as above, we would substitute a series

$$y(x) = x^\alpha \sum_{n=0}^{\infty} a_n x^n \tag{95}$$

and obtain the indicial equation and recurrence formula as in Eqs. (64), (65), and (66) with $p = 0$. If we ignore the indicial equation for the moment (that is, if we do not set $\alpha = 0$), then the recurrence formula gives

$$a_n = -\frac{1}{(\alpha + n)^2} a_{n-2} \tag{96}$$

and hence for $n = 2k$

$$a_{2k} = \frac{(-1)^k}{(\alpha + 2k)^2(\alpha + 2k - 2)^2 \cdots (\alpha + 2)^2} a_0 \tag{97}$$

Since the coefficients a_{2k} still depend on α, we shall designate them by $a_{2k}(\alpha)$, and we determine y as a function of the two variables x and α.

$$y(x,\alpha) = x^\alpha \sum_{k=0}^{\infty} a_{2k}(\alpha) x^{2k} \tag{98}$$

This is not yet a solution to Eq. (94), but if we substitute, we obtain

$$L[y(x,\alpha)] = a_0 \alpha^2 x^\alpha \; \dagger \tag{99}$$

All the other terms vanish since the recurrence formula (96) is satisfied.

If we now set $\alpha = 0$ in Eq. (99), we have

$$L[y(x,0)] = 0 \tag{100}$$

and in fact $y(x,0) = a_0 J_0(x)$. That is, we again obtain the solution $J_0(x)$.

However, since the right-hand side of Eq. (99) contains the factor α^2, it has a double root at $\alpha = 0$. Hence the derivative of this term with respect to α will also vanish at $\alpha = 0$, and so we have

$$\frac{\partial}{\partial \alpha} \{L[y(x,\alpha)]\} = a_0 x^\alpha (2\alpha + \alpha^2 \ln x) \tag{101}$$

However, since the operators $\partial/\partial\alpha$ and L commute,$\ddagger$ we have

$$L\left[\frac{\partial y(x,\alpha)}{\partial \alpha}\right] = a_0 x^\alpha (2\alpha + \alpha^2 \ln x) \tag{102}$$

and finally, when $\alpha = 0$,

$$L\left[\frac{\partial y(x,\alpha)}{\partial \alpha}\right]_{\alpha=0} = 0 \tag{103}$$

Therefore the function $[\partial y(x,\alpha)/\partial\alpha]_{\alpha=0}$ is a solution of Bessel's equation (94), and it can be shown (see Prob. 14) that this function is independent of $J_0(x)$ and hence furnishes the desired second solution.

In order to obtain an explicit representation of this function, we have from Eq. (98)

$$\frac{\partial y(x,\alpha)}{\partial \alpha} = x^\alpha \left[\sum_{k=0}^{\infty} \frac{\partial a_{2k}(\alpha)}{\partial \alpha} x^{2k} + (\ln x) \sum_{k=0}^{\infty} a_{2k}(\alpha) x^{2k} \right] \tag{104}$$

$\dagger$ The derivatives with respect to x in the operator L are to be interpreted here as partial derivatives with α held fixed.

$\ddagger$ For a proof, see Coddington and Levinson [11], chap. 1, sec. 7.

and, on setting $\alpha = 0$, we have

$$\left[\frac{\partial y(x,\alpha)}{\partial \alpha}\right]_{\alpha=0} = \sum_{k=0}^{\infty}\left[\frac{\partial a_{2k}(\alpha)}{\partial \alpha}\right]_{\alpha=0} x^{2k} + a_0 (\ln x) J_0(x) \tag{105}$$

The coefficients of the series in (105) can be computed from Eq. (97), and the result is

$$\left[\frac{\partial a_{2k}(\alpha)}{\partial \alpha}\right]_{\alpha=0} = \frac{(-1)^{k+1} a_0}{2^{2k}(k!)^2}\left(1 + \tfrac{1}{2} + \tfrac{1}{3} + \cdots + \frac{1}{k}\right) \qquad k \geq 1$$

Thus we have obtained a second solution to the Bessel equation of order zero and, for $a_0 = 1$, this is the solution obtained by Neumann and designated by $N_0(x)$.† Hence

$$N_0(x) = J_0(x) \ln x + \sum_{k=1}^{\infty}\frac{(-1)^{k+1}}{(k!)^2}\left(1 + \tfrac{1}{2} + \cdots + \frac{1}{k}\right)\left(\frac{x}{2}\right)^{2k} \tag{106‡}$$

and the general solution of Eq. (94) for $x > 0$ is then

$$y = c_1 J_0(x) + c_2 N_0(x) \tag{107}$$

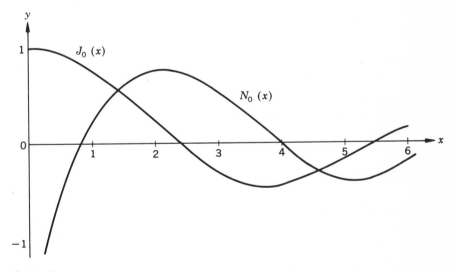

figure 2

† A graph of the function N_0 appears in Fig. 2.   Note that it becomes infinite as $x \to 0$.   This behavior is caused by the term $\ln x$ in Eq. (106).   There are several other methods of obtaining a second solution and many different expressions for the result.   E.g., see Watson [36].   Another function, designated by $Y_0(x)$, is often used as a second solution.   See Prob. 15.

‡ Compare with Eq. (86), page 210, and Theorem 1B, page 199.

An extension of the above method yields a second solution when the indicial roots differ by an integer (e.g., in Bessel's equation of order n where n is a positive integer). If the roots are α_1 and α_2, then the indicial equation can be written in the form

$$(\alpha - \alpha_1)(\alpha - \alpha_2) = 0 \tag{108}$$

If the differential equation is

$$L(y) \equiv P_1(x)y'' + P_2(x)y' + P_3(x)y = 0 \tag{109}$$

then substitution of the series $y(x,\alpha) = x^\alpha \Sigma a_n(\alpha)x^n$ as before yields

$$L[y(x,\alpha)] = a_0(\alpha - \alpha_1)(\alpha - \alpha_2)x^\alpha \tag{110}$$

if the recurrence formula is satisfied but the indicial equation is not.

A double root at $\alpha = \alpha_1$ can be introduced into Eq. (110) by multiplying by $(\alpha - \alpha_1)$. Thus

$$L[(\alpha - \alpha_1)y(x,\alpha)] = a_0(\alpha - \alpha_1)^2(\alpha - \alpha_2)x^\alpha \tag{111}$$

and then the desired second solution is given by

$$\frac{\partial}{\partial \alpha}[(\alpha - \alpha_1)y(x,\alpha)]_{\alpha=\alpha_1} \tag{112}$$

Problems 7-4

1. The equation $x^2y'' + xy' + (k^2x^2 - p^2)y = 0$, where k is a constant, appears frequently in applications. Prove that $J_p(kx)$ is a solution by changing the equation into the form of Bessel's equation of order p. Let $t = kx$ and apply the chain rule.

2. Derive the formula $d[x^2J_2(x)]/dx = x^2J_1(x)$ by differentiating the series for $x^2J_2(x)$.

3. Show that

$$\frac{d}{dx}[J_1^2(x)] = \frac{x}{2}[J_0^2(x) - J_2^2(x)]$$

4. Use the series (80) to compute the value of $J_0(1)$ to four decimal places. Estimate the error.

5. Show that if we define $1/\Gamma(-n) = 0$ for $n = 0, 1, 2, \ldots$, then formula (72) defines $J_{-n}(x)$. Show that $J_{-n}(x) = (-1)^n J_n(x)$.

6. By examining the appropriate series, show that $J_{\frac{1}{2}}(x) = \sqrt{2/\pi x}\, \sin x$ and $J_{-\frac{1}{2}}(x) = \sqrt{2/\pi x}\, \cos x$. Hence prove that $J_{\frac{1}{2}}$ and $J_{-\frac{1}{2}}$ are linearly independent.

7. Show that Bessel's equation (57) has an irregular singularity at infinity.

8. Use the facts that $J_0(0) = 1$ and $J_0(x)$ is continuous to prove that there exists a positive number x_0 such that the function $f(x) = xJ_0^2(x)$ does not vanish in the interval $0 < x \leq x_0$.

9. In Bessel's equation (57) substitute $y = u/\sqrt{x}$ to obtain the equation

$$u'' + \left(1 - \frac{4p^2 - 1}{4x^2}\right)u = 0$$

Assume that the solution of this equation behaves like the solution of the equation

$$u'' + u = 0$$

for large values of x. Hence show that, for large x, $J_p(x)$ behaves like $(A/\sqrt{x}) \times \sin(x + B)$.

10. a. The equation $4y'' + 9xy = 0$ can be changed into the form of Bessel's equation by setting

$$y = \sqrt{x}\,Y \qquad x = X^{\frac{2}{3}}$$

Hence show that a solution of the original equation is

$$y = \sqrt{x}\,J_{\frac{1}{3}}(x^{\frac{3}{2}})$$

b. Show that $y = x^{\frac{1}{2}}J_2(4x^{\frac{1}{4}})$ is a solution of the differential equation $y'' + x^{-\frac{3}{2}}y = 0$.

11. Find the second solution of Bessel's equation of order $p = 1$. See Eq. (112).

12. Let $y_1(x)$ and $y_2(x)$ be any two solutions of Bessel's equation (57) in some interval $a \le x \le b$ not including the origin. Prove that the Wronskian $W[y_1, y_2]$ is of the form $W = c/x$, where c is a constant. (Hint: Use Abel's formula, Sec. 5-8.) Interpret the result.

13. Given the function $F(x) = 1/\pi \displaystyle\int_0^\pi \cos(x \cos \theta)\,d\theta$, show that $xF'' + F' + xF = 0$ and $F(0) = 1$. Hence show that $F(x) \equiv J_0(x)$.

14. Prove that $J_0(x)$ and $N_0(x)$ are linearly independent.

15. If p is not an integer, the function $Y_p(x)$ is defined by

$$Y_p(x) = \frac{J_p(x) \cos p\pi - J_{-p}(x)}{\sin p\pi}$$

a. Show that J_p and Y_p are a pair of linearly independent solutions to Bessel's equation of order p.

b. When n is an integer, the function $Y_n(x)$ is defined by $Y_n(x) = \lim\limits_{p \to n} Y_p(x)$. Show that J_n and Y_n are a pair of linearly independent solutions to Bessel's equation of order n.

c. Show that $\lim\limits_{x \to 0^+} Y_n(x) = -\infty$ for $n = 0, 1, 2, \ldots$.

16. Use the method of the text [see Eq. (112)] to obtain an expression for the second solution $N_n(x)$ of Bessel's equation when n is a positive integer.

7-5   *Hypergeometric equation*

Another equation which is of considerable importance in the theory and applications of second-order equations is the hypergeometric equation

$$x(1 - x)y'' + [C - (A + B + 1)x]y' - ABy = 0 \tag{113}$$

where A, B, and C are any constants.† Many important equations can be written in the form of (113), and a surprising variety of functions are included among the solutions of (113) for special values of A, B, and C.   (See Prob. 1.)

Equation (113) has regular singularities at $x = 0$ and $x = 1$,‡ and hence the method of Frobenius can be applied to obtain solutions near these points.   Near $x = 0$ we assume a solution of the form

$$y = \sum_{n=0}^{\infty} a_n x^{\alpha+n} \tag{114}$$

and, on substituting into (113), we have

$$x(1 - x) \sum_{n=0}^{\infty} (\alpha + n)(\alpha + n - 1)a_n x^{\alpha+n-2} + [C - (A + B + 1)x]$$
$$\times \sum_{n=0}^{\infty} (\alpha + n)a_n x^{\alpha+n-1} - AB \sum_{n=0}^{\infty} a_n x^{\alpha+n} = 0 \tag{115}$$

On collecting terms with like powers of x, we have

$$\sum_{n=0}^{\infty} [(\alpha + n)(\alpha + n - 1) + C(\alpha + n)]a_n x^{\alpha+n-1}$$
$$- \sum_{n=0}^{\infty} [(\alpha + n)(\alpha + n - 1)$$
$$+ (A + B + 1)(\alpha + n) + AB]a_n x^{\alpha+n} = 0 \tag{116}$$

If we then adjust the index in the second sum, we have

$$\sum_{n=0}^{\infty} [(\alpha + n)(\alpha + n - 1) + C(\alpha + n)]a_n x^{\alpha+n-1}$$
$$- \sum_{n=1}^{\infty} [(\alpha + n - 1)(\alpha + n - 2)$$
$$+ (A + B + 1)(\alpha + n - 1) + AB]a_{n-1} x^{\alpha+n-1} \tag{117}$$

† We shall consider only real values of A, B, and C in this book.

‡ There is also a regular singularity at infinity (see Prob. 5), and it is interesting to note that *any* linear second-order equation with three distinct regular singularities can be transformed into the hypergeometric equation.   E.g., see Ince [19].

or

$$[\alpha(\alpha - 1) + C\alpha]a_0 x^{\alpha-1} + \sum_{n=1}^{\infty} \{[(\alpha + n)(\alpha + n - 1) + C(\alpha + n)]a_n$$
$$- [(\alpha + n - 1)(\alpha + n - 2) + (A + B + 1)$$
$$\times (\alpha + n - 1) + AB]a_{n-1}\}x^{\alpha+n-1} = 0 \quad (118)$$

This gives the *indicial equation*

$$\alpha(\alpha - 1) + C\alpha = 0 \quad (119)$$

and the *recurrence formula*

$$(\alpha + n)(\alpha + n + C - 1)a_n - [(\alpha + n - 1)$$
$$\times (\alpha + n + A + B - 1) + AB]a_{n-1} = 0 \quad (120)$$

From the indicial equation we have

$$\alpha = 0 \quad \text{or} \quad \alpha = 1 - C \quad (121)$$

In order to avoid the case where the indicial roots differ by an integer, we shall assume that C is not an integer.

For the case $\alpha = 0$, the recurrence formula becomes

$$n(C + n - 1)a_n - [(n - 1)(n + A + B - 1) + AB]a_{n-1} = 0 \quad (122)$$

and it is easily seen that this is equivalent to the equation

$$a_n = \frac{(A + n - 1)(B + n - 1)}{n(C + n - 1)} a_{n-1} \quad n = 1, 2, 3, \ldots \quad (123)$$

If we set $a_0 = 1$, then the solution that we have obtained is the so-called hypergeometric series, usually designated by $F(A,B;C;x)$. That is,

$$F(A,B;C;x) = 1 + \frac{AB}{C} x + \frac{A(A + 1)B(B + 1)}{2!C(C + 1)} x^2$$
$$+ \frac{A(A + 1)(A + 2)B(B + 1)(B + 2)}{3!C(C + 1)(C + 2)} x^3 + \cdots \quad (124)$$

The solution corresponding to the index $\alpha = 1 - C$ (when C is not an integer) is easily shown to be[†]

$$y_2(x) = x^{1-C}F(A - C + 1, B - C + 1; 2 - C; x) \quad (125)$$

Therefore the general solution of the hypergeometric equation (113) is given by a linear combination of the solutions in (124) and (125). This solution is valid at least in the interval $0 < x < 1$.

[†] See Prob. 6.

Problems 7-5

1. Use the series expression for $F(A,B;C;x)$ in Eq. (124) to prove that:

a. $F(1,B;B;x) = (1 - x)^{-1}$ b. $F(-n,B;B;-x) = (1 + x)^n$

c. $xF(1,1;2;-x) = \ln (1 + x)$ d. $e^x = \lim_{B \to \infty} F\left(1,B;1;\dfrac{x}{B}\right)$

2. Prove that $d\, F(A,B;C;x)/dx = (AB/C)F(A + 1, B + 1; C + 1; x)$
3. Find a solution in terms of the hypergeometric function F:

a. $x(1 - x)y'' + (\tfrac{1}{2} - 6x)y' = 0$
b. $2(x^2 - x)y'' + (7x - 3)y' - 2y = 0$

4. Show that

$$F(A, B + 1; C; x) - F(A,B;C;x) = \frac{Ax}{C} F(A + 1, B + 1; C + 1; x)$$

5. Prove that the hypergeometric equation (113) has a regular singularity at infinity.

6. Derive Eq. (125) in the text. Hint: Change variables in the hypergeometric equation (113) by setting $y = x^{1-C}Y$ and show that the equation in Y has the solution $F(A - C + 1, B - C + 1; 2 - C; x)$.

7-6 Legendre's equation

The differential equation

$$(1 - x^2)y'' - 2xy' + n(n + 1)y = 0 \tag{126}$$

where n is any constant (not necessarily an integer), is known as *Legendre's equation*. This equation and its solutions play an important role in many branches of engineering and mathematical physics.

Equation (126) has regular singularities at $x = \pm 1$ and at infinity (see Example 3, page 202). However, instead of examining the solutions near these singularities, we shall restrict our study to the solutions near the ordinary point $x = 0$. These solutions are the more interesting and are more important in the applications.

Since $x = 0$ is an ordinary point for Eq. (126), the general solution near $x = 0$ can be written in the form of a power series

$$y = \sum_{k=0}^{\infty} a_k x^k \tag{127}$$

On substituting this series into Eq. (126), we have

$$(1 - x^2) \sum_{k=0}^{\infty} k(k-1)a_k x^{k-2} - 2x \sum_{k=0}^{\infty} ka_k x^{k-1} + n(n+1) \sum_{k=0}^{\infty} a_k x^k = 0 \quad (128)$$

or

$$\sum_{k=0}^{\infty} k(k-1)a_k x^{k-2} + \sum_{k=0}^{\infty} [n(n+1) - k(k+1)]a_k x^k = 0 \quad (129)$$

and finally, after adjusting the index in the first sum,

$$\sum_{k=0}^{\infty} [(k+2)(k+1)a_{k+2} + (n^2 + n - k^2 - k)a_k]x^k = 0 \quad (130)$$

This equation yields a recurrence formula which can be written in the form

$$a_{k+2} = -\frac{(n-k)(n+k+1)}{(k+2)(k+1)} a_k \quad (131)$$

Thus the series (127) is the general solution to Legendre's equation within its interval of convergence $|x| < 1$ if the coefficients a_k satisfy (131). The coefficients a_0 and a_1 are of course arbitrary constants.

Legendre polynomials If n is zero or a positive integer, then one of the solutions of Legendre's equation (126) reduces to a polynomial. This is a result of the recurrence formula (131), where it is clear that for $k = n$ we have $a_{n+2} = 0$ and hence $a_{n+4} = a_{n+6} = \cdots = 0$. Also, if n is even, then by taking $a_1 = 0$, we have $a_3 = a_5 = \cdots = 0$, and the only terms remaining are

$$y = \sum_{k=0}^{n/2} a_{2k} x^{2k} \quad (132)$$

If n is odd, set $a_0 = 0$ and hence $a_2 = a_4 = \cdots = 0$, and the solution is

$$y = \sum_{k=0}^{(n-1)/2} a_{2k+1} x^{2k+1} \quad (133)$$

It is convenient to express these polynomials in *descending* powers of x since we may then represent them by a single formula which is valid for n even or odd.

In Eq. (131) set $k = n - 2$ and solve for a_{n-2} to get

$$a_{n-2} = -\frac{n(n-1)}{2(2n-1)} a_n$$

Similarly

$$a_{n-4} = -\frac{(n-2)(n-3)}{4(2n-3)} a_{n-2}$$

$$= \frac{n(n-1)(n-2)(n-3)}{2 \cdot 4(2n-1)(2n-3)} a_n \quad (134)$$

and so on. Hence, when n is zero or a positive integer, a particular solution of Legendre's equation is the polynomial

$$y = a_n\left[x^n - \frac{n(n-1)}{2(2n-1)}x^{n-2} + \frac{n(n-1)(n-2)(n-3)}{2\cdot 4(2n-1)(2n-3)}x^{n-4} - \cdots\right]$$

$$(135)$$

Note that the expression (135) terminates when the exponent of x reaches either 0 or 1, since the next term (and all succeeding terms) would contain the factor $n - n$ in the numerator of the coefficient and hence vanish.

The coefficient a_n in (135) is arbitrary, but if it is chosen in a special way we obtain the traditional form of the *Legendre polynomials*. Let

$$\begin{cases} a_0 = 1 \\ a_n = \dfrac{(2n)!}{2^n(n!)^2} = \dfrac{(2n-1)(2n-3)\cdots 1}{n!} \end{cases} \quad \text{if } n = 1, 2, \ldots \quad (136)$$

and then the functions in (135) become the *Legendre polynomials*

$$P_0(x) = 1$$
$$P_n(x) = \frac{(2n)!}{2^n(n!)^2}\left[x^n - \frac{n(n-1)}{2(2n-1)}x^{n-2}\right.$$
$$\left. + \frac{n(n-1)(n-2)(n-3)}{2\cdot 4(2n-1)(2n-3)}x^{n-4} - \cdots\right]$$
$$n = 1, 2, \ldots \quad (137)$$

Explicitly, the first few polynomials are

$$P_0(x) = 1$$
$$P_1(x) = x$$
$$P_2(x) = \tfrac{1}{2}(3x^2 - 1)$$
$$P_3(x) = \tfrac{1}{2}(5x^3 - 3x) \qquad\qquad\qquad (138)$$
$$P_4(x) = \tfrac{1}{8}(35x^4 - 30x^2 + 3)$$
$$P_5(x) = \tfrac{1}{8}(63x^5 - 70x^3 + 15x)$$

Rodrigues's formula   The Legendre polynomials can be expressed in the surprisingly compact form

$$P_n(x) = \frac{1}{2^n n!}\frac{d^n}{dx^n}(x^2 - 1)^n \qquad\qquad (139)$$

This is *Rodrigues's formula*, which can be obtained as follows.

By the binomial theorem we have

$$(x^2 - 1)^n = x^{2n} - nx^{2n-2} + \frac{n(n-1)}{2}x^{2n-4} - \cdots \qquad (140)$$

Differentiating, we have

$$\frac{d}{dx}(x^2 - 1)^n = 2nx^{2n-1} - n(2n - 2)x^{2n-3}$$

$$+ \frac{n(n - 1)(2n - 4)}{2!}x^{2n-5} - \cdots \quad (141)$$

and it is easy to show by induction that, after differentiating n times, we have

$$\frac{d^n}{dx^n}(x^2 - 1)^n = 2n(2n - 1) \cdots (n - 1)x^n - n(2n - 2)(2n - 3) \cdots$$

$$\times n(n - 1)x^{n-2} + \frac{n(n - 1)(2n - 4)(2n - 5)}{2!} \cdots$$

$$\times (n - 3)x^{n-4} - \cdots \quad (142)$$

The coefficient of x^n can be written in the form $(2n)!/n!$, and if this expression is factored out of the right-hand side of Eq. (142), we have

$$\frac{d^n}{dx^n}(x^2 - 1)^n = \frac{(2n)!}{n!}\left[x^n - \frac{n(n - 1)}{2(2n - 1)}x^{n-2}\right.$$

$$\left. + \frac{n(n - 1)(n - 2)(n - 3)}{2 \cdot 4(2n - 1)(2n - 3)}x^{n-4} - \cdots\right] \quad (143)$$

Finally, if we multiply both sides of (143) by $1/2^n n!$, the right side becomes $P_n(x)$ [see Eq. (137)] and Rodrigues's formula is established.

The generating function for Legendre polynomials   One of the most important applications of Legendre polynomials is in the field of potential theory, where the following formula plays a prominent role:

$$(1 - 2xr + r^2)^{-\frac{1}{2}} = \sum_{n=0}^{\infty} P_n(x)r^n \quad (144)$$

where $P_n(x)$ is the Legendre polynomial of degree n. The function $(1 - 2xr + r^2)^{-\frac{1}{2}}$ is called the *generating function* for the Legendre polynomials. It can be given the following geometrical interpretation (see Fig. 3). The distance R between A and B is given by the law of cosines

$$R = (1 + r^2 - 2r \cos \theta)^{\frac{1}{2}}$$

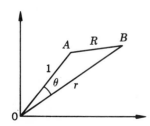

figure 3

and hence, if we set

$$x = \cos \theta \qquad 0 \leq \theta \leq \pi$$

then

$$\frac{1}{R} = (1 - 2xr + r^2)^{-\frac{1}{2}}$$

This quantity $1/R$ is called the "potential at B due to a unit charge at A."[†]
  The derivation of formula (144) is now indicated.[‡]   From the binomial theorem we have

$$(1 - 2xr + r^2)^{-\frac{1}{2}} = 1 - \tfrac{1}{2}(-2xr + r^2) + \frac{1 \cdot 3}{2^2 2!}(-2xr + r^2)^2$$

$$- \frac{1 \cdot 3 \cdot 5}{2^3 3!}(-2xr + r^2)^3 + \cdots$$

and, in ascending powers of r, this is

$$(1 - 2xr + r^2)^{-\frac{1}{2}} = 1 + xr + \left(-\tfrac{1}{2} + \frac{1 \cdot 3}{2!} x^2\right) r^2$$

$$+ \left[\frac{1 \cdot 3}{2^2 2!}(-4x) - \frac{1 \cdot 3 \cdot 5}{2^3 3!}(-2^3 x^3)\right] r^3 + \cdots$$

$$= 1 + xr + \tfrac{1}{2}(3x^2 - 1)r^2 + \tfrac{1}{2}(5x^3 - 3x)r^3 + \cdots$$

$$= P_0(x) + P_1(x)r + P_2(x)r^2 + P_3(x)r^3 + \ldots$$

This expansion is valid for $|2xr| + r^2 < 1$.
  If we set $x = \cos \theta$ (when $|x| < 1$) in (144), we have

$$(1 - 2r \cos \theta + r^2)^{-\frac{1}{2}} = \sum_{n=0}^{\infty} P_n (\cos \theta) r^n \tag{145}$$

When this formula is used in applications, it is often more convenient to express the functions $P_n (\cos \theta)$ in terms of $\cos k\theta$. By substituting $x = \cos \theta$ in Eqs. (138) and applying some simple trigonometric identities, we have

$$P_0 (\cos \theta) = 1$$
$$P_1 (\cos \theta) = \cos \theta$$
$$P_2 (\cos \theta) = \tfrac{1}{4}(3 \cos 2\theta + 1)$$
$$P_3 (\cos \theta) = \tfrac{1}{8}(5 \cos 3\theta + 3 \cos \theta) \tag{146}$$
$$P_4 (\cos \theta) = \tfrac{1}{64}(35 \cos 4\theta + 20 \cos 2\theta + 9)$$
$$P_5 (\cos \theta) = \tfrac{1}{128}(63 \cos 5\theta + 35 \cos 3\theta + 30 \cos \theta)$$

[†] See Sec. 15-2.
[‡] For a rigorous derivation see Ince [19], sec. 8.32.

The properties and applications of the Legendre polynomials are discussed further in Chaps. 13 and 15.

Problems 7-6

1. Find $P_6(x)$ by substituting in the general formula (137).

2. Substitute $x = 1$ in the expansion formula (144) and then, by identifying the coefficients in the series expansion of $(1 - r)^{-1}$, show that $P_n(1) = 1$, $n = 0, 1, 2, \ldots$.

3. Show that formula (137) for $P_n(x)$ can be written in the form

$$P_n(x) = \sum_{k=0}^{M} \frac{(-1)^k (2n - 2k)!}{2^n k! (n - k)! (n - 2k)!} x^{n-2k}$$

where $M = \frac{1}{2} n$ if n is even and $M = \frac{1}{2}(n - 1)$ if n is odd.

4. Prove $\int_{-1}^{1} P_n(x)\, dx = 0$ for $n = 1, 2, \ldots$. Use Rodrigues's formula (139).

5. Change the independent variable in Legendre's equation (126) by setting $x = 1 - 2t$. Show that the resulting equation is a hypergeometric equation with $A = n + 1$, $B = -n$, and $C = 1$, and hence show that

$$P_n(x) = F(n + 1, -n; 1; \tfrac{1}{2} - \tfrac{1}{2}x).$$

6. Use the expansion of the generating function, Eq. (145), to prove that

$$\frac{1 - r^2}{(1 - 2r \cos \theta + r^2)^{\frac{3}{2}}} = \sum_{n=0}^{\infty} (2n + 1) P_n (\cos \theta) r^n$$

Hint: Differentiate Eq. (145) with respect to r.

7. Use the result of Prob. 6 to prove that

$$P'_{n+1}(x) - P'_{n-1}(x) = (2n + 1) P_n(x) \qquad n = 1, 2, \ldots$$

8. Use Prob. 7 to prove that

$$\int_{x}^{1} P_n(s)\, ds = \frac{1}{2n + 1} [P_{n-1}(x) - P_{n+1}(x)] \qquad n = 1, 2, \ldots$$

9. Substitute $x = 0$ in the expansion formula (144) and then, by identifying the coefficients in the series expansion of $(1 + r^2)^{-\frac{1}{2}}$, show that

$$P_{2n}(0) = (-1)^n \frac{1 \cdot 3 \cdot 5 \cdots (2n - 1)}{2 \cdot 4 \cdot 6 \cdots (2n)}$$

$$P_{2n-1}(0) = 0 \qquad n = 1, 2, \ldots$$

Systems of Linear Differential Equations

8

8-1 Introduction

Many of the physical problems which lead to differential equations involve more than one unknown function. For example, in studying the motion of a particle in the xy plane, it is necessary to determine the two coordinates x and y of the particle as functions of time. For the motion of a point in three-dimensional space, three coordinates are needed. For the general motion of a rigid body in space, *six* coordinates are needed, three for the location of the center of mass and three more for the orientation of the body. In general, in a physically determined situation we have one differential equation for each unknown (or degree of freedom) in the system.

In this chapter we shall consider some of the theory and methods of solution for systems of differential equations. In Sec. 8-2 we illustrate these methods with some simple examples. A physical application is discussed in Sec. 8-3.

In Sec. 8-4 the notion of a matrix is introduced and some of the fundamental properties of vectors and matrices are established. The remainder of the chapter is devoted to the application of these concepts to the theory of linear systems and some general methods of solution.

8-2 Some illustrative examples

In this section we examine a few systems of equations in order to discover some of the basic properties of systems and their solutions. The theory underlying most of the intuitive discussions is developed carefully in Secs. 8-5 and 8-6.

Example 1 Find a pair of functions $x = x(t)$ and $y = y(t)$ so that the following pair of equations will be satisfied for all t in some interval I.

$$\frac{dx}{dt} = x - y$$
$$\frac{dy}{dt} = -2x \tag{1}$$

Since we have a pair of linear equations with constant coefficients, it is natural to attempt to find a solution pair, or, briefly, a *solution*, of the form

$$x = Ae^{\lambda t} \qquad y = Be^{\lambda t} \tag{2}$$

where the exponential $e^{\lambda t}$ is chosen to be the same in both x and y so that a common factor can be removed from the equations when the trial solutions are substituted.† On substituting (2) into (1) and canceling the nonzero factor $e^{\lambda t}$, we have

$$(1 - \lambda)A - B = 0 \tag{3a}$$

$$-2A - \lambda B = 0 \tag{3b}$$

Equations (3) are a pair of simultaneous equations which must be satisfied by A and B. One possibility, of course, is $A = B = 0$, which leads to the trivial solution $x \equiv y \equiv 0$ for the system (1). A nontrivial solution of Eqs. (3) will exist if and only if the determinant of the coefficients vanishes. That is,

$$\begin{vmatrix} 1 - \lambda & -1 \\ -2 & -\lambda \end{vmatrix} = 0 \tag{4}$$

Expanding the determinant, we obtain the *characteristic equation*

$$\lambda^2 - \lambda - 2 = 0 \tag{5}$$

with solutions

$$\lambda = 2 \qquad \text{or} \qquad \lambda = -1 \tag{6}$$

Now, if $\lambda = 2$, Eq. (3b) yields $B = -A$, whereas if $\lambda = -1$, we have $B = 2A$. Hence we have obtained a pair of solutions

$$x_1 = A_1 e^{2t} \qquad y_1 = -A_1 e^{2t} \tag{7a}$$
and
$$x_2 = A_2 e^{-t} \qquad y_2 = 2A_2 e^{-t} \tag{7b}$$

where A_1 and A_2 are arbitrary constants. The sum of the two solutions (7a) and (7b),

$$x = A_1 e^{2t} + A_2 e^{-t} \qquad y = -A_1 e^{2t} + 2A_2 e^{-t} \tag{8}$$

is also a solution of Eq. (1) as is easily verified by direct substitution. The solution given by (8) is, in fact, the general solution as will be shown in Sec. 8-5.

† A similar technique is used in finding the *normal modes* of a vibrating system. See Sec. 8-3.

Example 2   Solve:

$$\dot{x} = -x + 2y \tag{9a}$$

$$\dot{y} = -2x - y \tag{9b}$$

This problem, which could be solved by the method used in Example 1, can also be solved by a somewhat different approach.

Solving for y in (9a), we have

$$y = \tfrac{1}{2}(\dot{x} + x) \tag{10}$$

and hence, by differentiating,

$$\dot{y} = \tfrac{1}{2}(\ddot{x} + \dot{x}) \tag{11}$$

If (10) and (11) are now substituted into Eq. (9b), we have

$$\tfrac{1}{2}(\ddot{x} + \dot{x}) = -2x - \tfrac{1}{2}(\dot{x} + x) \tag{12}$$

or

$$\ddot{x} + 2\dot{x} + 5x = 0 \tag{13}$$

Thus the pair of first-order equations (9a) and (9b) has been reduced to a single second-order equation for x. The general solution (obtained by the methods of Chap. 5) is

$$x = e^{-t}(c_1 \cos 2t + c_2 \sin 2t) \tag{14}$$

The value of y is most easily found by differentiating (14) and substituting into (10).   We obtain

$$y = e^{-t}(-c_1 \sin 2t + c_2 \cos 2t) \tag{15}$$

Two features of this example require further comment.   In the first place the differentiation which produced Eq. (11) needs justification. One often introduces extraneous solutions by this procedure.† However, in this example we can verify by direct substitution that Eqs. (14) and (15) satisfy the system (9).

Finally in this example, after finding x, we could have repeated the same procedure to find y.   On eliminating x and $\dot{x}$ from (9), we would have

$$\ddot{y} + 2\dot{y} + 5y = 0 \tag{16}$$

and

$$y = e^{-t}(c_3 \cos 2t + c_4 \sin 2t) \tag{17}$$

This method seems to introduce two more arbitrary constants, but actually c_3 and c_4 depend on c_1 and c_2. This is seen by substituting (14) and (17) back into the system (9).   We would find that y in (17) is a solution only if $c_3 = c_2$ and $c_3 = -c_1$, thus reproducing Eq. (15).

Example 3   Solve:

$$\dot{x} = x - y \tag{18a}$$

$$\dot{y} = x + 3y \tag{18b}$$

† Compare the solutions of the equations $\dot{x} = 0$ and $\ddot{x} = 0$.

Substituting $x = Ae^{\lambda t}$, $y = Be^{\lambda t}$, we have

$$(1 - \lambda)A - B = 0$$
$$A + (3 - \lambda)B = 0$$

with nontrivial solutions only if

$$\lambda^2 - 4\lambda + 4 = 0$$

Since this characteristic equation has a double root at $\lambda = 2$, we obtain only one solution by this method. That is

$$x = Ae^{2t} \qquad y = -Ae^{2t} \tag{19}$$

A second solution can be obtained by the method of *variation of parameters*. That is, we try to find functions $u(t)$ and $v(t)$ so that the following is a solution of the system (18):

$$x = u(t)e^{2t} \qquad y = v(t)e^{2t} \tag{20}$$

If these expressions are substituted into Eqs. (18), we obtain a pair of equations for u and v

$$\dot{u} = -u - v \tag{21a}$$

$$\dot{v} = u + v \tag{21b}$$

Solving algebraically for v in the first of these equations, we have $v = -\dot{u} - u$, and on differentiating this result,† we have $\dot{v} = -\ddot{u} - \dot{u}$. Substituting these expressions for v and $\dot{v}$ into Eq. (21b), we obtain

$$\ddot{u} = 0$$

and hence

$$u = Bt + C \tag{22}$$

Differentiating this result and substituting back into Eq. (21a), we get

$$v = -Bt - B - C \tag{23}$$

Therefore the general solution of the original pair of Eqs. (18) is

$$x = (Bt + C)e^{2t} \qquad y = (-Bt - B - C)e^{2t} \tag{24}$$

where B and C are arbitrary constants.

It is instructive to note that if we set $B = 0$ in Eq. (24), we recapture the first solution (20). On the other hand, if we set $C = 0$ in Eq. (24), then we obtain the second solution

$$x = Bte^{2t} \qquad y = (-Bt - B)e^{2t} \tag{25}$$

Hence it was indeed necessary to use two different functions u and v in Eq. (20). Although one might have been tempted to infer from Eq. (19) that $v = -u$, we see from Eq. (25) that this is not the case.

† See Example 2.

One final comment on this example is in order. The fact that $u(t)$ and $v(t)$ are linear functions in Eqs. (22) and (23) is not accidental. In fact, for a pair of equations of the form $\dot{x} = ax + by$, $\dot{y} = cx + dy$, it is easy to show (see Prob. 7) that whenever the characteristic equation has a double root $\lambda = \lambda_1$, then the general solution will be of the *form* $x = (At + B)e^{\lambda_1 t}$, $y = (Ct + D)e^{\lambda_1 t}$. The relation between A, B, C, and D can be obtained by substituting x and y into the differential equations.

Example 4

$$2\dot{x} - 2\dot{y} = y \tag{26a}$$

$$-\dot{x} + \dot{y} = -2x \tag{26b}$$

This pair of equations differs from the pairs previously considered in that derivatives of both x and y occur in the same equation. If the second equation is multiplied by 2 and added to the first, we get

$$y = 4x \tag{27}$$

Substituting this result into (26b) yields

$$3\dot{x} + 2x = 0 \tag{28}$$

so that

$$x = ce^{-\frac{2}{3}t}$$

and by (27)

$$y = 4ce^{-\frac{2}{3}t} \tag{29}$$

Note that the solution involves only *one* arbitrary constant even though we had a pair of equations in the two unknowns x and y. This fact is closely associated with the fact that if we substitute $x = Ae^{\lambda t}$ and $y = Be^{\lambda t}$ into (26), the characteristic equation is of *first degree*, namely, $3\lambda + 2 = 0$. This yields the solution (29) but also indicates that there is no second linearly independent solution in this degenerate case.†

Problems 8-2

1. Find the general solution for each of the following systems of equations:

a. $\dot{x} = 4x + y \qquad \dot{y} = 3x + 2y$
b. $\dot{x} = 4x + 2y \qquad \dot{y} = 2x + y$
c. $\dot{x} = 5x - 2y \qquad \dot{y} = 2x + y$
d. $\dot{x} = x - 3y \qquad \dot{y} = 3x + y$
e. $\dot{x} + \dot{y} = x + y \qquad \dot{y} = x - y$
f. $3\dot{x} - 2\dot{y} = x - y \qquad -6\dot{x} + 4\dot{y} = x - y$
g. $\dot{x} = z \qquad \dot{y} = x \qquad \dot{z} = x - 2y + 2z$
h. $\dot{x} = 3x \qquad \dot{y} = 2y \qquad \dot{z} = -z$

† E.g., see Ince [19] for a more complete discussion of degenerate cases.

2. Show that the method of variation of parameters fails to produce a second solution in Example 4 of the text.

3. Using the result of Example 1 in the text, find the general solution of the nonhomogeneous system

$$\frac{dx}{dt} = x - y + 1$$

$$\frac{dy}{dt} = -2x + 4$$

Hint: Guess a "particular solution" and extend the method of Sec. 5-6.

4. Solve: $\dot{x} = Ax + By$ A, B, C, D const

$$\dot{y} = Cx + Dy$$

Consider all the different cases that arise depending on the sign of the discriminant $(A - D)^2 + 4BC$. (See Sec. 10-3 for a thorough discussion of this problem.)

5. Solve the following initial-value problems by first finding the general solution and then evaluating the arbitrary constants. (Compare with Probs. 1a and c.)

a. DE: $\dot{x} = 4x + y$ $\dot{y} = 3x + 2y$
   IC: $x(0) = -1$ $y(0) = 7$

b. DE: $\dot{x} = 5x - 2y$ $\dot{y} = 2x + y$
   IC: $x(0) = 1$ $y(0) = 0$

6. a. Solve the system $\ddot{x} = y$, $\dot{y} = x$ by assuming a solution of the form $x = Ae^{\lambda t}$, $y = Be^{\lambda t}$.

b. Solve the system in part a by substituting $y = \ddot{x}$ into the second equation and solving the resulting third-order equation for x.

7. Given the system $\dot{x} = ax + by$, $\dot{y} = cx + dy$, assume that the characteristic equation has a double root $\lambda = \lambda_1$ and, by substituting $x = u(t)e^{\lambda_1 t}$ and $y = v(t)e^{\lambda_1 t}$, show that $\ddot{u} = \ddot{v} = 0$. Hence show that $u(t)$ and $v(t)$ are linear functions. Find the general solution of the system.

8-3   *A two-degree-of-freedom vibration*

As an example of a physical problem which leads to a system of differential equations, consider a mechanical system consisting of two masses connected by three springs as shown in Fig. 1. The springs have spring constants of k, k', and k and the masses are equal. We assume that the motion of the masses takes place in a straight line on a smooth frictionless plane. The only forces on the masses are the spring forces. Let x_1 and x_2 denote the displacement of the masses from their equilibrium positions. A free-body diagram

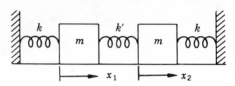

figure 1

of the masses is shown in Fig. 2.   From Newton's second law we obtain

$$m\ddot{x}_1 = -kx_1 + k'(x_2 - x_1)$$
$$m\ddot{x}_2 = -k'(x_2 - x_1) - kx_2$$

(30)

or

$$m\ddot{x}_1 + (k + k')x_1 - k'x_2 = 0$$
$$m\ddot{x}_2 - k'x_1 + (k + k')x_2 = 0$$

(31)

These are two simultaneous equations for the displacements x_1 and x_2. These equations could be readily solved by one of the methods of the preceding section.   However, it is instructive to proceed in a different manner, motivated by physical reasoning.

Because of the absence of damping and friction forces, we expect the motion to be combinations of simple harmonic motions.   We look for the simplest such motions, the so-called *normal modes of vibration*.   These are motions where both masses move with the same frequency, called the *natural frequency*, with possibly different amplitudes, and both masses go through their equilibrium positions together.   Therefore we look for a solution of the form

$$x_1 = a_1 \sin(\omega t + \epsilon)$$
$$x_2 = a_2 \sin(\omega t + \epsilon) \qquad \omega > 0$$

(32)

Substituting these into the differential equations, we obtain

$$(-m\omega^2 + k + k')a_1 - k'a_2 = 0$$
$$-k'a_1 + (-m\omega^2 + k + k')a_2 = 0$$

(33)

Equations (33) are homogeneous linear algebraic equations for a_1 and a_2, and in order for (33) to have a nontrivial solution, the determinant of the coefficients must equal zero:

$$\begin{vmatrix} (-m\omega^2 + k + k') & -k' \\ -k' & (-m\omega^2 + k + k') \end{vmatrix} = 0$$

(34)

figure 2

This equation simplifies to

$$(-m\omega^2 + k + k')^2 - k'^2 = 0 \tag{35}$$

Solving for ω, we find the two values

$$\omega_1 = \sqrt{\frac{k}{m}} \quad \text{and} \quad \omega_2 = \sqrt{\frac{k + 2k'}{m}} \tag{36}$$

For these values Eqs. (33) have nontrivial solutions for a_1 and a_2. Substituting $\omega = \omega_1$ into Eqs. (33), we find that

$$a_1 = a_2$$

Therefore the differential equations have the solutions

$$\begin{aligned} x_1 &= a_1 \sin(\omega_1 t + \epsilon_1) \\ x_2 &= a_1 \sin(\omega_1 t + \epsilon_1) \end{aligned} \tag{37}$$

where a_1 and ϵ_1 are arbitrary constants. This is the *first normal mode.* We see that $x_1 = x_2$, so that both masses move together with natural frequency ω_1. In this motion the middle spring remains unstretched.

Substituting $\omega = \omega_2$ into Eqs. (33), we obtain

$$a_1 = -a_2$$

This yields the solutions

$$\begin{aligned} x_1 &= a_2 \sin(\omega_2 t + \epsilon_2) \\ x_2 &= -a_2 \sin(\omega_2 t + \epsilon_2) \end{aligned} \tag{38}$$

where a_2 and ϵ_2 are arbitrary constants. This is the *second normal mode.* We see that $x_1 = -x_2$, so that the masses move oppositely.

The differential equations (31) are linear and homogeneous. Therefore, as we shall see in Sec. 8-5, the two solutions (37) and (38) may be superposed to yield the solution

$$\begin{aligned} x_1 &= a_1 \sin(\omega_1 t + \epsilon_1) + a_2 \sin(\omega_2 t + \epsilon_2) \\ x_2 &= a_1 \sin(\omega_1 t + \epsilon_1) - a_2 \sin(\omega_2 t + \epsilon_2) \end{aligned} \tag{39}$$

This, in fact, is the general solution of the system of differential equations.† By expanding the sine terms in (39) and renaming the constants, the solution can be written in the form

$$\begin{aligned} x_1 &= A_1 \sin \omega_1 t + B_1 \cos \omega_1 t + A_2 \sin \omega_2 t + B_2 \cos \omega_2 t \\ x_2 &= A_1 \sin \omega_1 t + B_1 \cos \omega_1 t - A_2 \sin \omega_2 t - B_2 \cos \omega_2 t \end{aligned} \tag{40}$$

† This will follow from Theorem 6, Sec. 8-5.

The solution contains four arbitrary constants. We can therefore expect to satisfy four initial conditions:

$$x_1(0) = s_1 \qquad x_2(0) = s_2$$
$$\dot{x}_1(0) = v_1 \qquad \dot{x}_2(0) = v_2 \tag{41}$$

It can easily be shown that these conditions uniquely determine the four constants in Eq. (39) or (40).

We have seen that the general motion of this system with two degrees of freedom consists of a superposition of the two normal modes. A similar statement holds for a system with n degree of freedom; that is, there are exactly n normal modes, with their corresponding natural frequencies. The general motion of the system can be obtained by superposing the n normal modes. (See Chap. 11.) In Chap. 14 we will see that a system with an infinite number of degrees of freedom (e.g., a vibrating string) has similar properties.

Problems 8-3

1. Determine the arbitrary constants A_1, B_1, A_2, B_2 in Eq. (40) for the following initial conditions:

$$x_1(0) = 0 \qquad x_2(0) = 1$$
$$\dot{x}_1(0) = 0 \qquad \dot{x}_2(0) = 0$$

2. Assume that $k' \ll k$, i.e., k' is much smaller than k. Then $\omega_2 = \omega_1 + \delta$ where δ is small compared to ω_1. Show that in this case the solution to Prob. 1 can be written approximately as:

$$x_1 = \sin \omega_1 t \sin \delta t$$
$$x_2 = \cos \omega_1 t \cos \delta t$$

Draw a sketch of x_1 and x_2. This illustrates the phenomenon of beats. See Sec. 6-2.

3. Derive the solutions (40) by one of the methods of Sec. 8-2.

4. Two beads of equal masses are equally spaced on a tightly stretched horizontal string. Assuming small vertical motions, derive the differential equations of motion. Find the natural frequencies and normal modes. Illustrate the normal modes geometrically.

5. Generalize Prob. 4 to three beads.

6. Two identical pendulums are suspended from a ceiling. The two pendulums are coupled with a spring connecting the midpoints of the pendulums. Assuming small motions in a plane, derive the differential equations of motion. Find the natural frequencies and normal modes.

8-4   *Vectors and matrices*†

In this section we shall introduce some elementary concepts from linear algebra since they will be of great assistance in the study of systems of differential equations. First we shall examine these concepts as they apply to the simplest case of a system of two equations in two unknowns, and then it will be fairly easy to extend the notions to the general case of n equations in n unknowns.

Consider the following pair of homogeneous linear differential equations:

$$\frac{dx}{dt} = a(t)x + b(t)y \tag{42a}$$

$$\frac{dy}{dt} = c(t)x + d(t)y \tag{42b}$$

Let $a(t)$, $b(t)$, $c(t)$, and $d(t)$ be given functions of t, continuous on some interval J, $t_0 \leq t \leq t_1$. A *solution* of the system (42) is defined to be a pair of continuously differentiable functions $x(t)$ and $y(t)$ which satisfy (42a) and (42b) simultaneously for all t in J.

Since we are dealing with a *pair* of functions $x(t)$ and $y(t)$, it is convenient to treat these quantities as *components* of a *vector*. We recall that we can define a vector v to be an ordered pair (x,y) of real numbers. For our purposes it is convenient to write the components of the vector in a column,

$$\mathbf{v} = \begin{bmatrix} x \\ y \end{bmatrix} \tag{43}$$

and call $\mathbf{v}$ a *column vector*. If x and y are functions of t, then so is $\mathbf{v}$ and we have

$$\mathbf{v}(t) = \begin{bmatrix} x(t) \\ y(t) \end{bmatrix} \tag{44}$$

In the general case where we have a system of equations which involves n unknown functions $x_1(t)$, $x_2(t)$, $\ldots$, $x_n(t)$ where n is any positive integer, we introduce the concept of an n-dimensional column vector as an ordered n-tuple of components $x_i(t)$:

$$\mathbf{x}(t) = \begin{bmatrix} x_1(t) \\ x_2(t) \\ \cdot \\ \cdot \\ \cdot \\ x_n(t) \end{bmatrix}$$

† See Sec. 11-6 for a further treatment of vectors.

In order to develop a useful algebra of vectors, we need the following definitions:

Definition 1   *The zero vector* **0** *is the vector with all components equal to zero, i.e.,*

$$\mathbf{0} = \begin{bmatrix} 0 \\ 0 \\ \cdot \\ \cdot \\ \cdot \\ 0 \end{bmatrix}$$

Definition 2   *If k is any real number and* **x** *is a vector, then we define*

$$k\mathbf{x} = k\begin{bmatrix} x_1 \\ x_2 \\ \cdot \\ \cdot \\ \cdot \\ x_n \end{bmatrix} = \begin{bmatrix} kx_1 \\ kx_2 \\ \cdot \\ \cdot \\ \cdot \\ kx_n \end{bmatrix}$$

The vector $k\mathbf{x}$ is called a scalar multiple of **x**.   *Also, we define* $-\mathbf{x} = (-1)\mathbf{x}$.

Definition 3   *If* **x** *and* **y** *are any two vectors, then we define the sum of* **x** *and* **y** *by*

$$\mathbf{x} + \mathbf{y} = \begin{bmatrix} x_1 \\ x_2 \\ \cdot \\ \cdot \\ \cdot \\ x_n \end{bmatrix} + \begin{bmatrix} y_1 \\ y_2 \\ \cdot \\ \cdot \\ \cdot \\ y_n \end{bmatrix} = \begin{bmatrix} x_1 + y_1 \\ x_2 + y_2 \\ \cdot \\ \cdot \\ \cdot \\ x_n + y_n \end{bmatrix}$$

Definition 4 $\mathbf{x} = \mathbf{y}$ *if and only if* $x_1 = y_1,\ x_2 = y_2,\ \ldots,\ x_n = y_n$.

It follows from these definitions that the following algebraic relations hold for all vectors **x**, **y**, and **z**, and for all scalars a and b:

$$\mathbf{x} + \mathbf{y} = \mathbf{y} + \mathbf{x} \qquad a(b\mathbf{x}) = (ab)\mathbf{x}$$

$$\mathbf{x} + (\mathbf{y} + \mathbf{z}) = (\mathbf{x} + \mathbf{y}) + \mathbf{z} \qquad a(\mathbf{x} + \mathbf{y}) = a\mathbf{x} + b\mathbf{y}$$

$$\mathbf{x} + \mathbf{0} = \mathbf{x} \qquad (a + b)\mathbf{x} = a\mathbf{x} + b\mathbf{x}$$

$$\mathbf{x} + (-\mathbf{x}) = \mathbf{0}$$

In the theory of the linear equation of order n in Chap. 5 the concept of linear independence played an important role. The same is true for systems.

Definition 5 *A set of m vectors $\mathbf{x}_1(t)$, $\mathbf{x}_2(t)$, ..., $\mathbf{x}_m(t)$ is said to be linearly dependent (LD) on an interval J, $t_0 \leq t \leq t_1$, if and only if there exists a set of m constants $k_1, \ldots, k_m$, not all zero, such that*

$$k_1\mathbf{x}_1(t) + k_2\mathbf{x}_2(t) + \cdots + k_m\mathbf{x}_m(t) \equiv \mathbf{0}$$

for all t in J.

Note that if a set of vectors is linearly dependent, then at least one of them can be expressed as a linear combination of the others. (See Prob. 1.)

Example 1 The vectors

$$\mathbf{x}_1 = \begin{bmatrix} 3t^2 - 3 \\ 3 + 6t \end{bmatrix} \quad \text{and} \quad \mathbf{x}_2 = \begin{bmatrix} 2 - 2t^2 \\ -2 - 4t \end{bmatrix}$$

are linearly dependent for all t since $2\mathbf{x}_1 + 3\mathbf{x}_2 \equiv \mathbf{0}$.

Example 2 The vectors

$$\mathbf{x}_1 = \begin{bmatrix} t \\ 0 \\ 0 \\ \cdot \\ \cdot \\ \cdot \\ 0 \end{bmatrix}, \mathbf{x}_2 = \begin{bmatrix} 0 \\ t \\ 0 \\ \cdot \\ \cdot \\ \cdot \\ 0 \end{bmatrix}, \ldots, \mathbf{x}_n = \begin{bmatrix} 0 \\ 0 \\ 0 \\ \cdot \\ \cdot \\ \cdot \\ t \end{bmatrix}$$

are not linearly dependent on *any* interval since the assumption that $k_1\mathbf{x}_1 + k_2\mathbf{x}_2 + \cdots + k_n\mathbf{x}_n = \mathbf{0}$ implies $k_1 = k_2 = \cdots = k_n = 0$. (Why?)

Definition 6 *The set of vectors $\mathbf{x}_1(t), \ldots, \mathbf{x}_m(t)$ defined on some interval J is said to be linearly independent (LI) on J if and only if it is not linearly dependent on J.*

In other words, $\mathbf{x}_1(t), \ldots, \mathbf{x}_m(t)$ are *linearly independent* if and only if the assumption that $k_1\mathbf{x}_1 + \cdots + k_m\mathbf{x}_m = \mathbf{0}$ on J implies that $k_1 = k_2 = \cdots = k_m = 0$. Note that the vectors of Example 2 are *linearly independent* on any interval J.

In the special case where the vectors $\mathbf{x}_1(t), \ldots, \mathbf{x}_m(t)$ are solution vectors of a system of linear differential equations, it is possible to give a simple and useful test for linear independence. This test is stated in Theorem 1, Sec. 8-5.

One further property of vectors is needed in this chapter.

Definition 7   *Given a vector*

$$\mathbf{x}(t) = \begin{bmatrix} x_1(t) \\ \cdot \\ \cdot \\ \cdot \\ x_n(t) \end{bmatrix}$$

where the components $x_1(t), \ldots, x_n(t)$ are differentiable functions of t on some interval J, the derivative $d\mathbf{x}(t)/dt$ is defined by

$$\frac{d\mathbf{x}(t)}{dt} = \begin{bmatrix} \dfrac{dx_1(t)}{dt} \\ \cdot \\ \cdot \\ \cdot \\ \dfrac{dx_n(t)}{dt} \end{bmatrix}$$

Note that differentiation of vectors is a linear operation, i.e.,

$$\frac{d}{dt}[c_1\mathbf{x}_1(t) + c_2\mathbf{x}_2(t)] = c_1\frac{d\mathbf{x}_1}{dt} + c_2\frac{d\mathbf{x}_2}{dt}$$

for any constants c_1 and c_2.   (See Prob. 3.)

Another algebraic concept that is quite useful in the study of systems of differential equations is the concept of a *matrix*.   It will be seen below that this notion will simplify and illuminate much of our work with systems.

Definition 8   *An $m \times n$ matrix is defined to be a rectangular array of real numbers*

$$\begin{bmatrix} a_{11} & a_{12} & \cdots & a_{1n} \\ a_{21} & a_{22} & \cdots & a_{2n} \\ \cdots\cdots\cdots\cdots\cdots \\ a_{m1} & a_{m2} & \cdots & a_{mn} \end{bmatrix}$$

containing m horizontal rows and n vertical columns of elements.   The element in the ith row and jth column is designated by a_{ij}, and if the matrix is denoted by A, the abbreviated notation $A = (a_{ij})$ is frequently used.   Note that a vector with m components can be considered as an $m \times 1$ or column matrix.

Definition 9   *Two $m \times n$ matrices $A = (a_{ij})$ and $B = (b_{ij})$ are equal if and only if corresponding elements are equal, i.e., $a_{ij} = b_{ij}$ for all i,j.   The* sum $A + B$ is the matrix formed by adding corresponding elements of A and B. If k is any real number, then the product kA is the matrix (ka_{ij}); i.e., each element is multiplied by k.

Example 3 $\quad \begin{bmatrix} 1 & 3 \\ -2 & 4 \end{bmatrix} + 3 \begin{bmatrix} 2 & -1 \\ 1 & 0 \end{bmatrix} = \begin{bmatrix} 1 & 3 \\ -2 & 4 \end{bmatrix} + \begin{bmatrix} 6 & -3 \\ 3 & 0 \end{bmatrix} = \begin{bmatrix} 7 & 0 \\ 1 & 4 \end{bmatrix}$

Definition 10   *Given an $m \times n$ matrix $A = (a_{ij})$ and an n-dimensional vector*

$$\mathbf{x} = \begin{bmatrix} x_1 \\ \cdot \\ \cdot \\ \cdot \\ x_n \end{bmatrix}$$

the product $A\mathbf{x}$ is defined to be the m-dimensional vector

$$A\mathbf{x} = \begin{bmatrix} a_{11}x_1 + a_{12}x_2 + \cdots + a_{1n}x_n \\ a_{21}x_1 + a_{22}x_2 + \cdots + a_{2n}x_n \\ \cdots\cdots\cdots\cdots\cdots\cdots\cdots \\ a_{m1}x_1 + a_{m2}x_2 + \cdots + a_{mn}x_n \end{bmatrix}$$

That is, if we set

$$A\mathbf{x} = \mathbf{y} \tag{45}$$

where

$$\mathbf{y} = \begin{bmatrix} y_1 \\ y_2 \\ \cdot \\ \cdot \\ \cdot \\ y_m \end{bmatrix}$$

then

$$\begin{aligned} y_1 &= a_{11}x_1 + a_{12}x_2 + \cdots + a_{1n}x_n \\ y_2 &= a_{21}x_1 + a_{22}x_2 + \cdots + a_{2n}x_n \\ &\;\;\cdot \\ &\;\;\cdot \\ &\;\;\cdot \\ y_m &= a_{m1}x_1 + a_{m2}x_2 + \cdots + a_{mn}x_n \end{aligned} \tag{45a}$$

or

$$y_i = \sum_{j=1}^{n} a_{ij}x_j \qquad i = 1, 2, \ldots, m$$

Note that the number of components of the vector must be the same as the number of *columns* of the matrix or else the product is not defined. The reader should also note that the vector-matrix notation that has been introduced in Eq. (45) provides a compact way of writing the system of simultaneous linear algebraic equations (45a).

$$\textit{Example 4}\quad \text{Given } A = \begin{bmatrix} 2 & 3 & -4 \\ 1 & 2 & 1 \\ -2 & 1 & 2 \end{bmatrix} \quad \mathbf{x} = \begin{bmatrix} x_1 \\ x_2 \\ x_3 \end{bmatrix}$$

then

$$A\mathbf{x} = \begin{bmatrix} 2x_1 + 3x_2 - 4x_3 \\ x_1 + 2x_2 + x_3 \\ -2x_1 + x_2 + 2x_3 \end{bmatrix}$$

that is, $A\mathbf{x}$ is a *vector* with components $2x_1 + 3x_2 - 4x_3$, $x_1 + 2x_2 + x_3$, and $-2x_1 + x_2 + 2x_3$.

Example 5   The square matrix (a_{ij}) with $a_{11} = a_{22} = \cdots = a_{nn} = 1$ and all other elements zero is called the *identity matrix I*, since

$$I\mathbf{x} = \begin{bmatrix} 1 & 0 & 0 & \cdots & 0 \\ 0 & 1 & 0 & \cdots & 0 \\ 0 & 0 & 1 & \cdots & 0 \\ \cdot & \cdot & \cdot & \cdots & \cdot \\ \cdot & \cdot & \cdot & \cdots & \cdot \\ \cdot & \cdot & \cdot & \cdots & \cdot \\ 0 & 0 & 0 & \cdots & 1 \end{bmatrix} \begin{bmatrix} x_1 \\ x_2 \\ x_3 \\ \cdot \\ \cdot \\ \cdot \\ x_n \end{bmatrix} = \begin{bmatrix} x_1 \\ x_2 \\ x_3 \\ \cdot \\ \cdot \\ \cdot \\ x_n \end{bmatrix} = \mathbf{x}$$

Problems 8-4

1. Prove that if two vectors are linearly dependent, then at least one of them can be written as a scalar multiple of the other.   Prove the generalization for n linearly dependent vectors.

2. Determine which of the following sets of vectors are linearly dependent and which are linearly independent.   Prove your answers.

a. $\begin{bmatrix} 1 \\ 0 \end{bmatrix}, \begin{bmatrix} t \\ 0 \end{bmatrix}$

b. $\begin{bmatrix} 1 \\ 0 \end{bmatrix}, \begin{bmatrix} t \\ 0 \end{bmatrix}, \begin{bmatrix} t^2 \\ 0 \end{bmatrix}$

c. $\begin{bmatrix} \sin t \\ \cos t \end{bmatrix}, \begin{bmatrix} \cos t \\ \sin t \end{bmatrix}$

d. $\begin{bmatrix} 1 \\ 1 \end{bmatrix}, \begin{bmatrix} e^t \\ e^t \end{bmatrix}$

e. $\begin{bmatrix} 3 \\ -2 \\ 4 \end{bmatrix}, \begin{bmatrix} -1 \\ 2 \\ -3 \end{bmatrix}, \begin{bmatrix} 1 \\ 2 \\ -2 \end{bmatrix}$

f. $\begin{bmatrix} 1 \\ 0 \\ 0 \end{bmatrix}, \begin{bmatrix} t \\ 0 \\ 0 \end{bmatrix}, \begin{bmatrix} t^2 \\ 0 \\ 0 \end{bmatrix}$

g. $\begin{bmatrix} \sin^2 t \\ \cos^2 t \\ 1 \end{bmatrix}, \begin{bmatrix} \cos^2 t \\ -\sin^2 t \\ -1 \end{bmatrix}, \begin{bmatrix} 1 \\ \cos 2t \\ 0 \end{bmatrix}$

3. If $x_1(t), x_2(t), \ldots, x_n(t)$ are differentiable, prove that

$$\frac{d}{dt} \sum_{i=1}^{n} a_i x_i(t) = \sum_{i=1}^{n} a_i \frac{dx_i(t)}{dt}$$

4. Solve for x: $\begin{bmatrix} x \\ 2 \end{bmatrix} + \begin{bmatrix} 1 & 3 \\ -1 & 2 \end{bmatrix} \begin{bmatrix} x \\ a \end{bmatrix} = \begin{bmatrix} 1 \\ 5 \end{bmatrix}$

8-5 *Theory of systems of linear differential equations*

Consider the system of differential equations

$$\frac{dx_1}{dt} = a_{11}x_1 + a_{12}x_2 + \cdots + a_{1n}x_n + f_1$$

$$\frac{dx_2}{dt} = a_{21}x_1 + a_{22}x_2 + \cdots + a_{2n}x_n + f_2 \qquad (46)$$

.

.

.

$$\frac{dx_n}{dt} = a_{n1}x_1 + a_{n2}x_2 + \cdots + a_{nn}x_n + f_n$$

This is a system of n linear nonhomogeneous first-order differential equations in the n unknown functions $x_1, x_2, \ldots, x_n$. The coefficients a_{ij} and f_i, $i, j = 1, 2, \ldots, n$, are assumed to be *continuous functions of t* in some interval J, $t_0 \leq t \leq t_1$. A solution of (46) on J is any set of continuously differentiable functions $x_1, \ldots, x_n$ which satisfy (46) identically on J.

The system (46) is somewhat special in that each equation contains only one derivative term. The more general system, in which derivatives of more than one function occur in the same equation, is treated, for example, in Wylie [38]. We also restrict our discussion to the case where the number of equations equals the number of unknown functions.

In order to write the system (46) in matrix form, we define an $n \times n$ matrix $A(t)$ and vectors $\mathbf{x}(t)$ and $\mathbf{f}(t)$ as follows:

$$A(t) = \begin{bmatrix} a_{11} & a_{12} & \cdots & a_{1n} \\ a_{21} & a_{22} & \cdots & a_{2n} \\ \multicolumn{4}{c}{\cdots\cdots\cdots\cdots\cdots} \\ a_{n1} & a_{n2} & & a_{nn} \end{bmatrix} \qquad \mathbf{x}(t) = \begin{bmatrix} x_1(t) \\ x_2(t) \\ \cdot \\ \cdot \\ \cdot \\ x_n(t) \end{bmatrix} \qquad \mathbf{f}(t) = \begin{bmatrix} f_1(t) \\ f_2(t) \\ \cdot \\ \cdot \\ \cdot \\ f_n(t) \end{bmatrix}$$

$$(47)$$

Then the system (46) can be written in the compact form

$$\frac{d\mathbf{x}(t)}{dt} = A(t)\mathbf{x}(t) + \mathbf{f}(t) \tag{48}$$

The initial-value problem for the system (46) would involve n initial conditions of the form

$$x_1(t_0) = c_1, \ . \ . \ . \ , \ x_n(t_0) = c_n \tag{49}$$

where $c_1, c_2, \ . \ . \ . \ , c_n$ are given constants.

If we define a constant vector

$$\mathbf{c} = \begin{bmatrix} c_1 \\ c_2 \\ \cdot \\ \cdot \\ \cdot \\ c_n \end{bmatrix} \tag{50}$$

then the initial condition becomes

$$\mathbf{x}(t_0) = \mathbf{c} \tag{51}$$

Example 1   Given the system

$$\frac{dx_1}{dt} = 3x_1 + t^2 x_2 + t$$
$$\tag{52}$$
$$\frac{dx_2}{dt} = -x_1 + 2tx_2 - t^3$$

and the initial conditions $x_1(0) = 1$, $x_2(0) = -2$.   In order to write (52) in matrix form, let

$$A = \begin{bmatrix} 3 & t^2 \\ -1 & 2t \end{bmatrix} \qquad \mathbf{x} = \begin{bmatrix} x_1 \\ x_2 \end{bmatrix} \qquad \mathbf{f} = \begin{bmatrix} t \\ -t^3 \end{bmatrix} \qquad \mathbf{c} = \begin{bmatrix} 1 \\ -2 \end{bmatrix} \tag{53}$$

and the system (52) becomes

$$\frac{d}{dt}\begin{bmatrix} x_1 \\ x_2 \end{bmatrix} = \begin{bmatrix} 3 & t^2 \\ -1 & 2t \end{bmatrix}\begin{bmatrix} x_1 \\ x_2 \end{bmatrix} + \begin{bmatrix} t \\ -t^3 \end{bmatrix} \tag{54}$$

or

$$\frac{d\mathbf{x}}{dt} = A\mathbf{x} + \mathbf{f} \tag{55}$$

with initial condition

$$\mathbf{x}(0) = \mathbf{c} \tag{56}$$

In order to state the existence theorem in matrix form, we need one more definition.

Definition 1   *A vector $\mathbf{v}(t)$ or a matrix $A(t)$ is said to be continuous on an interval J if and only if each component or element is continuous on J.*

The fundamental existence theorem for systems can now be stated as follows:†

Theorem 1‡   *Consider the initial-value problem*

$$\text{DE:} \frac{d\mathbf{x}(t)}{dt} = A(t)\mathbf{x}(t) + \mathbf{f}(t) \tag{57}$$

$$\text{IC:} \ \mathbf{x}(t_0) = \mathbf{c} \tag{58}$$

with $\mathbf{x}$, A, $\mathbf{f}$, and $\mathbf{c}$ defined as in (47) and (50). If $A(t)$ and $\mathbf{f}(t)$ are continuous on some interval J containing $t = t_0$, then the initial-value problem (57), (58) possesses a unique solution vector $\mathbf{x}(t)$ on the interval J.

In order to ensure the applicability of this theorem, A and $\mathbf{f}$ will be assumed continuous on some common interval J throughout the chapter.

Definition 2   *If $\mathbf{f}(t) \equiv 0$ throughout J, then the system (57) is called homogeneous. If any of the components of $\mathbf{f}$ differs from zero anywhere in J, the system is called nonhomogeneous.*

One important property of matrix multiplication is contained in the following theorem.

Theorem 2   *The multiplication of a vector by a matrix is a linear operation. That is, if A is an $m \times n$ matrix and $\mathbf{x}_1$ and $\mathbf{x}_2$ are n-dimensional vectors and c_1 and c_2 real constants, then*

$$A(c_1\mathbf{x}_1 + c_2\mathbf{x}_2) = c_1A\mathbf{x}_1 + c_2A\mathbf{x}_2 \tag{59}$$

Proof   See Prob. 1.

We leave it to the reader to show also that if A and B are matrices, then

$$(k_1A + k_2B)\mathbf{x} = k_1A\mathbf{x} + k_2B\mathbf{x}$$

The general theory for the system of linear differential equations (46) will now be given, expressed in the notation of the vector equation (48).

Theorem 3   *If $\mathbf{x}_1(t)$ and $\mathbf{x}_2(t)$ are any two solutions of the homogeneous equation $d\mathbf{x}(t)/dt = A(t)\mathbf{x}(t)$, then the vector $c_1\mathbf{x}_1 + c_2\mathbf{x}_2$ is also a solution.*

---

† See the appendix to Chap. 4 for a proof.

‡ In order to emphasize the similarity between the theory for systems and that developed for the nth-order equation, theorem numbers in this section correspond exactly to those of the corresponding theorems in Sec. 5-2 and also to those in Sec. 5-9.

Proof

$$\frac{d}{dt}(c_1\mathbf{x}_1 + c_2\mathbf{x}_2) = c_1\frac{d\mathbf{x}_1}{dt} + c_2\frac{d\mathbf{x}_2}{dt} = c_1 A\mathbf{x}_1 + c_2 A\mathbf{x}_2 = A(c_1\mathbf{x}_1 + c_2\mathbf{x}_2)$$

This theorem is easily extended to the case of n solutions $\mathbf{x}_1, \mathbf{x}_2, \ldots, \mathbf{x}_n$ (see Prob. 2). As in Chap. 5, we shall see that the *general solution* of a system (i.e., the set of all solutions) is the set of all linear combinations of n *linearly independent* solutions. As before, the concept of a *Wronskian* is useful in this connection.

Definition 3   *Given the vectors*

$$\mathbf{x}_1(t) = \begin{bmatrix} x_{11}(t) \\ x_{12}(t) \\ \cdot \\ \cdot \\ \cdot \\ x_{1n}(t) \end{bmatrix}, \qquad \mathbf{x}_2(t) = \begin{bmatrix} x_{21}(t) \\ x_{22}(t) \\ \cdot \\ \cdot \\ \cdot \\ x_{2n}(t) \end{bmatrix}, \qquad \ldots, \qquad \mathbf{x}_n(t) = \begin{bmatrix} x_{n1}(t) \\ x_{n2}(t) \\ \cdot \\ \cdot \\ \cdot \\ x_{nn}(t) \end{bmatrix}$$

the Wronskian of $\mathbf{x}_1, \mathbf{x}_2, \ldots, \mathbf{x}_n$ *is defined as the determinant*

$$W[\mathbf{x}_1, \mathbf{x}_2, \ldots, \mathbf{x}_n] = \begin{vmatrix} x_{11}(t) & x_{21}(t) & \cdots & x_{n1}(t) \\ x_{12}(t) & x_{22}(t) & \cdots & x_{n2}(t) \\ \cdots\cdots\cdots\cdots\cdots\cdots\cdots \\ x_{1n}(t) & x_{2n}(t) & \cdots & x_{nn}(t) \end{vmatrix} \tag{60}$$

The relationship between this definition of the Wronskian and that given in Sec. 5-8 is discussed in Prob. 3.

Theorem 4   *Let* $\mathbf{x}_1, \mathbf{x}_2, \ldots, \mathbf{x}_n$ *be functions of t defined on an interval J. If the Wronskian* $W[\mathbf{x}_1, \mathbf{x}_2, \ldots, \mathbf{x}_n]$ *is different from zero at any point $t = t_0$ in J, then* $\mathbf{x}_1, \mathbf{x}_2, \ldots, \mathbf{x}_n$ *are linearly independent on J.*

Proof by contradiction   If the theorem is false and $\mathbf{x}_1, \mathbf{x}_2, \ldots, \mathbf{x}_n$ are linearly *dependent*, then there exist constants $c_1, c_2, \ldots, c_n$, not all zero, such that

$$\sum_{i=1}^{n} c_i\mathbf{x}_i = \mathbf{0} \tag{61}$$

for all t in J. The vector equation (61) can be interpreted as a set of n linear homogeneous algebraic equations for $c_1, c_2, \ldots, c_n$. A nontrivial solution exists for all t only if the determinant of the coefficients vanishes for all t. But the determinant of the coefficients is just the Wronskian $W[\mathbf{x}_1, \mathbf{x}_2, \ldots, \mathbf{x}_n]$ and hence $W \equiv 0$ for all t in J contradicting our assumptions. Thus the theorem is proved.

Corollary *If* $\mathbf{x}_1, \mathbf{x}_2, \ldots, \mathbf{x}_n$ *are linearly dependent on* J, *then* $W[\mathbf{x}_1, \mathbf{x}_2, \ldots, \mathbf{x}_n] = 0$ *for all* t *in* J.

The converse of Theorem 4 is not true in general. (See Prob. 4.) However, if the $\mathbf{x}_i$ are solution vectors of a linear differential equation, the following modified converse will hold.

Theorem 5 *If* $\mathbf{x}_1, \mathbf{x}_2, \ldots, \mathbf{x}_n$ *are linearly independent solutions of* $d\mathbf{x}(t)/dt = A(t)\mathbf{x}(t)$ *on some interval* J, *then the Wronskian* $W[\mathbf{x}_1, \mathbf{x}_2, \ldots, \mathbf{x}_n]$ *is different from zero for all* t *in* J.

Proof by contradiction If $W[\mathbf{x}_1, \ldots, \mathbf{x}_n] = 0$ at some point $t = t_0$ in J, then the vector equation $\sum_{i=1}^{n} c_i \mathbf{x}_i(t_0) = \mathbf{0}$ has a nontrivial solution for $c_1, c_2, \ldots, c_n$ since the determinant of the coefficients is W, which is zero at t_0. Use these values of $c_1, c_2, \ldots, c_n$ to form the function $\sum_{i=1}^{n} c_i \mathbf{x}_i = \mathbf{x}$. This function is a solution of $d\mathbf{x}/dt = A\mathbf{x}$ (see Prob. 2), which vanishes at $t = t_0$. Hence by the uniqueness theorem (Theorem 1), $\mathbf{x} = \mathbf{0}$ for all t in J. That is, $\sum_{i=1}^{n} c_i \mathbf{x}_i \equiv \mathbf{0}$ and hence $\mathbf{x}_1, \ldots, \mathbf{x}_n$ are linearly *dependent* since $c_1, \ldots, c_n$ are not all zero. This contradicts the assumption of linear independence and hence the theorem is proved.

Corollary *The Wronskian of* n *solutions of* $d\mathbf{x}/dt = A\mathbf{x}$ *is either identically zero (if the solutions are linearly dependent) or never zero (if the solutions are linearly independent).*

We are now in a position to prove the following theorem.

Theorem 6 *If* $\mathbf{x}_1(t), \mathbf{x}_2(t), \ldots, \mathbf{x}_n(t)$ *are linearly independent solutions of* $d\mathbf{x}(t)/dt = A(t)\mathbf{x}(t)$ *on* J, *then* $\mathbf{x} = \sum_{i=1}^{n} c_i \mathbf{x}_i(t)$, *where* $c_1, c_2, \ldots, c_n$ *are arbitrary constants, is the general solution on* J.

Proof Let $\mathbf{x}(t)$ be *any* solution of $d\mathbf{x}/dt = A\mathbf{x}$. Let $t = t_0$ be any point in J and consider the possibility of solving the following system of linear algebraic equations for $c_1, \ldots, c_n$:

$$\sum_{i=1}^{n} c_i \mathbf{x}_i(t_0) = \mathbf{x}(t_0)$$

A unique solution for $c_1, c_2, \ldots, c_n$ exists since the determinant of coefficients is the Wronskian of linearly independent vectors and hence is not zero. Now the function $\sum_{i=1}^{n} c_i \mathbf{x}_i(t)$ is a solution of the differential equation which equals $\mathbf{x}$ at t_0. The uniqueness theorem then states that these two solutions are equal for all t in J and hence

$$\mathbf{x}(t) = \sum_{i=1}^{n} c_i \mathbf{x}_i(t)$$

Theorem 7   *There exist n linearly independent solutions of $dx/dt = Ax$ on J.*

Proof   Let t_0 be a point J and define constant vectors

$$
c_1 = \begin{bmatrix} 1 \\ 0 \\ 0 \\ \cdot \\ \cdot \\ \cdot \\ 0 \end{bmatrix}, \quad
c_2 = \begin{bmatrix} 0 \\ 1 \\ 0 \\ \cdot \\ \cdot \\ \cdot \\ 0 \end{bmatrix}, \quad \ldots, \quad
c_n = \begin{bmatrix} 0 \\ 0 \\ 0 \\ \cdot \\ \cdot \\ \cdot \\ 1 \end{bmatrix}
$$

Theorem 1 then ensures the existence of a unique set of solutions $x_1, x_2, \ldots, x_n$ of $dx/dt = Ax$ satisfying $x_1(t_0) = c_1$, $x_2(t_0) = c_2$, $\ldots$, $x_n(t_0) = c_n$.   The Wronskian of $x_1, \ldots, x_n$ at t_0 is

$$
W[x_1(t_0), \ldots, x_n(t_0)] = \begin{vmatrix} 1 & 0 & 0 & \cdots & 0 \\ 0 & 1 & 0 & \cdots & 0 \\ 0 & 0 & 1 & \cdots & 0 \\ \cdots & \cdots & \cdots & \cdots & \cdots \\ 0 & 0 & 0 & \cdots & 1 \end{vmatrix} = 1 \neq 0
$$

Since the Wronskian is different from zero at one point, the solutions are linearly independent on J by Theorem 4.

Theorem 8   *If x_p is any particular solution of the nonhomogeneous equation $dx/dt = Ax + f$ and x_h is the general solution of the homogeneous equation $dx/dt = Ax$, then the general solution of $dx/dt = Ax + f$ is $x = x_h + x_p$.*

Proof   The proof of this theorem is essentially the same as that of the corresponding Theorem 8 in Sec. 5-2.   See Prob. 5.

Example 2   Suppose

$$
A(t) = \begin{bmatrix} 1 & -1 \\ -2 & 0 \end{bmatrix} \quad \text{and} \quad f = \begin{bmatrix} t \\ 1 \end{bmatrix}
$$

The vectors

$$
x_1 = \begin{bmatrix} e^{2t} \\ -e^{2t} \end{bmatrix} \quad \text{and} \quad x_2 = \begin{bmatrix} e^{-t} \\ 2e^{-t} \end{bmatrix}
$$

are solutions of $dx/dt = Ax$ as may be verified by direct substitution (see also Example 1, Sec. 8-2).

The Wronskian

$$W[\mathbf{x_1},\mathbf{x_2}] = \begin{vmatrix} e^{2t} & e^{-t} \\ -e^{2t} & 2e^{-t} \end{vmatrix} = 3e^t$$

which never vanishes, and hence $\mathbf{x_1}$ and $\mathbf{x_2}$ are linearly independent. The general solution of $d\mathbf{x}/dt = A\mathbf{x}$ is therefore $\mathbf{x}_h = c_1\mathbf{x_1} + c_2\mathbf{x_2}$. The vector

$$\mathbf{x}_p = \begin{bmatrix} 0 \\ t \end{bmatrix}$$

satisfies $d\mathbf{x}/dt = A\mathbf{x} + \mathbf{f}$, and hence the general solution of this equation is $\mathbf{x} = \mathbf{x}_h + \mathbf{x}_p$.

Theorem 9   *Principle of superposition.*   *If $\mathbf{x_1}$ is a solution of $d\mathbf{x}/dt = A\mathbf{x} + \mathbf{f}_1$ and $\mathbf{x_2}$ is a solution of $d\mathbf{x}/dt = A\mathbf{x} + \mathbf{f}_2$, then $\mathbf{x} = \mathbf{x_1} + \mathbf{x_2}$ is a solution of $d\mathbf{x}/dt = A\mathbf{x} + \mathbf{f}_1 + \mathbf{f}_2$.*

Proof   See Prob. 6.

Reduction of an nth-order equation to a system†   Throughout this section the theory of systems of equations has been developed as an exact parallel to that of the nth-order equation in Chap. 5. The obvious similarity between the theories is of course not accidental. In fact, the theory of the nth-order equation is a special case of the theory of systems. This can be shown easily since it is possible to reduce the general nth-order linear differential equation to an equivalent system of the type considered above.

Consider the following equation for $x(t)$:

$$a_0(t)\frac{d^nx}{dt^n} + a_1(t)\frac{d^{n-1}x}{dt^{n-1}} + \cdots + a_{n-1}(t)\frac{dx}{dt} + a_n(t)x = f(t) \tag{62}$$

where $a_0, a_1, \ldots, a_n, f$ are continuous in some common interval J. We assume also that $a_0(t) \neq 0$ on J in order to ensure the existence of a solution. [See Eq. (64).]

Replace the single dependent variable x by n new variables defined as follows:

$$x_1 = x$$

$$x_2 = \frac{dx}{dt} = \frac{dx_1}{dt}$$

$$x_3 = \frac{d^2x}{dt^2} = \frac{dx_2}{dt}$$

$$\cdot$$
$$\cdot$$
$$\cdot$$

$$x_n = \frac{d^{n-1}x}{dt^{n-1}} = \frac{dx_{n-1}}{dt}$$

(63)

† See also Sec. 4-7.

These equations of definition provide differential equations for $x_1, x_2, \ldots,$ x_{n-1}, and the original differential equation (62) itself provides the equation for $dx_n/dt = d^n x/dt^n$. Thus we obtain the system

$$\frac{dx_1}{dt} = x_2$$

$$\frac{dx_2}{dt} = x_3$$

$$\cdot$$
$$\cdot \qquad\qquad\qquad\qquad\qquad\qquad\qquad (64)$$
$$\cdot$$

$$\frac{dx_{n-1}}{dt} = x_n$$

$$\frac{dx_n}{dt} = -\frac{a_n}{a_0} x_1 - \frac{a_{n-1}}{a_0} x_2 - \cdots - \frac{a_1}{a_0} x_n + \frac{f}{a_0}$$

Finally, if we let

$$\mathbf{x} = \begin{bmatrix} x_1 \\ x_2 \\ \cdot \\ \cdot \\ \cdot \\ x_n \end{bmatrix} \qquad A = \begin{bmatrix} 0 & 1 & 0 & 0 & \cdots & 0 \\ 0 & 0 & 1 & 0 & \cdots & 0 \\ 0 & 0 & 0 & 1 & \cdots & 0 \\ \cdots\cdots\cdots\cdots\cdots\cdots\cdots\cdots \\ -\dfrac{a_n}{a_0} & -\dfrac{a_{n-1}}{a_0} & \cdots\cdots\cdots\cdots & -\dfrac{a_1}{a_0} \end{bmatrix}$$

$$\qquad\qquad\qquad\qquad\qquad\qquad\qquad\qquad\qquad (65)$$

$$\mathbf{f}(t) = \begin{bmatrix} 0 \\ 0 \\ 0 \\ \cdot \\ \cdot \\ \cdot \\ \dfrac{f}{a_0} \end{bmatrix}$$

then the system (64) becomes

$$\frac{d\mathbf{x}}{dt} = A\mathbf{x} + \mathbf{f} \qquad\qquad\qquad\qquad\qquad (66)$$

Thus the system (66) is equivalent to Eq. (62) in the sense that if $\mathbf{x}$ is a solution of (66), then the first component $x_1(t) = x(t)$ is a solution of (62), and conversely.

Note that if initial conditions were specified for (62), they would be of the form

$$x(t_0) = c_1, \; x'(t_0) = c_2, \; \ldots, \; x^{(n-1)}(t_0) = c_n \qquad\qquad (67)$$

The equivalent vector form would be

$$\mathbf{x}(t_0) = \mathbf{c} \tag{68}$$

where $\mathbf{c}$ is the constant vector

$$\mathbf{c} = \begin{bmatrix} c_1 \\ c_2 \\ \cdot \\ \cdot \\ \cdot \\ c_n \end{bmatrix} \tag{69}$$

Problems 8-5

1. Prove Theorem 2.

2. State and prove the analog of Theorem 3 for the case of n solutions, $\mathbf{x}_1(t), \ldots, \mathbf{x}_n(t)$, of the equation $d\mathbf{x}(t)/dt = A(t)\mathbf{x}(t)$.

3. Suppose that the vectors $\mathbf{x}_1(t), \ldots, \mathbf{x}_n(t)$ are connected by the relations $\mathbf{x}_2 = d\mathbf{x}_1/dt$, $\mathbf{x}_3 = d\mathbf{x}_2/dt, \ldots, \mathbf{x}_n = d\mathbf{x}_{n-1}/dt$. Show that the Wronskian as defined in Eq. (60) reduces to the same form as in Eq. (90), Sec. 5-9.

4. a. Given

$$\mathbf{x}_1(t) = \begin{bmatrix} t \\ 1 \end{bmatrix} \qquad \mathbf{x}_2 = \begin{bmatrix} t^2 \\ t \end{bmatrix}$$

show that $\mathbf{x}_1$ and $\mathbf{x}_2$ are not linearly dependent on any interval even though the Wronskian $W[\mathbf{x}_1, \mathbf{x}_2] \equiv 0$.

b. Apply Theorem 5 to show that $\mathbf{x}_1$ and $\mathbf{x}_2$ cannot both satisfy the same linear differential equation $d\mathbf{x}/dt = A\mathbf{x}$.

5. Prove Theorem 8 by adapting the method used in proving Theorem 8 in Sec. 5-2.

6. Prove Theorem 9.

8-6   Homogeneous linear systems with constant coefficients

The general theory of Sec. 8-5 is valid for systems in which the coefficient matrix $A(t)$ varies with t. However, the task of actually solving such a system is in general quite difficult, and one must usually resort to approximate numerical methods. The difficulties encountered are similar to those involved in solving the nth-order equation with variable coefficients (see Chap. 7). A considerable simplification is effected if one restricts the investigation to systems in which A is a constant matrix. Fortunately many of the systems encountered in applications are of this type, or can at least be approximated by this type.

In this section, then, we shall study methods for solving the homogeneous equation†

$$\frac{d\mathbf{x}(t)}{dt} = A\mathbf{x}(t) \tag{70}$$

where A is the *constant matrix*

$$A = \begin{bmatrix} a_{11} & a_{12} & \cdots & a_{1n} \\ a_{21} & a_{22} & \cdots & a_{2n} \\ \cdots\cdots\cdots\cdots\cdots \\ a_{n1} & a_{n2} & \cdots & a_{nn} \end{bmatrix} \tag{71}$$

(i.e., the a_{ij} are all real constants).   Some of the methods that can be used to solve Eq. (70) have been described briefly in Sec. 8-2.   These methods will now be developed more generally.

Since each component dx_i/dt of $d\mathbf{x}/dt$ in (70) is to be equal to a linear combination of all the components $x_1, \ldots, x_n$ of $\mathbf{x}$, it is to be expected once again that all the functions must be multiples of the same exponential function $e^{\lambda t}$.   Hence we seek a solution in the form

$$x_1 = c_1 e^{\lambda t}, x_2 = c_2 e^{\lambda t}, \ldots, x_n = c_n e^{\lambda t} \tag{72}$$

or

$$\mathbf{x} = e^{\lambda t}\mathbf{c} \tag{73}$$

where $\mathbf{c}$ is the vector

$$\mathbf{c} = \begin{bmatrix} c_1 \\ c_2 \\ \cdot \\ \cdot \\ \cdot \\ c_n \end{bmatrix} \tag{74}$$

Substituting (73) into the differential equation (70), we have

$$\lambda e^{\lambda t}\mathbf{c} = e^{\lambda t}A\mathbf{c} \tag{75}$$

or

$$A\mathbf{c} = \lambda\mathbf{c} \tag{76}$$

If we introduce the identity matrix‡ I into (76) by using the fact that $\mathbf{c} = I\mathbf{c}$, then

$$A\mathbf{c} - \lambda I\mathbf{c} = \mathbf{0} \tag{77}$$

or

$$(A - \lambda I)\mathbf{c} = \mathbf{0} \tag{78}$$

† The nonhomogeneous equation is discussed in Sec. 8-7.
‡ See Sec. 8-4, Example 5.

Equation (78) can be interpreted as a set of n homogeneous linear equations for the constants $c_1, c_2, \ldots, c_n$. A nontrivial solution will exist if and only if the determinant of the matrix $A - \lambda I$ vanishes. That is,

$$|A - \lambda I| = 0 \tag{79}$$

This equation for λ is called the *characteristic equation* and is a polynomial equation of degree n with n solutions $\lambda_1, \lambda_2, \ldots, \lambda_n$.† These solutions of (79) are called the *eigenvalues* of the matrix A. For each λ_i, Eq. (78) has a corresponding nontrivial solution vector $\mathbf{c}_i \neq \mathbf{0}$. Hence these vectors, called *eigenvectors*, satisfy

$$A\mathbf{c}_i = \lambda_i \mathbf{c}_i \qquad i = 1, 2, \ldots, n \tag{80}$$

Note that any scalar multiple of an eigenvector is again an eigenvector. Finally, for each value of $\lambda = \lambda_i$, $i = 1, 2, \ldots, n$, satisfying Eq. (80), Eq. (78) will determine all the components of $\mathbf{c}_i$ in terms of one or more arbitrary constants. The vector $\mathbf{x}_i = e^{\lambda_i t}\mathbf{c}_i$ is then a solution of the differential equation (70). If all the eigenvalues $\lambda_1, \lambda_2, \ldots, \lambda_n$ are distinct, then the corresponding solutions $\mathbf{x}_i = e^{\lambda_i t}\mathbf{c}_i$, $i = 1, 2, \ldots, n$, are linearly independent (see Prob. 1), and the general solution of (70) is

$$\mathbf{x} = \sum_{i=1}^{n} e^{\lambda_i t}\mathbf{c}_i \tag{81}$$

where the constant vectors $\mathbf{c}_i$ each depend on one arbitrary constant. The solution in the case where two or more eigenvalues are equal is described in Prob. 2. See also Example 3, Sec. 8-2.

Example 1   Solve:

$$\frac{dx}{dt} = 4x + y$$

$$\frac{dy}{dt} = 3x + 2y \tag{82}$$

Let

$$\mathbf{x} = \begin{bmatrix} x \\ y \end{bmatrix} \quad \text{and} \quad A = \begin{bmatrix} 4 & 1 \\ 3 & 2 \end{bmatrix}$$

and the system becomes $d\mathbf{x}/dt = A\mathbf{x}$. Substituting $\mathbf{x} = e^{\lambda t}\mathbf{c}$ into the differential equation yields the equation

$$(A - \lambda I)\mathbf{c} = \mathbf{0} \tag{83}$$

† Note that the eigenvalues λ_i need not all be distinct and that some of them may be complex numbers. When complex eigenvalues occur the basic definitions and properties of vectors as developed in Secs. 8-4 and 8-5 must be extended to include complex scalars and complex components for vectors. The necessary modifications are straightforward.

with nontrivial solutions only if $|A - \lambda I| = 0$. That is,

$$\begin{vmatrix} 4 - \lambda & 1 \\ 3 & 2 - \lambda \end{vmatrix} = 0$$

or

$$\lambda^2 - 6\lambda + 5 = 0$$

The eigenvalues are therefore $\lambda_1 = 1$ and $\lambda_2 = 5$. Substituting $\lambda_1 = 1$ in (83) yields, for example, $3c_1 + c_2 = 0$, whence $c_2 = -3c_1$. For $\lambda_2 = 5$ we get $c_2 = c_1$. The eigenvectors corresponding to $\lambda_1 = 1$ and $\lambda_2 = 5$ are then

$$k_1 \begin{bmatrix} 1 \\ -3 \end{bmatrix} \qquad \text{and} \qquad k_2 \begin{bmatrix} 1 \\ 1 \end{bmatrix}$$

where k_1 and k_2 are arbitrary constants. Hence the general solution of the system (82) can be written in the form

$$\mathbf{x} = k_1 e^t \begin{bmatrix} 1 \\ -3 \end{bmatrix} + k_2 e^{5t} \begin{bmatrix} 1 \\ 1 \end{bmatrix}$$

The components x and y of $\mathbf{x}$ are then

$$x = k_1 e^t + k_2 e^{5t}$$
$$y = -3k_1 e^t + k_2 e^{5t} \tag{84}$$

Complex eigenvalues If any of the eigenvalues are complex numbers, they must occur in conjugate pairs (why?), and these can be combined to produce real functions as in Sec. 5-4.

Example 2 Solve:† $\dot{x} = -x + 2y$
$$\dot{y} = -2x - y$$

The characteristic equation is $\lambda^2 + 2\lambda + 5 = 0$ with solutions $\lambda_1 = -1 + 2i$ and $\lambda_2 = -1 - 2i$ with corresponding eigenvectors

$$k_1 \begin{bmatrix} 1 \\ i \end{bmatrix} \qquad \text{and} \qquad k_2 \begin{bmatrix} 1 \\ -i \end{bmatrix}$$

The general solution is then

$$\mathbf{x} = k_1 \begin{bmatrix} 1 \\ i \end{bmatrix} e^{(-1+2i)t} + k_2 \begin{bmatrix} 1 \\ -i \end{bmatrix} e^{(-1-2i)t}$$

If we use the fact that $e^{i\theta} = \cos \theta + i \sin \theta$ in this result, we can write the components x and y of $\mathbf{x}$ in the form

$$x = e^{-t}(C_1 \cos 2t + C_2 \sin 2t)$$
$$y = e^{-t}(-C_1 \sin 2t + C_2 \cos 2t)$$

where we have set $C_1 = k_1 + k_2$ and $C_2 = i(k_1 - k_2)$.

† This system was solved by a different method in Example 2, Sec. 8-2.

Problems 8-6

1. Given a constant matrix A with eigenvalues $\lambda_1, \ldots, \lambda_n$ and corresponding eigenvectors $\mathbf{c}_1, \ldots, \mathbf{c}_n$, prove that if the eigenvalues $\lambda_1, \ldots, \lambda_n$ are distinct, then the functions $\mathbf{x}_i = e^{\lambda_i t}\mathbf{c}_i$, $i = 1, \ldots, n$, are linearly independent solutions of the equation $d\mathbf{x}/dt = A\mathbf{x}$.

2. Find the general solution to the system $d\mathbf{x}/dt = A\mathbf{x}$ where

$$A = \begin{bmatrix} 1 & 1 \\ 0 & 1 \end{bmatrix}$$

by using variation of parameters to get the second solution. Describe a general method for finding linearly independent solutions when multiple eigenvalues occur. Assume that A is 2×2.

3. Solve Probs. 1b, c, d, and g of Sec. 8-2 by the methods of this section.

8-7   Solution by matrix methods

In order to generalize the above results and to extend the method of solution to nonhomogeneous equations, it will be advantageous to go a little deeper into the theory of matrices.

In what follows we shall restrict ourselves to square ($n \times n$) matrices $A(t)$ with real coefficients $a_{ij}(t)$,† defined and continuous on some common interval $J: a \le t \le b$.

Definition 1   *If $A = (a_{ij})$ and $B = (b_{ij})$ are $n \times n$ matrices then the product AB is an $n \times n$ matrix C with coefficients*

$$c_{ij} = \sum_{k=1}^{n} a_{ik}b_{kj}$$

That is, the element in the ith row and jth column of the product matrix is obtained by multiplying the terms in the ith row of A by the corresponding terms in the jth column of B and adding the results.

One of the most important facts about matrix multiplication that should always be kept in mind is the fact that, in general, it is *not commutative*.

Example 1   Let $A = \begin{bmatrix} 2 & 1 \\ -3 & 0 \end{bmatrix}$ $B = \begin{bmatrix} 6 & -2 \\ 0 & 5 \end{bmatrix}$

$$AB = \begin{bmatrix} 12 & 1 \\ -18 & 6 \end{bmatrix} \qquad BA = \begin{bmatrix} 18 & 6 \\ -15 & 0 \end{bmatrix}$$

The reader can easily verify these statements and *note that $AB \ne BA$.*

† The results of this section also hold for matrices with complex coefficients. See the footnote on page 249.

Even though matrix multiplication is in general noncommutative, the associative and distributive laws *do* hold. That is, for arbitrary $n \times n$ matrices A, B, C we have (see Prob. 1)

$$(AB)C = A(BC) \qquad \text{and} \qquad A(B+C) = AB + AC \tag{85}$$

The associative law allows us to define powers of a matrix.

Definition 2 $A^2 = A \cdot A, A^3 = A \cdot A^2 = A^2 \cdot A, \ldots, A^k = A \cdot A^{k-1} = A^{k-1} \cdot A$ *where* $k = 1, 2, \ldots,$ *and for convenience we define* $A^0 = I$, *the identity matrix. Note that* A *commutes with* A^k *where* k *is any nonnegative integer.*

Also, $A^m \cdot A^n = A^{m+n}$ for any nonnegative integers m and n.

Definition 3   *A polynomial* $P_m(A)$ *of degree* m *in the matrix* A *is any expression of the form*

$$P_m(A) = \sum_{0=k}^{m} a_k A^k = a_0 I + a_1 A + a_2 A^2 + \cdots + a_m A^m$$

where $a_0, a_1, \ldots, a_m$ *are real numbers. Note that* $P_m(A)$ *is itself an* $n \times n$ *matrix.*

Definition 4   *A matrix* D *is called diagonal if the elements* $d_{ij} = 0$ *for all* $i \neq j$. *That is,*

$$D = \begin{bmatrix} \lambda_1 & 0 & 0 & \cdots & 0 \\ 0 & \lambda_2 & 0 & \cdots & 0 \\ 0 & 0 & \lambda_3 & \cdots & 0 \\ \multicolumn{5}{c}{\cdots\cdots\cdots\cdots\cdots} \\ 0 & 0 & 0 & \cdots & \lambda_n \end{bmatrix} \tag{86}$$

It is easy to see that the elements λ_i on the main diagonal are the eigenvalues of D.

Theorem 1   *If* D *is a diagonal matrix and* $P_m(D) = \sum_{k=1}^{m} a_k D^k$ *is any polynomial in* D, *then* $P_m(D)$ *can be evaluated as follows:*

$$P_m(D) = \begin{bmatrix} P_m(\lambda_1) & & & \bigcirc \\ & P_m(\lambda_2) & & \\ & & \ddots & \\ \bigcirc & & & P_m(\lambda_n) \end{bmatrix}$$

Proof   See Prob. 3.

Definition 5   *The determinant $|A|$ of an $n \times n$ matrix A is the determinant whose elements are the elements of the matrix. A matrix whose determinant is different from zero is called a nonsingular matrix.*

Definition 6   *The inverse of a nonsingular matrix A is a matrix, denoted by A^{-1}, which satisfies*

$$AA^{-1} = I = A^{-1}A$$

It is fairly easy to show that the inverse of a nonsingular matrix always exists and is unique.† Furthermore the following simple procedure enables one to compute A^{-1} directly. If we denote by b_{ij} the element in the ith row and jth column of A^{-1}, then

$$b_{ij} = \frac{A_{ji}}{|A|} \tag{87}$$

where $|A|$ is the determinant of A (nonzero since A is nonsingular) and A_{ji} is the *cofactor*† of a_{ji}.

Example 2   If

$$A = \begin{bmatrix} 2 & 1 \\ -3 & 5 \end{bmatrix}$$

then $|A| = 13 \neq 0$ and

$$A^{-1} = \tfrac{1}{13} \begin{bmatrix} 5 & -1 \\ 3 & 2 \end{bmatrix}$$

Diagonalizing a matrix   We are interested in solving the differential equation

$$\frac{d\mathbf{x}(t)}{dt} = A\mathbf{x}(t) \tag{88}$$

In this section the solution of a very special case will lead us to the general procedure. If the constant matrix A is *diagonal*, then the solution is immediate. In this case the system is

$$\frac{d}{dt} \begin{bmatrix} x_1 \\ \cdot \\ \cdot \\ \cdot \\ x_n \end{bmatrix} = \begin{bmatrix} \lambda_1 & & \bigcirc \\ & \ddots & \\ \bigcirc & & \lambda_n \end{bmatrix} \begin{bmatrix} x_1 \\ \cdot \\ \cdot \\ \cdot \\ x_n \end{bmatrix} \tag{89}$$

and the eigenvalues of A are $\lambda_1, \ldots, \lambda_n$. We immediately obtain n *linearly*

† E.g., see Finkbeiner [15], chap. 5.

independent solution vectors $\mathbf{x}_i$

$$\mathbf{x}_1 = \begin{bmatrix} e^{\lambda_1 t} \\ 0 \\ 0 \\ \cdot \\ \cdot \\ \cdot \\ 0 \end{bmatrix}, \quad \mathbf{x}_2 = \begin{bmatrix} 0 \\ e^{\lambda_2 t} \\ 0 \\ \cdot \\ \cdot \\ \cdot \\ 0 \end{bmatrix}, \quad \ldots, \quad \mathbf{x}_n = \begin{bmatrix} 0 \\ 0 \\ 0 \\ \cdot \\ \cdot \\ \cdot \\ e^{\lambda_n t} \end{bmatrix} \tag{90}$$

Hence the general solution is $\mathbf{x} = \sum_{i=1}^{n} c_i \mathbf{x}_i$ or

$$\mathbf{x} = \begin{bmatrix} c_1 e^{\lambda_1 t} \\ c_2 e^{\lambda_2 t} \\ \cdot \\ \cdot \\ \cdot \\ c_n e^{\lambda_n t} \end{bmatrix} \dagger \tag{91}$$

Now, even if A is *not* diagonal, it is often possible to transform it into another matrix D which *is* diagonal. That is, we shall attempt to make a change of variables in the differential equation (88) in order that the new differential equation will have a diagonal matrix. Let

$$\mathbf{x} = P\mathbf{y} \tag{92}$$

where P is an $n \times n$ constant matrix. Then (see Prob. 7)

$$\frac{d\mathbf{x}}{dt} = P \frac{d\mathbf{y}}{dt} \tag{93}$$

and *if P is nonsingular*, substituting (92) and (93) into (88) yields the differential equation for $\mathbf{y}$

$$\frac{d\mathbf{y}}{dt} = P^{-1}AP\mathbf{y} \tag{94}$$

Let

$$D = P^{-1}AP \tag{95}$$

and thus, if we can choose P such that D is diagonal, then we can solve the system (94) and finally obtain the solution $\mathbf{x}$ of Eq. (88) from Eq. (92), i.e., $\mathbf{x} = P\mathbf{y}$.

The construction of a matrix P which *diagonalizes* A by the transformation (95)‡ is not always a simple matter. In fact it is not even always

† This result could also have been obtained by solving the equations $dx_i/dt = \lambda_i x_i$ to get $x_i = c_i e^{\lambda_i t}$.

‡ The transformation given by Eq. (95) is called a *similarity transformation* and the matrices A and D are said to be *similar*.

possible (see Prob. 5). However, *if the eigenvalues of A are distinct*, then a diagonalizing matrix P always exists and can be found by the following method.

Let $\lambda_1, \lambda_2, \ldots, \lambda_n$ be the (distinct) eigenvalues of A and $x_1, \ldots, x_n$ be corresponding eigenvectors. That is,

$$Ax_i = \lambda_i x_i \qquad i = 1, 2, \ldots, n \tag{96}$$

Now *construct the matrix P which has the eigenvectors* $x_1, \ldots, x_n$ *as its columns*. Thus if the eigenvectors are

$$x_1 = \begin{bmatrix} 1 \\ -3 \end{bmatrix} \quad \text{and} \quad x_2 = \begin{bmatrix} 2 \\ 0 \end{bmatrix}$$

then

$$P = \begin{bmatrix} 1 & 2 \\ -3 & 0 \end{bmatrix}$$

It is claimed that this matrix P will diagonalize A.

First of all, P is nonsingular since the determinant of P is just the Wronskian of the eigenvectors and this is not zero since the eigenvectors are linearly independent. Next, the product AP can be computed by successive application of Eq. (96) for $i = 1, 2, \ldots, n$. The result is a matrix C with columns $\lambda_1 x_1, \lambda_2 x_2, \ldots, \lambda_n x_n$. Now, if we define D to be the diagonal matrix

$$D = \begin{bmatrix} \lambda_1 & & & & \\ & \lambda_2 & & \bigcirc & \\ & & \cdot & & \\ & & & \cdot & \\ & \bigcirc & & & \cdot \\ & & & & \lambda_n \end{bmatrix}$$

then it is evident that the product PD produces exactly the same matrix C. Therefore

$$AP = PD \tag{97}$$

and, since P is nonsingular, P^{-1} exists, and, multiplying (97) by P^{-1}, we have

$$D = P^{-1}AP \tag{98}$$

Example 3 Find the general solution of $dx/dt = Ax$ by using matrix methods if

$$A = \begin{bmatrix} 4 & 1 \\ 3 & 2 \end{bmatrix}$$

From Example 1, Sec. 8-6, the eigenvalues of A are $\lambda_1 = 1$, $\lambda_2 = 5$ with corresponding eigenvectors

$$x_1 = \begin{bmatrix} 1 \\ -3 \end{bmatrix} \quad \text{and} \quad x_2 = \begin{bmatrix} 1 \\ 1 \end{bmatrix}$$

The matrix P is then

$$P = \begin{bmatrix} 1 & 1 \\ -3 & 1 \end{bmatrix}$$

and it follows that

$$P^{-1} = \tfrac{1}{4}\begin{bmatrix} 1 & -1 \\ 3 & 1 \end{bmatrix} \quad \text{and} \quad D = P^{-1}AP = \begin{bmatrix} 1 & 0 \\ 0 & 5 \end{bmatrix}$$

The general solution of $dy/dt = Dy$ is

$$\mathbf{y} = \begin{bmatrix} c_1 e^t \\ c_2 e^{5t} \end{bmatrix}$$

and hence the general solution of $d\mathbf{x}/dt = A\mathbf{x}$ is

$$\mathbf{x} = P\mathbf{y} = \begin{bmatrix} c_1 e^t + c_2 e^{5t} \\ -3c_1 e^t + c_2 e^{5t} \end{bmatrix}$$

Compare Eq. (84).

Cayley-Hamilton theorem   We are now in a position to state an extremely important theorem.

Theorem 2   *Cayley-Hamilton.   Every $n \times n$ matrix A satisfies its own characteristic equation.   That is, if the characteristic polynomial is*

$$p(\lambda) \equiv |A - \lambda I| \equiv (-1)^n \lambda^n + a_1 \lambda^{n-1} + \cdots + a_{n-1}\lambda + a_n \qquad (99)$$

then A satisfies the matrix equation

$$p(A) \equiv (-1)^n A^n + a_1 A^{n-1} + \cdots + a_{n-1}A + a_n I = 0 \qquad (100)$$

Proof   (for the case of *distinct eigenvalues*)†   If A has distinct eigenvalues $\lambda_1, \lambda_2, \ldots, \lambda_n$, there is a matrix P which diagonalizes A.   That is,

$$P^{-1}AP = D$$

where

$$D = \begin{bmatrix} \lambda_1 & & & \\ & \lambda_2 & & \bigcirc \\ & & \ddots & \\ & \bigcirc & & \ddots \\ & & & \lambda_n \end{bmatrix}$$

---

† For a proof in the general case see, for example, Finkbeiner [15], chap. 8.

Now since $A = PDP^{-1}$, we have $A^2 = PDP^{-1}PDP^{-1} = PD^2P^{-1}$ and, by induction, $A^n = PD^nP^{-1}$. Hence $p(A) = Pp(D)P^{-1}$, but, by Theorem 1, $p(D)$ is a diagonal matrix with the elements on the diagonal equal to $p(\lambda_1)$, $p(\lambda_2), \ldots, p(\lambda_n)$. These all vanish since $p(\lambda_i) = 0$ for all eigenvalues λ_i. Therefore $p(A) = 0$.

Besides having considerable theoretical importance, this theorem is also useful in carrying out computations with matrices. This stems from the fact that the coefficient of A^n in (100) is different from zero, and hence we can solve for A^n in terms of lower powers of A. Successive application of this technique enables one to express any power of A as a polynomial of degree $n - 1$ or less. In particular, we have the following theorem.

Theorem 3   *Let A be an $n \times n$ matrix with characteristic polynomial $p(\lambda)$ having distinct roots†* $\lambda_1, \lambda_2, \ldots, \lambda_n$. *If $f(A)$ is a polynomial of degree $m \geq n$, then $f(A)$ can be expressed as*

$$f(A) = \alpha_0 I + \alpha_1 A + \alpha_2 A^2 + \cdots + \alpha_{n-1}A^{n-1} \tag{101}$$

Furthermore the coefficients α_i can be found by solving the following system of simultaneous algebraic equations:

$$\begin{aligned}
f(\lambda_1) &= \alpha_0 + \alpha_1\lambda_1 + \cdots + \alpha_{n-1}\lambda_1^{n-1} \\
f(\lambda_2) &= \alpha_0 + \alpha_1\lambda_2 + \cdots + \alpha_{n-1}\lambda_2^{n-1} \\
& \vdots \\
f(\lambda_n) &= \alpha_0 + \alpha_1\lambda_n + \cdots + \alpha_{n-1}\lambda_n^{n-1}
\end{aligned} \tag{102}$$

Proof   Consider the scalar polynomial $f(\lambda)$ of degree $m \geq n$. A well-known theorem from elementary algebra states that if $f(\lambda)$ is divided by the characteristic polynomial $p(\lambda)$ of degree n, the result is a quotient polynomial $q(\lambda)$ plus a remainder polynomial $r(\lambda)$ of degree $n - 1$ or less divided by $p(\lambda)$. Thus we can write

$$f(\lambda) = p(\lambda)q(\lambda) + r(\lambda) \tag{103}$$

This polynomial identity in λ is also true for the matrix A (see Prob. 14). That is,

$$f(A) = p(A)q(A) + r(A) \tag{104}$$

But $p(A) = 0$ and hence

$$f(A) = r(A) \tag{105}$$

and the coefficients of r, and hence f, can be obtained by successive substitution of $\lambda_1, \lambda_2, \ldots, \lambda_n$ into Eq. (103) to yield Eqs. (102).

† See Probs. 18 and 19 for the case of multiple eigenvalues.

Example 4   Given

$$A = \begin{bmatrix} -3 & -1 \\ -8 & 4 \end{bmatrix}$$

compute $f(A) = A^{10}$. The characteristic polynomial is $p(\lambda) = \lambda^2 - \lambda - 20$ and the eigenvalues are $\lambda_1 = 5$ and $\lambda_2 = -4$. Since A is a 2×2 matrix, the remainder polynomial $r(\lambda)$ in Eq. (103) is of degree 1, i.e.,

$$r(\lambda) = \alpha_0 + \alpha_1 \lambda$$

Now, since $p(\lambda_i) = 0$ for the eigenvalues, substituting $\lambda_1 = 5$ and $\lambda_2 = -4$ into (103) yields the pair of equations

$$5^{10} = \alpha_0 + 5\alpha_1$$
$$(-4)^{10} = \alpha_0 - 4\alpha_1$$

with solutions

$$\alpha_0 = \tfrac{1}{9}(4 \times 5^{10} + 5 \times 4^{10})$$
$$\alpha_1 = \tfrac{1}{9}(5^{10} - 4^{10})$$

Hence equation (105) yields

$$f(A) = A^{10} = r(A) = \alpha_0 I + \alpha_1 A = \tfrac{1}{9}\begin{bmatrix} 5^{10} + 8 \times 4^{10} & -5^{10} + 4^{10} \\ -8 \times 5^{10} + 8 \times 4^{10} & 8 \times 5^{10} + 4^{10} \end{bmatrix}$$

Functions of matrices   Since we have defined a matrix polynomial $P_m(A) = \sum_{k=0}^{m} a_k A^k$ for any positive integer m, one is tempted to investigate the possibility of introducing the notion of a power series of matrices, since a power series is, after all, merely a sequence of polynomials.† To begin, we need the following definition.

Definition 7   *A sequence $\{B_m\}$ is a matrix-valued function of the nonnegative integers. That is, a unique matrix B_m is associated with each integer $m = 0, 1, 2, \ldots$. A sequence $\{B_m\}$ of matrices with elements $b_{ij}^{(m)}$ is said to converge to a matrix B with elements b_{ij} if all the sequences of elements $\{b_{ij}^{(m)}\}$ converge to the elements b_{ij}. We write $\{B_m\} \to B$ or $\lim_{m \to \infty} B_m = B$.*

Example 5   Given the sequence $\{B_m\}$ where

$$B_m = \begin{bmatrix} \dfrac{1}{m!} & m \\ e^{-m} & m+1 \\ & 6 \end{bmatrix}$$

then

$$\lim_{m \to \infty} B_m = \begin{bmatrix} 0 & 1 \\ 0 & 6 \end{bmatrix} = B$$

and the sequence $\{B_m\}$ converges to B.

† See Appendix B for a brief review of infinite series.

Definition 8 *Given a sequence* $\{B_k\}$, *form a new sequence* $\{S_m\}$ *of partial sums* S_m *by defining* $S_m = \sum_{k=0}^{m} B_k$. *The sequence* $\{S_m\}$ *is then called an infinite series. If the sequence* $\{S_m\}$ *converges to a matrix* S, *we say that the series converges, and* S *is its sum. That is,*

$$S = \lim_{m \to \infty} S_m = \lim_{m \to \infty} \sum_{k=0}^{m} B_k$$

Example 6 Given a matrix

$$A = \begin{bmatrix} 2 & 0 \\ 0 & 3 \end{bmatrix}$$

we know that

$$A^k = \begin{bmatrix} 2^k & 0 \\ 0 & 3^k \end{bmatrix}$$

and hence, if we define $P_m(A) = \sum_{k=0}^{m} (1/k!)A^k$, we see that

$$P_m(A) = \begin{bmatrix} \sum_{k=0}^{m} \dfrac{2^k}{k!} & 0 \\ 0 & \sum_{k=0}^{m} \dfrac{3^k}{k!} \end{bmatrix}$$

and hence

$$\lim_{m \to \infty} P_m(A) = \begin{bmatrix} \lim\limits_{m \to \infty} \sum_{k=0}^{m} \dfrac{2^k}{k!} & 0 \\ 0 & \lim\limits_{m \to \infty} \sum_{k=0}^{m} \dfrac{3^k}{k!} \end{bmatrix} = \begin{bmatrix} e^2 & 0 \\ 0 & e^3 \end{bmatrix}$$

In Example 6 we see that the $\lim_{m \to \infty} \sum_{k=0}^{m} (1/k!)A^k$ exists and is a matrix whose elements are exponentials. This suggests the following definition.

Definition 9 *Let* A *be a matrix and let* $\lim_{m \to \infty} \sum_{k=0}^{m} (1/k!)A^k$ *exist.*[†] *Then we define the matrix function* e^A *by*

$$e^A = \lim_{m \to \infty} \sum_{k=0}^{m} \frac{1}{k!} A^k \tag{106}$$

or as we sometimes write for convenience

$$e^A = I + A + \frac{A^2}{2!} + \cdots + \frac{A^m}{m!} + \cdots \tag{107}$$

[†] It can be shown (e.g., see Finkbeiner [15], chap. 10) that $\lim_{m \to \infty} \Sigma(1/k!)A^k$ will exist for *any* square matrix A.

Because of the importance of the exponential function e^x in analysis, it might be expected that the function e^A would play a central role in the theory of functions of matrices. This is so, at least in those parts of the theory which apply to differential equations. However, the reader must use caution in working with the function e^A since it retains *some but not all* of the properties of e^x. For example, $e^0 = I$, where 0 is the zero matrix, as can be seen from Eq. (106). Also $e^{-A} = (e^A)^{-1}$. But on the other hand, $e^A e^B$ is *not always* equal to e^{A+B}. In fact, it need not even be equal to $e^B e^A$, unless A and B commute. (See Prob. 17.)

Example 7   Compute e^A where

$$A = \begin{bmatrix} 4 & 1 \\ 3 & 2 \end{bmatrix}$$

From Example 3 we know that A can be diagonalized by the matrix

$$P = \begin{bmatrix} 1 & 1 \\ -3 & 1 \end{bmatrix}$$

That is,

$$P^{-1}AP = D = \begin{bmatrix} 1 & 0 \\ 0 & 5 \end{bmatrix}$$

Hence $A = PDP^{-1}$ and

$$
\begin{aligned}
e^A = e^{PDP^{-1}} &= \lim_{n \to \infty} \sum_{k=0}^{n} \frac{1}{k!} (PDP^{-1})^k \\
&= \lim_{n \to \infty} \sum_{k=0}^{n} \frac{1}{k!} PD^k P^{-1} \\
&= P \left(\lim_{n \to \infty} \sum_{k=0}^{n} \frac{1}{k!} D^k \right) P^{-1} \\
&= P e^D P^{-1}
\end{aligned}
$$

Now e^D is easily evaluated (see Example 6),

$$e^D = \begin{bmatrix} e & 0 \\ 0 & e^5 \end{bmatrix}$$

and therefore we have finally

$$
e^A = P e^D P^{-1} = \begin{bmatrix} 1 & 1 \\ -3 & 1 \end{bmatrix} \begin{bmatrix} e & 0 \\ 0 & e^5 \end{bmatrix} \frac{1}{4} \begin{bmatrix} 1 & -1 \\ 3 & 1 \end{bmatrix} = \frac{1}{4} \begin{bmatrix} e + 3e^5 & -e + e^5 \\ -3e + 3e^5 & 3e + e^5 \end{bmatrix}
$$

Another method of evaluating e^A which often involves less computation than the method of Example 7 is suggested by Theorem 3 and Example 4. In particular, if A is an $n \times n$ matrix with distinct eigenvalues $\lambda_1, \lambda_2, \ldots, \lambda_n$, then it can be shown† that

$$e^\lambda = p(\lambda)q(\lambda) + r(\lambda) \tag{108}$$

† See B. Friedman [16].

where λ is any complex number, $p(\lambda)$ is the characteristic polynomial (of degree n) of A, $q(\lambda)$ is an analytic function of λ, and $r(\lambda)$ is a polynomial of degree $\leq n-1$. The same identity holds for the matrix A, i.e.,

$$e^A = p(A)q(A) + r(A)$$

But $p(A) = 0$ and hence

$$e^A = r(A) = \alpha_0 I + \alpha_1 A + \cdots + \alpha_{n-1} A^{n-1}$$

As before, the coefficients of $r(A)$ can be found by substituting $\lambda_1, \lambda_2, \ldots, \lambda_n$ into Eq. (108).

Example 8   Find e^A if

$$A = \begin{bmatrix} 4 & 1 \\ 3 & 2 \end{bmatrix}$$

Set $e^{\lambda_i} = \alpha_0 + \alpha_1 \lambda_i$ and substitute the eigenvalues of A, i.e., $\lambda_1 = 1$, $\lambda_2 = 5$, to get

$$e = \alpha_0 + \alpha_1$$
$$e^5 = \alpha_0 + 5\alpha_1$$

and hence

$$\alpha_0 = \tfrac{1}{4}(5e - e^5) \qquad \alpha_1 = \tfrac{1}{4}(e^5 - e)$$

Finally

$$e^A = \alpha_0 I + \alpha_1 A = \tfrac{1}{4} \begin{bmatrix} e + 3e^5 & -e + e^5 \\ -3e + 3e^5 & 3e + e^5 \end{bmatrix}$$

If t is a real number and A is a constant square matrix, then tA is a square matrix and e^{tA} is defined as in Definition 9. That is,

$$e^{tA} = \lim_{m \to \infty} \sum_{k=0}^{m} \frac{1}{k!}(tA)^k = \lim_{m \to \infty} \sum_{k=0}^{m} \frac{1}{k!} t^k A^k$$

The derivative of e^{tA} is easily obtained from this equation. We have†

$$\frac{d}{dt}(e^{tA}) = \lim_{m \to \infty} \sum_{k=0}^{m} \frac{1}{k!}(kt^{k-1})A^k$$

$$= A \lim_{m \to \infty} \sum_{k=1}^{m} \frac{1}{(k-1)!} t^{k-1} A^{k-1}$$

$$= A \lim_{m \to \infty} \sum_{k=0}^{m-1} \frac{1}{k!} t^k A^k$$

or

$$\frac{d}{dt}(e^{tA}) = Ae^{tA} = e^{tA}A \tag{109}$$

† The justification for differentiating the series termwise is left as an exercise for the reader. (See Prob. 10.)

Thus the function e^{tA} retains the special property which made $e^{\lambda t}$ so important in the solution of linear differential equations. It follows immediately from Eq. (109) that if $\mathbf{c}$ is any constant vector, then the vector $\mathbf{x} = e^{tA}\mathbf{c}$ satisfies

$$\frac{d\mathbf{x}}{dt} = A\mathbf{x} \tag{110}$$

and hence we have a simple and suggestive representation for a solution of a system of linear differential equations. The following fundamental theorem summarizes and extends these remarks.

Theorem 4 *Let A be an $n \times n$ constant matrix. Then the general solution of the differential equation*

$$\frac{d\mathbf{x}(t)}{dt} = A\mathbf{x}(t) \tag{111}$$

is given by $\mathbf{x}(t) = e^{tA}\mathbf{c}$ where $\mathbf{c}$ is an arbitrary constant vector. The (unique) solution of the differential equation which also satisfies the initial condition $\mathbf{x}(t_0) = \mathbf{x}_0$ is given by $\mathbf{x}(t) = e^{(t-t_0)A}\mathbf{x}_0$.

Proof Let $\mathbf{x}(t)$ be *any* solution of (111) and consider the vector $\mathbf{v}(t)$ defined by

$$\mathbf{v}(t) = e^{-tA}\mathbf{x}(t) \tag{112}$$

Differentiating,† we have

$$\frac{d\mathbf{v}}{dt} = -e^{-tA}A\mathbf{x} + e^{-tA}\frac{d\mathbf{x}}{dt} = e^{-tA}\left(-A\mathbf{x} + \frac{d\mathbf{x}}{dt}\right)$$

and, since $\mathbf{x}$ satisfies (111), the term in parentheses vanishes and we have $d\mathbf{v}/dt = 0$. Thus $\mathbf{v}(t) = \mathbf{c} = \text{const}$ and then by multiplying both sides of Eq. (112) by e^{tA}, we have

$$\mathbf{x} = e^{tA}\mathbf{c} \tag{113}$$

If, further, we want to satisfy the initial condition $\mathbf{x}(t_0) = \mathbf{x}_0$, let $t = t_0$ in (113) and we have $\mathbf{x}_0 = e^{t_0 A}\mathbf{c}$ or $\mathbf{c} = e^{-t_0 A}\mathbf{x}_0$. Substituting this value of $\mathbf{c}$ into Eq. (113) yields the desired result.

Example 9 Use Theorem 4 to solve the initial-value problem

DE: $\dfrac{d\mathbf{x}}{dt} = A\mathbf{x}$ where $A = \begin{bmatrix} 4 & 1 \\ 3 & 2 \end{bmatrix}$

IC: $\mathbf{x}(0) = \mathbf{x}_0$ where $\mathbf{x}_0 = \begin{bmatrix} -1 \\ 7 \end{bmatrix}$

† The proof of the rule for differentiating the product of a matrix and a vector is left as an exercise. See Prob. 7.

From Example 3, A is diagonalized by the matrix

$$P = \begin{bmatrix} 1 & 1 \\ -3 & 1 \end{bmatrix}$$

and

$$D = \begin{bmatrix} 1 & 0 \\ 0 & 5 \end{bmatrix} = P^{-1}AP$$

Hence (as in Example 7)

$$e^{tA} = e^{tPDP^{-1}} = Pe^{tD}P^{-1} = P\begin{bmatrix} e^t & 0 \\ 0 & e^{5t} \end{bmatrix}P^{-1}$$

On multiplying by P and P^{-1}, we have

$$e^{tA} = \tfrac{1}{4}\begin{bmatrix} e^t + 3e^{5t} & -e^t + e^{5t} \\ -3e^t + 3e^{5t} & 3e^t + e^{5t} \end{bmatrix}$$

The solution of the initial-value problem is then found from $\mathbf{x} = e^{tA}\mathbf{x_0}$ and we obtain the *vector*

$$\mathbf{x} = \begin{bmatrix} -2e^t + e^{5t} \\ 6e^t + e^{5t} \end{bmatrix}$$

We leave it to the reader to carry out the calculations and to check the result.

The evaluation of e^{tA} in this example could also have been accomplished by using the method of Example 8 to evaluate e^B where $B = tA$. That is, in setting $e^\lambda = \alpha_0 + \alpha_1\lambda$, λ would be an eigenvalue of the matrix tA.

The nonhomogeneous equation   Matrix methods provide us with an elegant solution for nonhomogeneous systems. The essence of the familiar method of variation of parameters becomes apparent here for the first time.

Given a constant matrix A and a vector $\mathbf{f}(t)$ which is continuous on some interval J, consider the problem of solving the nonhomogeneous system

$$\frac{d\mathbf{x}}{dt} = A\mathbf{x} + \mathbf{f}(t) \tag{114}$$

The general solution $\mathbf{x}_h(t)$ of the corresponding homogeneous equation is given by

$$\mathbf{x}_h(t) = e^{tA}\mathbf{c} \tag{115}$$

where $\mathbf{c}$ is an arbitrary constant. The method of variation of parameters consists in replacing $\mathbf{c}$ by a function of t, say $\mathbf{v}(t)$, and determining $\mathbf{v}(t)$ so that

$$\mathbf{x}(t) = e^{tA}\mathbf{v}(t) \tag{116}$$

will be a solution of Eq. (114).

Differentiating (116), we have

$$\frac{d\mathbf{x}}{dt} = Ae^{tA}\mathbf{v} + e^{tA}\frac{d\mathbf{v}}{dt} = A\mathbf{x} + e^{tA}\frac{d\mathbf{v}}{dt}$$

but if this is to equal the right side of (114), we must have

$$e^{tA}\frac{d\mathbf{v}}{dt} = \mathbf{f}(t) \tag{117}$$

or

$$\frac{d\mathbf{v}}{dt} = e^{-tA}\mathbf{f}(t) \tag{118}$$

and therefore $\mathbf{x}(t)$ in Eq. (116) will be the solution of Eq. (114) provided that

$$\mathbf{v}(t) = \int e^{-tA}\mathbf{f}(t)\,dt \tag{119}$$

This method yields the general solution of Eq. (114) and, in fact, the same procedure will work even when A depends on t, provided that the homogeneous equation can be solved.   (See Prob. 11.)

If the initial condition $\mathbf{x}(t_0) = \mathbf{x}_0$ is to be satisfied by the solution of Eq. (114), then the above procedure yields the solution

$$\mathbf{x}(t) = e^{(t-t_0)A}\mathbf{x}_0 + \int_{t_0}^{t} e^{(t-\tau)A}\mathbf{f}(\tau)\,d\tau \tag{120}$$

It would be very instructive for the reader to compare the present discussion of the nonhomogeneous system with the discussion in Sec. 5-6. In particular, note that Eq. (118), for $\mathbf{v}$, which arose so naturally here, had to be "pulled out of the hat" in Sec. 5-6.

Problems 8-7

1. a. Prove the associative law for matrix multiplication.
   b. Prove the distributive law.   [See Eq. (85).]
2. Prove that $A^m A^k = A^k A^m = A^{m+k}$ for any square matrix A and any nonnegative integers m and k.
3. Prove Theorem 1.
4. Find a formula for the inverse of a general 2×2 matrix.
5. Given the matrix

$$A = \begin{bmatrix} 1 & 1 \\ 0 & 1 \end{bmatrix}$$

show that A *cannot* be diagonalized.   That is, show that there is no matrix P such that PAP^{-1} is diagonal.
6. The definition of the matrix function e^A by using the familiar power series suggests the possibility of defining other matrix functions in the same way.

For example, define $\sin A = A - A^3/3! + A^5/5! - \cdots$ and evaluate $\sin A$ if

$$A = \begin{bmatrix} \dfrac{\pi}{2} & 2 \\ 0 & \dfrac{3\pi}{2} \end{bmatrix}$$

See Example 7.

7. Let $A(t)$ be an $n \times n$ matrix and $\mathbf{x}(t)$ an n vector. Prove that if A and $\mathbf{x}$ are differentiable functions of t, then

$$\frac{d}{dt}[A(t)\mathbf{x}(t)] = A(t)\frac{d\mathbf{x}(t)}{dt} + \frac{dA(t)}{dt}\mathbf{x}(t)$$

8. Let $A(t)$ and $B(t)$ be differentiable $n \times n$ matrix functions of t. Prove that

$$\frac{d}{dt}[A(t)B(t)] = A(t)\frac{dB(t)}{dt} + \frac{dA(t)}{dt}B(t)$$

Note that the order of the terms in each product is important since the various matrices involved do not necessarily commute. See Prob. 7.

9. Let $A(t) = \begin{bmatrix} 1 & t \\ 0 & t^2 \end{bmatrix}$

Verify that

$$\frac{d[A^2(t)]}{dt} = A(t)\frac{dA(t)}{dt} + \frac{dA(t)}{dt}A(t)$$

Show also that $d[A^2(t)]/dt$ is *not* equal to $2A(t)\,dA(t)/dt$. Thus the power rule of differentiation does *not* hold for matrices since a matrix need not commute with its own derivative.

10. Complete the derivation of Eq. (109) by justifying the termwise differentiation of the series.

11. Let $\mathbf{x}_1(t), \ldots, \mathbf{x}_n(t)$ be n linearly independent vectors satisfying the differential equation $d\mathbf{x}(t)/dt = A(t)\mathbf{x}(t)$. Let $X(t)$ be the matrix whose columns are the vectors $\mathbf{x}_1(t), \ldots, \mathbf{x}_n(t)$.

a. Show that $dX(t)/dt = A(t)X(t)$.

b. Let $\mathbf{f}(t)$ be a given continuous vector function and use the method of variation of parameters to derive a formula for the general solution of the nonhomogeneous differential equation

$$\frac{d\mathbf{x}(t)}{dt} = A(t)\mathbf{x}(t) + \mathbf{f}(t) \tag{121}$$

c. Find the solution of Eq. (121) which satisfies $\mathbf{x}(t_0) = \mathbf{x}_0$.

12. Find the general solution by matrix methods.   See Example 8.

DE: $\dfrac{d\mathbf{x}}{dt} = A\mathbf{x}$ where $A = \begin{bmatrix} 1 & -1 \\ -2 & 0 \end{bmatrix}$

Compare Example 1, Sec. 8-2.

13. Solve by matrix methods.   See Example 9.

DE: $\dfrac{dx_1}{dt} = 9x_1 - 8x_2$

$\dfrac{dx_2}{dt} = 24x_1 - 19x_2$

IC: $x_1(0) = 1$

$x_2(0) = 0$

14. Establish the identity in Eq. (104).
15. Compute e^{tA} in Exercise 9 by using the method of Example 8.
16. Solve Prob. 13 by the method of Example 8.

17. Given $A = \begin{bmatrix} 0 & 1 \\ 0 & 0 \end{bmatrix}$ and $B = \begin{bmatrix} 1 & 0 \\ 0 & 0 \end{bmatrix}$

compute:

a. AB b. BA
c. e^A d. e^B
e. e^{A+B} f. $e^A e^B$
g. $e^B e^A$

Hence show $AB \neq BA$ and $e^A e^B \neq e^B e^A \neq e^{A+B}$.

18. If $\lambda = \lambda_0$ is a double root of the characteristic equation $p(\lambda_0) = 0$, then it is also true that $p'(\lambda_0) = 0$.   Use Eq. (103) to show that $f'(\lambda_0) = r'(\lambda_0)$.   This gives a second equation for the coefficients of $r(\lambda)$ in the case of a double root. Use this method to compute A^5 if

$A = \begin{bmatrix} 4 & -1 \\ 1 & 2 \end{bmatrix}$

19. Generalize the method of Prob. 18 to apply to the determination of $f(A)$ when A has a multiple eigenvalue.

The Laplace Transform

9

9-1 Introduction

The Laplace transform has been widely adopted by scientists and engineers as an efficient tool for solving linear differential equations. In this chapter we shall develop the fundamental properties of Laplace transforms, and we shall see how they are used to solve linear differential equations and certain other problems. However, a thorough treatment† of this subject requires more knowledge of analysis, particularly the theory of complex variables, than we presuppose here.

9-2 Improper integrals

We shall digress briefly to consider some facts about improper integrals that will be useful in the sequel.

We recall that if $f(x)$ is defined for $a \leq x < \infty$, then

$$\int_a^\infty f(x)\, dx \equiv \lim_{R \to \infty} \int_a^R f(x)\, dx \tag{1}$$

provided the limit exists. If the limit exists, we say the improper integral (1) *converges*; otherwise it *diverges*.

Theorem 1 *Comparison test. If*

$$0 \leq f(x) \leq g(x) \qquad a \leq x < \infty$$

and $\int_a^\infty g(x)\, dx$ *converges, then* $\int_a^\infty f(x)\, dx$ *converges.*

† See Churchill [9] or Widder [37].

Proof $F(R) = \int_a^R f(x)\,dx \le \int_a^R g(x)\,dx \le \int_a^\infty g(x)\,dx = A$

Therefore $F(R)$ is a nondecreasing function which is bounded by A.   By a theorem of calculus† $\lim_{R\to\infty} F(R)$ exists.

Definition 1   *If $\int_a^\infty |f(x)|\,dx$ converges, then $\int_a^\infty f(x)\,dx$ is said to converge absolutely.*

Theorem 2   *If $\int_a^\infty |f(x)|\,dx$ converges, then $\int_a^\infty f(x)\,dx$ converges.*

Proof $\int_a^\infty f(x)\,dx = \int_a^\infty |f(x)|\,dx - \int_a^\infty [|f(x)| - f(x)]\,dx$

The first integral on the right converges by hypothesis.   The second integral on the right converges by the comparison test since

$0 \le |f(x)| - f(x) \le 2\,|f(x)|$

Problems 9-2

1. If $f(x)$ is continuous and if there exist constants c, α such that $|f(x)| < ce^{\alpha x}$, $0 \le x < \infty$, show that $\int_0^\infty e^{-sx} f(x)\,dx$ converges absolutely (therefore converges) for $s > \alpha$.

2. Prove that if for $a \le x < \infty$ we have $0 \le f(x) \le g(x)$ and $\int_a^\infty f(x)\,dx$ diverges, then $\int_a^\infty g(x)\,dx$ diverges.

9-3   The Laplace transform

The Laplace transform of a function $f(x)$, $0 \le x < \infty$, is defined by the improper integral

$$\mathscr{L}\{f(x)\} = \hat{f}(s) = \int_0^\infty e^{-sx} f(x)\,dx \tag{2}$$

provided that the integral converges for at least one value of s.‡   From the defining equation we see that functions $f(x)$ are transformed into new functions $\hat{f}(s)$ (read "f roof of s") by the Laplace transformation.   We can consider that Eq. (2) defines an *operator* $\mathscr{L}$ which acts on *object functions* $f(x)$ to produce *image functions* or *transforms* $\hat{f}(s)$.   One advantage of this

† See Buck [5], pp. 48 and 81.

‡ It can be shown that if $\hat{f}(s)$ in Eq. (2) exists for $s = s_0$, then it exists for $s > s_0$.   See Agnew [1], p. 361.   See also Theorem 2 of this section.

operation is that in many cases the transform $\hat{f}(s)$ is simpler than the object function $f(x)$.

Example 1 $\mathscr{L}\{e^{ax}\} = \displaystyle\int_0^\infty e^{-sx}e^{ax}\, dx$

$$= \lim_{R\to\infty} \int_0^R e^{-(s-a)x}\, dx$$

$$= \lim_{R\to\infty} \frac{e^{-(s-a)R} - 1}{-(s-a)}$$

If $s > a$, $\displaystyle\lim_{R\to\infty} e^{-(s-a)R} = 0$; therefore

$$\mathscr{L}\{e^{ax}\} = \frac{1}{s-a} \qquad s > a \tag{3}$$

Thus the transform of the transcendental function e^{ax} is the rational function $1/(s-a)$.

The operator $\mathscr{L}$ has the important property of linearity, that is,

$$\mathscr{L}\{af(x) + bg(x)\} = a\mathscr{L}\{f(x)\} + b\mathscr{L}\{g(x)\} = a\hat{f}(s) + b\hat{g}(s) \tag{4}$$

for those values of s for which both $\hat{f}(s)$ and $\hat{g}(s)$ exist. The proof is direct (see Prob. 2).

It is sometimes convenient to consider the transform of a complex-valued function of a real variable x. If $f(x)$ is such a function, then by definition we have

$$\mathscr{L}\{f(x)\} = \int_0^\infty e^{-sx}f(x)\, dx = \lim_{R\to\infty} \int_0^R e^{-sx}f(x)\, dx$$

where now the limit is a limit of complex numbers (see Appendix B). If $f(x) = u(x) + iv(x)$ where $u(x)$ and $v(x)$ are real functions that possess transforms, then it is easy to show that

$$\mathscr{L}\{f(x)\} = \mathscr{L}\{u(x)\} + i\mathscr{L}\{v(x)\} \tag{5}$$

or

$$\mathscr{L}\{\operatorname{Re} f(x)\} = \operatorname{Re}\mathscr{L}\{f(x)\} \tag{5a}$$

$$\mathscr{L}\{\operatorname{Im} f(x)\} = \operatorname{Im}\mathscr{L}\{f(x)\} \tag{5b}$$

These relations are often quite useful.

Example 2   If $f(x) = e^{ibx}$, then

$$\mathscr{L}\{f(x)\} = \int_0^\infty e^{-sx}e^{ibx}\, dx = \lim_{R\to\infty} \int_0^R e^{(-s+ib)x}\, dx$$

$$= \lim_{R\to\infty} \frac{e^{-(s-ib)R} - 1}{-(s-ib)} = \lim_{R\to\infty} \frac{e^{-sR}e^{ibR} - 1}{-(s-ib)}$$

If $s > 0$, then $\lim_{R \to \infty} e^{-sR}e^{ibR} = 0$. (Note: $|e^{ibR}| = 1$.) Therefore for $s > 0$

$$\mathscr{L}\{e^{ibx}\} = \frac{1}{s - ib}$$

which corresponds to Eq. (3). Since $e^{ibx} = \cos bx + i \sin bx$, we have the useful formulas for $s > 0$

$$\mathscr{L}\{\cos bx\} = \mathscr{L}\{\operatorname{Re} e^{ibx}\} = \operatorname{Re} \mathscr{L}\{e^{ibx}\} = \operatorname{Re} \frac{1}{s - ib} = \frac{s}{s^2 + b^2} \tag{6a}$$

$$\mathscr{L}\{\sin bx\} = \mathscr{L}\{\operatorname{Im} e^{ibx}\} = \operatorname{Im} \mathscr{L}\{e^{ibx}\} = \operatorname{Im} \frac{1}{s - ib} = \frac{b}{s^2 + b^2} \tag{6b}$$

To each object function there corresponds a unique transform, assuming the transform exists. In order to guarantee that to each transform there exists a unique object function, we shall restrict ourselves to the set of *continuous object functions*. This is the content of the following theorem.

Theorem 1 *If $\mathscr{L}\{f(x)\} \equiv \mathscr{L}\{g(x)\}$ and $f(x)$ and $g(x)$ are continuous for $0 \le x < \infty$, then $f(x) \equiv g(x)$.*

The proof of this important theorem is deferred until the appendix to this chapter (page 303). If $F(s)$ is a given transform and $f(x)$ is a continuous function such that $\mathscr{L}\{f(x)\} = F(s)$, then no other continuous function has $F(s)$ for its transform; in this case we shall write

$$f(x) = \mathscr{L}^{-1}\{F(s)\}$$

and call $f(x)$ the *inverse Laplace transform*† of $F(s)$. For example, if $F(s) = 1/(s - 1)$, then

$$f(x) = \mathscr{L}^{-1}\left\{\frac{1}{s - 1}\right\} = e^x$$

and e^x is the only continuous function having the transform $1/(s - 1)$.

The Laplace transform of a continuous function will not exist if the function grows too rapidly as $x \to \infty$. For instance, $\mathscr{L}\{e^{x^2}\}$ does not exist. A useful set of functions for which the transform exists is the set of functions of *exponential order* defined as follows.

Definition 1 *$f(x)$ is of exponential order α if $f(x)$ is continuous for $0 \le x < \infty$ and*

$$|f(x)| < ce^{\alpha x} \qquad 0 \le x < \infty \tag{7}$$

where c and α are constants.

† A general formula for $\mathscr{L}^{-1}\{F(s)\}$ exists and is called the inversion integral. It is discussed in more advanced treatments. For example, see Churchill [9] or Widder [37], p. 63.

In other words, a continuous function is of exponential order if it does not grow more rapidly than an exponential as $x \to \infty$. We shall denote the set of all functions of exponential order α by $\mathscr{E}_\alpha$. If $f(x)$ is a function of exponential order α, we write

$$f(x) \in \mathscr{E}_\alpha \tag{8}$$

Example 3 $e^{\alpha x} \sin bx \in \mathscr{E}_\alpha$

since

$$|e^{\alpha x} \sin bx| \leq 1 \cdot e^{\alpha x}$$

Example 4 $x^n \in \mathscr{E}_1$, for since $e^x = 1 + x/2 + \cdots + x^n/n! + \cdots$ we can easily see that $|x^n| \leq n! \, e^x$. By a similar process we can see that if h is *any* number greater than zero, then $x^n \in \mathscr{E}_h$.

We now show that all functions of exponential order possess a Laplace transform.

Theorem 2   *If $f(x) \in \mathscr{E}_\alpha$, then $\hat{f}(s) = \mathscr{L}\{f(x)\}$ is absolutely convergent for $s > \alpha$.*

Proof $|e^{-sx} f(x)| = e^{-sx} |f(x)| \leq ce^{-sx} e^{\alpha x} = ce^{-(s-\alpha)x}$

Since for $s > \alpha, \displaystyle\int_0^\infty ce^{-(s-\alpha)x} \, dx$ converges, we have by the comparison test that $\hat{f}(s) = \displaystyle\int_0^\infty e^{-sx} f(x) \, dx$ converges absolutely.

Before proceeding with the development of the properties of Laplace transforms, we indicate how they are used to solve differential equations by means of a simple example.

Example 5   DE: $y' - 4y = e^x$
IC: $y(0) = 1$

We operate on both sides of the differential equations with the operator $\mathscr{L}$, assuming y and y' possess transforms

$$\mathscr{L}\{y' - 4y\} = \mathscr{L}\{e^x\}$$

$$\mathscr{L}\{y'\} - 4\mathscr{L}\{y\} = \frac{1}{s-1}$$

In Probs. 5, 9, 10, and 11 below we shall show that the solutions of linear differential equations with constant coefficients are of exponential order if the right-hand sides are of exponential order; therefore the solutions will possess transforms.

Consider

$$\mathscr{L}\{y'\} = \int_0^\infty e^{-sx} y'(x) \, dx$$

Integrating by parts, we obtain

$$\mathscr{L}\{y'\} = e^{-sx}y(x)\,\Big|_0^\infty + s\int_0^\infty e^{-sx}y(x)\,dx$$

$$= \lim_{R\to\infty} e^{-sR}y(R) - y(0) + s\mathscr{L}\{y(x)\}$$

In Prob. 4 below we show that $\lim_{R\to\infty} e^{-sR}y(R) = 0$ if y is of exponential order. Therefore

$$\mathscr{L}\{y'\} = s\hat{y}(s) - y(0) = s\hat{y}(s) - 1$$

The transformed differential equation now becomes

$$s\hat{y}(s) - 1 - 4\hat{y}(s) = \frac{1}{s-1}$$

or

$$\hat{y}(s) = \frac{1}{s-4} + \frac{1}{(s-1)(s-4)}$$

In order to find the solution $y(x)$ which is the inverse transform of $\hat{y}(s)$, we replace the second term by its partial-fraction expansion

$$\frac{1}{(s-1)(s-4)} = \frac{1}{3}\frac{1}{s-4} - \frac{1}{3}\frac{1}{s-1}$$

Therefore

$$\hat{y}(s) = \frac{4}{3}\frac{1}{s-4} - \frac{1}{3}\frac{1}{s-1}$$

and (see Prob. 3)

$$y(x) = \mathscr{L}^{-1}\{\hat{y}(s)\} = \tfrac{4}{3}e^{4x} - \tfrac{1}{3}e^x$$

We can verify directly that this is the unique solution of the initial-value problem.

Problems 9-3

1. Find:

a. $\mathscr{L}\{5 + 6e^{-x}\}$ b. $\mathscr{L}\{xe^{ax}\}$

2. Prove that $\mathscr{L}$ is a linear operator [Eq. (4)].
3. Prove that $\mathscr{L}^{-1}$ is a linear operator.
4. If $f(x) \in \mathscr{E}_\alpha$, show that $\lim_{x\to\infty} e^{-sx}f(x) = 0$, $(s > \alpha)$.
5. Show that $p(x)e^{ax}$, where $p(x)$ is a polynomial, is of exponential order.
6. Using Laplace transforms, solve

DE: $y' - 5y = e^{3x} + 4$
IC: $y(0) = 0$

7. If $f(x) \in \mathcal{E}_\alpha$, show that $F(x) = \int_0^x f(t)\, dt \in \mathcal{E}_\alpha$.

8. If $f_1 \in \mathcal{E}_\alpha$ and $f_2 \in \mathcal{E}_\alpha$, show that:

a. $(\alpha f_1 + \beta f_2) \in \mathcal{E}_\alpha$ b. $f_1 f_2 \in \mathcal{E}_{2\alpha}$

9. Prove that if $f(x)$ is of exponential order and

$$y' - ay = f(x)$$

then y and y' are of exponential order. Show also that if f' is of exponential order, then y'' is of exponential order.

10. Show that if $f(x)$ is of exponential order and y is a solution of

$$y'' + ay' + by = f(x)$$

where a and b are constants, then y, y', y'' are of exponential order.

11. Generalize Prob. 10 to an nth-order differential equation.

12. Using the method of Example 2, find $\mathcal{L}\{e^{ax} \cos bx\}$ and $\mathcal{L}\{e^{ax} \sin bx\}$.

9-4   Properties of the Laplace transform

We shall develop those properties of Laplace transforms which will permit us to find the transforms of many functions and to solve linear differential equations with constant coefficients. The first theorem concerns the transform of a derivative. This is the key to the use of transforms in the solution of differential equations (see Sec. 9-3, Example 5).

Theorem 1   *If $f(x) \in \mathcal{E}_\alpha$ and $f'(x)$ is continuous, then $\mathcal{L}\{f'(x)\}$ exists for* $s > \alpha$ *and*

$$\mathcal{L}\{f'(x)\} = s\mathcal{L}\{f(x)\} - f(0) \tag{9}$$

Proof $\mathcal{L}\{f'(x)\} = \displaystyle\int_0^\infty e^{-sx} f'(x)\, dx$

Integrating by parts, we obtain

$$\mathcal{L}\{f'(x)\} = e^{-sx} f(x)\, \Big|_0^\infty + s\int_0^\infty e^{-sx} f(x)\, dx$$

$$= \lim_{R \to \infty} \{e^{-sR} f(R)\} - f(0) + s\mathcal{L}\{f(x)\}$$

From Prob. 4 of Sec. 9-3 we see that $\lim\limits_{R \to \infty} e^{-sR} f(R) = 0$. Therefore

$$\mathcal{L}\{f'(x)\} = s\mathcal{L}\{f(x)\} - f(0)$$

By repeated application of Theorem 1 we may find the transform of derivatives of any order.   If $f, f', \ldots, f^{(n)}$ belong to $\mathscr{E}_\alpha$, we have for $s > \alpha$

$$\mathscr{L}\{f'(x)\} = s\hat{f}(s) - f(0)$$
$$\mathscr{L}\{f''(x)\} = s^2\hat{f}(s) - sf(0) - f'(0)$$

$$\qquad\qquad\vdots \tag{10}$$

$$\mathscr{L}\{f^{(n)}(x)\} = s^n\hat{f}(s) - s^{n-1}f(0) - s^{n-2}f'(0) - \cdots - f^{(n-1)}(0)$$

Although the main use of (10) is in solving linear differential equations, we can also use the formulas to find $\mathscr{L}(f')$ if $\mathscr{L}(f)$ is known.

Example 1   From Example 2, Sec. 9-3, we have

$$\mathscr{L}\{\sin bx\} = \frac{b}{s^2 + b^2}$$

Therefore

$$\mathscr{L}\{\cos bx\} = \frac{1}{b}\mathscr{L}\{(\sin bx)'\} = \frac{s}{s^2 + b^2}$$

Theorem 2   *If* $f \in \mathscr{E}_\alpha$ *and* $F(x) = \displaystyle\int_0^x f(t)\, dt$, *then* $F(x) \in \mathscr{E}_\alpha$ *and*

$$\mathscr{L}\left\{\int_0^x f(t)\, dt\right\} = \frac{1}{s}\hat{f}(s) \qquad s > \alpha \tag{11}$$

Proof   In Prob. 7 of Sec. 9-3 it was shown that $F(x) \in \mathscr{E}_\alpha$.   Applying Theorem 1 to $F(x)$ and noting $F(0) = 0$, we obtain (11).

Example 2   We have seen that $\mathscr{L}\{1\} = 1/s$; therefore

$$\mathscr{L}\{x\} = \mathscr{L}\int_0^x 1 \cdot dt = \frac{1}{s}\mathscr{L}\{1\} = \frac{1}{s^2} \tag{12}$$

Theorem 3   *If* $f \in \mathscr{E}_\alpha$, *then*

$$\mathscr{L}\{e^{-ax}f(x)\} = \hat{f}(s + a) \qquad s > \alpha - a \tag{13}$$

Proof   See Prob. 1 below.

Example 3   Since $\mathscr{L}\{\sin bx\} = \dfrac{b}{s^2 + b^2} \qquad s > 0$
from (13) we have

$$\mathscr{L}\{e^{-ax}\sin bx\} = \frac{b}{(s + a)^2 + b^2} \qquad s > -a$$

Example 4   Since $\mathscr{L}\{x\} = \dfrac{1}{s^2} \qquad s > 0$
we have

$$\mathscr{L}\{xe^{ax}\} = \frac{1}{(s - a)^2} \qquad s > a$$

We now consider what happens to the object function when we differentiate the transform function $\hat{f}(s)$. We have

$$\frac{d}{ds}\hat{f}(s) = \frac{d}{ds}\int_0^\infty e^{-sx}f(x)\,dx$$

If it were possible to differentiate under the integral sign, we would obtain

$$\hat{f}'(s) = \int_0^\infty e^{-sx}[-xf(x)]\,dx = \mathscr{L}\{-xf(x)\}$$

The following theorem shows that this result is correct if $f \in \mathscr{E}_\alpha$.

Theorem 4 *If $f \in \mathscr{E}_\alpha$, then $\hat{f}(s)$ has derivatives of all orders and for $s > \alpha$*

$$\hat{f}'(s) = \mathscr{L}\{-xf(x)\}$$
$$\hat{f}''(s) = \mathscr{L}\{x^2 f(x)\}$$
$$\cdot$$
$$\cdot \qquad\qquad\qquad\qquad\qquad\qquad (14)$$
$$\cdot$$
$$\hat{f}^{(n)}(s) = \mathscr{L}\{(-x)^n f(x)\}$$

This theorem is proved in the appendix to this chapter. The theorem enables us to find many transforms as illustrated below.

Example 5 $\mathscr{L}\{e^{ax}\} = \dfrac{1}{s-a} \qquad s > a$

By (14) we have

$$\mathscr{L}\{xe^{ax}\} = -\frac{d}{ds}\frac{1}{s-a} = \frac{1}{(s-a)^2} \qquad s > a$$

$$\mathscr{L}\{x^2 e^{ax}\} = \frac{d^2}{ds^2}\frac{1}{s-a} = \frac{2}{(s-a)^3} \qquad s > a$$

and by repeated application we have for n, a positive integer,

$$\mathscr{L}\{x^n e^{ax}\} = \frac{n!}{(s-a)^{n+1}} \qquad s > a \qquad\qquad (15)$$

In particular we have

$$\mathscr{L}\{x^n\} = \frac{n!}{s^{n+1}} \qquad s > 0 \qquad\qquad (16)$$

In all of the examples of transforms that we have obtained, it was true that $\lim_{s\to\infty}\hat{f}(s) = 0$. This is true quite generally.

Theorem 5 *If $f(x) \in \mathscr{E}_\alpha$, then $\lim\limits_{s \to \infty} \hat{f}(s) = 0$. Furthermore $|s\hat{f}(s)|$ is bounded as $s \to \infty$.*

Proof Since $f \in \mathscr{E}_\alpha$, we have for $s > \alpha$

$$|f(x)| < ce^{\alpha x} \qquad \text{and} \qquad |e^{-sx}f(x)| < ce^{-(s-\alpha)x}$$

Therefore

$$|\hat{f}(s)| = \left| \int_0^\infty e^{-sx}f(x)\,dx \right| \leq \int_0^\infty e^{-sx}|f(x)|\,dx \leq c\int_0^\infty e^{-(s-\alpha)x}\,dx = \frac{c}{s-\alpha}$$

so that $\hat{f}(s) \to 0$ as $s \to \infty$. We also have

$$|s\hat{f}(s)| \leq \frac{sc}{s-\alpha} = \frac{c}{1-\alpha/s} \leq 2c$$

if s is big enough.

It follows immediately from this theorem that the functions 1, s, $s/(s-1)$ cannot be transforms of functions of exponential order.

Also $1/\sqrt{s}$ cannot be the transform of a function of exponential order since $s(1/\sqrt{s}) = \sqrt{s}$ is unbounded as $s \to \infty$.

Problems 9-4

1. Prove Theorem 3.

2. If $\hat{f}(s) = \displaystyle\int_0^\infty e^{-sx}f(x)\,dx$ and $a > 0$, show that

$$\frac{1}{a}\hat{f}\left(\frac{s}{a}\right) = \int_0^\infty e^{-sx}f(ax)\,dx$$

and verify for $f(x) = \sin x$.

3. Find: $\mathscr{L}\{\sinh x\}$ $\mathscr{L}\{x \sinh x\}$

$\qquad\qquad \mathscr{L}\{\cosh x\}$ $\mathscr{L}\{x \cosh x\}$

4. Find: $\mathscr{L}^{-1}\left\{ \dfrac{d^k}{ds^k}\dfrac{1}{s^2+a^2} \right\}$ $k = 1, 2, 3, \ldots$

5. Find: $\mathscr{L}^{-1}\left\{ \dfrac{d^k}{ds^k}\dfrac{s}{s^2+a^2} \right\}$ $k = 1, 2, 3, \ldots$

6. Find: $\mathscr{L}\{xe^{ax}\sin bx\}$

7. Show that

$$\mathscr{L}^{-1}\left\{ \frac{1}{(s-a)^c} \right\} = \frac{x^{c-1}e^{ax}}{\Gamma(c)} \ \dagger$$

if $c > 0$ and $s > a$.

8. Derive all of the formulas listed in the table of Laplace transforms on page 278.

† See page 28 for a definition of the gamma function $\Gamma(c)$.

9. If $Y(s) = 1/(as^2 + bs + c)$, $a \neq 0$, and $\mathscr{L}^{-1}\{Y(s)\} = W(x)$, show that:

a. $W(0) = 0$ b. $W'(0) = 1/a$

10. If $Y(s) = 1/p(s)$ where $p(s)$ is a polynomial of degree n with leading coefficient a_0 and $\mathscr{L}^{-1}\{Y(s)\} = W(x)$, show that $W(0) = W'(0) = \ldots = W^{(n-2)}(0) = 0$, $W^{(n-1)}(0) = 1/a_0$.

11. a. If $f(x)$ is continuous and bounded in the interval $0 \leq x < p$ and has period p [i.e., $f(x + p) = f(x)$ for all x], show that for $s > 0$

$$\mathscr{L}\{f(x)\} = (1 - e^{-sp})^{-1} \int_0^p f(x)e^{-sx}\,dx.$$

b. Verify the above for $f(x) = \sin x$, $p = 2\pi$.

12. Find the transforms of the periodic functions:

a. $f(x) = \begin{cases} 1 & 0 \leq x < 1 \\ -1 & 1 \leq x < 2 \end{cases}$ period 2

b. $f(x) = x$ $0 \leq x < 1$ period 1

13. *Laplace transforms from Taylor series.* In the problems below we assume that the transform of a power series may be computed term by term.

a. Using the series $e^x = \sum_0^\infty x^n/n!$, find $\mathscr{L}\{e^x\}$.

Using the Taylor's series for the functions, show:

b. $\mathscr{L}\{\sin x\} = \dfrac{1}{s^2 + 1}$

c. $\mathscr{L}\left\{\dfrac{\sin x}{x}\right\} = \arctan \dfrac{1}{s}$

d. $\mathscr{L}\left\{\displaystyle\int_0^x \dfrac{\sin u}{u}\,du\right\} = \dfrac{1}{s} \arctan \dfrac{1}{s}$

e. $\mathscr{L}\{\sin \sqrt{x}\} = \dfrac{\sqrt{\pi}e^{-1/4s}}{2s^{\frac{3}{2}}}$

f. Let $\operatorname{Erf} x = (2/\sqrt{\pi}) \displaystyle\int_0^x e^{-u^2}\,du$. Show that

$$\mathscr{L}\{\operatorname{Erf} \sqrt{x}\} = \dfrac{1}{s\sqrt{s+1}}$$

g. Recall the Bessel function of order zero is

$$J_0(x) = \sum_{n=0}^\infty \dfrac{(-1)^n x^{2n}}{2^{2n}(n!)^2}$$

Show that

$$\mathcal{L}\{J_0(x)\} = \frac{1}{\sqrt{s^2 + 1}}$$

h. Recalling $J_0'(x) = -J_1(x)$, show that

$$\mathcal{L}\{J_1(x)\} = \frac{\sqrt{s^2 + 1} - s}{\sqrt{s^2 + 1}}$$

Table of Laplace Transforms

	Transform function	Object function	Conditions
1.	$\dfrac{1}{s}$	1	$s > 0$
2.	$\dfrac{1}{s^n}$	$\dfrac{x^{n-1}}{(n-1)!}$	$s > 0$ $n = 1, 2, 3, \ldots$
3.	$\dfrac{1}{s - a}$	e^{ax}	$s > a$
4.	$\dfrac{1}{(s - a)^n}$	$\dfrac{x^{n-1}e^{ax}}{(n-1)!}$	$s > a$ $n = 1, 2, 3, \ldots$
5.	$\dfrac{1}{(s - a)^c}$	$\dfrac{x^{c-1}e^{ax}}{\Gamma(c)}$	$c > 0 \quad s > a$
6.	$\dfrac{1}{s^2 + a^2}$	$\dfrac{1}{a} \sin ax$	$a \neq 0 \quad s > 0$
7.	$\dfrac{s}{s^2 + a^2}$	$\cos ax$	$s > 0$
8.	$\dfrac{1}{s^2 - a^2}$	$\dfrac{1}{a} \sinh ax = \dfrac{e^{ax} - e^{-ax}}{2a}$	$s > |a|$
9.	$\dfrac{s}{s^2 - a^2}$	$\cosh ax = \dfrac{e^{ax} + e^{-ax}}{2}$	$s > |a|$
10.	$\dfrac{1}{(s - a)(s - b)}$	$\dfrac{e^{ax} - e^{bx}}{a - b}$	$a \neq b \quad s > \max(a,b)$
11.	$\dfrac{s}{(s - a)(s - b)}$	$\dfrac{ae^{ax} - be^{bx}}{a - b}$	$a \neq b \quad s > \max(a,b)$
12.	$\dfrac{1}{(s - a)^2 + b^2}$	$\dfrac{e^{ax} \sin bx}{b}$	$s > a$
13.	$\dfrac{s - a}{(s - a)^2 + b^2}$	$e^{ax} \cos bx$	$s > a$
14.	$\dfrac{s}{(s^2 + a^2)^2}$	$\dfrac{1}{2a} x \sin ax$	$s > 0$
15.	$\dfrac{s^2 - a^2}{(s^2 + a^2)^2}$	$x \cos ax$	$s > 0$
16.	$\dfrac{1}{(s^2 + a^2)^2}$	$\dfrac{1}{2a^3}(\sin ax - ax \cos ax)$	$s > 0$

9-5   *Solution of linear equations with constant coefficients*

In the table on page 278 we summarize the Laplace transforms of some common functions. We shall find these useful in the solution of linear differential equations with constant coefficients. Consider the initial-value problem

$$\text{DE: } ay'' + by' + cy = f(x)$$
$$\text{IC: } y(0) = \alpha \tag{17}$$
$$y'(0) = \beta$$

where a, b, c are constants, $a \neq 0$, and $f(x)$ is of exponential order. We know† from Prob. 10, Sec. 9-3, that y, y', and y'' are of exponential order and therefore possess transforms. Taking the transforms of both sides, we obtain

$$a\mathscr{L}\{y''\} + b\mathscr{L}\{y'\} + c\mathscr{L}\{y\} = \mathscr{L}\{f(x)\}$$

$$a[s^2\hat{y}(s) - s\alpha - \beta] + b[s\hat{y}(s) - \alpha] + c\hat{y}(s) = \hat{f}(s)$$

$$(as^2 + bs + c)\hat{y}(s) - (as + b)\alpha - a\beta = \hat{f}(s)$$

$$\hat{y}(s) = \frac{\hat{f}(s) + (as + b)\alpha + a\beta}{as^2 + bs + c} \tag{18}$$

Note that the denominator of $\hat{y}(s)$ is $p(s) = as^2 + bs + c$, the characteristic polynomial of the differential equation. To find $y(x)$, we must find the inverse transform of $\hat{y}(s)$. In many cases $\hat{f}(s)$ is a rational function and therefore $\hat{y}(s)$ is also. If $\hat{y}(s)$ does not appear directly in the table, we can express $\hat{y}(s)$ as a sum of partial fractions which do appear in the table. The following examples illustrate the technique.

Example 1   DE: $y'' + 5y' + 6y = e^{7x}$
$$y(0) = 0$$
$$y'(0) = 2$$

Taking transforms, we obtain

$$\hat{y} = \frac{2}{s^2 + 5s + 6} + \frac{1}{(s - 7)(s^2 + 5s + 6)}$$

$$= \frac{2}{(s + 3)(s + 2)} + \frac{1}{(s - 7)(s + 3)(s + 2)}$$

† Even without this knowledge, we could *assume* that the solution possessed a transform and, after finding a tentative solution, check to see that it satisfied the differential equation.

The inverse transform of the first term appears directly in the table, line 10:

$$\mathscr{L}^{-1}\left\{\frac{2}{(s+3)(s+2)}\right\} = -2(e^{-3x} - e^{-2x}) \tag{i}$$

We split the second term up into its partial-fraction expansion

$$\frac{1}{(s-7)(s+3)(s+2)} = \frac{A}{s-7} + \frac{B}{s+3} + \frac{C}{s+2}$$

This is an identity if

$$1 \equiv (s+3)(s+2)A + (s-7)(s+2)B + (s-7)(s+3)C$$

This equation must hold for all s. Setting $s = 7$, we obtain $A = \frac{1}{90}$; setting $s = -3$, we obtain $B = \frac{1}{10}$; and setting $s = -2$, we obtain $C = -\frac{1}{9}$. Therefore

$$\frac{1}{(s-7)(s+3)(s+2)} = \frac{1}{90}\frac{1}{s-7} + \frac{1}{10}\frac{1}{s+3} - \frac{1}{9}\frac{1}{s+2}$$

and

$$\mathscr{L}^{-1}\left\{\frac{1}{(s-7)(s+3)(s+2)}\right\} = \frac{1}{90}e^{7x} + \frac{1}{10}e^{-3x} - \frac{1}{9}e^{-2x} \tag{ii}$$

The solution $y(x)$ is the sum of (i) and (ii):

$$y(x) = \frac{1}{90}e^{7x} - \frac{19}{10}e^{-3x} + \frac{17}{9}e^{-2x}$$

Example 2   DE: $y'' + y = 3$

IC: $y(0) = 1$

$$y'(0) = 2$$

$$(s^2\hat{y} - s - 2) + \hat{y} = \frac{3}{s}$$

$$\hat{y} = \frac{3}{s(s^2+1)} + \frac{s+2}{s^2+1}$$

Splitting the first term into partial fractions,

$$\frac{3}{s(s^2+1)} = \frac{A}{s} + \frac{Bs+C}{s^2+1}$$

or

$$3 \equiv A(s^2+1) + (Bs+C)s$$

Equating coefficients of s on both sides, we obtain $A = 3$, $B = -3$, $C = 0$. Therefore

$$\hat{y}(s) = \frac{3}{s} - \frac{2s}{s^2+1} + \frac{2}{s^2+1}$$

Using the table, lines 1, 6, and 7, we obtain

$$y(x) = 3 - 2\cos x + 2\sin x$$

Example 3 $y'' + y = e^x \sin 2x$

$$y(0) = 1$$

$$y'(0) = 3$$

$$(s^2 \hat{y} - s - 3) + \hat{y} = \frac{2}{(s-1)^2 + 4}$$

$$\hat{y} = \frac{s+3}{s^2+1} + \frac{2}{(s^2+1)[(s-1)^2+4]}$$

We must find the partial-fraction expansion for the second term:

$$\frac{2}{(s^2+1)[(s-1)^2+4]} = \frac{As+B}{s^2+1} + \frac{Cs+D}{(s-1)^2+4}$$

or

$$2 = (As+B)[(s-1)^2+4] + (Cs+D)(s^2+1)$$

Equating coefficients of like powers of s on both sides,† we obtain

$$A = \tfrac{1}{5} \qquad B = \tfrac{2}{5} \qquad C = -\tfrac{1}{5} \qquad D = 0$$

Therefore

$$\hat{y} = \frac{6}{5}\frac{s}{s^2+1} + \frac{17}{5}\frac{1}{s^2+1} - \frac{1}{5}\frac{(s-1)+1}{(s-1)^2+4}$$

Using the table, lines 6, 7, 12, and 13, we obtain

$$y(x) = \tfrac{6}{5} \cos x + \tfrac{17}{5} \sin x - \tfrac{1}{10} e^x \sin 2x - \tfrac{1}{5} e^x \cos 2x$$

Partial fractions   We have seen that it is often necessary to expand a transform into partial fractions before finding the inverse transform. We shall develop an efficient formula for expanding a rational function into partial fractions provided that the zeros of the denominator are distinct. The rational function $P(s)/Q(s)$, where $Q(s)$ is a polynomial with n distinct zeros $\alpha_1, \ldots, \alpha_n$ and $P(s)$ is a polynomial of degree less than n, can be written in the form

$$\frac{P(s)}{Q(s)} = \frac{A_1}{s - \alpha_1} + \frac{A_2}{s - \alpha_2} + \cdots + \frac{A_n}{s - \alpha_n} \tag{19}$$

where the A_i are constants. To find A_i, multiply both sides by $s - \alpha_i$ and let s approach α_i. We obtain

$$A_i = \lim_{s \to \alpha_i} \frac{(s - \alpha_i)P(s)}{Q(s)}$$

$$= \lim_{s \to \alpha_i} \frac{P(s)}{[Q(s) - Q(\alpha_i)]/(s - \alpha_i)} = \frac{P(\alpha_i)}{Q'(\alpha_i)} \tag{20}$$

† Alternatively we could set $s = \pm i$, $s = 1 \pm 2i$, and solve for A, B, C, D.

Therefore we have

$$\frac{P(s)}{Q(s)} = \frac{P(\alpha_1)}{Q'(\alpha_1)} \frac{1}{s - \alpha_1} + \cdots + \frac{P(\alpha_n)}{Q'(\alpha_n)} \frac{1}{s - \alpha_n} \tag{21}$$

and the companion formula

$$\mathcal{L}^{-1}\left\{\frac{P(s)}{Q(s)}\right\} = \frac{P(\alpha_1)}{Q'(\alpha_1)} e^{\alpha_1 x} + \cdots + \frac{P(\alpha_n)}{Q'(\alpha_n)} e^{\alpha_n x} \tag{22}$$

These formulas are often called the Heaviside expansions.

Example 4   DE: $y'' - 5y' + 6y = e^x$

IC: $y(0) = 1$

$y'(0) = 1$

Taking transforms, we get

$$s^2\hat{y} - s - 1 - 5(s\hat{y} - 1) + 6\hat{y} = \frac{1}{s - 1}$$

or

$$\hat{y} = \frac{s^2 - 5s - 5}{(s - 1)(s - 2)(s - 3)}$$

We have $P(s) = s^2 - 5s - 5$, $Q(s) = (s - 1)(s - 2)(s - 3) = s^3 - 6s^2 + 11s - 6$, and $Q'(s) = 3s^2 - 12s + 11$.   Using Eq. (21), we obtain

$$\hat{y} = \frac{P(1)}{Q'(1)} \frac{1}{s - 1} + \frac{P(2)}{Q'(2)} \frac{1}{s - 2} + \frac{P(3)}{Q'(3)} (s - 3)$$

$$= -\frac{9}{2} \frac{1}{s - 1} + \frac{11}{s - 2} - \frac{11}{2} \frac{1}{s - 3}$$

and

$$y(x) = -\tfrac{9}{2} e^x + 11e^{2x} - \tfrac{11}{2} e^{3x}$$

Example 5   Find $\mathcal{L}^{-1}\left\{\dfrac{s^3}{s^4 - 1}\right\}$

$$\frac{s^3}{s^4 - 1} = \frac{A_1}{s - 1} + \frac{A_2}{s + 1} + \frac{A_3}{s - i} + \frac{A_4}{s + i}$$

We find from the Heaviside expansion that $A_1 = A_2 = A_3 = A_4 = \tfrac{1}{4}$; therefore

$$\mathcal{L}^{-1}\left\{\frac{s^3}{s^4 - 1}\right\} = \frac{e^{-x} + e^x + e^{ix} + e^{-ix}}{4} = \tfrac{1}{2}\cosh x + \tfrac{1}{2}\cos x$$

Problems 9-5

Solve by using Laplace transforms:

1. $y' - y = e^{ax}$ $y(0) = 1$ for the cases $a \neq 1$, $a = 1$
2. $y'' + 3y' + 2y = x$ $y(0) = 1$, $y'(0) = -1$

3. $y'' + 4y' + 4y = e^{-2x}$ $y(0) = y'(0) = 0$
4. $y'' + 2y' + 2y = \sin x$ $y(0) = 0, y'(0) = 1$
5. $y''' + y = 0$ $y(0) = 1, y'(0) = y''(0) = 0$
6. $y^{(iv)} - y = 0$ $y(0) = y'(0) = y''(0) = 0, y'''(0) = 1$
7. $y^{(iv)} + y = 0$ $y(0) = 1, y'(0) = y''(0) = y'''(0) = 0$
8. $y'' - 4y' + 5y = e^{2x} \cos x$ $y(0) = y'(0) = 0$
9. Find the general solution of $y'' + y' = x$.
10. Solve the harmonic-oscillator equation $m\ddot{x} + kx = F(t)$, $x(0) = x_0$, $\dot{x}(0) = v_0$ when:

a. $F(t) \equiv 0$ b. $F(t) \equiv 1$

c. $F(t) \equiv F_0 \cos wt$ $w \neq \sqrt{k/m}$
                            $w = \sqrt{k/m}$

Systems of differential equations†

11. Using Laplace transforms, solve the following systems of differential equations:

a. $\dfrac{dx}{dt} = x - 2y$ $\dfrac{dy}{dt} = -3x + 2y$ $x(0) = 3, y(0) = 8$

b. $\dfrac{dx}{dt} + \dfrac{dy}{dt} + x + y = 1$ $\dfrac{dx}{dt} + 2\dfrac{dy}{dt} + y = 0$ $x(0) = 0, y(0) = 1$

c. $\ddot{x} - x + y = 0$ $\ddot{y} - y + x = e^t$ $x(0) = y(0) = 0$
                                                 $\dot{x}(0) = \dot{y}(0) = 1$

12. Find the conditions, if any, on a and b so that the following system has a solution, and find the solution:

$$\dot{x} + \dot{y} + 2x + 3y = 0 \qquad \dot{x} + \dot{y} + 2x - 3y = 0 \qquad \begin{aligned} x(0) &= a \\ y(0) &= b \end{aligned}$$

13. Consider the system of first-order equations

$$\dot{\mathbf{x}} = A\mathbf{x} + \mathbf{b}(t) \qquad \mathbf{x}(0) = \mathbf{x}^0$$

where $\mathbf{x}$ is an n-dimensional column vector, A is a constant $n \times n$ matrix, and $\mathbf{b}(t)$ is a column vector with components $b_i(t) \in \mathscr{E}_\alpha$. Define $\mathscr{L}\{\mathbf{x}\} = \hat{\mathbf{x}}(s)$ as the n vector whose components are $\hat{x}_i(s) = \mathscr{L}\{x_i(t)\}$. Show that $\hat{\mathbf{x}}$ satisfies

$$\hat{\mathbf{x}} = (sI - A)^{-1}[\mathbf{x}^0 + \hat{\mathbf{b}}(s)]$$

where I is the $n \times n$ identity matrix, provided s is greater than the real part of

† See Chap. 8 for basic theory of systems of linear differential equations.

all the eigenvalues of A.   Use the above to solve the systems for which:

a. $A = \begin{bmatrix} 1 & 2 \\ 0 & 2 \end{bmatrix}$ $\mathbf{b} = \begin{bmatrix} 1 \\ 0 \end{bmatrix}$ $\mathbf{x}^0 = \begin{bmatrix} 1 \\ 0 \end{bmatrix}$

b. $A = \begin{bmatrix} 1 & 2 & 3 \\ 0 & 2 & 0 \\ 0 & 0 & 3 \end{bmatrix}$ $\mathbf{b} = \begin{bmatrix} 0 \\ 0 \\ 0 \end{bmatrix}$ $\mathbf{x}^0 = \begin{bmatrix} 0 \\ 1 \\ 0 \end{bmatrix}$

c. $A = \begin{bmatrix} 4 & 2 \\ -1 & 1 \end{bmatrix}$ $\mathbf{b} = \begin{bmatrix} e^t \\ 0 \end{bmatrix}$ $\mathbf{x}^0 = \begin{bmatrix} 0 \\ 0 \end{bmatrix}$

14. a. Solve by using Laplace transforms:

$$f'(x) + \int_0^x f(t)\, dt = 0 \qquad f(0) = 1$$

b. Solve also by first differentiating.

15. The current in an RLC circuit satisfies

$$L\frac{dI}{dt} + RI + \frac{1}{C}\int_0^t I(\tau)\, d\tau = E(t) - \frac{Q_0}{C} \qquad I(0) = I_0$$

where $E(t)$ is the impressed voltage and Q_0 initial charge on the capacitor (see Chap. 6, page 174).   Show by using Laplace transforms that the solution is

$$I(t) = \mathcal{L}^{-1}\left\{\frac{\hat{E}}{p(s)}\right\} + \mathcal{L}^{-1}\left\{\frac{I_0 - Q/sC}{p(s)}\right\}$$

where $p(s) = sL + R + 1/sC$ is called the *impedance* of the circuit.   Show that if R, L, $C > 0$, the second term on the right is a transient, i.e., approaches zero as $t \to \infty$.   Show that if $E(t) = E_0 \cos \omega t$ or $E_0 \sin \omega t$, the steady-state solution can be obtained by taking the real or imaginary part of $E_0 e^{i\omega t}/p(i\omega)$.

16. *Differential equations with variable coefficients:*

a. To solve

$$y'' + xy' - y = 0 \qquad y(0) = 0,\, y'(0) = 1$$

take the transform of both sides (assuming $\hat{y}$ exists) to obtain

$$\frac{d\hat{y}}{ds} + \frac{2 - s^2}{s}\hat{y} = -\frac{1}{s}$$

Then solve this first-order differential equation to obtain

$$\hat{y} = \frac{1}{s^2} + \frac{ce^{s^2/2}}{s^2}$$

Note that if $c \neq 0$, then $\hat{y} \to \infty$ as $s \to \infty$ and $\hat{y}$ could not be the transform of any function. Therefore set $c = 0$ and obtain $y(t) = t$. Check this in the differential equation to see that it is the desired solution.

b. Show by the above method that a solution of

$$xy'' + y' + xy = 0 \qquad y(0) = c, y'(0) = 0$$

is $y = cJ_0(x)$. (See Prob. 13, Sec. 9-4.)   Why is only one solution obtained?

c. Show that the solution of

$$xy'' + 2y' + xy = 0 \qquad y(0) = 1, y(\pi) = 0$$

is $y = (\sin x)/x, x \neq 0; y(0) = 1$. (See Prob. 13, Sec. 9-4.)

9-6   *Product of transform functions; convolutions*

When Laplace transforms are used to solve differential equations, it is often necessary to find the inverse transform of the product of two transform functions $\hat{f}(s)\hat{g}(s)$. If $\hat{f}(s)$ and $\hat{g}(s)$ are both rational functions, the inverse transform could be found by the method of partial fractions. However, it would be useful to know whether or not the product of two transform functions is itself a transform function and if so, what is the relation between the corresponding object functions.

Let $f(x)$ and $g(x)$ be continuous functions that possess the transforms $\hat{f}(s)$ and $\hat{g}(s)$, respectively. Let $h(x)$ be a continuous function such that

$$\mathscr{L}\{h(x)\} = \hat{f}(s)\hat{g}(s) \tag{23}$$

assuming such a function exists. Equation (23) is equivalent to

$$\mathscr{L}\{h(x)\} = \int_0^\infty e^{-su}f(u)\, du \int_0^\infty e^{-st}g(t)\, dt \tag{24}$$

where both improper integrals exist for, say, $s > \alpha$. We *assume* that the right hand side can be written as an iterated integral:

$$\mathscr{L}\{h(x)\} = \int_0^\infty \left[\int_0^\infty e^{-s(u+t)}f(u)g(t)\, du \right] dt$$

Making the transformation $u + t = x, t = t$, we obtain

$$\mathscr{L}\{h(x)\} = \int_0^\infty \left[\int_t^\infty e^{-sx}f(x - t)g(t)\, dx \right] dt$$

Assuming that it is possible to change the order of integration, we obtain (see Fig. 1):

$$\mathscr{L}\{h(x)\} = \int_0^\infty e^{-sx} \left[\int_0^x f(x - t)g(t)\, dt \right] dx \tag{25}$$

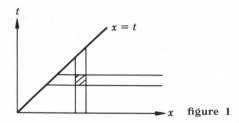

figure 1

Therefore from the uniqueness theorem we obtain

$$h(x) = \int_0^x f(x - t)g(t)\, dt \tag{26}$$

The expression on the right-hand side is called the *convolution*† of f and g and is symbolized by $f(x) * g(x)$, that is,

$$f(x) * g(x) = \int_0^x f(x - t)g(t)\, dt \tag{27}$$

We see therefore that multiplication of transform functions $\hat{f}(s)\hat{g}(s)$ corresponds to convolution of the object functions $f(x) * g(x)$, that is,

$$\mathscr{L}\{f(x) * g(x)\} = \hat{f}(s)\hat{g}(s) \tag{28}$$

The above derivation was purely formal. However, in the appendix to this chapter we shall prove the theorem that follows.

Theorem 1   *If $f(x) \in \mathscr{E}_\alpha$ and $g(x) \in \mathscr{E}_\alpha$, then $\{f * g\}$ exists and*

$$\mathscr{L}\{f(x) * g(x)\} = \mathscr{L}\{f(x)\}\mathscr{L}\{g(x)\}$$

The conditions that f and g belong to $\mathscr{E}_\alpha$ can be weakened simply to the requirement that $\mathscr{L}\{f(x)\}$ and $\mathscr{L}\{g(x)\}$ be absolutely convergent.‡

Example 1   Let $f(x) = \sin x$, $g(x) = 1$.   Then

$$f(x) * g(x) = \int_0^x \sin (x - t)\, dt = 1 - \cos x$$

We know that

$$\mathscr{L}\{1 - \cos x\} = \mathscr{L}\{1\} - \mathscr{L}\{\cos x\} = \frac{1}{s} - \frac{s}{s^2 + 1} = \frac{1}{s(s^2 + 1)}$$

On the other hand, from Theorem 1

$$\mathscr{L}\{1 - \cos x\} = \mathscr{L}\{1 * \sin x\} = \mathscr{L}\{1\}\mathscr{L}\{\sin x\} = \frac{1}{s}\frac{1}{s^2 + 1}$$

† The words *resultant* and *faltung* are often used for *convolution*.
‡ See Widder [37], p. 84.

The notation $f * g$ for convolution suggests that convolution can be thought of as a new kind of multiplication of functions; in fact, convolution has the following properties similar to ordinary multiplication:

$$f * g = g * f \qquad \text{commutative law}$$
$$(f * g) * h = f * (g * h) \qquad \text{associative law}$$
$$f * (g + h) = f * g + f * h \qquad \text{distributive law} \tag{29}$$
$$(cf) * g = f * (cg) = c(f * g) \qquad c = \text{const}$$

where all the functions are assumed to be continuous. These properties can be easily proved using Eq. (28) (see the problems below).

Example 2
$$\mathscr{L}^{-1}\left\{\frac{1}{(s^2 + a^2)^2}\right\} = \mathscr{L}^{-1}\left\{\frac{1}{s^2 + a^2}\frac{1}{s^2 + a^2}\right\}$$

$$= \frac{1}{a}\sin ax * \frac{1}{a}\sin ax$$

$$= \frac{1}{a^2}\sin ax * \sin ax = \frac{1}{a^2}\int_0^x \sin a(x - t)\sin at\, dt$$

$$= \frac{1}{2a^3}(\sin ax - ax \cos ax)$$

Convolution is useful in solving differential equations with arbitrary right-hand sides as in the following example.

Example 3   DE: $y'' + y = f(x)$
              IC: $y(0) = y'(0) = 0$

where $f(x)$ is assumed to have a Laplace transform. Taking transforms, we get

$$(s^2 + 1)\hat{y} = \hat{f}$$

$$\hat{y} = \frac{1}{s^2 + 1}\hat{f}$$

Therefore

$$y = \sin x * f(x)$$

$$= \int_0^x f(t)\sin(x - t)\, dt$$

or, since convolution is commutative, we also have

$$y = \int_0^x f(x - t)\sin t\, dt$$

Example 4   Solve the *integral equation* of convolution type

$$y(x) = x^2 + \int_0^x y(u) \sin (x - u) \, du$$

or

$$y(x) = x^2 + y(x) * \sin x$$

Assuming $y(x)$ has a transform, we find

$$\hat{y} = \frac{2}{s^3} + \frac{\hat{y}}{s^2 + 1}$$

or

$$\hat{y} = \frac{2}{s^3} + \frac{2}{s^5}$$

therefore

$$y(x) = x^2 + \tfrac{1}{12}x^4$$

To see that this is a solution, we substitute in the integral equation and find that it satisfies the equation.

Problems 9-6

1. Using convolutions, find:

a. $\mathscr{L}^{-1}\left\{\dfrac{1}{s^2(s - a)}\right\}$ $a \neq 0$

b. $\mathscr{L}^{-1}\left\{\dfrac{s^2}{(s^2 + 4)^2}\right\}$

2. Solve DE: $y'' - y = f(x)$
   IC: $y(0) = y'(0) = 0$

3. Using Laplace transforms, find a solution of the *integral equations of convolution type*:

a. $f(x) = 1 + \displaystyle\int_0^x f(t) \sin (x - t) \, dt$

b. $f(x) = \sin x - 2\displaystyle\int_0^x f(x) \cos (x - t) \, dt$

4. Prove formulas (29), using (28) and the uniqueness theorem.
5. Show that

$$\mathscr{L}^{-1}\left\{\frac{1}{\sqrt{s}(s - 1)}\right\} = e^t \operatorname{Erf}(\sqrt{t}) = e^t \frac{2}{\sqrt{\pi}} \int_0^x e^{-t^2} \, dt$$

(See Prob. 7, Sec. 9-4.)

6. Show, using Prob. 5, that $\mathscr{L}^{-1}\left\{\dfrac{1}{s\sqrt{s+1}}\right\} = \text{Erf}\,(\sqrt{t})$

7. Using Laplace transforms, show that

$$\int_0^x \left[\int_0^t f(u)\,du\right] dt = \int_0^x f(t)(x-t)\,dt$$

8. Generalize Prob. 7 to an n-fold integral.

9. Using Laplace transforms, find a function $y(x)$ satisfying

$$y'(x) + y(x) = a - \int_0^x e^{(x-u)} y(u)\,du \qquad y(0) = 0$$

10. Find a function $g(x)$, if it exists, such that:

a. $x * g(x) = x^4$ 　　　　　　　　　　b. $1 * 1 * g(x) = \dfrac{x^2}{2}$

c. $1 * g(x) = 1$

11. Construct examples to show:

a. $f * 1$ is not necessarily equal to f.
b. $f * f$ is not always greater than zero.

9-7 *Discontinuous functions*

In the preceding sections we assumed, for simplicity, that the functions considered were continuous in $0 \le x < \infty$. There is no particular difficulty in taking Laplace transforms of functions with finite or jump discontinuities, and the transform method is very efficient in solving linear differential equations with a discontinuous forcing term; such problems often occur in practical applications. We shall generalize our treatment of transforms to piecewise continuous functions defined below.

Definition 1 *$f(x)$ is said to be piecewise continuous on every finite interval, if, for every A, $f(x)$ is continuous on $0 \le x \le A$ except for a finite number of points x_i at which $f(x)$ possesses a right- and left-hand limit. The difference $f(x_i + 0) - f(x_i - 0)$ is called the jump of $f(x)$ at $x = x_i$.*

A graph of a typical piecewise continuous function is shown in Fig. 2.

A very simple and very useful discontinuous function is the *Heaviside function* or *unit-step function* shown in Fig. 3 and defined by

$$u(x - x_0) = \begin{cases} 1 & x > x_0 \\ 0 & x < x_0 \end{cases} \tag{30}$$

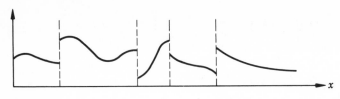

figure 2

The value assigned to this function at x_0 depends on the problem at hand; often it can be left undefined.   The transform of this function is

$$\mathscr{L}\{u(x - x_0)\} = \int_0^\infty u(x - x_0)e^{-sx}\, dx = \int_{x_0}^\infty e^{-sx}\, dx = \frac{e^{-sx_0}}{s} \qquad (31)$$

If $f(x)$ is defined on $0 \leq x < \infty$, then the function (see Fig. 4a and b)

$$f_s(x) = f(x - x_0)u(x - x_0) = \begin{cases} f(x - x_0) & x > x_0 \\ 0 & x < x_0 \end{cases}$$

is the function $f(x)$ *shifted to the right by an amount* x_0 [here we have assumed $f(x) = 0$, $x < 0$].   If $\mathscr{L}\{f\}$ exists, then $\mathscr{L}\{f_s\}$ exists, and is given by (see Prob. 2)

$$\mathscr{L}\{f_s(x)\} = \mathscr{L}\{f(x - x_0)u(x - x_0)\} = e^{-x_0 s}\hat{f}(s) \qquad (32)$$

This is known as the *shifting theorem*.   The function which agrees with $f(x)$ for $x > x_0$ but is zero for $x < x_0$ (see Fig. 4c) is given by

$$f_r(x) = f(x)u(x - x_0) \qquad (33)$$

and the function which is zero for $x > x_0$ but equal to $f(x)$ for $x < x_0$ (see Fig. 4d) is

$$f_l(x) = f(x)[1 - u(x - x_0)] = f(x)u(x_0 - x) \qquad (34)$$

In Prob. 3 we shall see that, if $\mathscr{L}\{f\}$ exists,

$$\mathscr{L}\{f_r(x)\} = \mathscr{L}\{f(x)u(x - x_0)\} = e^{-x_0 s}\mathscr{L}\{f(x + x_0)\} \qquad (35)$$

$$\mathscr{L}\{f_l(x)\} = \mathscr{L}\{f(x)[1 - u(x - x_0)]\} = \mathscr{L}\{f\} - e^{-x_0 s}\mathscr{L}\{f(x + x_0)\} \qquad (36)$$

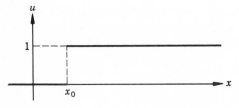

figure 3

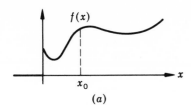

(a)

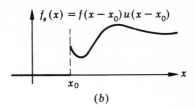

(b)

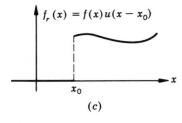

(c)

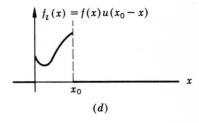

(d)

figure 4

If a function is given by different analytic expressions in different domains, i.e.,

$$F(x) = \begin{cases} f(x) & x < x_0 \\ g(x) & x > x_0 \end{cases} \tag{37}$$

we can conveniently write $F(x)$ using the Heaviside function

$$F(x) = f(x)u(x_0 - x) + g(x)u(x - x_0) \tag{38}$$

We now consider an nth-order linear equation with constant coefficients and a discontinuous right-hand side

$$a_0 y^{(n)}(x) + a_1 y^{(n-1)}(x) + \cdots + a_n y(x) = f(x)$$

$$y(0) = b_0,\ y'(0) = b_1,\ \ldots,\ y^{(n-1)}(0) = b_{n-1} \tag{39}$$

where $a_0 \neq 0$ and initial conditions are given at $x = 0$.

For simplicity assume $f(x)$ is continuous except for a finite discontinuity at $x = x_0 > 0$. First of all we must generalize our notion of solution, for it is clear that $y^{(n)}(x)$ does not exist at x_0. However, a unique solution exists in $0 \leq x \leq x_0$; therefore the left-hand limits $y(x_0-), \ldots, y^{(n-1)}(x_0-)$ exist. Using these values as new initial conditions at $x = x_0$, a unique solution for $x \geq x_0$ can be found. We define the function obtained by this piecing-together process to be the solution. Note that $y(x), \ldots, y^{(n-1)}(x)$ are continuous for $x \geq 0$ (even at $x = x_0$), whereas $y^{(n)}(x)$ has a finite discontinuity at $x = x_0$.

Example 1 $L(y) = y'' + 5y' + 6y = f(x)$

$$y(0) = y'(0) = 0$$

where $f(x) = 1$, $x < 2$, $f(x) = -1$, $x > 2$.   Solving $L(y) = 1$, $y(0) = y'(0) = 0$
for $0 \le x < 2$, we obtain

$$y = \tfrac{1}{6} - \tfrac{1}{2}e^{-2x} + \tfrac{1}{3}e^{-3x} \qquad 0 \le x \le 2$$

We now solve $L(y) = -1$, $y(2) = \tfrac{1}{6} - \tfrac{1}{2}e^{-4} + \tfrac{1}{3}e^{-6}$, $y'(2) = e^{-4} - e^{-6}$ for $x \ge 2$.
The general solution of the differential equation is $y = c_1 e^{-2x} + c_2 e^{-3x} - \tfrac{1}{6}$.
Satisfying the initial conditions at $x = 2$, we find:

$$y(x) = \tfrac{1}{2}e^{-2x}(e^4 - 1) + \tfrac{1}{3}e^{-3x}(1 - e^6) \qquad x \ge 2$$

Note that $y(x)$ and $y'(x)$ are continuous at $x = 2$ whereas $y''(2+) - y''(2-) = -2$.

The Laplace transform is an efficient tool for the solution of Eq. (39)
when the right-hand side $f(x)$ is discontinuous, provided $f(x)$ still possesses a
transform.   Equations (10), page 274, for the transforms of derivatives still
hold if $y(x), \ldots, y^{(n-1)}(x) \in \mathscr{E}_\alpha$ but $y^{(n)}(x)$ is piecewise continuous (see Prob. 4).

Example 2   We solve Example 1 using transforms.   We write $f(x) = u(x) -$
$2u(x - 2)$ so the differential equation is

$$y'' + 5y' + 6y = u(x) - 2u(x - 2) \qquad y(0) = y'(0) = 0$$

Taking transforms, we obtain

$$(s^2 + 5s + 6)\hat{y} = \frac{1}{s} - \frac{2e^{-2s}}{s}$$

$$\hat{y} = \frac{1}{s(s^2 + 5s + 6)} - \frac{2e^{-2s}}{s(s^2 + 5s + 6)}$$

We find

$$\mathscr{L}^{-1}\left\{\frac{1}{s(s + 3)(s + 2)}\right\} = \tfrac{1}{6} - \tfrac{1}{2}e^{-2x} + \tfrac{1}{3}e^{-3x} \equiv g(x)$$

Therefore

$$\mathscr{L}^{-1}\left\{\frac{e^{-2s}}{s(s + 3)(s + 2)}\right\} = g(x - 2)u(x - 2) \qquad \text{from Eq. (32)}$$

and

$$y = \tfrac{1}{6} + \tfrac{1}{2}e^{-2x} + \tfrac{1}{3}e^{-3x} + (\tfrac{1}{6} - \tfrac{1}{2}e^{-2(x-2)} + \tfrac{1}{3}e^{-3(x-2)})u(x - 2)$$

which agrees with the solution obtained in Example 1.

Problems 9-7

1. If $f(x)$ is defined for $0 \le x < \infty$, what is the difference, if any, between
the two following statements?

a. $f(x)$ has a finite number of discontinuities.
b. $f(x)$ has a finite number of discontinuities in every finite interval.

2. Derive Eq. (32).

3. Derive Eqs. (35) and (36).

4. Prove that if $y, y', \ldots, y^{(n-1)} \in \mathscr{E}_\alpha$ and $y^{(n)}(x)$ is piecewise continuous, then $\mathscr{L}\{y^n(x)\}$ exists and

$$\mathscr{L}\{y^n(x)\} = s^n\hat{y}(s) - [s^{n-1}y(0) + \cdots + sy^{n-2}(0) + y^{(n-1)}(0)]$$

5. Write the following functions as linear combinations of Heaviside functions:

a. $\operatorname{sgn} x \equiv \begin{cases} -1 & x < 0 \\ 1 & x > 0 \end{cases}$

b. $\delta_\epsilon(x) = \begin{cases} \dfrac{1}{\epsilon} & 0 \leq x < \epsilon \\ 0 & x > \epsilon \end{cases}$

6. If $f(x)$ is a given function, sketch the graph of

$$F(x) = f(x_0)u(x - x_0) + \sum_{i=1}^{n} [f(x_i) - f(x_{i-1})]u(x - x_i)$$

where $\{x_i\}$ is an increasing sequence of real numbers.

7. Solve $y'' + y = f(x)$ \qquad $y(0) = y'(0) = 0$

where

$$f(x) = \begin{cases} 1 & 0 \leq x < 5 \\ 0 & x > 5 \end{cases}$$

9-8 Linear systems analysis

We shall consider the use of Laplace transforms in *linear systems analysis.* Consider a *physical device* or *system* that is governed by an nth-order linear differential equation with constant coefficients:

$$\text{DE: } a_0 y^{(n)}(t) + a_1 y^{(n-1)}(t) + \cdots + a_n y(t) = f(t) \qquad t \geq 0$$
$$\text{IC: } y(0) = b_0,\ y'(0) = b_1,\ \ldots,\ y^{(n-1)}(0) = b_{n-1}$$

$$(40)$$

where the a_i and b_i are given constants $(a_0 \neq 0)$ and $f(t) \in \mathscr{E}_\alpha$ is a given function of the time t. The function $f(t)$ can be thought of as an *input* or *excitation* to the system and $y(t)$ as the *output* or *response* of the system. We indicate this by the simple block diagram shown in Fig. 5. The set of functions $\{y(t), y'(t), \ldots, y^{(n-1)}(t)\}$ is called a set of *state variables* for the system; the values of this set of functions at time t_0 is called the *state of the system* at time t_0. The output $y(t)$ of the system at time t is uniquely determined by the state of the system at, say, $t = 0$ (the initial conditions) and knowledge of the input $f(\tau)$ in the time interval $0 \leq \tau \leq t$. This important

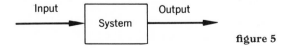

figure 5

property of the state variables is nothing but a restatement of the existence and uniqueness theorem for linear differential equations (see page 149).

The output of the system for a given input and initial state can conveniently be obtained by Laplace transforms. Taking transforms† of (40), we get

$$a_0[s^n\hat{y} - (s^{n-1}b_0 + \cdots + b_{n-1})] + a_1[s^{n-1}\hat{y} - (s^{n-2}b_0 + \cdots + b_{n-2})]$$
$$+ \cdots + a_{n-1}(s\hat{y} - b_0) + a_n\hat{y} = \hat{f}(s) \qquad (41)$$

Letting

$$p(s) = a_0 s^n + a_1 s^{n-1} + \cdots + a_{n-1}s + a_n \qquad (42)$$

$$q(s) = a_0 b_0 s^{n-1} + (a_0 b_1 + a_1 b_0)s^{n-2} + \cdots$$
$$+ (a_0 b_{n-2} + a_1 b_{n-3} + \cdots + a_{n-2}b_0)s + (a_0 b_{n-1} + \cdots + a_{n-1}b_0) \qquad (43)$$

we can rewrite (41) as

$$p(s)\hat{y}(s) = \hat{f}(s) + q(s) \qquad (44)$$

We note that $p(s)$ is a polynomial in s of degree n, namely, the characteristic polynomial of the differential equation, $q(s)$ is a polynomial of degree $\leq n - 1$ which depends on the b_i, that is, on the initial state of the system; if the system is in the *zero-initial state* (all $b_i = 0$), then $q(s) \equiv 0$.

From (44) we find that the transform $\hat{y}$ is

$$\hat{y}(s) = \frac{\hat{f}(s)}{p(s)} + \frac{q(s)}{p(s)} \qquad (45)$$

It is customary to define the *transfer function* $Y(s)$ of the system [see Probs. 6, 8, and 9 below for properties of $Y(s)$] by

$$Y(s) = \frac{1}{p(s)} \qquad (46)$$

Therefore (45) becomes

$$\hat{y}(s) = Y(s)\hat{f}(s) + Y(s)q(s) \qquad (47)$$

If we define

$$y_0(t) = \mathscr{L}^{-1}\{Y(s)\hat{f}(s)\} \qquad (48)$$

$$y_1(t) = \mathscr{L}^{-1}\{Y(s)q(s)\} \qquad (49)$$

then the output $y(t)$ is given by

$$y(t) = y_0(t) + y_1(t)$$

It is easy to see that $y_0(t)$ is a particular solution of the differential equation (40) and $y_1(t)$ is the solution of the corresponding homogeneous differential

† Note that we are considering the time t to be the independent variable, instead of x as in the previous portions of this chapter.

equation.   The function $y_0(t)$ is the output due to the input $f(t)$ *when the
initial state is the zero state*; we call $y_0(t)$ the *zero-state response* or *the response
due to the input*.   The function $y_1(t)$ is the response of the system if the input
$f(t) \equiv 0$ and is called the *zero-input response* or the *response due to the initial
state*.

The zero-state response and the weighting function   The zero-
state response $y_0(t)$ is given by

$$y_0(t) = \mathscr{L}^{-1}\{Y(s)\hat{f}(s)\} \tag{50}$$

Since $Y(s)$ is a rational function of s, this inverse transform could be evaluated
by partial fractions provided that $\hat{f}(s)$ is a rational function of s.   Even if
$\hat{f}(s)$ is not a rational function, we can proceed using convolutions.   For this
purpose we define the *weighting function* by

$$w(t) = \mathscr{L}^{-1}\{Y(s)\} \tag{51}$$

This function certainly exists since $Y(s)$ is a proper rational function (see
page 279).   The zero-state response can now be given by any of the forms

$$y_0(t) = w(t) * f(t) \tag{52}$$

$$y_0(t) = \int_0^t w(\tau)f(t - \tau)\,d\tau \tag{53}$$

$$y_0(t) = \int_0^t w(t - \tau)f(\tau)\,d\tau \tag{54}$$

From (53) we see that the zero-state output at time t is the *weighted
integral of the input*; the input τ units in the past, namely, $f(t - \tau)$, is weighted
by $w(\tau)$.   Knowledge of the weighting function completely determines the
zero-state response for a given input.

A graph of $w(\tau)$ gives useful information about the response of the system.
For instance, if $w(\tau)$ is as in Fig. 6a, the weighting function is almost zero for
$\tau \geq 2$; therefore the values of the input more than two time units in the past

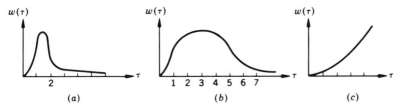

(a)   (b)   (c)

figure 6

do not appreciably affect the output.   We say that the system *remembers* the input for about two units of time or that the weighting function has a *memory* of about two time units.   The weighting function of Fig. 6b has a memory of about five units; the function in Fig. 6c has an infinite memory. In general we do not expect reasonable practical systems to have weighting functions like that in Fig. 6c; rather, we expect that inputs far in the past do not appreciably affect the output.

Example 1   Consider the system governed by the differential equation

$$2y''' + 9y'' + 13y' + 6y = f(t)$$

with zero-initial state.   The characteristic polynomial is

$$p(s) = 2s^3 + 9s^2 + 13s + 6 = (2s + 3)(s + 2)(s + 1)$$

The transfer function is

$$Y(s) = \frac{1}{p(s)} = \frac{1}{(2s + 3)(s + 2)(s + 1)} = \frac{-2}{s + \frac{3}{2}} + \frac{1}{s + 2} + \frac{1}{s + 1}$$

The weighting function is

$$w(t) = \mathscr{L}^{-1}\{Y(s)\} = -2e^{-3t/2} + e^{-2t} + e^{-t}$$

Thus the zero-state response is

$$y_0(t) = \int_0^t (-2e^{-3\tau/2} + e^{-2\tau} + e^{-\tau})f(t - \tau)\, d\tau$$

The zero-state response for a given input can be obtained from the above; for instance, if $f(t)$ is the *unit-step function* $u(t)$, then

$$y_0(t) = \int_0^t (-2e^{-3\tau/2} + e^{-2\tau} + e^{-\tau})\, d\tau = \frac{1}{6} + \frac{4e^{-3t/2}}{3} - \frac{e^{-2t}}{2} - e^{-t}$$

A sketch of $w(\tau)$ is shown in Fig. 7.

We see that $w(\tau)$ approaches zero as $\tau \to \infty$, so that inputs far in the past have little effect on the output.

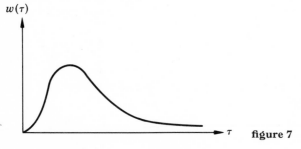

$w(\tau)$

τ **figure 7**

The zero-input response   The response $y_1(t)$ due to the initial state is given by

$$y_1(t) = \mathscr{L}^{-1}\{Y(s)q(s)\} \tag{55}$$

Now $q(s)$ is a polynomial of degree $n-1$ or less given by Eq. (43). For convenience, we write $q(s)$ in the form

$$q(s) = \alpha_0 s^{n-1} + \alpha_1 s^{n-2} + \cdots + \alpha_{n-2} s + \alpha_{n-1} \tag{56}$$

where

$$\alpha_0 = a_0 b_0, \ \alpha_1 = a_0 b_1 + a_1 b_0, \ \ldots, \ \alpha_{n-2} = a_0 b_{n-2} + \cdots + a_{n-2} b_0,$$
$$\alpha_{n-1} = a_0 b_{n-1} + \cdots + a_{n-1} b_0$$

Letting $\hat{y}_1(s) = \mathscr{L}\{y_1(t)\}$, Eq. (55) can be written

$$\hat{y}_1(s) = \alpha_0 s^{n-1} Y(s) + \alpha_1 s^{n-2} Y(s) + \cdots + \alpha_{n-2} s Y(s) + \alpha_{n-1} Y(s)$$

or

$$y_1(t) = \alpha_0 \mathscr{L}^{-1}\{s^{n-1} Y(s)\} + \alpha_1 \mathscr{L}^{-1}\{s^{n-2} Y(s)\}$$
$$+ \cdots + \alpha_{n-2} \mathscr{L}^{-1}\{s Y(s)\} + \alpha_{n-1} \mathscr{L}^{-1}\{Y(s)\} \tag{57}$$

We know that $\mathscr{L}^{-1}\{Y(s)\} = w(t)$. Furthermore it is shown in Sec. 9-4, Probs. 9 and 10, that

$$w(0) = w'(0) = \cdots = w^{(n-2)}(0) = 0 \tag{58}$$

Using the above and Eq. (10), page 274, we have

$$\mathscr{L}^{-1}\{s Y(s)\} = w'(t)$$
$$\mathscr{L}^{-1}\{s^2 Y(s)\} = w''(t)$$
$$\mathscr{L}^{-1}\{s^{n-1} Y(s)\} = w^{(n-1)}(t)$$

Using these results in Eq. (57), we see that the zero-input response is given by

$$y_1(t) = \alpha_{n-1} w(t) + \alpha_{n-2} w'(t) + \cdots + \alpha_0 w^{(n-1)}(t) \tag{59}$$

Therefore the zero-input response is given entirely in terms of the weighting function $w(t)$ and its derivatives.

From (59) we see that $y_1(t) = w(t)$ if $\alpha_{n-1} = 1$ and $\alpha_0 = \alpha_1 = \alpha_2 = \cdots = \alpha_{n-2} = 0$. This will be the case if the initial state of the system is

$$y(0) = 0 \qquad y'(0) = y^{(n-2)}(0) = 0 \qquad y^{(n-1)}(0) = \frac{1}{a_0} \tag{60}$$

Thus the *weighting function is the zero-input response of the system, the initial state being*

$$w(0+) = 0, \ w'(0+) = 0, \ \ldots, \ w^{(n-2)}(0+) = 0, \ w^{(n-1)}(0+) = \frac{1}{a_0} \tag{60a}$$

where, since $w(t)$ is defined only for $t \geq 0$, the values at $t = 0$ are really right-hand limits.

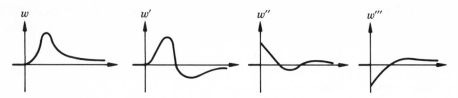

w w' w'' w'''

figure 8

Example 2   Consider the system of Example 1.   The weighting function is

$$w(t) = -2e^{-3t/2} + e^{-2t} + e^{-t}$$

We can easily verify that $w(0) = 0$, $w'(0) = 0$, $w''(0) = \frac{1}{2}$ in accordance with (60) (see Fig. 8).   If the initial state is

$$y(0) = 1 \qquad y'(0) = 0 \qquad y''(0) = 1$$

then we find that $\alpha_0 = 2$, $\alpha_1 = 9$, $\alpha_2 = 15$, and the zero-input response is

$$y_1(t) = 15w(t) + 9w'(t) + 2w''(t)$$
$$= -12e^{-3t/2} + 5e^{-2t} + 8e^{-t}$$

Weighting function as the unit-impulse response   If we try to find the input $f(t)$ such that the zero-state response is the weighting function, we are led to the equation

$$w(t) = w(t) * f(t)$$

or, taking transforms,

$$Y(s) = Y(s)\hat{f}(s)$$

Therefore $\hat{f}(s) \equiv 1$; *this is impossible* (see page 276).   However, let us consider an input $f(t)$ whose Laplace transform is "close to" 1.   Such an input is the ϵ *pulse*

$$\delta_\epsilon(t) = \begin{cases} \dfrac{1}{\epsilon} & 0 \leq t < \epsilon \\ 0 & t \geq \epsilon \end{cases} \tag{61}$$

and is shown in Fig. 9.   If ϵ is small, $\delta_\epsilon(t)$ represents a function that is very large for a small time interval but with area equal to 1.   This is an approximation to what is often called a *unit impulse*.†   It is easy to calculate the transform of $\delta_\epsilon(t)$:

$$\mathscr{L}\{\delta_\epsilon(t)\} = \int_0^\infty e^{-st}\delta_\epsilon(t)\, dt = \frac{1}{\epsilon}\int_0^\epsilon e^{-st}\, dt = \frac{1 - e^{-s\epsilon}}{s\epsilon}$$

Furthermore we find that as ϵ approaches zero (see Prob. 10),

$$\lim_{\epsilon \to 0} \mathscr{L}\{\delta_\epsilon(t)\} = \lim_{\epsilon \to 0} \frac{1 - e^{-s\epsilon}}{s\epsilon} = 1 \tag{62}$$

† The unit impulse is often defined as the limit of $\delta_\epsilon(t)$ as $\epsilon \to 0$.   However, such a limit does not exist at $t = 0$.   (See Probs. 13 and 14.)

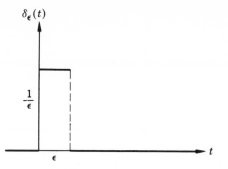

figure 9

Therefore, for small ϵ, $\mathscr{L}\{\delta_\epsilon(t)\}$ is "close to" 1. Now let $w_\epsilon(t)$ be the zero-state response to the input $\delta_\epsilon(t)$. We have

$$
\begin{aligned}
w_\epsilon(t) &= w(t) * \delta_\epsilon(t) \\
&= \int_0^t w(t-\tau)\delta_\epsilon(\tau)\,d\tau \\
&= \frac{1}{\epsilon}\int_0^\epsilon w(t-\tau)\,d\tau
\end{aligned}
\tag{63}
$$

It is easy to show (see Prob. 11) that

$$
\lim_{\epsilon\to 0} w_\epsilon(t) = w(t)
\tag{64}
$$

Therefore the weighting function is the limit of the zero-state response to the ϵ pulse $\delta_\epsilon(t)$ as ϵ approaches zero. Briefly but less precisely we shall state that *the weighting function is the zero-state response*† *to a unit impulse.*

† Strictly speaking, $w(t)$ is not itself a zero-state response, but the limit of the zero-state response $w_\epsilon(t)$ as ϵ approaches zero. For, as we saw in the last subsection, $w(t)$ satisfies the initial conditions (60a) and not the zero-initial conditions. However, if we consider the entire t interval $-\infty < t < \infty$ and define $\bar{w}(t)$ by

$$
\bar{w}(t) = \begin{cases} w(t) & t > 0 \\ 0 & t \le 0 \end{cases}
$$

then $\bar{w}(t)$ has the following properties:

1. $\bar{w}(t)$ satisfies the differential equation (40) for all t in $-\infty < t < \infty$ *except at $t = 0$.*

2. $\bar{w}(t)$ satisfies the conditions $\bar{w}(0-) = \bar{w}'(0-) = \cdots = \bar{w}^{(n-1)}(0-) = 0$.

3. $\bar{w}(t), \ldots, w^{(n-2)}(t)$ are continuous for $-\infty < t < \infty$.

4. $\bar{w}^{(n-1)}(t)$ is continuous except at $t = 0$, where it has a finite jump of $1/a_0$, i.e., $\bar{w}^{(n-1)}(0+) - \bar{w}^{(n-1)}(0-) = (1/a_0)$.

The reader should have no trouble verifying these statements. Thus $w(t)$ for $t > 0$ can be thought of as the response of a system which has been in the zero state for $t \le 0$ when a unit impulse is applied at $t = 0$.

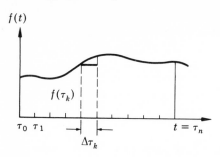

figure 10

There is no difficulty in showing that $w(t - \tau)$ is the zero-state response of the system to a unit impulse at time τ in the same sense as above.

The above considerations show that $w(t)$ can be found experimentally by observing the output of a system in the zero state when an impulse is applied at $t = 0$. Once $w(t)$ is known, the zero-state response to any input can be predicted by estimating the integral

$$y_0(t) = \int_0^t w(t - \tau)f(\tau)\,d\tau \tag{65}$$

This is of great practical value when the system is too complicated to find $w(t)$ analytically.

A useful interpretation of (65) can now be given. We subdivide the time interval $0 \leq \tau \leq t$ by the points $t_0 = 0 \leq \tau_1 \leq \cdots \leq \tau_n = t$ and let $\Delta\tau_k = \tau_k - \tau_{k-1}$. Then $y_0(t)$ is the limit of

$$\sum_{k=0}^{n-1} w(t - \tau_k)f(\tau_k)\,\Delta\tau_k \tag{66}$$

as the subdivision is refined. Since $w(t - \tau_k)$ is the zero-state response at time t due to a unit impulse at time τ_k, the term $w(t - \tau_k)f(\tau_k)\,\Delta\tau_k$ is the zero-state response at time t due to an impulse of weight $f(\tau_k)\,\Delta\tau_k$. The sum (66) of these responses approximates $y_0(t)$. Thus if $f(t)$ is approximated by a sum of pulses of width $\Delta\tau_k$ and height $f(\tau_k)$ (see Fig. 10), then each term in (66) is approximately the response due to this input pulse. By the principle of superposition the total response is the sum of the responses due to the individual input pulses. Thus (66) is an expression of the principle of superposition and the *integral* (65) *is a limiting expression of the principle of superposition.*

Problems 9-8

1. For each of the systems whose differential equations are given below, find the transfer function, weighting function, zero-state response, $A(t) \equiv$ zero-state response for a unit-step-function input, zero-input response for given

initial state using the weighting function. Verify that the weighting function and its first $(n - 1)$ derivatives satisfy the proper initial conditions.

a. $y'' + y = f(t)$ $y(0) = y'(0) = 1$
b. $y''' + y = f(t)$ $y(0) = 0, y'(0) = 1, y''(0) = 0$
c. $y^{(iv)} + y = f(t)$ $y(0) = 1, y'(0) = y''(0) = y'''(0) = 0$
d. $y''' - y'' + y' - y = f(t)$ $y(0) = 1, y'(0) = y''(0) = 0$
e. $y' = f(t)$ $y(0) = 1$

2. If $p(s)$ has distinct zeros $\alpha_1, \ldots, \alpha_n$, show that

$$w(t) = \sum_{k=1}^{n} \frac{e^{\alpha_k t}}{p'(\alpha_k)}$$

3. A system is called *stable* if the zero-input response approaches zero as $t \to \infty$ for all possible initial states.

a. Which of the systems in Prob. 1 are stable?
b. Show that if the roots of the characteristic equation all have negative real parts, then the system is stable.
c. Show that if a system is stable, $w(t) \to 0$ as $t \to \infty$.

4. If a system has a transfer function $Y(s)$, we define the *admittance function* $A(t) = \mathcal{L}^{-1}\{Y(s)/s\}$.

a. Find a relation between $A(t)$ and $w(t)$.

b. Show that $y_0(t) = \int_0^t \frac{\partial A}{\partial t}(t - \tau)f(\tau)\,d\tau$.

c. Show that $y_0(t) = \int_0^t A(t - \tau)f'(\tau)\,d\tau + A(t)f(0)$.

d. Show that $A(t)$ is the zero-state response to a unit-step-function input.

5. If $p(s)$ has distinct nonzero zeros $\alpha_1, \ldots, \alpha_n$, show that

$$A(t) = \frac{1}{p(0)} + \sum_{k=1}^{n} \frac{e^{\alpha_k t}}{\alpha_k p'(\alpha_k)}$$

6. a. If the output of a system with transfer function $Y_1(s)$ is used as an input to a system with transfer function $Y_2(s)$, the systems are said to be *cascaded*. Find the transfer function of the *cascaded* system (considered as a single system).
b. What is the relation between the weighting functions of the individual systems and that of the cascaded system?

7. Suppose a system has the weighting function t; find the weighting function of a second system which, when cascaded with the first, will produce the weighting function t^4.

8. If a stable system (see Prob 3) with transfer function $Y(s)$ has a sinusoidal input $f(t) = ce^{i\omega t}$, show that the output is

$$y(t) = Y(i\omega)ce^{i\omega t} + y_r(t)$$

where the transient $y_r(t) \to 0$ as $t \to \infty$. Thus $Y(i\omega)ce^{i\omega t}$ is the *steady-state* output. $Y(i\omega)$ as a function of ω is called the *frequency-response function*. If $Y(i\omega) = A(\omega)e^{i\varphi(\omega)}$, then the steady-state response is $A(\omega)ce^{c[\omega t + \varphi(\omega)]}$ so that an input sinusoid of frequency ω has its amplitude multiplied by the *amplification factor* $A(\omega)$ and its phase angle increased by $\varphi(\omega)$. A plot of $A(\omega)$ vs ω gives at a glance the effect the system has on various frequencies. (See page 170.)

9. Given the transfer functions:

a. $Y(s) = \dfrac{1}{s^2 + 5s + 6}$

b. $Y(s) = \dfrac{1}{s^2 + s + 1}$

c. $Y(s) = \dfrac{1}{s^2 + s}$

In each case find: (i) the frequency-response function and the amplification factor; (ii) the steady-state response due to an input $5 \sin 4t$; (iii) a plot of $A(\omega)$ vs ω; (iv) the value of ω for which $A(\omega)$ is a maximum.

10. Verify Eq. (62).

11. Prove Eq. (64).

12. Verify directly that $\omega_\epsilon(t)$ as given by (63) satisfies the differential equation (40).

13. *Generalized functions.*† We have seen that the limit of the ϵ pulse, $\delta_\epsilon(t) = [u(t) - u(t - \epsilon)]/\epsilon$, does not exist at $t = 0$. However, we can introduce a *generalized function* or unit impulse denoted by $\delta(t)$ which will behave like the ϵ pulse for small ϵ. We define operations on $\delta(t)$ by means of appropriate limiting operations involving $\delta_\epsilon(t)$.

a. Define $\mathscr{L}\{\delta(t)\} \equiv \lim_{\epsilon \to 0} \mathscr{L}\{\delta_\epsilon(t)\}$ and show $\mathscr{L}\{\delta(t)\} = 1$.

b. Define $\mathscr{L}\{\delta(t - t_0)\} \equiv \lim_{\epsilon \to 0} \mathscr{L}\{\delta_\epsilon(t - t_0)\}$ and show $\mathscr{L}\{\delta(t - t_0)\} = e^{-st_0}$.

c. Define $\displaystyle\int_{-\infty}^{\infty} \delta(t - t_0)f(t)\,dt \equiv \lim_{\epsilon \to 0} \int_{-\infty}^{\infty} \delta_\epsilon(t - t_0)f(t)\,dt = f(t_0)$. This is known as the *sifting* property of $\delta(t)$. Show that

$$\mathscr{L}\{\delta(t - t_0)f(t)\} = e^{-st_0}f(t_0)$$

d. The weighting function $w(t)$ is the limit as $\epsilon \to 0$ of the solution of $L(y) = \delta_\epsilon(t)$ with zero-initial state. We therefore define the solution of $L(y) = \delta(t)$ with zero-initial state to be $w(t)$. Show that $w(t)$ can be obtained formally by taking

† For an elementary treatment of generalized functions see Lighthill [24].

the transform of the differential equation using the usual rules and part a above. Why is this procedure not rigorous?

14. Solve *formally* by using Prob. 13.

a. $y'' - 4y = \delta(t - 1) + \delta(t - 2)$ $y(0) = y'(0) = 0$
b. $y'' + 4y' + 4y = \delta(t - 2)$ $y(0) = 0, y'(0) = 1$
c. $y'' + y = \delta(t) + u(t - 2)$ $y(0) = y'(0) = 0$
d. $y'' + y = \delta(t - 3)e^{2t}$ $y(0) = 0, y'(0) = 1$

Appendix

1. Uniqueness theorem. We now prove Theorem 1 of Sec. 9-3.

If $\mathscr{L}\{f(x)\} \equiv \mathscr{L}\{g(x)\}$ and $f(x)$ and $g(x)$ are continuous for $0 \leq x < \infty$, then $f(x) \equiv g(x)$.

Proof We must show that $h(x) = f(x) - g(x)$ is identically zero. We have

$$\hat{h}(s) = \hat{f}(s) - \hat{g}(s) \equiv 0 \qquad \text{for } s \geq \alpha \tag{i}$$

Setting $s = \alpha + n$, then $\hat{h}(s) = 0$ for $n \geq 0$. In particular $\hat{h}(s) = 0$ for $n = 0$, $1, 2, \ldots$. That is,

$$\hat{h}(s) = \int_0^\infty e^{-(\alpha+n)x}h(x)\, dx = 0 \qquad n = 0, 1, 2, \ldots \tag{ii}$$

Letting $v(x) = \int_0^x e^{-\alpha t}h(t)\, dt$ and integrating (ii) by parts, we obtain

$$0 = e^{-\alpha x}v(x)\, \Big|_0^\infty + n\int_0^\infty e^{-nx}v(x)\, dx$$

or since $v(0) = 0$

$$\int_0^\infty e^{-nx}v(x)\, dx = 0 \qquad n = 0, 1, 2, \ldots \tag{iii}$$

Since $v'(x) = e^{-\alpha x}h(x)$, if we show that $v(x) \equiv 0$, it will follow that $h(x) \equiv 0$. In (iii) we make the substitution $x = -\ln t$ and let $u(t) = v(-\ln t)$. We have $u(1) = v(0) = 0$ and in order to make $u(t)$ continuous in the closed interval $0 \leq t \leq 1$, we define $u(0) = \lim_{t \to 0^+} v(-\ln t) = v(\infty) = \hat{h}(\alpha) = 0$. With this substitution (iii) becomes

$$\int_0^1 t^n u(t)\, dt = 0 \qquad n = 0, 1, 2, \ldots \tag{iv}$$

The fact that (iv) implies that $u(t) \equiv 0$ is a basic nontrivial theorem in analysis.

We indicate the proof of this fact in the problem below. Since $u(t) \equiv 0$, then $v \equiv 0$, and the theorem follows.

Problem   If (iv) holds and the continuous function $u(t)$ is not identically zero, there exists a point in $(0,1)$ where $u(\xi) \neq 0$, say $u(\xi) = 2a > 0$. We can also find an interval $[\xi - \delta, \xi + \delta]$ where $u(t) > a$. (Why?) Below we shall construct a polynomial $P_N(t)$ such that

$$\int_0^1 P_N(t)u(t)\, dt > a \tag{v}$$

However, if (iv) holds, then $\int_0^1 P_N(t)u(t)\, dt = 0$. This contradiction shows that $u(t) \equiv 0$.

Construction of $P_N(t)$.   To find a $P_N(t)$ satisfying (v), we construct a polynomial which is large in $[\xi - \delta, \xi + \delta]$ and small outside this interval. This is done by finding a quadratic polynomial which is greater than 1 in $(\xi - \delta, \xi + \delta)$ and less than 1 in magnitude outside $[\xi - \delta, \xi + \delta]$ and raising this quadratic to high powers. (See Fig. 11.) The quadratic is

$$y(t) = 1 + t(2\xi - t) + (\delta^2 - \xi^2) \qquad 0 \leq t \leq 1$$

Show that:

a. $y'(\xi) = 0$ b. $y(\xi - \delta) = y(\xi + \delta) = 1$
c. $y(t) \geq 1$ in $[\xi - \delta, \xi + \delta]$ d. $|y(t)| \leq 1$ outside of $[\xi - \delta, \xi + \delta]$
e. $|y(t)| \geq 1 + \frac{3}{4}\delta^2$ in $[\xi - \frac{1}{2}\delta, \xi + \frac{1}{2}\delta]$

Now define $P_N(t) = [y(t)]^N$. Show that:

f. $P_N(t) \geq 1$ in $[\xi - \delta, \xi + \delta]$
g. $|P(t)| \leq 1$ outside of $[\xi - \delta, \xi + \delta]$
h. $P(t) > (1 + \frac{3}{4}\delta^2)^N > 1 + \frac{3}{4}N\delta^2$ in $[\xi - \frac{1}{2}\delta, \xi + \frac{1}{2}\delta]$
i. By picking N large enough, show that Eq. (v) holds.

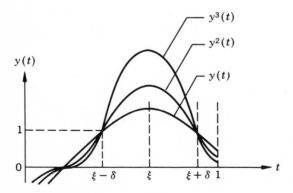

figure 11

2. Derivative of the Laplace transform. We prove Theorem 4 of Sec. 9-4:

If $f(x) \in \mathcal{E}_\alpha$, then $\hat{f}(s)$ has derivatives of all orders and $\hat{f}^{(n)}(s) = \mathcal{L}\{(-x)^n f(x)\}$.

Proof Let $I = \int_0^\infty -xf(x)e^{-sx}\,dx$. We must show that $\hat{f}'(s) = \lim\limits_{\Delta s \to 0} \Delta\hat{f}/\Delta s = I$; i.e., we must show that, for $\epsilon > 0$,

$$\left|\frac{\Delta\hat{f}}{\Delta s} - I\right| < \epsilon$$

for sufficiently small Δs. Now since $\hat{f}(s) = \int_0^\infty e^{-sx}f(x)\,dx$ for $s > \alpha$

$$\frac{\Delta\hat{f}}{\Delta s} = \int_0^\infty f(x)\,\frac{e^{-(s+\Delta s)x} - e^{-sx}}{\Delta s}\,dx$$

provided Δs is small enough so that $s + \Delta s > \alpha$. We certainly ensure this by requiring $|\Delta s| \leq \beta = (s - \alpha)/2$. Now

$$\frac{\Delta\hat{f}}{\Delta s} - I = \int_0^\infty e^{-sx}f(x)\left(\frac{e^{-x\,\Delta s} - 1}{\Delta s} + x\right)dx$$

In Prob. 1 below we show that

$$\left|\frac{e^{-x\,\Delta s} - 1}{\Delta s} + x\right| \leq |\Delta s|\,x^2 e^{|\Delta s|\,x} \leq |\Delta s|\,x^2 e^{\beta x} \tag{i}$$

Therefore

$$\left|\frac{\Delta\hat{f}}{\Delta s} - I\right| \leq |\Delta s|\int_0^\infty x^2\,|f(x)|\,e^{-(s-\beta)x}\,dx$$

Now $f(x) \in \mathcal{E}_\alpha$ and therefore $x^2 f(x) \in \mathcal{E}_{\alpha'}$ for every $\alpha' > \alpha$ (see Prob. 2 below), and $|x^2 f(x)| \leq ce^{\alpha'x}$. In particular we can take $\alpha' = \alpha + \frac{1}{2}\beta$. Therefore

$$\left|\frac{\Delta\hat{f}}{\Delta s} - I\right| \leq |\Delta s|\int_0^\infty ce^{-(s-\alpha'-\beta)x}\,dx$$

Since $s - \alpha' - \beta = \frac{1}{2}\beta > 0$, the integral on the right exists, and therefore the left-hand side can be made as small as desired by taking $|\Delta s|$ sufficiently small. Thus $\hat{f}'(s) = I = \mathcal{L}\{-xf(x)\}$ as we wished to prove. Since $(-x)f(x) \in \mathcal{E}_{\alpha'}$, a similar proof can be carried through for higher derivatives.

Problem 1. Prove (i) above by expanding $e^{-x\,\Delta s}$ in a Taylor series.

Problem 2. Show that if $f \in \mathcal{E}_\alpha$, then $x^n f(x) \in \mathcal{E}_{\alpha'}$ for every $\alpha' > \alpha$. (See Example 4, Sec. 9–3.)

3. The convolution theorem. We now prove Theorem 1; Sec. 9-6.
If $f(x) \in \mathcal{E}_\alpha$ and $g(x) \in \mathcal{E}_\alpha$, then $\mathcal{L}\{f * g\}$ exists and

$$\mathcal{L}\{f(x) * g(x)\} = \mathcal{L}\{f(x)\}\mathcal{L}\{g(x)\}.$$

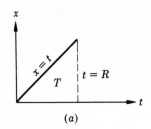
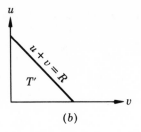

(a)   (b)

figure 12

Proof $\mathscr{L}\{f(x) * g(x)\} = \int_0^\infty e^{-sx}\left[\int_0^x f(x - t)g(t)\, dt\right] dx = \lim_{R\to\infty} I_R$ where

$$I_R = \int_0^R e^{-sx}\left[\int_0^x f(x - t)g(t)\, dt\right] dx \tag{i}$$

provided the limit exists. I_R can be written as a double integral

$$I_R = \iint_T e^{-sx} f(x - t)g(t)\, dt\, dx \tag{ii}$$

where T is the triangle shown in Fig. 12a. We make a transformation to new coordinates u, v by means of

$$x = u + v$$

$$t = v$$

The triangle T is transformed into the triangle T' in the uv plane, and since the Jacobian† of the transformation is 1, we have

$$I_R = \iint_{T'} e^{-s(u+v)} f(u)g(v)\, du\, dv = \int_0^R e^{-su} f(u)\int_0^{R-u} e^{-sv}g(v)\, dv\, du$$

Let

$$J_R = \int_0^R e^{-su} f(u)\, du \int_0^\infty e^{-sv} g(v)\, dv$$

We know that $\lim_{R\to\infty} J_R$ exists since f and g possess Laplace transforms and $\lim_{R\to\infty} J_R = \mathscr{L}\{f\}\mathscr{L}\{g\}$. We shall show that I_R differs from J_R by an arbitrarily small amount and therefore $\lim I_R = \mathscr{L}\{f\}\mathscr{L}\{g\}$. Now

$$|I_R - J_R| = \left|\int_0^R e^{-su} f(u)\int_{R-u}^\infty e^{-sv}g(v)\, dv\, du\right|$$

$$\leq \int_0^R e^{-su}|f(u)|\left|\int_{R-u}^\infty e^{-sv}g(u)\, dv\right|$$

† See Courant [12], vol. 2, p. 247.

Since $f,\, g \in \mathscr{E}_\alpha$, we have $|f(x)| \le ce^{\alpha x}$, $|g(x)| \le ce^{\alpha x}$. For $s > \alpha$ we have

$$\left| \int_{R-u}^{\infty} e^{-sv}g(v)\, dv \right| \le c \int_{R-u}^{\infty} e^{-(s-\alpha)v}\, dv = \frac{ce^{-(s-\alpha)(R-u)}}{s-\alpha}$$

Therefore

$$|I_R - J_R| \le \int_0^R e^{-su}ce^{\alpha u}\, \frac{ce^{-(s-\alpha)(R-u)}}{s-\alpha}\, du \le \frac{c^2 R e^{-(s-\alpha)R}}{s-\alpha}$$

and as $R \to \infty$, $Re^{-(s-\alpha)R} \to 0$ (since $s > \alpha$). This completes the proof.

Nonlinear Differential Equations

10

10-1 Introduction

When a differential equation is *linear*, the problem of finding the solution may be reduced to several simpler problems. That is, solutions of the homogeneous equation may be found one at a time and then combined to obtain its general solution. A single solution of the nonhomogeneous equation is then appended to produce the desired general solution. The success of this method is completely dependent upon the linearity of the equation. (See Sec. 5-2.) If an equation is *nonlinear*, this method is no longer available and the entire equation must be considered as a whole. For example, both e^x and e^{-x} are solutions of the nonlinear equation $(y' - y)(y' + y) = 0$, but $c_1 e^x + c_2 e^{-x}$ is *not* a solution for arbitrary c_1 and c_2. (Prove this.)

For nonlinear equations there is unfortunately no general procedure for obtaining solutions in closed form. In fact, when a physical problem leads to a nonlinear differential equation (which is usually the case), one must often be content with a qualitative description of the solutions together with a numerical approximation. Some useful information can often be obtained by neglecting the nonlinear part of the equation, but this procedure sometimes gives misleading results since a nonlinear system can behave quite differently from a linear approximation. (E.g., see Sec. 10-4.)

In addition to the practical difficulties encountered in seeking solutions of nonlinear equations, there are other difficulties which are even more fundamental. The very existence of solutions is sometimes in doubt and, even when they do exist, they may not be unique. Still another unpleasant feature of nonlinear equations is the fact that the solutions may vary *discontinuously* with respect to some parameter in the system. That is, a slight change in the conditions of the problem may produce radical changes in the behavior of the solution. (E.g., see Sec. 10-3.) Such instability is often troublesome in a physical problem since physical parameters can never be measured with absolute precision. These important questions of existence, uniqueness, and stability are easily settled for linear equations (see Sec. 4-7) but can cause considerable difficulty for nonlinear equations.

In this chapter several methods are developed for obtaining information about the solutions of nonlinear equations, and the results are then compared with the corresponding linear case.

10-2   The pendulum

One of the simplest nonlinear systems is the pendulum. In this section we shall study this system in order to illustrate some of the techniques which are used in studying nonlinear equations. These techniques will then be developed more formally in Sec. 10-3.

The differential equation governing the motion of a *free undamped pendulum* is

$$m\ddot{x} + mk^2 \sin x = 0 \tag{1}$$

Here x is the angular displacement (in radians), counterclockwise from the equilibrium position (see Fig. 1), $k^2 = g/l$ is a constant, g is the acceleration of gravity, and l is the length of the pendulum arm. The mass m of the pendulum bob could be canceled out in Eq. (1) since we have neglected the mass of the rod, friction, and outside forces. However, it is often helpful to consider the physical meaning of the terms in the equation, and hence we shall sometimes retain the m (see below).

Since the dependent variable x in Eq. (1) appears in the sine term, the equation is nonlinear. However, a *linear* approximation to the equation can be obtained as follows. Sin x can be expanded in a Maclaurin series,

$$\sin x = x - \frac{x^3}{3!} + \frac{x^5}{5!} - \cdots \tag{2}$$

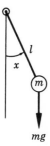

figure 1

and if x is assumed to be small at all times, we can approximate sin x by x, the first term in the series.   If we make this approximation, then Eq. (1) becomes

$$\ddot{x} + k^2 x = 0 \tag{3}$$

This equation is now *linear* with constant coefficients and has the solution

$$x = c_1 \cos kt + c_2 \sin kt \tag{4}$$

where $k = \sqrt{g/l}$ is the *constant circular frequency*.   The motion indicated by this result is a fairly good approximation to the actual motion of a pendulum for a short period of time, provided the initial displacement and velocity are small.

If greater accuracy is desired, two terms in the sine series can be used to yield the equation

$$\ddot{x} + k^2 \left(x - \frac{x^3}{3!} \right) = 0 \tag{5}$$

However, this equation is again nonlinear (because of the x^3 term) and is really no simpler than the original equation (1).

We return now to Eq. (1).   Since the independent variable t does not appear explicitly, we can reduce Eq. (1) to a first-order equation by making the substitution

$$v = \dot{x} \tag{6}$$

By the chain rule of differentiation we have then

$$\ddot{x} = \frac{dv}{dt} = \frac{dv}{dx}\frac{dx}{dt} = \frac{dv}{dx} v$$

Thus

$$\ddot{x} = v \frac{dv}{dx} \tag{7}$$

and Eq. (1) becomes

$$mv \frac{dv}{dx} + mk^2 \sin x = 0 \tag{8}$$

After separating variables and integrating, we have

$$\tfrac{1}{2}mv^2 - \tfrac{1}{2}mv_0^2 = mk^2 (\cos x - \cos x_0) \tag{9}$$

where $v = v_0$ when $x = x_0$.

Since $\tfrac{1}{2}mv^2$ represents the kinetic energy of the pendulum, Eq. (9) states the law of conservation of energy.   The left side is the change in kinetic energy, and the right side represents the work done by the "restoring force."

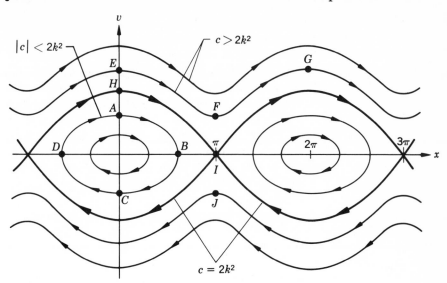

figure 2.   *Energy curves for the undamped pendulum* $\ddot{x} + k^2 \sin x = 0$.

Equation (9) can be written in the form

$$\tfrac{1}{2}mv^2 - mk^2 \cos x = \tfrac{1}{2}mc \tag{10}$$

where

$$c = v_0^2 - 2k^2 \cos x_0 \tag{11}$$

A surprising amount of valuable information about the pendulum can
be obtained by plotting Eq. (10) in the xv plane.   The displacement x
and velocity v are called *phases* of the motion, and the xv plane is known
as the *phase plane*.   If Eq. (10) is plotted for several values of the constant c,
a set of solution curves is obtained (see Fig. 2).   In this case of undamped
motion these curves are called *energy curves* since the energy is constant on a
given curve.†   That is, each curve represents a particular motion resulting
from a given set of initial conditions.

The entire history of a motion can be read from the energy curves in
the phase plane.   For example, if the pendulum is set in motion at $x = 0$
with small positive initial velocity v_0, then these initial conditions are
represented by the point A in Fig. 2 and the curve through A represents the
subsequent motion of the pendulum.   Since the velocity is positive at A,
the displacement x will increase as time goes on and hence, to trace the
history of the motion, we move to the right along the curve to B.   Here x

† The left-hand side of Eq. (10) represents the sum of the kinetic and potential
energies.

reaches its maximum and v is zero. This occurs when the pendulum has reached its highest position in its swing to the right and is just beginning its return swing. As the pendulum returns, x decreases; that is, v is negative until we pass through the equilibrium position again at point C, where the velocity has reached its greatest negative value. The pendulum is now swinging to the left ($x < 0$, $v < 0$) until we reach D, where the pendulum again stops ($v = 0$) to reverse its motion and return to equilibrium at A. If there is no damping present, the pendulum will return with the same velocity v_0 that it had initially and hence will continue to carry out the same oscillation as described above. Thus the motion is *periodic*. In fact, periodic motion is always represented by a *closed curve* in the phase plane. (Why?)

Another type of motion results if the pendulum is set in motion at equilibrium with a large initial velocity (point E in Fig. 2). In this case the pendulum is moving with such speed that it reaches the vertical position ($x = \pi$) with positive velocity (point F). Hence it continues over the top of its swing and on around again to its starting position $x = 2\pi$ with the same velocity that it had initially (point G). Hence in this case the pendulum will continue to revolve in the same direction about its point of suspension. Note that although v is evidently a periodic function of x, the motion of the pendulum itself is *not* periodic since the angular displacement x is a monotonically increasing function of time. It is also important to realize that this type of motion could not be explained by the linearized equation (3). (Why not?)

The motion represented by the energy curve EFG is characterized by the fact that the velocity v never vanishes. From Eq. (10) we have, solving for v,

$$v = \pm\sqrt{c + 2k^2 \cos x} \tag{11a}$$

Hence v will never vanish if $c > 2k^2$ or, by (11), if $v_0 > \sqrt{2k}$ for $x_0 = 0$.

Separating the two motions which have just been described is a special curve (heavy line through H in Fig. 2) called a *separatrix*. This is the curve for $c = 2k^2$ and hence it has the equation

$$v^2 = 2k^2 (\cos x + 1) \tag{12}$$

This curve represents the motion of the pendulum which starts with just the proper initial velocity (point H) to make it approach the vertical (unstable) equilibrium $x = \pi$ as a limit. It is easily shown (see Prob. 1) that this point I could not be reached in finite time.

Returning now to Eq. (10), since $v = dx/dt$, we have

$$\frac{dx}{dt} = \pm\sqrt{2k^2 \cos x + c} \tag{13}$$

where the sign of the square root must be chosen appropriately. When dx/dt is positive, we can write

$$\frac{dx}{\sqrt{2k^2 \cos x + c}} = dt \tag{14}$$

and it might be expected that an integration would yield x as a function of t. Unfortunately, however, the left side of (14) cannot be integrated in terms of elementary functions. Nevertheless, we can get some useful information by integrating (14) in the case of small initial velocity ($c < 2k^2$). In this case the energy curve is like $ABCD$ in Fig. 2, and hence the motion is periodic with period T and amplitude M. We can compute M from (13) by setting $x = M$ and $dx/dt = 0$ to give

$$2k^2 \cos M = -c \tag{15}$$

Next we note that x goes from 0 to M in time $\frac{1}{4}T$ so that

$$\int_0^M \frac{dx}{\sqrt{2k^2 \cos x + c}} = \frac{T}{4} \tag{16}$$

On substituting for c from (15), we have

$$T = \frac{2\sqrt{2}}{k} \int_0^M \frac{dx}{\sqrt{\cos x - \cos M}} \tag{17}$$

This result can be put into a standard form by introducing θ as a new variable of integration by means of the substitution $\sin \frac{1}{2} x = (\sin \frac{1}{2}M) \sin \theta$. We get

$$T = \frac{4}{k} \int_0^{\pi/2} \frac{d\theta}{\sqrt{1 - b^2 \sin^2 \theta}} \tag{18}$$

where $b = \sin \frac{1}{2}M$. The integral in (18) is called a "complete elliptic integral of the first kind" and is tabulated as a function of the parameter b.† One interesting consequence of formula (18) is the fact that the period T varies as the amplitude M (and hence b) varies. It is easily shown (see Prob. 2) that T increases steadily with M as M approaches π. This is in marked contrast with the constant period predicted by the linear analysis.

It will be shown in the next section that much of the knowledge which we have obtained about the behavior of the solutions could have been obtained directly from the differential equation (8). Solving (8) for dv/dx, we have

$$\frac{dv}{dx} = \frac{-k^2 \sin x}{v} \tag{19}$$

† See Jahnke, Emde, and Lösch [20].

This equation can be interpreted as the definition of a *direction field.*†
That is, the slope dv/dx of an energy curve through the point (x,v) is given
by the right side of (19).

For example, if $x = 0$ and $v \neq 0$, then $dv/dx = 0$ and hence the energy
curves are horizontal as they cross the v axis. (see Fig. 2).   If $v = 0$, then
dv/dx is undefined, but if $\sin x \neq 0$, we have $dx/dv = 0$ and hence the energy
curve is vertical at such a point.   This occurs on the x axis in Fig. 2 except
where $\sin x = 0$.   It should be noted that dv/dx is undefined when $\sin x = 0$
and $v = 0$, that is, when $x = \pm n\pi$, $v = 0$.   Such points are called *singular
points* or *critical points*; they are discussed in detail in the next section.

In order to facilitate the sketching of the energy curves as in Fig. 2,
it is often helpful to sketch first the *isoclines* or *curves of constant slope*
obtained by setting $dv/dx = S$ (a constant) in (19).   This leads to a set of
equations

$$v = - \frac{k^2}{S} \sin x \tag{20}$$

For any given value of $S \neq 0$, Eq. (20) gives the locus of all points at which
the lineal elements have slope S.   These curves are plotted in Fig. 3 for
several values of S.   (For simplicity we assume $k = 1$.)   By drawing short
line segments with the indicated slopes, one obtains a fair picture of the
direction field.   It is then possible to make rough sketches of energy curves
by following this direction field.   (Compare Figs. 2 and 3.)

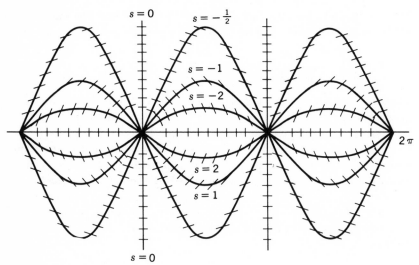

figure 3

† See Sec. 4-2.

Problems 10-2

1. Show that a free undamped pendulum cannot reach an unstable equilibrium position in finite time and remain there. Hint: Use Eq. (12) to show that the time T_1 needed for x to go from 0 to π on the separatrix is

$$T_1 = \frac{2\sqrt{2}}{k} \int_0^\pi \frac{dx}{\sqrt{1 + \cos x}}$$

Evaluate the integral to show that T_1 is infinite.

2. Show that the period T of a pendulum is a monotonically increasing function of the amplitude M for $0 < M < \pi$. Hint: Compute dT/db from Eq. (18) and show that dT/db (and hence dT/dM) is positive.

3. Describe the motion associated with the energy curve through the point J in Fig. 2.

10-3   Singularities and the phase plane

Consider the motion of a particle of mass m which is acted on by forces which do not depend explicitly on the time.† The differential equation describing such motion can be written in the form

$$m\ddot{x} = F(x,\dot{x}) \tag{21}$$

If we substitute $\dot{x} = v$, $\ddot{x} = v\, dv/dx$, and divide by mv, we can write the equation in the form

$$\frac{dv}{dx} = G(x,v) \tag{22}$$

where $G(x,v) = F(x,v)/mv$ Now, assuming that $G(x,v)$ can be written as a quotient of two analytic functions,‡ we write (22) in the form

$$\frac{dv}{dx} = \frac{g(x,v)}{f(x,v)} \tag{23}$$

This last equation defines a direction field at every point *except* where *both* g and f vanish.§ These exceptional points are called *singular points*. That is, we define a singular point to be a point (x_0,v_0) at which both $f(x_0,v_0) = 0$ and $g(x_0,v_0) = 0$.

---

† Such a system is called *autonomous*.
‡ See Sec. 7-2 for a definition of analytic function.
§ Note that if $f = 0$ but $g \neq 0$ at some point, dv/dx is undefined but $dx/dv = 0$, and hence the direction field is vertical at such a point.

For example, in the previous section the equation for the pendulum was

$$\frac{dv}{dx} = -\frac{k^2 \sin x}{v} \tag{24}$$

Here $f(x,v) = v$, $g(x,v) = -k^2 \sin x$ and the points $(\pm n\pi, 0)$, $n = 0, 1, \ldots$, are the *singular points* or *singularities*.

We shall see below that once we know the behavior of the solution curves near the singularities, we can predict the behavior throughout the plane.

Since both x and v are functions of the parameter t, it is often convenient to separate Eq. (23) into an equivalent pair of simultaneous equations

$$\frac{dx}{dt} = f(x,v) \qquad \frac{dv}{dt} = g(x,v) \tag{25}$$

That is, if x and v satisfy Eqs. (25), then they will also satisfy (23).† The advantage of working with the system (25) is that the solution curves in the phase plane are given in terms of the parameter t, and physical intuition is then helpful in interpreting the results.‡

In order to simplify the following analysis of singularities of Eq. (23), we shall assume that:

1. The origin is a singular point,§ i.e.,

$$f(0,0) = 0 \qquad g(0,0) = 0$$

2. There are no other singular points near the origin.
3. f and g are analytic functions of x and v near the origin

These assumptions imply that f and g can be expanded in Taylor series of the form

$$\begin{aligned}
f(x,v) &= ax + bv + P_2(x,v) \\
g(x,v) &= cx + dv + Q_2(x,v)
\end{aligned} \tag{26}$$

where a, b, c, and d are constants and P_2 and Q_2 contain terms of at least the second order in x and v. Hence, if x and v are small, f and g are given

† The converse is not necessarily true since f and g might contain a common factor which would cancel out in (23).

‡ In *this section*, however, we shall analyze the system of Eqs. (25) from a purely mathematical point of view and we shall *not* impose the physical restriction that $v = dx/dt$.

§ This is not a serious restriction since, if a singularity occurs at a point (x_1,v_1), the transformation $x' = x - x_1$, $v' = v - v_1$ produces a new equation with the singularity shifted to the origin.

approximately by

$$f(x,v) = ax + bv$$
$$g(x,v) = cx + dv\dagger \tag{27}$$

Thus, in order to determine the behavior of the nonlinear system (25) in the vicinity of a singularity, we shall first investigate the solutions of the *linearized system*

$$\frac{dx}{dt} = ax + bv \qquad \frac{dv}{dt} = cx + dv \tag{28}$$

It is shown in more advanced texts‡ that the solutions of the two systems (25) and (28) have *essentially* the same behavior near singular points, although some differences can occur. (E.g., see Sec. 10-6, Example 1.)

Equations (28) are a pair of homogeneous linear equations with constant coefficients which can be solved by the methods of Chap. 8.

Substitute a trial solution

$$x = Ae^{\lambda t} \qquad v = Be^{\lambda t}$$

into (28) and the following equations result:

$$(a - \lambda)A + bB = 0$$
$$cA + (d - \lambda)B = 0 \tag{29}$$

This pair of equations can be satisfied by nonzero A and B only if the determinant of the coefficients vanishes. That is,

$$\begin{vmatrix} a - \lambda & b \\ c & d - \lambda \end{vmatrix} = 0 \tag{30}$$

This yields the following *characteristic equation* for λ

$$\lambda^2 - (a + d)\lambda + (ad - bc) = 0 \tag{31}$$

with solutions

$$\lambda = \frac{a + d}{2} \pm \tfrac{1}{2}\sqrt{(a - d)^2 + 4bc} \tag{32}$$

Several different cases arise depending on the value of the discriminant

$$\Delta = (a - d)^2 + 4bc \tag{33}$$

† In order to simplify the subsequent discussion, we assume that the quantity $ad - bc \neq 0$. Otherwise the singularity would be *removable* in the linearized equations and its character would be determined by the higher-order terms. For more details on this exceptional case see, for example, Kaplan [21], chap. 11.

‡ E.g., see Coddington and Levinson [11], chap. 15.

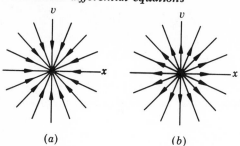

(a)   (b)

figure 4.   (*a*) *Stable node.* (*b*) *Unstable node.*

Since a detailed analysis of all the cases can become somewhat tedious, we shall content ourselves here with an analysis of some special cases which illustrate the general results.   For a more complete treatment see Kaplan [21].

Case I. $\Delta = 0$.   This could occur, for example, if $a = d \neq 0$ and $b = c = 0$.†   Eqs. (28) would then be $\dot{x} = ax$ and $\dot{v} = av$ with solutions

$$x = c_1 e^{at} \qquad v = c_2 e^{at} \tag{34}$$

We can eliminate t in (34) to get

$$c_2 x = c_1 v \tag{35}$$

This is a straight line through the origin in the phase plane for any choice of c_1 and c_2, not both zero.

In a case such as this, where all the solution curves go through the singularity, the singularity is called a *node* (see Fig. 4).   From Eqs. (34) we see that if $a < 0$, then, as $t \to \infty$, both x and $v \to 0$.   In this case the node is said to be *stable* (see Fig. 4a).   If $a > 0$ so that $x \to \infty$ and $v \to \infty$, then we have an *unstable node*, as in Fig. 4b.

Case II. $\Delta > 0$.   This case is typified by the situation that arises when $b = c = 0$ and $a \neq d$.   Eqs. (28) become $\dot{x} = ax$ and $\dot{v} = dv$ with solutions

$$x = c_1 e^{at} \qquad v = c_2 e^{dt} \tag{36}$$

or, eliminating t,

$$c_4 v^a = c_3 x^d \tag{37}$$

If a and d have the same sign, we again have a node since all the curves will go through the origin (see Fig. 5a).

However, if a and d have different signs so that when $x \to 0$, $v \to \infty$, and when $x \to \infty$, $v \to 0$, the curves will look like those in Fig. 5b and the singularity is called a *saddle point*.

† General criteria for each case will be stated later.   See page 320.

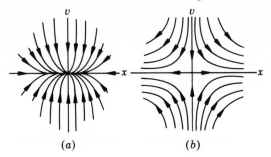

figure 5.   (*a*) *Stable node.* (*b*) *Saddle point.*

Whether we have a node or a saddle, there are two special curves which are obtained when $c_3 = 0$ or when $c_4 = 0$. These curves are then $v = 0$ and $x = 0$. In the node the curve $x = 0$ is the only one which enters the origin vertically. In the saddle the curves $x = 0$ and $v = 0$ are the only ones which enter the singularity.

Case III. $\Delta < 0$. This occurs, for example, if $d = a$ and $c = -b$. The differential equations (28) become

$$\dot{x} = ax + bv$$
$$\dot{v} = -bx + av$$

These equations have a solution $x = e^{at} \sin bt$, $v = e^{at} \cos bt$, and hence

$$\sqrt{x^2 + v^2} = e^{at}$$

If $a < 0$, the distance $\sqrt{x^2 + v^2}$ from the origin to a point moving on the solution curve will decrease steadily as the point moves around the origin. The resulting curves will be similar to those in Fig. 6*a*. In such a case the

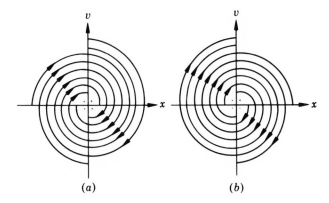

figure 6.   (*a*) *Stable spiral.* (*b*) *Unstable spiral.*

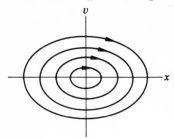

figure 7.   *Center.*

singularity is called a *stable spiral*.   If $a > 0$, the curves spiral away from the origin and we have an *unstable spiral*, as in Fig. 6*b*.

In the above example, if $a = 0$, the solution would be $x = \sin bt$, $v = \cos bt$, and the integral curves would be circles or, more generally, ellipses. In this case the origin is called a *center* and the curves look like those in Fig. 7.   It is of considerable importance in the application of these techniques that solution curves near a *center* are *closed*, and hence a point traversing such a curve returns to its starting position in some finite time T.   Physically this implies periodic motion with period T.

In the above examples we have illustrated the various types of singularities that can occur.   It is shown in more advanced texts† that there are no other types of singularities under the restrictions that we have imposed. The general classification criteria are summarized in Table 1.

Table 1   *Classification of Singularities*

Equation: $\dfrac{dv}{dx} = \dfrac{cx + dv}{ax + bv}$ or $\begin{cases} \dot{x} = ax + bv \\ \dot{v} = cx + dv \end{cases}$

Fundamental quantities: $\begin{cases} \Delta = (a - d)^2 + 4bc \\ p = a + d \\ q = ad - bc \neq 0 \end{cases}$

Case	Singularity	Stability
1. $\Delta = 0$	Node	$\begin{cases} \text{Stable if } p < 0 \\ \text{Unstable if } p > 0 \end{cases}$
2. $\Delta > 0$	$\begin{cases} \text{Node if } q > 0 \\ \\ \text{Saddle if } q < 0 \end{cases}$	$\begin{cases} \text{Stable if } p < 0 \\ \text{Unstable if } p > 0 \end{cases}$
3. $\Delta < 0$	$\begin{cases} \text{Center if } p = 0 \\ \text{Spiral if } p \neq 0 \end{cases}$	$\begin{cases} \text{Stable if } p < 0 \\ \text{Unstable if } p > 0 \end{cases}$

†E.g., see Coddington and Levinson [11].

Example 1   The damped pendulum.   Suppose that the pendulum described in Sec. 10-2 is subjected to a small viscous damping force such as air resistance. If we assume that this resistance is proportional to the angular velocity, the equation of motion becomes

$$m\ddot{x} + r\dot{x} + mk^2 \sin x = 0 \tag{38}$$

where r is a small positive constant.   When we set $\dot{x} = v$, this equation becomes

$$\frac{dv}{dx} = \frac{-k^2 \sin x - (r/m)v}{v} \tag{39}$$

and the corresponding system is

$$\dot{x} = v$$
$$\dot{v} = -k^2 \sin x - \frac{r}{m} v \tag{40}$$

The singularities occur at $v = 0$, $x = n\pi$, $n = 0, \pm 1, \pm 2, \ldots$.   We shall first examine the behavior near the singularity at $(0,0)$.   Expanding $\sin x$ about $x = 0$, we have $\sin x = x - x^3/3! + \cdots$ and hence Eqs. (40) become

$$\dot{x} = v$$
$$\dot{v} = -k^2 x - \frac{r}{m} v + \text{higher-order terms} \tag{41}$$

In the notation of the classification scheme developed above

$$a = 0 \qquad b = 1 \qquad c = -k^2 \qquad d = \frac{-r}{m} \qquad p = a + d = \frac{-r}{m}$$

$$q = ad - bc = k^2 \qquad \Delta = (a - d)^2 + 4bc = \left(\frac{r}{m}\right)^2 - 4k^2$$

Now, if r is small ($r < 2mk$), then we have $\Delta < 0$, and since $p < 0$, we have a *stable spiral* at $(0,0)$.   The solution curves near $(0,0)$ must look like those in Fig. 8.

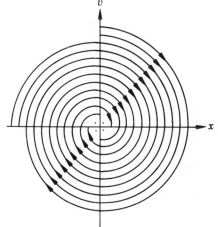

figure 8.   *Stable spiral at* $(0,0)$.

Next we examine the singularity at $(\pi,0)$. We can shift this singularity to the origin by making the change of variable

$$x_1 = x - \pi$$

so that

$$\sin x = \sin (x_1 + \pi) = -\sin x_1$$

In terms of the new variable, the linearized version of (40) becomes

$$\dot{x}_1 = v$$
$$\dot{v} = +k^2 x_1 - \frac{r}{m} v \tag{42}$$

and hence we have now

$$a = 0 \qquad b = 1 \qquad c = k^2 \qquad d = -\frac{r}{m} \qquad p = -\frac{r}{m} \qquad q = -k^2$$

$$\Delta = \left(\frac{r}{m}\right)^2 + 4k^2$$

Therefore at $(\pi,0)$ we have $\Delta > 0$, $q < 0$, and by Table 1 we have a *saddle* (see Fig. 9).

By a similar analysis it is easily shown that the singularities at the points $(2n\pi,0)$ are stable spirals and at $([2n + 1]\pi, 0)$ saddles, for $n = 0, \pm1, \pm2, \ldots$ (see Prob. 1).

From this information on the singularities it is possible to infer the qualitative behavior of the solution curves in the phase plane, and a sketch such as Fig. 9 can then be drawn. A careful study of the curves in Fig. 9 reveals a vivid picture of all the possible motions of a pendulum. For instance, if the bob is set in motion at $x = 0$ with a fairly high initial velocity, as at A, then the displacement increases although the velocity decreases until the bob passes through the unstable vertical position (B) with positive velocity. As the bob falls down it picks up speed until it passes $x = 2\pi$ and begins to rise and slow down. The resistance has retarded the bob sufficiently so that it no longer has enough speed to carry it over the top; it reaches its maximum displacement

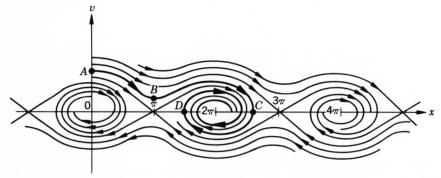

figure 9

at point C and then swings back toward the position $x = 2\pi$. The subsequent motion consists of damped oscillations about the stable equilibrium position at $x = 2\pi$.

If there is any question as to the form of any section of the phase picture, it is often useful to sketch the isoclines first and then use the information on the singularities.

Problems 10-3

1. Find and classify all singularities of Eq. (39) for the damped pendulum.

2. The displacement x of a body satisfies the differential equation $\ddot{x} + 2\dot{x} - \sin x = 0$. Discuss the motion of the body if it is released from equilibrium $(x = 0)$ with:

    a. small initial velocity   b. large initial velocity

3. Discuss the singularities and then sketch the integral curves for the equation

$$\frac{dv}{dx} = \frac{x(x-1) + v}{x(x-1) + 2v}$$

4. Locate and classify the singularities of the equation $\ddot{x} + k^2 \sin x = 0$ for the undamped pendulum. Sketch the energy curves in the phase plane and compare with Fig. 2.

5. The equation $\ddot{x} + k^2(x - x^3/3!) = 0$ [Eq. (5) in the text] is an approximation to the equation for the undamped pendulum. Locate and classify the singularities, sketch the solution curves, and compare with Fig. 2.

6. Given the equation $\dfrac{dy}{dx} = \dfrac{-x + y}{x + y}$

    a. Locate and classify the singularities and sketch the solution curves in the xy plane.

    b. Solve the equation by one of the methods of Chap. 2.

    c. Introduce polar coordinates into your solution in part b and show that the solution curves are the spirals $r = ce^{-\theta}$ for arbitrary constants $c \geq 0$.

7. The linear equation for free vibrations is $m\ddot{x} + r\dot{x} + kx = 0$. Let $v = \dot{x}$ and analyze the singularities of the resulting equation by using Table 1. Compare the various cases with those discussed in Sec. 6-2.

8. A pendulum is subjected to a damping force which is proportional to the *square* of the velocity.

    a. Show that the equation governing the motion is of the form

$$\ddot{x} + c\dot{x}\,|\dot{x}| + k \sin x \neq 0$$

where c is a positive constant.

    b. Examine the integral curves in the phase plane and describe the motion.

c. Solve for $v = \dot{x}$ in terms of x.

d. For what range of values of the initial velocity v_0 will the pendulum make exactly n complete revolutions before tending to the equilibrium position?

10-4   Van der Pol's equation

In a classical paper† in 1927 van der Pol showed that certain properties of vacuum tubes could be predicted from the differential equations of the circuit only if the nonlinear terms in the equation were retained. In fact, the oversimplified linear theory predicted that no periodic oscillations in the current would occur under certain conditions when experimental evidence showed that they did occur. But, by retaining the nonlinear terms in the equations, van der Pol was able to give a complete explanation of the phenomenon. Since that time many similar situations have arisen, in which physical systems were found to have certain properties which could be explained only in terms of the nonlinearity of the associated differential equations.

In this section we shall study the differential equation‡

$$\ddot{x} - \mu[\dot{x} - \tfrac{1}{3}(\dot{x})^3] + x = 0 \tag{43}$$

with positive parameter μ, which is closely related to van der Pol's equation. The equation which was actually studied by van der Pol is essentially the one obtained by differentiating (43) and substituting $y = \dot{x}$, i.e.,

$$\ddot{y} + \mu(y^2 - 1)\dot{y} + y = 0 \tag{44}$$

In particular we shall examine the possibility of periodic solutions of Eq. (43) or (44) since such solutions are observed experimentally.

Before beginning a mathematical analysis of Eqs. (43) and (44), some indication of the effect of the nonlinearity can be obtained by physical reasoning. The quantity y in Eq. (44) is proportional to the current in the electrical circuit studied by van der Pol, and hence it is instructive to compare (44) with the equation for the current in an RLC series circuit§

$$L\ddot{I} + R\dot{I} + \frac{1}{C}I = 0 \tag{45}$$

On comparing Eqs. (44) and (45), we see that the quantity $\mu(y^2 - 1)$ is equivalent to the resistance in the circuit. Thus the linearized version of (44) would contain a *negative* resistance term $-\mu$. This would produce a

† B. van der Pol, Forced Oscillations in a Circuit with Non-linear Resistance, *Phil. Mag.*, Vol. 3., pp. 65–80, Jan., 1927.

‡ Equation (43) is sometimes referred to as Rayleigh's equation.

§ See Sec. 6-4.

current with steadily increasing amplitude (see Prob. 4) and hence, according to the *linear* theory, no periodic solution could exist. However, with a resistance $\mu(y^2 - 1)$ in the circuit, as the current y becomes greater than 1, the resistance becomes *positive*, thus limiting the growth of the current and providing at least the possibility for the existence of a periodic solution.

For a phase-plane analysis of Eq. (43) let $v = \dot{x}$ to obtain the system

$$\dot{x} = v$$
$$\dot{v} = -x + \mu(v - \tfrac{1}{3}v^3) \tag{46}$$

This system has a singular point at (0,0), and near (0,0) the associated linear system is

$$\dot{x} = v$$
$$\dot{v} = -x + \mu v \tag{47}$$

Thus the fundamental quantities for classifying the singularity by Table 1 are $\Delta = \mu^2 - 4$, $p = \mu$, $q = 1$. Hence the singularity is an unstable node if $\mu \geq 2$ and an unstable spiral if $\mu < 2$. In either case the solution curves near (0,0) are all directed away from the origin. On the other hand, it is possible to show that the curves far from the origin are approaching the origin.† It was first proved by Liénard‡ that all these solution curves are approaching a unique closed solution curve. This closed curve represents the periodic solution to van der Pol's equation.

Although the analytical proof of the preceding remarks is too involved to be included here, it is possible to gain some insight into the behavior of the solutions by means of a very simple geometrical technique.

Liénard's construction The following method was devised by Liénard for studying the geometrical character of the solution curves for van der Pol's and related equations.

The system (46) can be combined to give an equation for the slope dv/dx of the solution curve through the point (x,v), i.e.,

$$\frac{dv}{dx} = \frac{\mu(v - \tfrac{1}{3}v^3) - x}{v} \tag{48}$$

In the more general case we consider equations of the form

$$\frac{dv}{dx} = \frac{F(v) - x}{v} \tag{49}$$

for an arbitrary function $F(v)$. A line with slope given by (49) is easily constructed. First sketch the so-called *characteristic curve* $x = F(v)$. [See

† E.g., see Stoker [33], chap. 5A.
‡ A. Liénard, Étude des oscillations entretenues, *Revue Générale de l'Électricité*, vol. 23, 1928, pp. 901–912, 946–954.

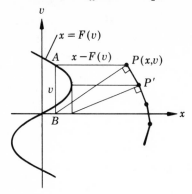

figure 10

Fig. 10 for the special case where $F(v) = \mu(v - \frac{1}{3}v^3)$.] Next, from any point $P(x,v)$ a horizontal line AP is drawn to the characteristic. This segment has length $x - F(v)$. The vertical line AB has length v and hence the segment BP has slope $v/[x - F(v)]$. Hence the line PP' perpendicular to PB has the desired slope and represents a tangent to the solution curve through P. If the process is repeated at P', etc., the resulting chain of

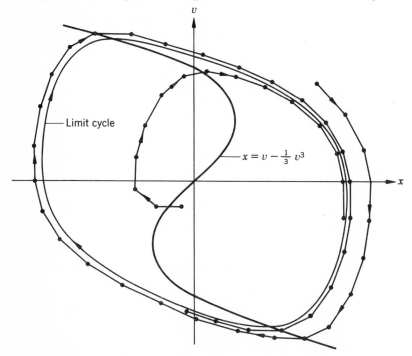

figure 11.   *Liénard's construction applied to van der Pol's equation*

$$\frac{dv}{dx} = \frac{v - \frac{1}{3}v^3 - x}{v}.$$

line segments is a rough approximation to a solution curve. By drawing several such curves, it is possible to obtain a very good approximation to the closed periodic solution curve or *limit cycle*. See Fig. 11 for the case $\mu = 1$ in Eq. (48).

Problems 10-4

1. Suppose that the motion of a body is governed by the equation

$$\ddot{x} + (\dot{x}^2 - \dot{x}) + x = 0$$

a. Using physical arguments, discuss the possible motions.

b. Use Liénard's construction to sketch the solution curves in the phase plane.

c. Does the result in part b agree with your results in part a?

2. Using Liénard's construction, obtain a graphical approximation to the limit cycle for van der Pol's equation (48) when:

a. $\mu = 0.1$ b. $\mu = 10$

Compare with Fig. 11.

3. Use Eq. (45) for the RLC series circuit to show that if R is assumed to be a negative constant, then the amplitude of the resulting current I is an increasing function of time. See Sec. 6-4.

10-5 Piecewise linear system

Consider an object of mass m attached to a wall by means of a spring as shown in Fig. 12. Suppose that the object is sliding along a rough surface which exerts a force opposing the motion. If, in addition, this friction force has a *constant magnitude* $r > 0$, then it is known as *coulomb friction* and can be represented by a function $f(\dot{x})$ where

$$f(\dot{x}) = \begin{cases} +r & \text{if} \quad \dot{x} < 0 \\ -r & \text{if} \quad \dot{x} > 0 \end{cases} \tag{50}$$

and x is the displacement from equilibrium. ($\dot{x}$ is then the velocity.)

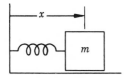

figure 12

The differential equation governing the motion of the sliding object is

$$m\ddot{x} = -kx + f(\dot{x}) \tag{51}$$

where k is the spring constant. This equation can be written in the form

$$m\ddot{x} + kx = r \quad \text{if} \quad \dot{x} < 0 \tag{52a}$$

$$m\ddot{x} + kx = -r \quad \text{if} \quad \dot{x} > 0 \tag{52b}$$

These equations are both linear but the problem is itself nonlinear for $t > 0$ since it cannot be represented by a *single linear equation* which is valid for all $t > 0$.

The qualitative behavior of the solution of Eqs. (52) is unaffected by changes in the relative magnitudes of the parameters m, k, r. Hence we can simplify the following discussion, without restricting the range of applicability of the result, by setting $m = k = r = 1$. Equations (52) then become

$$\ddot{x} + x = 1 \quad \dot{x} < 0 \tag{53a}$$

$$\ddot{x} + x = -1 \quad \dot{x} > 0 \tag{53b}$$

or, on substituting $\dot{x} = v$,

$$v\frac{dv}{dx} + x = 1 \quad v < 0 \tag{54a}$$

$$v\frac{dv}{dx} + x = -1 \quad v > 0 \tag{54b}$$

The integral curves in the phase plane can be obtained explicitly by integration. We have

$$v^2 + x^2 = 2x + C_1 \quad v < 0 \tag{55a}$$

$$v^2 + x^2 = -2x + C_2 \quad v > 0 \tag{55b}$$

or

$$v^2 + (x - 1)^2 = C_3 \quad v < 0 \tag{56a}$$

$$v^2 + (x + 1)^2 = C_4 \quad v > 0 \tag{56b}$$

These curves are semicircles with centers at $(1,0)$ and $(-1,0)$ respectively (see Fig. 13).

Some typical motions of the system are shown in Fig. 13. For example, suppose that the weight is set in motion at its equilibrium position with a large positive velocity (point A in Fig. 13). Since the velocity is positive, the motion is governed by the differential equation (53b) and the corresponding solution curve is of the form (56b), a semicircle through A with center at

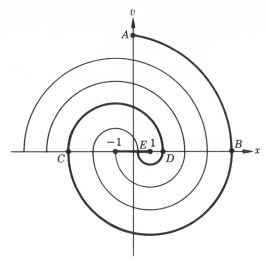

figure 13

$(-1,0)$. The combined action of the spring and friction forces reduces the speed of the weight until it is brought to rest $(v = 0)$ at point B. The extended spring now reverses the direction of the motion, and since the velocity is now negative, Eq. (53a) describes the motion and the integral curve is given by (56a), a semicircle through B with center at $(1,0)$. The motion continues in the manner indicated by the curve $ABCDE$. At point E the displacement from equilibrium $(x = 0)$ is so small that the spring force is not strong enough to overcome the friction force and the motion stops. This will happen at any point on the x axis in the interval $-1 \le x \le 1$.

Problems 10-5

1. Discuss the motion of the general piecewise linear system as determined by Eqs. (52) for arbitrary positive m, r, and k.

2. Use Liénard's construction (see Sec. 10-4) to sketch the solution curves for Eq. (51). Compare with Fig. 13.

3. Sketch the solution curves for Eq. (51) if the friction force $f(\dot{x})$ is given by

$$f(\dot{x}) = \begin{cases} +r & \text{if} \quad \dot{x} < -r \quad \text{or} \quad 0 < \dot{x} < r \\ -r & \text{if} \quad \dot{x} > r \quad \text{or} \quad -r < \dot{x} < 0 \end{cases}$$

What happens when a solution curve reaches one of the segments $v = \pm r$ for $-r < x < r$?

4. For the motion governed by Eqs. (53), assume initial conditions $x(0) = 0$, $v(0) = 2\sqrt{2}$, and determine the time at which the system finally comes to rest.

10-6   *Liapunov's second method*

In the previous sections we have discussed several methods of obtaining
information about the behavior of solutions of equations without actually
solving the equations.   Another such method, which has received considerable attention in recent years, is one due to A. M. Liapunov.†   It is known
as his *second*, or *direct*, *method*.

A simple example will illustrate the basic idea, and then the method
will be developed more formally.

> *Example* 1   Determine the behavior of the solutions of the system
>
> $$\dot{x} = -y - x^3$$
> $$\dot{y} = x - y^3 \tag{57}$$
>
> in the vicinity of the point (0,0) in the xy plane.   The technique of linearizing
> the system, as described in Sec. 10-3, leads to the prediction that the origin
> is a *center*.   That is, the solution curves should be *closed* curves encircling the
> origin, such as those in Fig. 7.   It is easy to show, however, that this is *not* the
> case.
>
> Consider the set of circles about the origin, given by the equation,
>
> $$x^2 + y^2 = c^2 \tag{58}$$
>
> for all $c > 0$.   See Fig. 14.   Now, if we could show that every time a solution
> curve of the system (57) crosses one of these circles, the tangent vector $\mathbf{T}$ of the

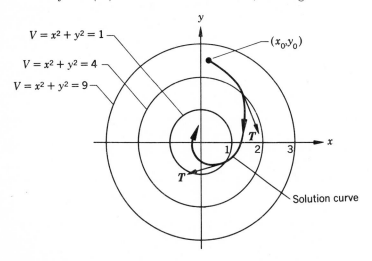

figure 14

† A. M. Liapunov, *Problème générale de la stabilité du mouvement*, Annals of
Mathematics Studies, No. 17, Princeton University Press, Princeton, N.J., 1947.

solution is pointing into the *interior* of the circle, then the solution could never leave the circle. We shall show, in fact, that a point P: $(x(t),y(t))$ on a solution curve continues to move closer and closer to the origin as t increases.

Let $V = V(x,y) = x^2 + y^2$. That is, V is the square of the distance from the origin to the point (x,y). As the point (x,y) moves along a solution curve, the rate of change of V in the direction tangent to the curve will indicate whether or not the curve is approaching the origin. The unit tangent vector $\mathbf{T}$ to the solution curve is given by[†]

$$\mathbf{T}(t) = \frac{\dot{x}(t)\mathbf{i} + \dot{y}(t)\mathbf{j}}{\sqrt{\dot{x}^2 + \dot{y}^2}}$$

and hence the derivative of $V(x,y)$ in the direction of $\mathbf{T}$ is the dot product of $\mathbf{T}$ with the gradient of V, i.e.,

$$\boldsymbol{\nabla} V \cdot \mathbf{T} = \frac{V_x \dot{x} + V_y \dot{y}}{\sqrt{\dot{x}^2 + \dot{y}^2}}. \tag{59}$$

Since the denominator is always positive, the sign of $\boldsymbol{\nabla} V \cdot \mathbf{T}$ is determined by the sign of $V_x \dot{x} + V_y \dot{y}$ which is just $(d/dt)[V(x(t),y(t))]$. Thus if $(d/dt)[V(x(t),y(t))]$ is always negative, then the point $(x(t),y(t))$ is approaching the origin.

Using (57), we have

$$\begin{aligned}
\frac{d}{dt}[V(x(t),y(t))] &= V_x \dot{x} + V_y \dot{y} \\
&= 2x(-y - x^3) + 2y(x - y^3) \\
&= -2(x^4 + y^4) \\
&< 0 \qquad \text{for} \qquad (x,y) \neq (0,0)
\end{aligned}$$

Therefore we have shown that the solution curve through any point remains inside a circle through that point.

To show, further, that the solutions actually approach the origin, we first note that, for any $\delta > 0$, if $x^2 + y^2 \geq \delta^2$, then $x^4 + y^4 \geq \delta^4/2$ (why?), and hence

$$\frac{dV}{dt} \leq -\delta^4 \tag{60}$$

Integrating both sides of (60) from some t_0 to $t > t_0$, we have

$$V(x(t),y(t)) - V(x(t_0),y(t_0)) \leq -\delta^4(t - t_0)$$

or

$$V(x(t),y(t)) \leq V(x(t_0),y(t_0)) + \delta^4 t_0 - \delta^4 t \tag{61}$$

Now, if Eq. (61) held for arbitrarily large t, then the right side would ultimately become negative. But then we would have $V = x^2 + y^2 < 0$, which is impossible. Hence (61) cannot hold for all t; this means that $x^2 + y^2$ cannot remain always $\geq \delta^2$. Thus the point (x,y) must ultimately enter every circle, no matter how small, and therefore the solution curve must approach the origin as $t \to \infty$.

[†] E.g., see Courant [12], vol. 2.

The method which was described in Example 1 will now be generalized in order to apply it to the system of equations†

$$\dot{x} = f(x,y)$$
$$\dot{y} = g(x,y)$$

(62)

where f and g are assumed to be analytic and nonzero near the origin, but $f(0,0) = g(0,0) = 0$.

Definition 1   *The origin,* $x = 0$, $y = 0$, *is called a stable solution of* (62) *in the sense of Liapunov (or L stable) if, given any* $\epsilon > 0$, *there exists a* $\delta > 0$ *such that, whenever* $x^2(t_0) + y^2(t_0) < \delta^2$, $x^2(t) + y^2(t) < \epsilon^2$ *for all* $t > t_0$.

Geometrically, this means that any solution that starts inside a circle of radius δ must remain inside a circle of radius ϵ.   See Fig. 15.   If, in addition, $\lim\limits_{t \to \infty} [x^2(t) + y^2(t)] = 0$, then the origin is called *asymptotically stable*.

For example, we have proved in Example 1 that the origin is asymptotically stable for the system (57).

Liapunov's second method consists in constructing functions $V(x,y)$, called *Liapunov functions*, that play a role similar to that played by the distance function V in Example 1.

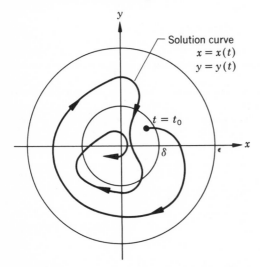

figure 15

† The present discussion will be restricted to *autonomous systems*, i.e., systems in which the independent variable t does not appear explicitly.   Also, we shall consider only pairs of equations in two unknown functions.   For the extension of our results to systems of n equations, see Probs. 7 and 8.

Definition 2   *A function* $V(x,y)$ *is called a Liapunov function for the system* (62) *if there exists some neighborhood of the origin in which the following conditions hold*:

1. V *is a differentiable function of* x *and* y.
2. $V > 0$ *except at the origin, where* $V(0,0) = 0$.
3. *For any solution* $x(t)$, $y(t)$ *of* (62) *there exists a* t_0 *such that*

$$\frac{d}{dt}[V(x(t),y(t))] \le 0 \qquad \text{for all } t \ge t_0$$

The following two theorems indicate the way that Liapunov functions are applied.

Theorem 1   *If there exists a Liapunov function for the system* (62), *then the origin is L-stable.*

Theorem 2   *If, for any* $\delta > 0$, *there exists a Liapunov function* $V(x,y)$ *and a number* $\beta(\delta) > 0$, *and a* T_0, *such that* $(d/dt)[V(x(t),y(t))] \le -\beta$ *outside the circle* $x^2 + y^2 = \delta^2$ *for all* $t \ge T_0$, *then the origin is asymptotically stable.*

The proofs of Theorems 1 and 2 are left as exercises since they follow exactly the same kind of arguments that were used in establishing the results in Example 1.   See Prob. 1.

For further information on Liapunov's second method see the problems at the end of this section.   See also, for example, La Salle and Lefschetz [23].

Problems 10-6

1. Using arguments similar to those in Example 1 of the text:

a. Prove Theorem 1.   b. Prove Theorem 2.

2. Show that the function $V = x^4 + y^4$ is a Liapunov function for the system $\dot{x} = -xy^4$, $\dot{y} = yx^4$.   What conclusion can be drawn?

3. Discuss the stability of the origin for the system $\dot{x} = -4y - x^3$, $\dot{y} = 3x - y^3$.

4. Given the system $\dot{x} = Ax - xy^2$, $\dot{y} = Ay - yx^2$:

a. For what values of A is $V = x^2 + y^2$ a Liapunov function?   Compare with the result of applying the method of Sec. 10-3.

b. What can be said about asymptotic stability?

5. a. Is $V = x^2 - y^2$ a Liapunov function for the system $\dot{x} = -x - x^3y^2$, $\dot{y} = -y - x^4y$?

b. What does the answer to part a show about the stability of the origin?

6. Consider the second-order equation $\ddot{x} + f(x)\dot{x} + g(x) = 0$, where f and g are continuous. Define $F(x) = \int_0^x f(s)\,ds$, $G(x) = \int_0^x g(s)\,ds$. Assume that in some neighborhood of $x = 0$:

     a. $g(x)F(x) \geq 0$ b. $xg(x) > 0$ if $x \neq 0$

First show that the differential equation is equivalent to the system $\dot{x} = y - F(x)$, $\dot{y} = -g(x)$, and then prove that $V(x,y) = \frac{1}{2}y^2 + G(x)$ is a Liapunov function for the system. What conclusions can be drawn about the solutions of the original second-order equation?

7. Given the system $\dot{\mathbf{x}} = \mathbf{f}(\mathbf{x})$ where $\mathbf{x} = \mathbf{x}(t)$ is a vector function of t with components $x_1(t), x_2(t), \ldots, x_n(t)$ and $\mathbf{f}$ is a vector with components $f_1, f_2, \ldots, f_n$, which are analytic functions of $x_1, x_2, \ldots, x_n$ such that $\mathbf{f}(\mathbf{0}) = \mathbf{0}$:

     a. Define *stability* and *asymptotic stability* of the origin $\mathbf{x} \equiv (0, 0, \ldots, 0)$.
     b. State the analogs of Theorems 1 and 2 of the text for this system.
     c. Prove the theorems stated in part b.

8. Find a Liapunov function for the system $\dot{\mathbf{x}} = A(t)\mathbf{x}$, where $A(t)$ is an $n \times n$ matrix with coefficients $a_{ij}(t)$ satisfying $a_{ij}(t) = -a_{ji}(t)$, if $i \neq j$ and $a_{ii}(t) \leq 0$.

Linear Difference Equations

11

11-1 Introduction

A *recurrence relation of order* p is an equation of the form

$$y_{n+p} = f(n, y_n, y_{n+1}, \ldots, y_{n+(p-1)}) \qquad n = 0, 1, 2, \ldots \tag{1}$$

where p is a fixed positive integer. This equation determines the $(n + p)$th term of a sequence $\{y_n\}$ in terms of the previous p terms. We have already met examples of recurrence relations in Chap. 7 in finding the nth coefficient of the series solution to a differential equation. Recurrence relations occur in many parts of mathematics and are particularly important in the numerical solution of differential equations (see Chap. 12).

It is customary to call Eq. (1) a *difference equation of order* p. The reason is that Eq. (1) can be expressed in terms of *differences of order* p or less defined by

$$\Delta y_n = y_{n+1} - y_n$$
$$\Delta^2 y_n = \Delta(\Delta y_n) = y_{n+2} - 2y_{n+1} + y_n$$
$$\vdots \tag{2}$$
$$\Delta^p y_n = \Delta(\Delta^{p-1} y_n)$$

However, we will have no need for the difference notation in our discussion.

By *a solution of the difference equation* (1) *we mean a sequence* y_n *which satisfies Eq. (1) for all n under consideration. The general solution is the set of all solutions.* In this chapter we will concern ourselves with the theory and methods of solution of linear difference equations. For simplicity we restrict ourselves to equations of the first and second orders.

335

11-2   First-order linear difference equations

Consider the initial-value problem

$$\Delta E :\dagger\ y_{n+1} = p_n y_n + q_n \qquad n = 0, 1, 2, \ldots$$
$$\text{IC: } y_0 = a \tag{3}$$

where p_n and q_n are known sequences. We see immediately that once $y_0 = a$ is given, then the difference equation uniquely determines y_1; and if y_n is known, the difference equation uniquely determines y_{n+1}. Therefore, by the principle of mathematical induction, y_n is uniquely determined for all n. This proves that *the initial-value problem* (3) *possesses a unique solution.*

We now turn to solving the difference equation and first consider the homogeneous equation

$$y_{n+1} = p_n y_n \qquad n \geq 0 \tag{4}$$

The solution is easily obtained by successive substitutions:

$$y_1 = p_0 y_0$$
$$y_2 = p_1 y_1 = (p_1 p_0) y_0$$
$$y_3 = p_2 y_2 = (p_2 p_1 p_0) y_0$$

$$\cdot$$
$$\cdot$$
$$\cdot$$

$$y_n = p_{n-1} y_{n-1} = (p_0 p_1 p_2 \cdots p_{n-1}) y_0 \qquad n > 0$$

It is convenient to introduce the product notation defined by

$$\prod_{k=0}^{n-1} a_k \equiv a_0 \cdot a_1 \cdot a_2 \cdots a_{n-1}$$

In this notation the solution can be written

$$y_n = \left(\prod_{k=0}^{n-1} p_k \right) y_0 \qquad n > 0 \tag{5}$$

This is a solution of the homogeneous equation (4) for any initial value y_0.

Example 1   Solve $y_{n+1} = a y_n$, $n \geq 0$, y_0 given. By successive substitution or by Eq. (5) we find

$$y_n = a^n y_0$$

† ΔE stands for difference equation.

We now consider the nonhomogeneous problem

$$\Delta E: y_{n+1} = p_n y_n + q_n \qquad n \geq 0 \tag{6}$$

$$IC: y_0 = a$$

We proceed by successive substitutions:

$$y_1 = p_0 y_0 + q_0$$

$$y_2 = p_1 y_1 + q_1 = p_1 p_0 y_0 + p_1 q_0 + q_1$$

$$y_3 = p_2 y_2 + q_2 = p_2 p_1 p_0 y_0 + p_2 p_1 q_0 + p_2 q_1 + q_2$$

.

.

.

$$y_n = p_{n-1} y_{n-1} + q_{n-1}$$

$$= y_0 \prod_{k=0}^{n-1} p_k + q_0 \prod_{k=1}^{n-1} p_k + q_1 \prod_{k=2}^{n-1} p_k + \cdots + q_{n-2} p_{n-1} + q_{n-1}$$

The solution can be written

$$y_n = a \prod_{k=0}^{n-1} p_k + \sum_{j=0}^{n-2} q_j \prod_{k=j+1}^{n-1} p_k + q_{n-1} \qquad n > 0 \tag{7}$$

$$y_0 = a$$

We note that the first term in (7) is the general solution of the homogeneous equation (for arbitrary a) and the second term is a particular solution of the nonhomogeneous equation. This is analogous to the situation for linear differential equations.

Example 2 $\Delta E: y_{n+1} = (n + 1)y_n + (n + 1)! \qquad n \geq 0$

             $IC: y_0 = a$

From Eq. (7) we have

$$y_n = a \prod_{k=0}^{n-1} (k + 1) + \sum_{j=0}^{n-2} (j + 1)! \prod_{k=j+1}^{n-1} (k + 1) + n!$$

$$= an! + \sum_{j=0}^{n-1} n! = an! + n! \sum_{j=0}^{n-1} 1$$

$$= an! + (n!)n$$

$$= n!(a + n) \qquad n \geq 0$$

Example 3 $\Delta E: y_{n+1} = ay_n + f_n$

             $IC: y_0$ given

From Eq. (7) we have

$$y_n = y_0 \prod_{k=0}^{n-1} a + \sum_{j=0}^{n-2} f_j \prod_{k=j+1}^{n-1} a + f_{n-1}$$

$$= a^n y_0 + \sum_{j=0}^{n-1} f_j a^{n-j+1}$$

Consider the special case $a = 2, f_n = 3^n$. We obtain from above

$$y_n = 2^n y_0 + \sum_{j=0}^{n-1} 3^j 2^{n-j+1} = 2^n y_0 + 2^{n-1} \sum_{j=0}^{n-1} (\tfrac{3}{2})^j$$

Summing the geometric series

$$\sum_{j=0}^{n-1} (\tfrac{3}{2})^j = \frac{1 - (\tfrac{3}{2})^n}{1 - \tfrac{3}{2}}$$

and simplifying, we obtain

$$y_n = 3^n + (y_0 - 1)2^n$$

Another procedure for solving

$$y_{n+1} = 2y_n + 3^n$$

is as follows. First solve the homogeneous equation to obtain $y_n^{(h)} = 2^n c$, where c is any constant. Next find a particular solution of the nonhomogeneous equation by undetermined coefficients. Assume a particular solution

$$y_n^{(p)} = A3^n$$

Substituting into the equation, we easily find $A = 1$. The general solution is therefore

$$y_n = 3^n + c2^n$$

Putting $n = 0$, we obtain $c = y_0 - 1$, which reproduces the solution obtained above.

> *Example 4* ΔE: $S_{n+1} - S_n = (n + 1)^2$ $n \geq 0$
>
> IC: $S_0 = 0$

The solution to this problem is easily obtained by successive substitutions:

$$S_n = 1^2 + 2^2 + 3^2 + \cdots + n^2 \tag{8}$$

Although this is an explicit solution, it is desirable to obtain a solution in closed form. This can be done by solving the difference equation using undetermined coefficients. To obtain a particular solution of the nonhomogeneous equation, we assume

$$S_n = An^3 + Bn^2 + Cn\dagger \tag{9}$$

$\dagger$ A second-degree polynomial in n is not suitable since the n^2 term would cancel out on substitution into the difference equation.

Substituting into the difference equation, we obtain

$$A(3n^2 + 3n + 1) + B(2n + 1) + C = n^2 + 2n + 1$$

Setting coefficients of like powers of n equal, we obtain

$$A = \tfrac{1}{3} \qquad B = \tfrac{1}{2} \qquad C = \tfrac{1}{6}$$

Therefore a particular solution is

$$S_n = \tfrac{1}{3}n^3 + \tfrac{1}{2}n^2 + \tfrac{1}{6}n = \tfrac{1}{6}n(n + 1)(2n + 1) \tag{10}$$

We see that this also satisfies $S_0 = 0$ and therefore must be the desired solution. Thus we have derived a closed form for the sum of the first n squares

$$1^2 + 2^2 + \cdots + n^2 = \tfrac{1}{6}n(n + 1)(2n + 1) \tag{11}$$

Problems 11-2

Solve

1. $y_{n+1} + 5y_n = 0 \qquad n \geq 0$
   $y_0 = 5$
2. $y_{n+1} = 3y_n + 5 + n \qquad n \geq 0$
   $y_0 = 2$
3. $y_{n+2} = ny_{n+1} + 1 \qquad n \geq 0$
   $y_1 = 3$
4. Discuss the solution of

$$y_{n+2} = p_n y_n \qquad n \geq 0$$
$$y_0 = a$$
$$y_1 = b$$

5. Solve the recurrence relation obtained in the discussion of Bessel functions in Sec. 7-4:

$$a_n = \frac{1}{2pn + n^2} a_{n-2} \qquad n \geq 2$$

where $a_1 = 0$ and a_0 is given.

6. Discuss the solution of

$$a_n y_{n+1} = b_n y_n \qquad b_n \neq 0$$

if $a_n = 0$ for exactly one value of n, say $a_{n_0} = 0$.

7. Solve: $ny_{n+1} = y_n \qquad n \geq 1, y_1 = a$
8. Solve: $y_{n+1} = a^n y_n \qquad n \geq 0, y_0 = 1$
9. Find a closed form for

$$S_n = 1^3 + 2^3 + \cdots + n^3$$

Hint: See Example 4.

10. Solve: $y_{n+1} = ay_n^2 \qquad n \geq 0, y_0 = a \neq 0$

11-3   Second-order linear difference equations

The basic theory for linear difference equations closely parallels that for linear differential equations.† We consider the initial-value problem

$$\Delta E: a_n y_{n+2} + b_n y_{n+1} + c_n y_n = f_n \qquad n \geq 0$$
$$IC: y_0 \text{ and } y_1 \text{ given} \tag{12}$$

Theorem 1   *If $a_n \neq 0$ for $n \geq 0$, the initial-value problem* (12) *possesses a unique solution.*

Proof   See Prob. 1.

It is convenient to introduce an operator L defined by

$$L(y_n) \equiv a_n y_{n+2} + b_n y_{n+1} + c_n y_n \qquad a_n \neq 0, \, n \geq 0 \tag{13}$$

Theorem 2   *The operator L is linear, that is,*

$$L(\alpha u_n + \beta v_n) = \alpha L(u_n) + \beta L(v_n) \tag{14}$$

where α and β are constants.

Proof   See Prob. 2.

Theorem 3   *If u_n and v_n are solutions of $L(y_n) = 0$, then $y_n = \alpha u_n + \beta v_n$ is also a solution.*

Proof $L(\alpha u_n + \beta v_n) = \alpha L(u_n) + \beta L(v_n) = 0$

Before proceeding further with properties of solutions, we need the notion of linear independence and dependence. This notion is the same for sequences as for functions. Actually a sequence $\{f_n\}$ or $f(n)$ is nothing but a function whose domain is the set of integers $n = 0, 1, 2, \ldots$.

Definition 1   *Two sequences u_n and v_n are called linearly dependent (LD) if it is possible to find two constants c_1 and c_2, not both zero, so that $c_1 u_n + c_2 v_n \equiv 0$ for $n \geq 0$. Two sequences are called linearly independent (LI) if they are not linearly dependent, i.e., if $c_1 u_n + c_2 v_n \equiv 0$ for $n \geq 0$ only when $c_1 = c_2 = 0$.*

Example 1 $u_n = n$, $v_n = n^2$ are linearly independent.

Example 2 $u_n = 2n$, $v_n = 5n$ are linearly dependent.

† To emphasize the similarity to linear differential equations, Theorems 1 through 7 are numbered to correspond to the Theorems in Secs. 5-2, 5-9, and 8-5.

Example 3   If $u_n \equiv 0$, then u_n and v_n are linearly dependent.

Example 4   Let u_n be the solution of $L(y_n) = 0$ satisfying $u_0 = 1$, $u_1 = 0$, and let v_n be the solution of $L(y_n) = 0$ satisfying $v_0 = 0$, $v_1 = 1$. Then u_n and v_n are linearly independent. (See Prob. 3.)

A useful test for linear independence can be given in terms of the *Wronskian* defined by

$$W[u_n,v_n] = \begin{vmatrix} u_n & v_n \\ u_{n+1} & v_{n+1} \end{vmatrix} = u_n v_{n+1} - v_n u_{n+1} \tag{15}$$

Theorem 4   *If $W[u_n,v_n]$ is different from zero for at least one value of $n \geq 0$, then u_n and v_n are linearly independent.*

Proof   Assume u_n and v_n are linearly dependent. Then there exist c_1 and c_2, not both zero, so that

$$c_1 u_n + c_2 v_n \equiv 0 \qquad n \geq 0 \tag{16a}$$

Therefore

$$c_1 u_{n+1} + c_2 v_{n+1} \equiv 0 \qquad n \geq 0 \tag{16b}$$

Equations (16a) and (16b) are two homogeneous equations for c_1 and c_2 which have, by assumption, a nontrivial solution. Therefore the determinant of the coefficients must be zero for each $n \geq 0$. However, the determinant is precisely $W[u_n,v_n]$, which is assumed to be different from zero for at least one value of $n \geq 0$. Therefore u_n and v_n cannot be linearly dependent.

Corollary   *If u_n and v_n are linearly dependent, then $W[u_n,v_n] \equiv 0$ for $n \geq 0$.*

If we restrict ourselves to sequences u_n and v_n which are solutions of $L(y_n) = 0$, a partial converse of Theorem 4 holds.

Theorem 5   *If u_n and v_n are linearly independent solutions of $L(y_n) = 0$, then $W[u_0,v_0] \neq 0$.†*

Proof   Suppose $W[u_0,v_0] = 0$. Then the equations

$$\alpha u_0 + \beta v_0 = 0$$

$$\alpha u_1 + \beta v_1 = 0$$

† If in addition to $a_n \neq 0$ in Eq. (13) we have $c_n \neq 0$ for $n \geq 0$, then $W[u_n,v_n] \neq 0$ for $n \geq 0$. See Prob. 8.

whose determinant is the Wronskian, have a nontrivial solution.   Consider

$$y_n = \alpha u_n + \beta v_n$$

where α and β are determined above.   We have that y_n is a solution of $L(y_n) = 0$ and $y_0 = y_1 = 0$.   Therefore $y_n \equiv 0$.   (See Prob. 4.)   This means that u_n and v_n are linearly dependent, contrary to assumption.

We are now in a position to prove the following theorem.

Theorem 6   *If u_n and v_n are linearly independent solutions of $L(y_n) = 0$, $n \geq 0$, then $y_n = \alpha u_n + \beta v_n$ is the general solution.*†

Proof   Let y_n be *any* solution of $L(y_n) = 0$.   We determine two numbers α and β by solving the equations

$$\alpha u_0 + \beta b_0 = y_0$$
$$\alpha u_1 + \beta v_1 = y_1$$

These equations have a unique solution since the coefficient determinant is $W[u_n, v_n]$ evaluated at $n = 0$, which is not zero from Theorem 5.   Consider now the sequence

$$w_n = \alpha u_n + \beta v_n$$

w_n is a solution of $L(y_n) = 0$ and $w_0 = y_0$, $w_1 = y_1$.   Since the solution is unique, we must have $w_n \equiv y_n$.   Therefore any solution y_n can be expressed as

$$y_n = \alpha u_n + \beta v_n$$

Example 5   By substitution we see that $(-3)^n$ and $(-2)^n$ are solutions of $y_{n+2} + 5y_{n+1} + 6y_n = 0$.   Therefore $y_n = c_1(-3)^n + c_2(-2)^n$ is the general solution.

Theorem 7   *If w_n is any particular solution of the nonhomogeneous equation $L(y_n) = f_n$ and u_n is the general solution of $L(y_n) = 0$, then $y_n = u_n + w_n$ is the general solution of $L(y_n) = f_n$.*

Proof   See Prob. 6 below.

Example 6   Consider

$$y_{n+2} + 5y_{n+1} + 6y_n = 4^n$$

To find a particular solution, we assume

$$w_n = A \cdot 4^n$$

† "General solution" means the set of all solutions.

Substituting into the equation, we obtain

$$A4^n(4^2 + 5 \cdot 4 + 6) = 4^n$$

Therefore $A = \frac{1}{42}$ and $w_n = \frac{1}{42}(4^n)$ is a particular solution. Combining this with the result of the previous example, we see that the general solution is

$$y_n = \frac{1}{42}(4^n) + c_1(-3)^n + c_2(-2)^n$$

Problems 11-3

1. Use mathematical induction to prove Theorem 1.
2. Prove Theorem 2.
3. Prove the statement in Example 4.
4. Using the uniqueness theorem, prove that if a solution y_n of Eq. (12) (where $a_n \neq 0$) satisfies $y_0 = y_1 \equiv 0$, then $y_n \equiv 0$.
5. Prove that two solutions of $a_n y_{n+2} + b_n y_{n+1} + c_n y_n = 0$ that agree for two successive values of n are identical, provided a_n and c_n are never zero.
6. Prove Theorem 7.
7. Using the Wronskian criterion for linear independence, show that the following pairs of functions are linearly independent:

    a. α_1^n, α_2^n where $\alpha_1 \neq \alpha_2 \neq 0$
    b. α^n, $n\alpha^n$ where $\alpha \neq 0$
    c. $\rho^n \cos n\theta$, $\rho^n \sin n\theta$ where $0 < \theta < \pi$, $\rho \neq 0$

8. Let u_n and v_n be any two solutions of

$$a_n y_{n+2} + b_n y_{n+1} + c_n y_n = 0 \qquad n \geq 0$$

where $a_n \neq 0, c_n \neq 0$ for $n \geq 0$. Derive the following formula for the Wronskian $W_n = W[u_n, v_n]$:

$$W_n = \prod_{k=0}^{n-1}\left(\frac{c_k}{a_k}\right) W_0$$

and therefore show that W_n is either identically zero (if u_n and v_n are linearly dependent) or never zero (if u_n and v_n are linearly independent).

Systems of first-order difference equations

9. Consider the system of two first-order linear difference equations for y_n and z_n:

$$y_{n+1} = a_n y_n + b_n z_n + f_n \qquad n > 0$$

$$z_{n+1} = c_n y_n + d_n y_n + g_n$$

where the initial values y_0 and z_0 are known. State and prove theorems analogous to Theorems 1 through 7.

10. a. Write the system of equations in Prob. 9 in the matrix form

ΔE: $\mathbf{x}_{n+1} = A_n\mathbf{x}_n + \mathbf{F}_n$ $n > 0$

IC: $\mathbf{x}_0$ given

by suitably defining the vectors $\mathbf{x}_n$, $\mathbf{F}_n$ and the matrix A_n.

b. Find a formula for the solution of the homogeneous equation ($\mathbf{F}_n \equiv 0$).

c. Find a formula for the solution of the nonhomogeneous equation.

11. Show that the second-order equation

$$a_n y_{n+2} + b_n y_{n+1} + c_n y_n = f_n \qquad n \geq 0$$

where $a_n \neq 0$, can be written as a system of two first-order equations.   Hint: Let $y_{n+1} = z_n$.

12. Extend Probs. 9 and 10 to a system of N equations.

11-4   Homogeneous linear difference equations with constant coefficients

The solution of linear difference equations with constant coefficients is very similar to the solution of linear differential equations with constant coefficients.

Consider the homogeneous difference equation

$$ay_{n+2} + by_{n+1} + cy_n = 0 \qquad n \geq 0 \tag{17}$$

where $a \neq 0$.   We assume a solution of the form

$$y_n = \alpha^n \tag{18}$$

where α must be determined.   Substituting into Eq. (17), we obtain

$$\alpha^n(a\alpha^2 + b\alpha + c) = 0 \tag{19}$$

If $\alpha \neq 0$ ($\alpha = 0$ provides the trivial solution $y_n \equiv 0$), we must have α satisfying the *characteristic equation*

$$a\alpha^2 + b\alpha + c = 0 \tag{20}$$

The roots α_1 and α_2 of this equation are given by

$$\alpha_{1,2} = \frac{-b \pm \sqrt{b^2 - 4ac}}{2a} \tag{21}$$

If $b^2 - 4ac > 0$, the roots α_1, α_2 are real and distinct.   Since α_1^n and α_2^n are linearly independent, the general solution is

$$y_n = c_1\alpha_1^n + c_2\alpha_2^n \tag{22}$$

If $b^2 - 4ac = 0$, the roots are real and equal. We obtain only one solution α_1^n. It can be verified (see Prob. 1) that a second linearly independent solution is $n\alpha_1^n$ and the general solution is

$$y_n = (c_1 + nc_2)\alpha_1^n \tag{23}$$

If $b^2 - 4ac < 0$, the roots are conjugate complex, say

$$\begin{aligned}
\alpha_1 &= \xi + i\eta = \rho e^{i\theta} \\
\alpha_2 &= \xi - i\eta = \rho e^{-i\theta}
\end{aligned} \tag{24}$$

where $\rho = \sqrt{\xi^2 + \eta^2}$ is the modulus of the roots and θ is the argument. We can write the general solution in the form

$$y_n = c_1(\rho e^{i\theta})^n + c_2(\rho e^{-i\theta})^n \tag{25}$$

To obtain a more convenient form, we write

$$y_n = \rho^n(c_1 e^{in\theta} + c_2 e^{-in\theta}) \tag{26}$$

Using Euler's formulas $e^{\pm i\theta} = \cos\theta \pm i\sin\theta$, we obtain

$$y_n = \rho^n[(\cos n\theta)(c_1 + ic_2) + (\sin n\theta)(c_1 - ic_2)]$$

Letting

$$\begin{aligned}
c_1 + ic_2 &= A \\
c_1 - ic_2 &= B
\end{aligned}$$

we obtain the final form for the solution

$$y_n = \rho^n(A\cos n\theta + B\sin n\theta) \tag{27}$$

Example 1 $y_{n+2} + 5y_{n+1} + 6y_n = 0$

The characteristic equation is $\alpha^2 + 5\alpha + 6 = 0$ with roots $\alpha = -3$, $\alpha = -2$. The general solution is

$$y_n = c_1(-3)^n + c_2(-2)^n$$

Example 2 $y_{n+2} + 2y_{n+1} + y_n = 0$

The characteristic equation is $\alpha^2 + 2\alpha + 1 = 0$ with a double root $\alpha = -1$. The general solution is

$$y_n = c_1(-1)^n + c_2 n(-1)^n$$

Example 3 $y_{n+2} - \sqrt{2}y_{n+1} + y_n = 0$

The characteristic equation is $\alpha^2 - \sqrt{2}\alpha + 1 = 0$ with roots

$$\alpha = \tfrac{1}{2}(\sqrt{2} \pm i\sqrt{2}) = 1 \cdot e^{\pm i\pi/4}$$

The general solution is

$$y_n = c_1 \cos \tfrac{1}{4}n\pi + c_2 \sin \tfrac{1}{4}n\pi$$

Problems 11-4

1. In the case of equal roots of the characteristic equation (20), show that $n\alpha_1^n$ is a second, linearly independent solution of (17).

2. Find the general solution:

a. $y_{n+2} + y_n = 0$ b. $y_{n+2} + 2y_{n+1} - 3y_n = 0$

c. $y_{n+2} - 6y_{n+1} + 9y_n = 0$ d. $y_{n+1} - 2y_n \cos \beta + y_{n-1} = 0$

e. $y_{n+2} + y_{n+1} + y_n = 0$ f. $y_{n+1} - 2y_n = 0$

3. Solve: $\Delta\text{E:}\; y_{n+2} + 2y_{n+1} - 3y_n = 0$

         IC: $y_0 = 1$ $y_1 = 0$

4. Find the second-order difference equations with constant coefficients having the following as their general solution:

a. $y_n = A(-1)^n + B3^n$

b. $y_n = (A + Bn)4^n$

c. $y_n = A \cos n\beta + B \sin n\beta$ $\sin \beta \neq 0$

5. Find the so-called *Fibonacci numbers* defined as the solution of

$$\Delta\text{E:}\; y_{n+2} = y_n + y_{n+1}$$

$$\text{IC:}\; y_0 = 0 \qquad y_1 = 1$$

6. If y_n is the nth Fibonacci number, show that

$$\lim_{n \to \infty} \frac{y_n}{y_{n+1}} = \tfrac{1}{2}(\sqrt{5} - 1)$$

(This number is called the *golden mean*; it was thought by the ancient Greeks to be the ratio of sides of that rectangle which is most pleasing to the eye.)

7. The following problem concerning the breeding of rabbits was the origin of the Fibonacci numbers. Find the number of pairs of rabbits that are produced from one pair at the end of n months, assuming that each month every pair of rabbits produce another pair and that the rabbits begin to bear young 2 months after their own birth.

8. Show that the following boundary-value problem has only the trivial solution

$$\Delta\text{E:}\; y_{n+2} - y_n = 0$$

$$\text{BC:}\; y_0 = 0 \qquad y_N = 0$$

where N is some given positive integer.

9. Find the values of β for which the following boundary-value problem has a nontrivial solution and find the corresponding solutions

ΔE: $y_{n+1} - 2y_n \cos \beta + y_{n-1} = 0$ $0 < \beta < \pi$

BC: $y_0 = 0$ $y_N = 0$

10. Discuss the conditions under which the general solution y_n of a second-order difference equation with constant coefficients has the property that

$$\lim_{n \to \infty} y_n = 0$$

11. Solve the following third-order equations by extending the method in the text:

a. $y_{n+3} + y_n = 0$
b. $y_{n+3} - 7y_{n+2} + 16y_{n+1} - 12y_n = 0$
c. $y_{n+3} - 3y_{n+2} + 3y_{n+1} - y_n = 0$

Systems of difference equations with constant coefficients

12. Consider the system of two homogeneous linear difference equations with constant coefficients

$y_{n+1} = ay_n + bz_n$
$z_{n+2} = cy_n + dz_n$ $n \geq 0$

Where y_0 and z_0 are given, discuss how to find solutions by assuming $y_n = A\alpha^n$, $z_n = B\alpha^n$. Be sure to discuss the case of *multiple roots*. (See Sec. 11-3, Prob. 9.)

13. Solve: $y_{n+1} = y_n - 2z_n$

$$z_{n+1} = y_n + 4z_n \qquad n \geq 0$$

14. Write the system of equations in Prob. 12 in the matrix form

ΔE: $\mathbf{x}_{n+1} = A\mathbf{x}_n$

IC: $\mathbf{x}_0$ given

Show the solution is given by $\mathbf{x}_n = A^n\mathbf{x}_0$.

15. Using the method of Sec. 8-7 to evaluate the matrix A^n, solve Prob. 13 by the matrix method.

16. a. Write the equations of Prob. 2 as systems of first-order equations.
b. Solve these equations by evaluating the matrix A^n.

17. Solve the systems of two first-order equations $\mathbf{x}_{n+1} = A\mathbf{x}_n$ where:

a. $A = \begin{bmatrix} 3 & 4 \\ 1 & 3 \end{bmatrix}$

b. $A = \begin{bmatrix} \cos\theta & \sin\theta \\ -\sin\theta & \cos\theta \end{bmatrix}$

c. $A = \begin{bmatrix} \cosh a & b\sinh a \\ \dfrac{\sinh a}{b} & \cosh a \end{bmatrix}$

d. $A = \begin{bmatrix} 1 & 1 \\ -1 & 3 \end{bmatrix}$

18. Solve the following systems of three first-order equations where the matrix of coefficients is:

a. $A = \begin{bmatrix} 1 & 3 & -3 \\ 3 & 1 & -3 \\ 0 & 0 & 2 \end{bmatrix}$

b. $A = \begin{bmatrix} 6 & 2 & -2 \\ -2 & 2 & 2 \\ 2 & 2 & 2 \end{bmatrix}$

c. $A = \begin{bmatrix} 2 & 1 & 1 \\ 0 & 2 & 1 \\ 0 & 0 & 2 \end{bmatrix}$

19. a. Show that if p_n and q_n satisfy

$$p_{n+1} = ap_n + bq_n$$
$$q_{n+1} = cp_n + dq_n$$

then $R_n = p_n/q_n$ satisfies

$$R_{n+1} = \frac{aR_n + b}{cR_n + d}$$

b. Solve: $R_{n+1} = \dfrac{R_n - 2}{R_n + 4}$ $R_0 = 2$

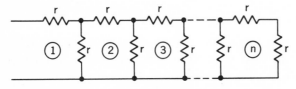

20. Show that the resistance R_n between the terminals P and Q of the n-section network shown in the figure satisfies

$$R'_{n+1} = \frac{1 + 2R'_n}{1 + R'_n}   R'_1 = 2   \text{where } R'_n = \frac{R_n}{r}$$

Find R_n and find $\lim R_n$ as $n \to \infty$.

11-5   The nonhomogeneous equation—undetermined coefficients

Particular solutions of the nonhomogeneous equation may sometimes be obtained by the method of undetermined coefficients in a manner similar to the corresponding method for differential equations. The following example illustrates the procedure.

Example 1   Find a particular solution of

$$y_{n+2} + 5y_{n+1} + 6y_n = (-2)^n$$

We are tempted to try a particular solution of the form $w_n = A(-2)^n$. However, we see from Example 1 of the last section that $(-2)^n$ is a solution of the homogeneous equation and cannot therefore also be a solution of the nonhomogeneous equation. We therefore try

$$w_n = An(-2)^n$$

Substituting into the difference equation, we obtain

$$A(-2)^n[(n+2)(-2)^2 + 5(n+1)(-2) + 6n] = (-2)^n$$

or

$$A = -\tfrac{1}{2}$$

A particular solution is therefore $-\tfrac{1}{2}n(-2)^n = +n(-2)^{n-1}$ and the general solution is

$$y_n = c_1(-3)^n + c_2(-2)^n + n(-2)^{n-1}$$

Variation of parameters   A particular solution of any nonhomogeneous linear difference equation

$$a_n y_{n+2} + b_n y_{n+1} + c_n y_n = f_n \tag{28}$$

can be obtained by the method of variation of parameters provided that two linearly independent solutions of the corresponding homogeneous equation are known. We seek a solution of the form

$$y_n = \alpha_n u_n + \beta_n v_n \tag{29}$$

where u_n and v_n are linearly independent solutions of the homogeneous equation and α_n and β_n are to be determined so that Eq. (28) is satisfied. Calculating y_n, y_{n+1}, y_{n+2} from Eq. (29) and substituting into Eq. (30) will result in one equation for α_n and β_n. We can expect that it is possible to put some other condition on α_n and β_n. This condition can be rather arbitrary but should serve to simplify the calculation of α_n and β_n. From (29) we have

$$y_{n+1} = \alpha_{n+1} u_{n+1} + \beta_{n+1} v_{n+1} \tag{30}$$

A convenient condition to put on α_n and β_n is

$$\alpha_{n+1} u_{n+1} + \beta_{n+1} v_{n+1} \equiv \alpha_n u_{n+1} + \beta_n v_{n+1} \tag{31}$$

With this condition we obtain

$$y_{n+2} = \alpha_{n+1} u_{n+2} + \beta_{n+1} v_{n+2} \tag{32}$$

Substituting (29), (30), and (31) into (28), we obtain

$$\alpha_{n+1}(a_n u_{n+2} + b_n u_{n+1}) + \alpha_n c_n u_n + \beta_{n+1}(a_n v_{n+2} + b_n v_{n+1})$$
$$+ \beta_n c_n v_n = f_n \tag{32a}$$

Since u_n and v_n satisfy the homogeneous equation, we have

$$a_n u_{n+2} + b_n u_{n+1} = -c_n u_n$$

and

$$a_n v_{n+2} + b_n v_{n+1} = -c_n v_n$$

Equation (32a) now becomes [assuming $(c_n \neq 0)$]

$$(\alpha_{n+1} - \alpha_n)u_n + (\beta_{n+1} - \beta_n)v_n = -\frac{f_n}{c_n} \tag{33}$$

Rewriting (31), we have also

$$(\alpha_{n+1} - \alpha_n)u_{n+1} + (\beta_{n+1} - \beta_n)v_{n+1} = 0 \tag{34}$$

Solving (33) and (34) for $(\alpha_{n+1} - \alpha_n)$ and $(\beta_{n+1} - \beta_n)$, we obtain

$$\alpha_{n+1} - \alpha_n = -\frac{v_{n+1}f_n}{c_n W_n} \tag{35}$$

$$\beta_{n+1} - \beta_n = \frac{u_{n+1}f_n}{c_n W_n} \tag{36}$$

where $W_n = W(u_n, v_n)$ is the Wronskian of u_n and v_n. We have seen in Sec. 11-3, Prob. 8, that if $c_n \neq 0$ and u_n and v_n are linearly independent, then $W_n \neq 0$. Equations (35) and (36) are first-order difference equations for α_n and β_n respectively which can be solved as discussed in Sec. 11-2.

Example 2 $y_{n+2} - y_n = 2^n$

Linearly independent solutions of the homogeneous equation are $u_n = 1^n = 1$, $v_n = (-1)^n$. The Wronskian $W[u_n, v_n]$ is

$$W[u_n, v_n] = \begin{vmatrix} 1 & (-1)^n \\ 1 & (-1)^{n+1} \end{vmatrix} = (-1)^{n+1} - (-1)^n = -2(-1)^n$$

We have from above that a particular solution exists in the form $y_n = \alpha_n u_n + \beta_n v_n$, where

$$\alpha_{n+1} - \alpha_n = \frac{(-1)^{n+1}2^n}{-2(-1)^n} = \tfrac{1}{2}2^n$$

$$\beta_{n+1} - \beta_n = \frac{-2^n}{-2(-1)^n} = \tfrac{1}{2}(-2)^n$$

Solving these equations as in Sec. 11-2, using the initial conditions $\alpha_0 = 0$, $\beta_0 = 0$ for convenience, we obtain

$$\alpha_n = \tfrac{1}{2} \sum_{k=0}^{n-1} 2^k = \frac{1}{2}\frac{1-2^n}{1-2} = \frac{2^n-1}{2}$$

$$\beta_n = \tfrac{1}{2} \sum_{k=0}^{n-1} (-2)^k = \frac{1}{2}\frac{1-(-2)^n}{1-(-2)} = \frac{1-(-2)^n}{6}$$

where we have used the formula for the partial sum of a geometric series. A particular solution is therefore

$$y_n = \tfrac{1}{2}(2^n - 1) + \tfrac{1}{6}(-1)^n[1 - (-2)^n]$$

Since the terms $\tfrac{1}{2}$ and $\tfrac{1}{6}(-1)^n$ are solutions of the homogeneous equation, we may omit them and still obtain the particular solution

$$y_n = \tfrac{1}{2}(2^n) - \tfrac{1}{6}(-1)^n(-2)^n = \tfrac{1}{3}(2^n)$$

The general solution is

$$y_n = A + B(-1)^n + \tfrac{1}{3}(2^n)$$

Problems 11-5

1. ΔE: $y_{n+2} - 2y_{n+1} + y_n = 1$

   IC: $y_0 = 0$ $y_1 = 1$

2. Find the general solution of

$$y_{n+2} - 2y_{n+1} + y_n = f_n$$

where:

a. $f_n = n$ b. $f_n = a^n$ $a \neq 1$
c. $f_n = \sin(\alpha n)$

3. Show that the general solution of

$$y_{n+1} - 2y_n \cos \alpha + y_{n-1} = f_n \qquad n \geq 1$$

is

$$y_n = \frac{1}{\sin \alpha} \sum_{k=1}^{n} f_k \sin(n-k)\alpha + c_1 \cos n\alpha + c_2 \sin n\alpha$$

where $0 < \alpha < \pi$. Note that the above difference equation can also be written

$$y_{n+2} - 2y_{n+1} \cos \alpha + y_n = f_{n+1} \qquad n \geq 0$$

4. Consider the difference equation

$$\Delta E: y_n + a_n y_{n-1} + b_n y_{n-2} = f_n \qquad n \geq 0$$

IC: $y_n = 0$ $n < 0$

where $\{f_n\}$ is an arbitrary sequence. The *weighting sequence* w_n is defined to be the solution of the above problem when f_n is the *unit sequence*, $\delta_n = 1$ for $n = 0$, $\delta_n = 0$ otherwise. Show that the solution for an arbitrary f_n is

$$y_n = \sum_{k=0}^{n} f_k w_{n-k} = \sum_{k=0}^{n} w_k f_{n-k} \qquad n \geq 0$$

5. Show that the weighting sequence is the solution of

ΔE: $y_n + a_n y_{n-1} + b_n y_{n-2} = 0 \qquad n \geq 2$

IC: $y_0 = 1 \qquad y_1 = -a_1$

6. Find the weighting sequence for each of the following and express the solution in terms of the weighting sequence:

a. $y_n + 5y_{n-1} + 6y_{n-2} = f_n \qquad n \geq 0$
$\quad y_n = 0 \qquad n < 0$

b. $y_n - 2y_{n-1} + y_{n-2} = f_n \qquad n \geq 0$
$\quad y_n = 0 \qquad n < 0$

c. $y_n - 2y_{n-1} \cos \alpha + y_{n-2} = f_n \qquad n \geq 0$
$\quad y_n = 0 \qquad n < 0$

The Z transform

7. The Z transform of a sequence is analogous to the Laplace transform (see Chap. 9) of a function. If f_n is a sequence, the Z transform of f_n is defined by

$$Z(f_n) = \tilde{f}(z) = \sum_{n=0}^{\infty} f_n z^{-n}$$

provided the series converges for $|z| > R$ for some R. Show:

a. $Z(\delta_n) = 1$, where $\delta_n = 1$, $n = 0$; $\delta_n = 0$, $n \neq 0$.

b. $Z(1) = \dfrac{z}{z - 1}$

Hint: Sum the geometric series.

c. $Z(a^n) = \dfrac{z}{z - a}$

d. If $b_n = 0$, $n = 0$, $b_n = a^{n-1}$ for $n > 0$, show

$Z(b_n) = \dfrac{1}{z - a}$

e. $Z(\cos n\alpha) = \dfrac{z(z - \cos \alpha)}{z^2 - 2z \cos \alpha + 1}$

f. $Z(\sin n\alpha) = \dfrac{z \sin \alpha}{z^2 - 2z \cos \alpha + 1}$

g. $Z\left(\dfrac{\rho^n \sin n\theta}{b}\right) = \dfrac{z}{(z - a)^2 + b^2}$

where a, b, ρ, θ are such that $a + ib = \rho e^{i\theta}$.

8. a. Prove: If $F(z)$ is a given function, analytic for $|z| > R$, then there exists a unique sequence f_n such that $Z(f_n) = F(z)$. [We call f_n the inverse Z transform of $F(z)$ and write $f_n = Z^{-1}(F(z))$.]

b. Find: $Z^{-1}\left(\dfrac{z}{z^2 - 5z + 6}\right)$

Hint: Use partial fractions on $1/(z^2 - 5z - 6)$ to obtain

$$\frac{z}{z^2 - 5z + 6} = \frac{z}{(z - 3)} - \frac{z}{(z - 2)}$$

and use Prob. 7c.

c. Find: $Z^{-1}\left(\dfrac{1}{z^2 - 5z + 6}\right)$

Hint: Use partial fractions and Prob. 7d.

d. Find: $Z^{-1}\left(\dfrac{z}{z^2 + z + 1}\right)$

9. Derive the following properties of Z transforms. Assume that all sequences possess transforms.

a. $Z(f_{n-r}) = z^{-r}[\tilde{f}(z) + \displaystyle\sum_{k=1}^{r} f_{-k}z^{k}]$ r = fixed positive integer

b. $Z(f_{n+r}) = z^{r}[\tilde{f}(z) - \displaystyle\sum_{k=0}^{r-1} f_{k}z^{-k}]$ r = fixed positive integer

c. If $g_n = \displaystyle\sum_{k=0}^{n} f_k$, then $Z(g_n) = z\tilde{f}(z)/(z - 1)$.

d. If $\tilde{f}(z) = Z(f_n)$, then $Z(nf_n) = -z^{-1}\tilde{f}\,'(z)$.

e. Define the convolution of f_n and g_n to be $f_n * g_n = \displaystyle\sum_{k=0}^{n} f_k g_{n-k} = \displaystyle\sum_{k=0}^{n} g_k f_{n-k}$.

Show that $Z(f_n * g_n) = \tilde{f}(z)\tilde{g}(z)$.

10. Derive the following, using Probs. 7 and 9:

a. $Z(n) = \dfrac{2}{(z - 1)^2}$

b. $Z(na^n) = \dfrac{az}{(z - a)^2}$

c. $Z(n^2) = \dfrac{z(z + 1)}{(z - 1)^3}$

d. $Z(\delta_{n+j}) = z^j$ j = fixed positive integer

e. $Z(\delta_{n-j}) = z^{-j}$

f. Consider the sequence H_n defined by

$H_n = 0$ $n < 0$

$H_n = 1$ $n \geq 0$

Show that

$$Z(a^{n-1}H_{n-1}) = \frac{1}{z - a} \qquad Z(a^{n-j}H_{n-j}) = \frac{z^{1-j}}{z - a}$$

11. Define $n^{[m]}$ by $n^{[m]} = n(n + 1) \cdots (n + m - 1)$.   Show that:

a. $Z(n^{[m]}) = \dfrac{m!z^m}{(z - a)^{m+1}}$

b. $Z(n^{[m]}f_n) = (-1)^m z^{-m} \dfrac{d^m}{dz^m} \tilde{f}(z)$

c. $Z(n^{[m]}a^n) = \dfrac{am!z^m}{(z - a)^{m+1}}$

12. a. Using Z transforms, show that

$$\sum_{k=0}^{n} k = \tfrac{1}{2}n(n + 1)$$

b. Find: $\displaystyle\sum_{k=0}^{n} k^2$

c. Find: $\displaystyle\sum_{k=0}^{n} k^3$

d. Find: $\displaystyle\sum_{k=0}^{n} k^4$

13. Consider $y_{n+2} + 5y_{n+1} + 6y_n = 4^n$, $n \geq 0$, $y_0 = 1$, $y_1 = 2$.   This problem can be solved by taking Z transforms of both sides to obtain

$$z^2[\tilde{y}(z) - (y_0 + y_1 z^{-1})] + 5z[\tilde{y}(z) - y_0] + 6\tilde{y}(z) = \frac{z}{z - 4}$$

or

$$(z^2 + 5z + 6)\tilde{y}(z) = \frac{z}{z - 4} + z^2 + 7z$$

$$\tilde{y}(z) = \frac{z}{(z - 4)(z + 3)(z + 2)} + \frac{z^2 + 7z}{(z + 3)(z + 2)}$$

$$= z\left(\frac{1}{42}\frac{1}{z - 4} + \frac{1}{7}\frac{1}{z + 3} - \frac{1}{6}\frac{1}{z + 2}\right) + 1 + \frac{12}{z + 3} - \frac{10}{z + 2}$$

$$y_n = \tfrac{1}{42}(4)^n + \tfrac{1}{7}(-3)^n - \tfrac{1}{6}(-2)^n + \delta_n$$
$$+ 12(-3)^{n-1}H_{n-1} - 10(-2)^{n-1}H_{n-1}$$

$$\begin{cases} y_n = \tfrac{1}{42}(4)^n + \tfrac{8}{7}\tfrac{1}{}(-3)^{n-1} - \tfrac{2}{3}\tfrac{9}{}(-2)^{n-1} & n \geq 1 \\ y_0 = 1 \end{cases}$$

Using this method, solve:

a. $y_{n+2} + 4y_{n+1} - 5y_n = 2^n$ $y_0 = 0, y_1 = 2$
b. $y_{n+2} + 4y_{n+1} - 5y_n = 1$ $y_0 = 1, y_1 = -1$
c. $y_{n+2} + 4y_{n+1} + 4y_n = 1$ $y_0 = y_1 = 0$
d. $y_{n+2} + y_{n+1} + y_n = 0$ $y_0 = y_1 = 1$

14. Using Z transforms, show that (assuming f_n has a Z transform) the solution of

$$y_n + ay_{n-1} + by_{n-2} = f_n \qquad n \geq 0$$
$$y_n = 0 \qquad n < 0$$

is

$$y_n = \sum_{k=0}^{n} w_{n-k}f_k \qquad \text{where } w_n = Z^{-1}\left(\frac{1}{z^2 + az + b}\right)$$

(w_n is the weighting sequence defined in Prob. 4.)

15. Using Z transforms, find the weighting sequence for each of the equations in Prob. 6.

11-6   The vector space E_N

We have already introduced N-dimensional vectors in Sec. 8-4. We now discuss some further properties of such vectors in order to present some interesting applications in the next two sections. This material will also serve as an excellent background for the study of Fourier series, boundary-value problems, and partial differential equations which are taken up in Chaps. 13 to 15.

If N is a positive integer, we have the following definition.

Definition 1 *An N-dimensional vector is an ordered set of N real numbers. A typical vector is denoted by*

$$\mathbf{v} = (v_1, v_2, \ldots, v_N)$$

where the v_i are called the components of $\mathbf{v}$. The set of all such N-tuples $\mathbf{v}$ is called the N-dimensional vector space E_N. If $\mathbf{v} \in E_N$ and $w \in E_N$, we define the sum of two vectors by

$$\mathbf{v} + \mathbf{w} = (v_1 + w_1, v_2 + w_2, \ldots, v_N + w_N)$$

If α is a real number (a scalar), we define multiplication by a scalar by

$$\alpha\mathbf{v} = (\alpha v_1, \alpha v_2, \ldots, \alpha v_N)$$

The zero vector is denoted by

$$\mathbf{0} = (0, 0, \ldots, 0)$$

and the negative of a vector by

$$-\mathbf{v} = (-1) \cdot \mathbf{v}$$

We say $\mathbf{v} = \mathbf{w}$ if and only if $v_i = w_i$, $i = 1, \ldots, N$.

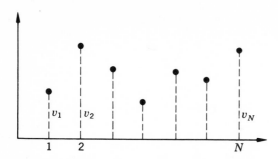

figure 1. *The vector* **v**.

We have seen in Sec. 8-4 that the following algebraic properties hold for all **v**, **w**, **z** $\in E_N$ and all scalars α, β.

$$\begin{array}{ll}
\mathbf{v} + \mathbf{w} = \mathbf{w} + \mathbf{v} & \alpha(\beta\mathbf{v}) = (\alpha\beta)\mathbf{v} \\
\mathbf{v} + (\mathbf{w} + \mathbf{z}) = (\mathbf{v} + \mathbf{w}) + \mathbf{z} & \alpha(\mathbf{v} + \mathbf{w}) = \alpha\mathbf{v} + \alpha\mathbf{w} \\
\mathbf{v} + \mathbf{0} = \mathbf{v} & (\alpha + \beta)\mathbf{v} = \alpha\mathbf{v} + \beta\mathbf{v} \\
\mathbf{v} + (-\mathbf{v}) = \mathbf{0} &
\end{array} \qquad (37)$$

The usual geometrical interpretation of a vector as a directed line segment is not possible if $N > 3$. We should think of vectors simply as objects on which certain algebraic operations are defined. There is a useful geometric interpretation of an N-dimensional vector given in Fig. 1. Here we represent the components v_1, v_2, ..., v_n as ordinates corresponding to the abscissas 1, 2, ..., N respectively. The vector is then simply the collection of these ordinates in their natural order. We leave it to the reader to interpret addition and scalar multiplication of vectors using this model.

The concept of linear independence and dependence of a set of vectors plays a fundamental role in the study of vector spaces. We have the following definition.

Definition 2 *The set of vectors* $\mathbf{v}^1$, $\mathbf{v}^2$, ..., $\mathbf{v}^k \in E_N$ *is called linearly dependent if there exist* k *scalars* α_1, ..., α_k, *not all zero, such that* $\alpha_1\mathbf{v}^1 + \alpha_2\mathbf{v}^2 + \cdots + \alpha_k\mathbf{v}^k = \mathbf{0}$. *A set that is not linearly dependent is called linearly independent. That is* $\mathbf{v}^1$, ..., $\mathbf{v}^k$ *are linearly independent if* $\alpha_1\mathbf{v}^1 + \alpha_2\mathbf{v}^2 + \cdots + \alpha_k\mathbf{v}^k = \mathbf{0}$ *only when* $\alpha_1 = \alpha_2 = \cdots = \alpha_k = 0$.

Example 1 Any set of vectors including the zero vector is linearly dependent. Call the vectors $\mathbf{0}$, $\mathbf{v}^1$, ..., $\mathbf{v}^k$; then we may write

$$\alpha \cdot \mathbf{0} + 0 \cdot \mathbf{v}^1 + 0 \cdot \mathbf{v}^2 + \cdots + 0 \cdot \mathbf{v}^k = \mathbf{0}$$

where $\alpha \neq 0$.

Example 2   Consider the following vectors in E_N:

$\mathbf{e}^1 = (1, 0, 0, \ldots, 0, 0)$
$\mathbf{e}^2 = (0, 1, 0, \ldots, 0, 0)$
.
.
.
$\mathbf{e}^N = (0, 0, 0, \ldots, 0, 1)$

We have

$$\alpha_1 \mathbf{e}^1 + \alpha_2 \mathbf{e}^2 + \cdots + \alpha_N \mathbf{e}^N = (\alpha_1, \alpha_2, \ldots, \alpha_N)$$

which is equal to the zero vector if and only if

$$\alpha_1 = \alpha_2 = \cdots = \alpha_N = 0$$

Therefore $\mathbf{e}^1, \ldots, \mathbf{e}^N$ are linearly independent.

The above example shows that in E_N there exists at least one linearly independent set containing N vectors. The following theorem shows that no linearly independent set in E_N can contain more than N vectors. This is basically the reason that E_N is called N-dimensional.

Theorem 1   *In E_N the maximum number of vectors in a linearly independent set is N.*

Proof   Let $\mathbf{v}^1, \ldots, \mathbf{v}^M$, $M > N$, be any M vectors in E_N. If these are linearly independent, then the only possibility for the coefficients α_i in the equation $\sum_{i=1}^{M} \alpha_i \mathbf{v}^i = \mathbf{0}$ is $\alpha_i = 0$, $i = 1, \ldots, M$. However, this vector equation is equivalent to the system

$$\sum_{i=1}^{M} \alpha_i v_k^i = 0 \qquad k = 1, 2, \ldots, N$$

where v_k^i is the kth component of $\mathbf{v}^i$. We can consider the above as N homogeneous linear equations for the unknown $\alpha_1, \ldots, \alpha_M$.

An important theorem of algebra† states that N homogeneous equations in M unknowns with $M > N$ must always possess a nontrivial solution. (See Prob. 1.) Thus some of the α_i can be taken to be different from zero and the set must be linearly dependent. Therefore if M is the maximum number of linearly independent vectors in E_N, we must have $M \leq N$. However from Example 2 we see that $M = N$.

A set of N linearly independent vectors $\mathbf{u}^1, \ldots, \mathbf{u}^N$ in E_N is called a *basis* for E_N. The importance of a basis is shown by the following theorem.

† See Birkhoff and MacLane [3], pp. 46–48.

Theorem 2   *If* $\mathbf{u}^1, \ldots, \mathbf{u}^N$ *is a basis for* E_N *and* $\mathbf{v}$ *is any vector in* E_N, *then* $\mathbf{v}$ *can be expressed uniquely as a linear combination of the basis vectors. That is,*

$$\mathbf{v} = \sum_{i=1}^{N} \alpha_i \mathbf{u}^i$$

where the α^i *are uniquely determined.*

Proof   Existence.   The set of $N + 1$ vectors $\mathbf{v}, \mathbf{u}^1, \ldots, \mathbf{u}^N$ is linearly dependent.   Therefore there exist constants $\alpha_0, \alpha_1, \ldots, \alpha_N$, not all zero, such that

$$\alpha_0 \mathbf{v} + \alpha_1 \mathbf{u}^1 + \alpha_2 \mathbf{u}^2 + \cdots + \alpha_N \mathbf{u}^N = \mathbf{0}$$

We cannot have $\alpha_0 = 0$ since this would imply that

$$\alpha_1 \mathbf{u}^1 + \alpha_2 \mathbf{u}^2 + \cdots + \alpha_N \mathbf{u}^N = \mathbf{0}$$

with at least one of the $\alpha_i \neq 0$.   This, however, contradicts the fact that the $\mathbf{u}^i$ are linearly independent.   Thus we can divide by α_0 and solve for $\mathbf{v}$.

Uniqueness.   Suppose there are two representations

$$\mathbf{v} = \sum_{i=1}^{N} \alpha_i \mathbf{u}^i \qquad \mathbf{v} = \sum_{i=1}^{N} \beta_i \mathbf{u}^i$$

This implies that

$$\sum_{i=1}^{N} \alpha_i \mathbf{u}^i = \sum_{i=1}^{N} \beta_i \mathbf{u}^i$$

or

$$\sum_{i=1}^{N} (\alpha_i - \beta_i) \mathbf{u}^i = \mathbf{0}$$

Since the $\mathbf{u}^i$ are linearly independent, we must have

$$\alpha_i - \beta_i = 0 \qquad \text{or} \qquad \alpha_i = \beta_i$$

so that the two representations are identical.

If $\mathbf{v}$ is a given vector and $\mathbf{u}^1, \ldots, \mathbf{u}^N$ are linearly independent, then the coefficients α_i in the representation

$$\mathbf{v} = \sum_{i=1}^{N} \alpha_i \mathbf{u}^i$$

are uniquely determined.   In order to find the α_i, we replace this single vector equation by the N scalar equation,

$$v_k = \sum_{i=1}^{N} \alpha_i u_k^i \tag{38}$$

where u_k^i represents the kth component of the vector $\mathbf{u}^i$. Since the v_k are known, this is a system of N linear equations for the N unknowns $\alpha_1, \ldots, \alpha_N$. To find the α_i's, we must solve this system of equations. The above theorem guarantees the existence of unique solutions for the α_i.

Sets of orthogonal vectors in E_N As an immediate generalization of the dot product of two vectors in E_3, we define the product of two vectors in E_N by

$$\langle \mathbf{v}, \mathbf{w} \rangle = \sum_{k=1}^{N} v_k w_k \tag{39}$$

where the v_k and w_k are the components of $\mathbf{v}$ and $\mathbf{w}$. This product is called by any of several names: *dot product, scalar product,* or *inner product*. We shall use the term inner product. We note that the inner product of two vectors is a number or a scalar and *not* a vector. It is easily verified that the following properties of the inner product hold:

$$\begin{aligned}
\langle \mathbf{v}, \mathbf{v} \rangle &> 0 \quad \text{if} \quad \mathbf{v} \neq 0 \\
\langle \mathbf{v}, \mathbf{v} \rangle &= 0 \quad \text{if} \quad \mathbf{v} = 0 \\
\langle \mathbf{v}, \mathbf{w} \rangle &= \langle \mathbf{w}, \mathbf{v} \rangle \\
\langle \alpha\mathbf{v} + \beta\mathbf{w}, \mathbf{z} \rangle &= \alpha\langle \mathbf{v}, \mathbf{z} \rangle + \beta\langle \mathbf{w}, \mathbf{z} \rangle
\end{aligned} \tag{40}$$

The last property is the distributive law. The *length* of a vector in E_N is denoted by $\|\mathbf{v}\|$ and defined by

$$\|\mathbf{v}\| = \sqrt{\langle \mathbf{v}, \mathbf{v} \rangle} = \sqrt{\sum_{k=1}^{N} v_k^2} \tag{41}$$

If $\|\mathbf{v}\| = 1$, we say that $\mathbf{v}$ is a *unit vector* or that $\mathbf{v}$ is *normalized*. We see that only the zero vector has zero length.

Two vectors $\mathbf{u}$ and $\mathbf{v}$ in E_N are called orthogonal to each other if and only if

$$\langle \mathbf{v}, \mathbf{w} \rangle = 0 \tag{42}$$

Again this is a generalization of the corresponding definition in three dimensions. We note that

$$\langle \mathbf{0}, \mathbf{v} \rangle = \sum_{k=1}^{N} 0 \cdot v_k = 0$$

so that the zero vector is orthogonal to all vectors in E_N.

We now consider a set of N *nonzero, mutually orthogonal* vectors $\mathbf{u}^1, \ldots, \mathbf{u}^N$, that is,

$$\langle \mathbf{u}^i, \mathbf{u}^j \rangle = \begin{cases} 0 & i \neq j \\ \|\mathbf{u}^i\|^2 \neq 0 & i = j \end{cases} \tag{43}$$

For instance, the vectors $\mathbf{e}^1, \ldots, \mathbf{e}^N$ defined in Example 2 are mutually orthogonal. A set of N vectors in E_N satisfying Eq. (43) is called a *complete orthogonal set* in E_N; if also each $\mathbf{u}^i$ is a unit vector, the set is called a *complete orthonormal set*.

Theorem 3   *If* $\mathbf{u}^1, \ldots, \mathbf{u}^N$ *is a complete orthogonal set in* E_N, *then the* $\mathbf{u}^i$ *are linearly independent*.

Proof   Suppose

$$\alpha_1\mathbf{u}^1 + \alpha_1\mathbf{u}^2 + \cdots + \alpha_N\mathbf{u}^N = \mathbf{0}$$

Taking the inner product of both sides with $\mathbf{u}^1$, we find

$$\alpha_1\langle\mathbf{u}^1,\mathbf{u}^1\rangle + \alpha_2\langle\mathbf{u}^2,\mathbf{u}^1\rangle + \cdots + \alpha_N\langle\mathbf{u}^N,\mathbf{u}^1\rangle = 0$$

Taking into account the orthogonality conditions, we find

$$\alpha_1 = 0$$

In a similar way we find $\alpha_2 = \alpha_3 = \cdots = \alpha_N = 0$. Therefore the $\mathbf{u}^i$ are linearly independent.

Suppose now that $\mathbf{v}$ is any vector in E_N and $\mathbf{u}^1, \ldots, \mathbf{u}^N$ is a complete orthogonal set. Since the $\mathbf{u}^i$ are linearly independent, we can express $\mathbf{v}$ as a linear combination of the $\mathbf{u}^i$:

$$\mathbf{v} = \sum_{r=1}^{N} \alpha_r\mathbf{u}^r \tag{44}$$

One of the reasons why orthogonal sets of vectors are important is that the coefficients α_r in Eq. (44) are easy to find. Taking the inner product of both sides of Eq. (44) with $\mathbf{u}^j$, we find

$$\langle\mathbf{v},\mathbf{u}^j\rangle = \sum_{r=1}^{N}\alpha_r\langle\mathbf{u}^r,\mathbf{u}^j\rangle = \alpha_j\langle\mathbf{u}^j,\mathbf{u}^j\rangle$$

since all the inner products $\langle\mathbf{u}^r,\mathbf{u}^j\rangle$ are zero except where $r = j$. Therefore the coefficients α_j are given by the simple formula

$$\alpha_j = \frac{\langle\mathbf{v},\mathbf{u}^j\rangle}{\langle\mathbf{u}^j,\mathbf{u}^j\rangle} \qquad j = 1, \ldots, N \tag{45}$$

The pair of equations (44) and (45) can be written

$$v_n = \sum_{r=1}^{N} \alpha_r u_n^r \qquad n = 1, \ldots, N \tag{46a}$$

$$\alpha_r = \sum_{k=1}^{N}\left(\frac{v_k u_k^r}{\|\mathbf{u}^r\|^2}\right) \qquad r = 1, \ldots, N \tag{46b}$$

Equation (44) or (46a) is sometimes called a finite Fourier series, and the coefficients given by Eq. (45) or (46b) are called the Fourier coefficients.

Example 3 Consider the following vectors in E_4:

$$\mathbf{u}^1 = (1,1,1,1)$$
$$\mathbf{u}^2 = (-1,-1,1,1)$$
$$\mathbf{u}^3 = (-1,1,-1,1)$$
$$\mathbf{u}^4 = (-1,1,1,-1)$$

It is easily verified that $\langle \mathbf{u}^i, \mathbf{u}^j \rangle = 0$, $i \neq j$, and $\langle \mathbf{u}^i, \mathbf{u}^i \rangle = 4$. Therefore $\mathbf{u}^1$, $\mathbf{u}^2$, $\mathbf{u}^3$, $\mathbf{u}^4$ is a complete orthogonal set. Let

$$\mathbf{v} = (2,3,-1,4)$$

Then the coefficient α_1 in the representation $\mathbf{v} = \sum_{i=1}^4 \alpha_i \mathbf{v}^i$ is given by

$$\alpha_1 = \frac{\langle \mathbf{v}, \mathbf{u}^1 \rangle}{\langle \mathbf{u}^1, \mathbf{u}^1 \rangle} = \frac{2 \cdot 1 + 3 \cdot 1 + (-1) \cdot 1 + 4 \cdot 1}{1 \cdot 1 + 1 \cdot 1 + 1 \cdot 1 + 1 \cdot 1} = \frac{8}{4} = 2$$

Similarly $\alpha_2 = -\frac{1}{2}$, $\alpha_3 = \frac{3}{2}$, $\alpha_4 = -1$. Therefore

$$\mathbf{v} = 2\mathbf{u}^1 - \tfrac{1}{2}\mathbf{u}^2 + \tfrac{3}{2}\mathbf{u}^3 - \mathbf{u}^4$$

Example 4 An important example of a complete orthogonal set in E_N is given by

$$\mathbf{u}^1 = \left(\sin \frac{\pi}{N+1} , \sin \frac{2\pi}{N+1} , \ldots , \sin \frac{N\pi}{N+1} \right)$$

$$\mathbf{u}^2 = \left(\sin \frac{2\pi}{N+1} , \sin \frac{4\pi}{N+1} , \ldots , \sin \frac{2N\pi}{N+1} \right)$$

.
.
.

$$\mathbf{u}^N = \left(\sin \frac{N\pi}{N+1} , \sin \frac{2N\pi}{N+1} , \ldots , \sin \frac{N^2\pi}{N+1} \right)$$

In Prob. 8 we show that (see also page 366)

$$\langle \mathbf{u}^i, \mathbf{u}^j \rangle = \begin{cases} 0 & i \neq j \\ \dfrac{N+1}{2} & i = j \end{cases}$$

If $\mathbf{v}$ is any vector in E_N, then we can write Eqs. (46a) and (46b) as

$$v_k = \sum_{r=1}^N \alpha_r \sin \frac{rk\pi}{N+1} \qquad k = 1, \ldots, N \tag{47a}$$

$$\alpha_r = \frac{2}{N+1} \sum_{k=1}^N \sin \frac{rk\pi}{N+1} \qquad r = 1, \ldots, N \tag{47b}$$

The reader familiar with the ordinary Fourier sine series (see Chap. 13) will note the remarkable similarity they bear to the above formulas.

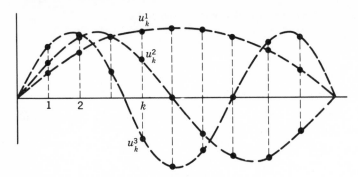

figure 2.   *The vectors* $\mathbf{u}^1$, $\mathbf{u}^2$, $\mathbf{u}^3$.

The components $u_k^r = \sin rk\pi/(N+1)$ of the vector $\mathbf{u}^r$ are sketched in Fig. 2 for $r = 1, 2, 3$. We see that these components are points on the sine curve $\sin r\pi x/(N+1)$ for the integer values $x = 1, 2, \ldots, N$. Our example shows that if $v_1, v_2, \ldots, v_N$ is an arbitrary set of ordinates, then they can be uniquely expressed as a linear combination of the *discrete sinusoidal harmonics*

$$[\sin r\pi/(N+1), \ldots, \sin rN\pi/(N+1)]$$

Problems 11-6

1. a. Consider two homogeneous equations in the three unknowns x_1, x_2, x_3 with coefficients a_{ij}:

$$a_{11}x_1 + a_{12}x_2 + a_{13}x_3 = 0$$
$$a_{21}x_1 + a_{22}x_2 + a_{23}x_3 = 0$$

Prove that a nontrivial solution exists. Hint: If all $a_{ij} = 0$, then obviously the x_i are arbitrary; otherwise at least one coefficient is not zero. Say $a_{11} \neq 0$. Eliminate x_1 and show the resulting equation has a nontrivial solution.

b. Generalize the above to N homogeneous equations in M unknowns with $M > N$. Prove the theorem by induction.

2. If $\{\mathbf{u}^i\}$ is a complete orthonormal set in E_N and $\mathbf{v} \in E_N$, show that $\|\mathbf{v}\|^2 = \sum_{k=1}^N \alpha_k^2$ where $\alpha_k = \langle \mathbf{v}, \mathbf{u}^k \rangle$.

3. In Example 4 put $N = 3$ to obtain

$$\mathbf{u}^1 = \left(\frac{\sqrt{2}}{2}, 1, \frac{\sqrt{2}}{2} \right) \qquad \mathbf{u}^2 = (1, 0, -1) \qquad \mathbf{u}^3 = \left(\frac{\sqrt{2}}{2}, -1, \frac{\sqrt{2}}{2} \right)$$

Verify that $\langle \mathbf{u}^i, \mathbf{u}^j \rangle = 0$, if $i \neq j$ and $\langle \mathbf{u}^i, \mathbf{u}^i \rangle = 2$. If $\mathbf{v} = (-1, +2, -3)$, find the Fourier coefficients α_r in $\mathbf{v} = \sum_{r=1}^3 \alpha_r \mathbf{u}^r$ and check formulas (47a) and (47b).

4. Establish the following identities which will be useful in Prob. 5:

a. $$\sum_{k=1}^N e^{ik\alpha} = \frac{\sin \frac{1}{2}N\alpha \, e^{i[N+1/2]\alpha}}{\sin \frac{1}{2}\alpha} \qquad \alpha \neq 2n\pi$$

Hint: Sum the geometric series.

b. Take real and imaginary parts in part a to obtain

$$C_N(\alpha) = \sum_1^N \cos k\alpha = \frac{\sin \frac{1}{2}N\alpha \cos \frac{1}{2}(N+1)\alpha}{\sin \frac{1}{2}\alpha} \qquad \alpha \neq 2n\pi$$

$$S_N(\alpha) = \sum_1^N \sin k\alpha = \frac{\sin \frac{1}{2}N\alpha \sin \frac{1}{2}(N+1)\alpha}{\sin \frac{1}{2}\alpha} \qquad \alpha \neq 2n\pi$$

c. $\displaystyle\sum_1^N \sin k\alpha \sin k\beta = \frac{1}{2}[C_N(\alpha - \beta) - C_N(\alpha + \beta)]$

d. $\displaystyle\sum_{k=1}^N \sin k\alpha \cos k\beta = \frac{1}{2}[S_N(\alpha + \beta) + S_N(\alpha - \beta)]$

e. $\displaystyle\sum_{k=1}^N \cos k\alpha \cos k\beta = \frac{1}{2}[C_N(\alpha + \beta) + C_N(\alpha - \beta)]$

f. $\displaystyle\sum_{k=1}^N \sin^2 k\alpha = \frac{1}{2}N - \frac{1}{2}C_N(2\alpha)$

g. $\displaystyle\sum_{k=1}^N \cos^2 k\alpha = \frac{1}{2}N + \frac{1}{2}C_N(2\alpha)$

5. a. Establish that the vectors $\langle \mathbf{u}^1, \ldots, \mathbf{u}^N \rangle$ in Example 4 satisfy $\langle \mathbf{u}^r, \mathbf{u}^s \rangle = 0$, $r \neq s$. (Use Prob. 4.)

   b. Establish that $\langle \mathbf{u}^r, \mathbf{u}^r \rangle = \frac{1}{2}(N+1)$. (Use Prob. 4.)

11-7 A boundary-value problem

Let us consider the following boundary-value problem involving a difference equation:

$$\begin{aligned} &\Delta\text{E: } a_{n+1} + (\lambda - 2)a_n + a_{n-1} = 0 \qquad n = 1, 2, \ldots, N \\ &\text{BC: } a_0 = 0 \qquad a_{N+1} = 0 \end{aligned} \tag{48}$$

The solution to this problem is a certain set of numbers $a_1, a_2, \ldots, a_N$ which may be thought of as the components of an N vector. Since the difference equation and the boundary conditions are homogeneous, the boundary-value problem possesses the trivial solution

$$a_n \equiv 0 \qquad n = 1, 2, \ldots, N$$

no matter what value the parameter λ has. We seek those values of λ for which a nontrivial solution of the boundary-value problem exists. These values of λ are called *eigenvalues*, and the corresponding solutions are called *eigenvectors*. We shall see below that there will exist exactly N distinct eigenvalues for this problem and that the corresponding set of N eigenvectors will be an orthogonal set. Our reasons for taking up this problem at this time are twofold. First, it furnishes an example of the generation of a set of orthogonal vectors and, second, this same problem will arise in the solution of the problem of finding the motion of N beads on a tightly stretched string considered in the next section. In this problem a physical interpretation of eigenvalues and eigenvectors will appear. We now proceed to find the eigenvalues and eigenvectors for the boundary-value problem.

We look for a solution of the difference equation of the form

$$a_n = \alpha^n \qquad \alpha \neq 0$$

On substitution into the difference equation, we obtain

$$\alpha^2 + (\lambda - 2)\alpha + 1 = 0$$

or

$$\alpha = \tfrac{1}{2}[2 - \lambda \pm \sqrt{\lambda(\lambda - 4)}]$$

We see that if $\lambda > 4$ or $\lambda < 0$, the roots are real and distinct; if $\lambda = 0$ or $\lambda = 4$, the roots are real and equal; if $0 < \lambda < 4$, the roots are complex. We consider these three cases separately. If the roots are real and distinct, say $\alpha_1 \neq \alpha_2$, the general solution of the difference equation is

$$a_n = c_1\alpha_1^n + c_2\alpha_2^n$$

However, in order to satisfy the boundary conditions, we must have

$$0 = c_1 + c_2$$
$$0 = c_1\alpha_1^{N+1} + c_2\alpha_2^{N+1}$$

It can be shown that these equations have only the trivial solution $c_1 = c_2 = 0$. Therefore in this case the boundary-value problem has only the trivial solution $a_n \equiv 0$. Thus the values $\lambda > 4$ and $\lambda < 0$ are not eigenvalues.

In the case of real, equal roots it again turns out that only the trivial solution exists, and therefore $\lambda = 0$ and $\lambda = 4$ are not eigenvalues. (See Prob. 1.)

We now consider the third case of complex roots, where $0 < \lambda < 4$. We write the roots in the form

$$\alpha = \tfrac{1}{2}[2 - \lambda \pm i\sqrt{\lambda(4 - \lambda)}] \tag{49}$$

or equivalently

$$\alpha_1 = 1 \cdot e^{i\beta}, \; \alpha_2 = 1 \cdot e^{-i\beta}$$

where (see Fig. 3) $-\pi < \beta < +\pi$ and

$$\cos \beta = \tfrac{1}{2}(2 - \lambda) \tag{50}$$

The general solution of the difference equation in this case can therefore be written

$$a_n = A \cos (n\beta) + B \sin (n\beta)$$

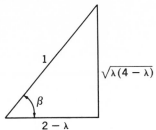

figure 3

In order to satisfy the boundary condition $a_0 = 0$, we must have $A = 0$. To satisfy the second boundary condition $a_{N+1} = 0$, we must have

$$0 = B \sin (N + 1)\beta$$

If $B = 0$, we again obtain the trivial solution. However, we may also have

$$\sin (N + 1)\beta = 0$$

provided that $(N + 1)\beta$ is a multiple of π. Therefore if we determine β as one of the numbers β_r given by

$$\beta_r = \frac{r\pi}{N + 1} \qquad r = 1, 2, \ldots, N$$

we will have nontrivial solutions. There will be a nontrivial solution for a_n corresponding to each value of $r = 1, \ldots, N$. We denote these solutions by a_n^r which are given by

$$a_n^r = B_r \sin \left(\frac{nr}{N + 1}\right)\pi \qquad \begin{array}{l} r = 1, \ldots, N \\ n = 1, \ldots, N \end{array} \tag{51}$$

where the B_r are arbitrary constants.

The eigenvalues λ_r can be determined from Eq. (50). We have

$$\cos \beta_r = \tfrac{1}{2}(2 - \lambda_r)$$

Therefore

$$\lambda_r = 2(1 - \cos \beta_r) = 4 \sin^2 \tfrac{1}{2}\beta_r$$

or

$$\lambda_r = 4 \sin^2 \frac{r\pi}{2(N + 1)} \qquad r = 1, 2, \ldots, N \tag{52}$$

We have therefore determined the N eigenvalues [Eq. (52)] and the corresponding eigenvectors

$$\mathbf{a}^r = (a_1^r, a_2^r, a_3^r, \ldots, a_n^r) \qquad r = 1, 2, \ldots, N \tag{53}$$

where the components of $\mathbf{a}^r$ are given by Eq. (51).

It was shown in Prob. 5 of Sec. 11-6 that these eigenvectors were mutually orthogonal.   We give here an independent proof of this fact.   Let λ_r and λ_s be two eigenvalues and a_n^r and a_n^s the corresponding eigenvectors.   By the definition of these terms we have

$$a_{n+1}^r + (\lambda_r - 2)a_n^r + a_{n-1}^r = 0 \qquad n = 1, 2, \ldots, N$$
$$a_{n+1}^s + (\lambda_s - 2)a_n^s + a_{n-1}^s = 0 \qquad n = 1, 2, \ldots, N$$

and

$$a_0^r = a_0^s = a_{N+1}^r = a_{N+1}^s = 0$$

Multiplying the first equation by a_n^s and the second equation by a_n^r and subtracting, we obtain (after rearranging terms)

$$(a_{n+1}^r a_n^s - a_{n+1}^s a_n^r) - a_{n-1}^s a_n^r + a_{n-1}^r a_n^s = (\lambda_s - \lambda_r)a_n^r a_n^s \tag{54}$$

Since we seek to prove that

$$\langle \mathbf{a}^r, \mathbf{a}^s \rangle = \sum_{n=1}^N a_n^r a_n^s = 0$$

we sum both sides of Eq. (54) from 1 to N:

$$\sum_1^N (a_{n+1}^r a_n^s - a_{n+1}^s a_n^r) - \sum_1^N (a_{n-1}^s a_n^r - a_{n-1}^r a_n^s) = (\lambda_r - \lambda_s)\Sigma a_n^r a_n^s \tag{55}$$

In the second summation we change dummy variables by means of the substitution $n = k + 1$, and we obtain

$$\sum_{n=1}^N (a_{n-1}^s a_n^r - a_{n-1}^s a_n^r) = \sum_{k=0}^{N-1} (a_k^s a_{k+1}^r - a_k^s a_{k+1}^r)$$

Because of the boundary conditions we do not change this sum if we make the limits 1 and N.   Therefore the left-hand side of Eq. (55) becomes

$$\sum_{k=1}^N (a_{k+1}^r a_k^s - a_{k+1}^s a_k^r) - \sum_{k=1}^N (a_k^s a_{k+1}^r - a_k^s a_{k+1}^r) = 0$$

where we have changed the dummy variable from n to k in the first summation. We therefore have shown that

$$(\lambda_r - \lambda_s) \sum_{n=1}^N a_n^r a_n^s = 0$$

and if $r \neq s$, we have $\lambda_r - \lambda_s \neq 0$ and consequently

$$\langle \mathbf{a}^r, \mathbf{a}^s \rangle = \sum_{n=1}^N a_n^r a_n^s = 0 \qquad r \neq s$$

which completes the proof of orthogonality.

Problems 11-7

1. Show that $\lambda = 0$ and $\lambda = 4$ are not eigenvalues for Eq. (48).
2. Consider ΔE: $a_{n+1} + (\lambda - 2)a_n + a_{n-1} = 0 \qquad n = 1, 2, \ldots, N$
   BC: $a_0 = 0 \qquad a_N = a_{N+1}$

a. Show that $\lambda = 0$ and $\lambda = 4$ are not eigenvalues.
b. Show that $\lambda < 0$ and $\lambda > 4$ are not eigenvalues.
c. Find the eigenvalues λ_r and the corresponding eigenvectors $\mathbf{a}^r$.
d. Show that the eigenvectors are orthogonal.   Hint: Eq. (55) still holds.
e. Find $\langle \mathbf{a}^r, \mathbf{a}^r \rangle$.   Hint: Use Sec. 11-6, Prob. 4f.
f. Write the Fourier expansions analogous to Eqs. (47a) and (47b).

3. ΔE: $a_{n+1} + (\lambda - 2)a_n + a_{n-1} = 0 \qquad n = 1, 2, \ldots, N$

   BC: $a_0 = a_1 \qquad a_N = a_{N+1}$

a. Show $\lambda = 0$ is an eigenvalue and the corresponding eigenvector has components $a_n^{(0)} = 1$, $n = 1, 2, \ldots, N$. Show that $\lambda = 4$ is not an eigenvalue.
b. Do parts b, c, d, e, and f, of Prob. 2 for this problem.

4. Consider: ΔE: $a_{n+1} + (\lambda - 2)a_n + a_{n-1} = 0 \qquad n = 1, 2, \ldots, N$

   BC: $a_0 = 0 \qquad a_{N+1} - ha_N = 0 \qquad 0 \leq h \leq 1$

a. Show that the rth eigenvector has components given by

$a_n^r = \beta_r \sin n\beta_r \qquad n = 1, 2, \ldots, N$

where β_r is the rth root of the transcendental equation

$\sin (N + 1)\beta = h \sin (N\beta)$

b. Show that the eigenvectors are orthogonal.

11-8   N beads on a tightly stretched string

We consider the problem of analyzing the motion of N beads on a tightly stretched string. We assume that all the beads have the same mass m and that the beads are equally spaced a distance L apart as shown in Fig. 4. The end points of the string are fixed. In addition to the above we make some further assumptions:

1. We assume the tension T in the string is large and is the same value at all points in the string during the motion.
2. We neglect all forces on the masses except for the tension (i.e., we neglect friction forces, gravity, etc.).
3. We assume each mass moves along a vertical line.
4. We assume that the angle θ made between the string and the x axis is small, and we shall replace $\sin \theta$ by $\tan \theta$ for such angles.

figure 4.   *Beads on a tightly stretched string.*

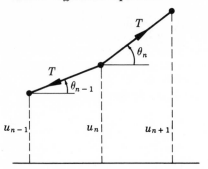

figure 5

We now set up the differential equation for this system, using Newton's second law. The forces on the nth bead are due to the tension in the string connecting the nth bead to the beads on either side as shown in Fig. 5. Referring to the figure, we set the sum of the vertical components of the forces on the nth bead equal to the mass times the acceleration:

$$m\ddot{u}_n = T\,(\sin\theta_n - \sin\theta_{n-1})$$

where $u_n(t)$ is the displacement of the nth mass and θ_n and θ_{n-1} are the angles shown in the figure. Using assumption 4 above, we have

$$\sin\theta_{n-1} \approx \frac{u_n - u_{n-1}}{L} \qquad \sin\theta_n \approx \frac{u_{n+1} - u_n}{L}$$

Therefore the differential equation becomes

$$m\ddot{u}_n = \frac{T}{L}\,(u_{n+1} - 2u_n + u_{n-1}) \qquad n = 1, 2, \ldots, N \tag{56}$$

with the boundary conditions

$$u_0 = 0 \qquad u_{N+1} = 0$$

To complete the statement of the problem, we also need to know the initial displacement and velocity of each bead:

$$\begin{aligned} u_n(0) &= f_n \qquad n = 1, 2, \ldots, N \\ \dot{u}_n(0) &= g_n \qquad n = 1, 2, \ldots, N \end{aligned} \tag{57}$$

where f_n and g_n are given numbers.
Equation (56) is actually a system of differential equations. Our method of solution will be to look for simple solutions of Eq. (56) called normal modes and get the general solution as a superposition of these normal modes.

Because of the absence of friction and damping forces, we expect the solution to be composed of sinusoids. The simplest such sinusoidal motions are the normal modes, which we now define. *A normal mode is a motion*

where each mass vibrates with simple harmonic motion with the same frequency
ω, but with possibly different amplitudes, and all masses go through the equi-
librium position together.   *A frequency ω as defined above is called a natural*
frequency of the system.   We shall see that there are exactly N distinct
natural frequencies for this system and N corresponding normal modes.

According to the above, we look for solutions of the form

$$u_n = a_n \sin(\omega t + \epsilon) \qquad n = 1, 2, \ldots, N$$
$$u_0 = a_0 = 0 \tag{58}$$
$$u_{N+1} = a_{N+1} = 0$$

Substituting into the differential equation, we obtain

$$-m\omega^2 a_n = \frac{T}{L}(a_{n+1} - 2a_n + a_{n-1}) \qquad n = 1, 2, \ldots, N \tag{59}$$

We let

$$\lambda = \frac{Lm\omega^2}{T} \tag{60}$$

so that Eq. (59) becomes

$$a_{n+1} + (\lambda - 2)a_n + a_{n-1} = 0 \qquad n = 1, 2, \ldots, N \tag{60a}$$

with

$$a_0 = a_{N+1} = 0 \tag{61}$$

This is a boundary-value problem involving a difference equation and is the
exact problem considered in Sec. 11-7. Using the results obtained there,
we see that nontrivial solutions will exist if and only if λ is equal to one of the
N eigenvalues

$$\lambda_r = 4\sin^2\frac{r\pi}{2(N+1)} \qquad r = 1, 2, \ldots, N \tag{62}$$

For each λ_r there is a corresponding solution

$$a_n^r = B_r \sin\frac{nr}{N+1} \qquad \begin{array}{l} n = 1, 2, \ldots, N \\ r = 1, 2, \ldots, N \end{array} \tag{63}$$

We can determine the N natural frequencies ω_r from the equation

$$\omega_r = \sqrt{\frac{T\lambda_r}{mL}} \qquad r = 1, 2, \ldots, N \tag{64}$$

Using Eq. (62), we have

$$\omega_r = 2\sqrt{\frac{T}{mL}}\sin\frac{r\pi}{2(N+1)} \qquad r = 1, 2, \ldots, N \tag{65}$$

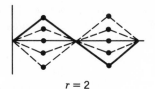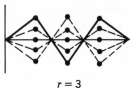

figure 6.   *Normal modes; three beads on a string.*

The corresponding N normal modes, denoted by u_n^r, are given by

$$u_n^r = a_n^r \sin{(\omega_r t + \epsilon_r)} \qquad \begin{aligned} r &= 1, 2, \ldots, N \\ n &= 1, 2, \ldots, N \end{aligned}$$

$$= \left(B_r \sin \frac{nr}{N+1} \right) \sin{(\omega_r t + \epsilon_r)} \tag{66}$$

where the B_r and ϵ_r are arbitrary constants.

These normal modes are shown graphically for $r = 1, 2, 3$ in Fig. 6 for $N = 3$. We see that in each normal mode the beads lie on a sine curve at any given time; the maximum of this sine curve changes with time as shown by the dotted lines on the figure. Because of this physical behavior the normal modes are often called *standing waves*.

Because the differential equation (56) is linear and homogeneous, we may add the solutions (66) to obtain a more general solution:

$$u_n = \sum_{r=1}^{N} u_n^r$$

or

$$u_n = \sum_{r=1}^{N} B_r \sin \frac{nr}{N+1} \sin{(\omega_r t + \epsilon_r)} \qquad n = 1, 2, \ldots, N \tag{67}$$

We shall see that this solution is sufficiently general† to allow us to satisfy the initial conditions

$$\begin{aligned} u_n(0) &= f_n \qquad n = 1, 2, \ldots, N \\ \dot{u}_n(0) &= g_n \qquad n = 1, 2, \ldots, N \end{aligned} \tag{68}$$

where f_n and g_n are arbitrarily given initial positions and velocities respectively. Before proceeding to do this, it is convenient to use the familiar identity

$$\sin{(\omega_r t + \epsilon_r)} = \sin{\omega_r t} \cos{\epsilon_r} + \cos{\omega_r t} \sin{\epsilon_r}$$

and to write Eq. (67) in the form

$$u_n = \sum_{r=1}^{N} \left(\alpha_r \sin \frac{nr}{N+1} \sin{\omega_r t} + \beta_r \sin \frac{nr}{N+1} \cos{\omega_r t} \right) \tag{69}$$

† It can be shown that it is indeed the general solution.   See Sec. 8-5.

where we have set

$$
\begin{aligned}
\alpha_r &= B_r \cos \epsilon_r \\
\beta_r &= B_r \sin \epsilon_r
\end{aligned} \tag{70}
$$

According to the first initial condition, we have

$$
f_n = \sum_{r=1}^{N} \beta_r \sin \frac{nr\pi}{N+1} \tag{71}
$$

whence [see Sec. 11-6, Eq. (47)]

$$
\beta_r = \frac{2}{N+1} \sum_{n=1}^{N} f_n \sin \frac{nr\pi}{N+1} \tag{72}
$$

According to the second initial condition, we have

$$
g_n = \sum_{r=1}^{N} \omega_r \alpha_r \sin \frac{nr\pi}{N+1} \tag{73}
$$

whence [see Sec. 11-6, Eq. (47)]

$$
\alpha_r = \frac{2}{\omega_r(N+1)} \sum_{n=1}^{N} g_n \sin \frac{n\pi r}{N+1} \tag{74}
$$

This determines the constants α_r and β_r in the solution (69).

We see therefore that the solution satisfying arbitrary initial conditions can always be obtained by a superposition of the N normal modes.

Problems 11-8

1. Solve: DE: $\dfrac{du_n}{dt} = \alpha^2(u_{n+1} - 2u_n + u_{n-1}) \qquad n = 1, 2, \ldots, N$

   BC: $u_0 = u_{N+1} = 0$

   IC: $u_n(0) = f_n \qquad n = 1, 2, \ldots, N$

2. *The vibrating string*

   a. The partial differential equation for the small motions of a tightly stretched continuous string with constant density ρ and constant tension T is (see Sec. 14-2)

$$
\frac{\rho}{T} \frac{\partial^2 u}{\partial t^2} = \frac{\partial^2 u}{\partial x^2}
$$

where $u(x,t)$ is the displacement of the string at position x at time t. Obtain this equation, formally, as a limiting case of N beads on a string, Eq. (56), as $N \to \infty$ and $L \to 0$. Hint: Let $L = \Delta x$; assume $(N + 1)L = l = \text{const}$ (the length of the string); assume $m \to 0$ as $\Delta x \to 0$ but that $m/\Delta x = \rho = \text{const}$; write $u_n(t) = u(x,t)$, where $x = nL$.

b. Obtain the natural frequencies of the continuous string using Eq. (65).

c. Obtain the solution of the partial differential equation satisfying the boundary condition $u(0,t) = u(l,t) = 0$ and the initial condition $u(x,0) = f(x)$, $\partial u/\partial t(x,0) = 0$, using Eqs. (69) and (72). (Note that g_n should be taken equal to zero.)

d. Verify, formally, that the the solution obtained in part c satisfies the partial differential equation and the boundary conditions.

e. Find the limiting case of Eq. (71).

Numerical Methods

12

12-1 Introduction

Consider the initial-value problem

$$\text{DE}: y' = f(x,y)$$
$$\text{IC}: y(x_0) = y_0 \tag{1}$$

where f and f_y are continuous in some rectangle R containing the point (x_0, y_0) in its interior. The existence of a unique solution in some interval about x_0 was proved in Chap. 4. In this chapter we consider methods for obtaining numerical approximations of the solution. In simple cases we may of course solve the equation analytically and use the analytic solution to calculate numerical values. However, analytic solutions are often not known, and direct numerical solutions of the initial-value problem are needed. Moreover, even when an analytic solution is known, it may be so complicated that it is easier to find a numerical solution of the differential equation than it is to evaluate the analytic solution.

We shall consider methods of approximating the solution of (1) at the discrete points $x_0 < x_1 < x_2 \cdots x_n \cdots$. The approximate (calculated) value of the solution at x_k will be denoted by y_k, whereas the exact value will be denoted by $y(x_k) = z_k$ (see Fig. 1).

The methods we shall consider can be divided into two classes: *one-step* or *starting* methods and *multistep* or *continuing* methods. In order to calculate y_{n+1} by a one-step method, only the one previous value y_n is needed. Since y_0 is known, y_1 can be determined, and knowing y_1, we can determine y_2, and so on. A typical example of a one-step method is the Euler method discussed in Chap. 4 and below in Secs. 12-2 and 12-3. More refined one-step methods are the Runge-Kutta methods discussed in Sec.

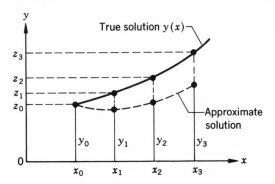

figure 1.   *True and approximate solutions.*

12-7.   Multistep methods require values at more than one previous point, say the values y_n, y_{n-1}, y_{n-2}, in order to calculate y_{n+1}.   Such methods are discussed in Secs. 12-4 through 12-6.   In order to use a multistep method, it is necessary to use a one-step method to start the solution for the first few points.   The multistep method can then be used to continue the solution. Of course, a starting method can be used for the entire solution; however, as we shall see, there are many advantages to using a continuing method.

Most of this chapter is devoted to numerical methods for the initial-value problem for a single differential equation.   Methods for higher-order equations and systems of equations will be discussed briefly in Sec. 12-8.

For more information on numerical methods, we refer the reader to the many recent excellent books on numerical analysis.†

12-2   The Euler method

The Euler approximations to the solution of

$$\text{DE:}\ y' = f(x,y)$$
$$\text{IC:}\ y(x_0) = y_0 \tag{2}$$

at the equally spaced points

$$x_0,\ x_1 = x_0 + h,\ x_2 = x_0 + 2h,\ \ldots,\ x_n = x_0 + nh,\ \ldots \tag{3}$$

are given by

$$y_{n+1} = y_n + hy'_n \tag{4}$$

where

$$y'_n = f(x_n, y_n) \tag{5}$$

Equation (4) is obtained simply by replacing the integral curve through (x_n, y_n) by a straight line of slope $y'_n = f(x_n, y_n)$ (see Fig. 2).   Since y_0 is known, y_1, y_2, $\ldots$ can be successively calculated using (4).

† See Ralston [30] or Hamming [17].

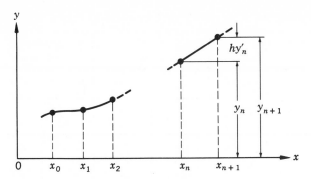

figure 2. *Euler's method.*

To illustrate Euler's method, we shall use the following three examples:

$$y' = y \qquad y(0) = 1 \qquad (y = e^x) \tag{6}$$

$$y' = -y \qquad y(0) = 1 \qquad (y = e^{-x}) \tag{7}$$

$$y' = 2xy^2 \qquad y(0) = 1 \qquad [y = (1 - x^2)^{-1}] \tag{8}$$

These same examples will be used for the other methods developed in this chapter in order to compare the various methods. Each of the examples has a known analytic solution, shown in parentheses, which can be used to compare with the calculated solution.

The results of the numerical calculations are shown in Tables 1, 2, and 3. In all cases the calculations were made to six significant digits and the results rounded to four significant digits. In Tables 1 and 2 the calculations were performed with two different step sizes, $h = 0.1$ and $h = 0.05$. We note that

Table 1 *Euler's Method for* $y' = y$, $y(0) = 1$

x_n	y_n $h = 0.1$	$h = 0.05$	Exact $y = e^x$	Error $h = 0.1$	$h = 0.05$	Percent relative error $h = 0.1$	$h = 0.05$
0.0	1.000	1.000	1.000	—	—	—	—
0.05	—	1.050	—	—	—	—	—
0.1	1.100	1.103	1.105	0.005	0.002	0.45	0.18
0.2	1.210	1.216	1.221	0.011	0.005	0.90	0.41
0.3	1.331	1.340	1.350	0.019	0.010	1.4	0.74
0.4	1.464	1.477	1.492	0.028	0.015	1.9	1.0
0.5	1.611	1.629	1.649	0.038	0.020	2.3	1.2
0.6	1.722	1.796	1.822	0.050	0.026	2.7	1.4
0.7	1.949	1.980	2.014	0.065	0.034	3.2	1.6
0.8	2.144	2.183	2.226	0.082	0.043	3.7	1.9
0.9	2.358	2.407	2.460	0.102	0.053	4.1	2.2
1.0	2.594	2.653	2.718	0.124	0.065	5.7	2.4

Table 2 *Euler's Method for* $y' = -y$, $y(0) = 1$

x_n	y_n $h = 0.1$	$h = 0.05$	Exact $y = e^{-x}$	Error $h = 0.1$	$h = 0.05$	Percent relative error $h = 0.1$	$h = 0.05$
0.0	1.0000	1.000	1.000	—	—	—	—
0.1	0.9000	0.9025	0.9048	0.00481	0.0023	0.53	0.25
0.2	0.8100	0.8145	0.8187	0.0087	0.0042	1.1	0.51
0.3	0.7290	0.7351	0.7408	0.0118	0.0057	1.6	0.77
0.4	0.6561	0.6634	0.6703	0.0142	0.0069	2.1	1.0
0.5	0.5905	0.5987	0.6065	0.0160	0.0078	2.6	1.3
0.6	0.5314	0.5404	0.5488	0.0174	0.0084	3.2	1.5
0.7	0.4783	0.4877	0.4966	0.0183	0.0089	3.7	1.8
0.8	0.4305	0.4401	0.4493	0.0188	0.0092	4.2	2.0
0.9	0.3874	0.3972	0.4066	0.0192	0.0094	4.7	2.3
1.0	0.3487	0.3585	0.3679	0.0192	0.0094	5.2	2.6

both the error $z_n - y_n$ and the relative error $(z_n - y_n)/z_n$ are decreased by reducing h, but of course twice as many calculations are necessary when h is reduced by a factor of 2. In Table 1 the errors are much larger than in Table 2. However, the relative errors are about the same. The reason for this behavior will be seen in the next section. In Table 3 we see that the errors grow rapidly and are actually infinite at $x = 1$. This indicates one of the dangers in the careless numerical solution of differential equations. For, if only the calculated values are looked at, the calculation appears to be going well, with no hint of the fact that the actual solution is infinite at $x = 1$. This indicates the need for more sophisticated methods.

Table 3 *Euler's Method for* $y' = 2xy^2$, $y(0) = 1$ (when $h = 0.1$)

x_n	y_n	Exact $y = (1 - x^2)^{-1}$	Error	Percent relative error
0.0	1.000	1.000	0.000	0.0
0.1	1.000	1.010	0.010	1.0
0.2	1.020	1.042	0.022	2.1
0.3	1.062	1.099	0.037	3.4
0.4	1.130	1.190	0.060	5.0
0.5	1.231	1.333	0.102	7.7
0.6	1.383	1.563	0.180	12
0.7	1.612	1.961	0.349	18
0.8	1.976	2.778	0.802	29
0.9	2.601	5.263	2.662	51
1.0	3.819	(∞)	(∞)	(∞)

Problems 12-2

1. Check the first few computations in Tables 1, 2, and 3.
2. Given $y' = 1 + y^2$, $y(0) = 0$, find $y(0.2)$ by Euler's method, using the step sizes:

   a. $h = 0.1$ b. $h = 0.05$

Work with three decimal places and round results to two decimal places.

12-3   Error analysis

Three sources of error can be distinguished in the numerical solution of differential equations.

Round-off error   This is the error resulting from the necessity of rounding each number to a certain number of decimal digits when calculations are performed.   In solving a differential equation, we use some formula such as Euler's method, Eq. (4), to determine the successive values $y_1, y_2, \ldots,$ $y_n, \ldots$.   However, because of rounding errors, the actual calculated values are certain other numbers $y_1^*, y_2^*, \ldots, y_n^*, \ldots$.   The difference

$$R_n = y_n - y_n^* \tag{9}$$

is called the *accumulated round-off error*.   It is quite difficult to analyze the round-off error, for it depends not only on the numbers and operations that are performed, but even on the order in which the operations are performed (see Probs. 1, 2, 3).   We shall ignore round-off error in most of what follows and assume that enough decimal places are carried so that the round-off does not affect the number of decimal places that are desired in the result.

One rule of thumb can be given if round-off is suspected to be a problem: repeat the calculation with several more places and see whether any difference occurs in the decimal digits retained in the result.

Truncation error   All numerical methods calculate an approximation y_{n+1} to the exact solution $z_{n+1} = y(x_{n+1})$ based on previously calculated approximations $y_n, y_{n-1}, y_{n-2}, \ldots$.   If the exact solution were known, and the values $z_n, z_{n-1}, z_{n-2}, \ldots$ were used in place of $y_n, y_{n-1}, y_{n-2}, \ldots$, the numerical method would still not give the exact solution, but rather an approximation $\tilde{y}_{n+1}$.   The difference

$$T_n = z_{n+1} - \tilde{y}_{n+1} \tag{10}$$

is called the truncation error.

To estimate the truncation error for Euler's method, we expand the exact solution z_{n+1} in a Taylor series about x_n, using two terms plus remainder:

$$z_{n+1} = z_n + hz_n' + \tfrac{1}{2}h^2 z''(\xi) \qquad x_n \leq \xi \leq x_{n+1} \tag{11}$$

From Eq. (4) we get

$$\tilde{y}_{n+1} = z_n + hz_n' \tag{12}$$

Therefore the truncation error is

$$T_n = z_{n+1} - \tilde{y}_{n+1} = \tfrac{1}{2}h^2 z''(\xi) \qquad x_n \leq \xi \leq x_{n+1} \tag{13}$$

In all of the methods considered in this chapter, the truncation error will be of the form

$$T_n = Kh^{r+1} z^{(r+1)}(\xi) \tag{14}$$

where K is a constant and r is a positive integer. Such methods are said to have *order of accuracy* r, or simply to be of order r. Euler's method is a first-order method.

Knowledge of the truncation error allows us to estimate the error committed during *one step* of the calculation, assuming some knowledge of $z_n^{(r+1)}$ is at hand. This gives a clue to picking the step size h in order to obtain a solution accurate to a certain number of places.

Consider the use of Euler's method to calculate $y(1)$ for $y' = -y$, $y(0) = 1$, using $h = 0.05$ (see Table 1). Since $y'' = -y' = y$, the true solution is decreasing and $|y''| \leq 1$. Therefore the error at each step is $|T_n| \leq (0.05)^2 \cdot \tfrac{1}{2} = \frac{0.0025}{2}$. The error after the 20 steps required to reach $x = 1$ is therefore less than $20 \cdot \frac{0.0025}{2} = 0.025$. The actual error is much smaller (0.0094) since we have not taken account of the fact that the second derivative of the true solution is decreasing.

Accumulated error   This is the error in y_n produced by the errors in all previous steps. Neglecting round-off, this error is

$$E_n = z_n - y_n \tag{15}$$

We now estimate the accumulated error for Euler's method. Since

$$z_{n+1} = z_n + hf(x_n, z_n) + T_n \tag{16}$$

$$y_{n+1} = y_n + hf(x_n, y_n) \tag{17}$$

we have, by subtraction,

$$E_{n+1} = E_n + h[f(x_n, z_n) - f(x_n, y_n)] + T_n \tag{18}$$

By the mean-value theorem

$$f(x_n, z_n) - f(x_n, y_n) = (z_n - y_n)f_y(x_n, \bar{y}_n)$$

where $\bar{y}_n$ is between y_n and z_n. If we let $A_n = f_y(x_n, \bar{y}_n)$, Eq. (18) becomes

$$E_{n+1} = (1 + hA_n)E_n + T_n \tag{19}$$

This difference equation describes the propagation of error throughout the computation. The values T_n and A_n depend on n and are not completely known in advance. However, in most cases the step size h is such that T_n and A_n vary slowly with n and some idea of the behavior of E_n can be obtained by assuming T_n and A_n are constants. Calling these constants T and A respectively, we obtain

$$E_{n+1} = (1 + hA)E_n + T \tag{20}$$

a difference equation with constant coefficients. This can be readily solved (see Sec 11-2) to obtain

$$E_n = \frac{T}{hA} [(1 + hA)^n - 1] \tag{20a}$$

where we have used the fact that $E_0 = z_0 - y_0 = 0$. If $|1 + hA| \leq 1$, that is, if $-2/h < A < 0$, the error remains bounded as n increases. However, if $|1 + hA| > 1$, that is, if $A > 0$ or if $A < -2/h$, the error increases exponentially. This explains why the error in Table 1 ($A = 1$) increases more rapidly than in Table 2 ($A = -1$).

Relative error In many cases the *relative error* is more significant than the error itself. We can often tolerate large errors, if the exact solution is large enough so that the relative error E_n/z_n is small. Even if the error is small, if the exact solution is also small, the relative error can be so large as to make the calculated solution useless.

It is difficult to determine the relative error of a given numerical method for a general differential equation $y' = f(x,y)$. However, a good idea of the relative error can often be obtained by considering the simple special case $y' = Ay$. For if the step size is picked properly, f_y varies slowly in the region of the xy plane involved in the calculation. If A is a typical value of f_y, then the relative error can be expected to behave approximately as for the simple case $y' = Ay$.

We illustrate this approach for Euler's method. The exact solution of $y' = Ay$, $y(x_0) = y_0$, is $y = y_0 e^{A(x-x_0)}$. Therefore

$$z_n = y(x_n) = e^{A(x_n-x_0)}y_0 = e^{Anh}y_0 \tag{21}$$

since $x_n - x_0 = nh$. The approximate solution y_n satisfies

$$y_{n+1} = y_n + hy'_n = (1 + hA)y_n$$

since $y'_n = Ay_n$. Solving this difference equation, we get

$$y_n = (1 + hA)^n y_0 \tag{22}$$

The relative error is therefore

$$\frac{E_n}{z_n} = \frac{z_n - y_n}{z_n} = 1 - \left(\frac{1 + hA}{e^{hA}}\right)^n \tag{23}$$

Since $e^{hA} \geq 1 + hA$ (see Prob. 4), we see that the relative error remains bounded as n increases. This holds regardless of the sign of A. In Tables 1 and 2 we see that the relative error grows rather slowly in both cases $A = 1$ (Table 1) and $A = -1$ (Table 2), whereas the error itself grows much more rapidly for $A = 1$. In Table 3 the relative error grows quite rapidly. Here, however, $A = f_y = 4xy$, and since the exact solution y gets quite large in absolute value near $x = 1$, A changes rapidly, and the above analysis does not apply.

Problems 12-3

1. The distributive law $a(b - c) = ab - ac$ does not necessarily hold if multiplications are rounded. To see this, let $a = 0.110$, $b = 2.080$, $c = 1.040$, and after each multiplication, round off to three decimal digits.

2. A common source of rounding errors occurs when nearly equal numbers are subtracted. If b and c are nearly equal is it better to calculate $a(b - c)$ or $ab - ac$ (assuming products are rounded)? Illustrate with a numerical example.

3. Sometimes the subtraction of nearly equal numbers can be avoided. Show that

$$\frac{-b + \sqrt{b^2 - 4ac}}{2a} = \frac{-2c}{b + \sqrt{b^2 - 4ac}}$$

If b^2 is much greater than $4ac$ the left side involves subtraction of nearly equal numbers, whereas the right side does not. Illustrate with a numerical example that (assuming products and quotients are rounded) use of the right-hand side is preferable under the above conditions.

4. Prove that $e^x \geq 1 + x$. Illustrate the inequality graphically.

5. A variation of Euler's method called *Nystrom's method* is given by

$$y_{n+1} = y_{n-1} + 2hy_n'$$

This uses the slope at the midpoint x_n between x_{n-1} and x_{n+1} instead of at the end point as in Euler's method. Nystrom's method is not a starting method, since both y_0 and y_1 must be known before y_2 can be found. One way of starting the solution is to use the first few Taylor-series terms to calculate y_1. That is, $y_1 = y_0 + hy_0' + \frac{1}{2}h^2y_0''$, where y_0'' can be calculated from the differential equation. Use this method to solve Eq. (6) with $h = 0.1$ for $0 \leq x \leq 1$. Use four decimal places and round results to three places.

6. Show that the truncation error in the use of Nystrom's method is $\frac{1}{3}h^3y'''(\xi)$, $x_n \leq \xi \leq x_{n+1}$. Hint: T_n is defined by $z_{n+1} = z_{n-1} + 2hz_n' + T_n$; expand z_{n+1} and z_{n-1} in Taylor series about x_n (three terms plus remainder) to show

$$T_n = \frac{1}{3}h^3 \cdot \frac{1}{2}[z'''(\xi_1) + z'''(\xi_2)] = \frac{1}{3}h^3z'''(\xi)$$

7. For Nystrom's method show that the accumulated error is given by

$$E_{n+1} = E_{n-1} + 2hA_nE_n + T_n$$

where $A_n = f_y(x_n,\bar{y}_n)$. Investigate the behavior of E_n assuming A_n and T_n are constant.

8. Investigate the growth of the relative error for Nystrom's method when the differential equation is $y' = Ay$.

9. The method known as the *trapezoidal rule* is given by

$$y_{n+1} = y_n + \tfrac{1}{2}h(y'_n + y'_{n+1}) \tag{i}$$

where $y'_n = f(x_n,y_n)$, $y'_{n+1} = f(x_{n+1},y_{n+1})$. Here the average of the slopes at x_n and x_{n+1} is used, rather than the slope at x_n as in Euler's method. Note that the trapezoidal rule is a starting method. However, the equation is not solved for y_{n+1} since y_{n+1} also appears in the right-hand side in $y'_{n+1} = f(x_{n+1},y_{n+1})$. In general this equation must be solved by iteration. That is, a value $y_{n+1}^{\{0\}}$ is assumed (this could be obtained, say, by Euler's method) and substituted into the right-hand side to obtain

$$y_{n+1}^{\{1\}} = y_n + \tfrac{1}{2}h[f(x_n,y_n) + f(x_{n+1},y_{n+1}^{\{0\}})]$$

Then $y_{n+1}^{\{1\}}$ can be used as the new estimate. Thus the iteration is given by

$$y_{n+1}^{\{i+1\}} = y_n + \tfrac{1}{2}h[f(x_n,y_n) + f(x_{n+1},y_{n+1}^{\{i\}})] \qquad i = 0, 1, 2, \ldots \tag{ii}$$

Use this method to find approximate values of $y(0.1)$ and $y(0.2)$ for $y' = 1 + y^2$, $y(0) = 0$. (Use $h = 0.1$.)

10. The following sequence of problems shows that the sequence $y_{n+1}^{\{i\}}$, $i = 0, 1, 2, \ldots$, determined by the iteration in (ii) converges to a number y_{n+1} satisfying Eq. (i) if $|f_y| \le A$ and the *convergence factor* $\theta = |\tfrac{1}{2}hA| < 1$.

   a. Let $\delta_{i+1} = y_{n+1}^{\{i+1\}} - y_{n+1}^{\{i\}}$ and show that

$$\delta_{i+1} = \tfrac{1}{2}\delta_i h f_y(x_{n+1},\bar{y}_{n+1}^{\{i\}})$$

where $\bar{y}_{n+1}^{\{i\}}$ is between $y_{n+1}^{\{i+1\}}$ and $y_{n+1}^{\{i\}}$.

   b. Show that $|\delta_i| \le \theta^i |\delta_0|$.

   c. If $|\theta| < 1$, show that, if $j > i$,

$$
\begin{aligned}
|y_{n+1}^{\{j\}} - y_{n+1}^{\{i\}}| &\le |\delta_j| + |\delta_{j-1}| + \cdots + |\delta_{i+1}| \\
&\le (\theta^j + \theta^{j-1} + \cdots + \theta^{i+1}) |\delta_0| \\
&\le |\delta_0| \, \theta^{i+1}/(1 - \theta)
\end{aligned}
$$

Thus, by the Cauchy test, the sequence $y_{n+1}^{\{i\}}$ converges to some number u_{n+1} if $\theta < 1$.

   d. If y_{n+1} satisfies (i), show that

$$|y_{n+1} - y_{n+1}^{\{i\}}| \le \theta^i |y_{n+1} - y_{n+1}^{\{0\}}|$$

   e. Show that, if $\theta < 1$ and $\epsilon > 0$,

$$|u_{n+1} - y_{n+1}| \le |u_{n+1} - y_{n+1}^{\{i\}}| + |y_{n+1}^{\{i\}} - y_{n+1}| < \epsilon$$

for i large enough. Since the left side is a fixed number which can be made less than any $\epsilon > 0$, we must have $|u_{n+1} - y_{n+1}| = 0$ or $u_{n+1} = y_{n+1}$.

11. Find the truncation error for the trapezoidal rule. Hint: T_n is defined by

$$z_{n+1} = z_n + \tfrac{1}{2}h(z_n' + z_{n+1}') + T_n$$

Expand z_{n+1} and z_{n+1}' in a Taylor series about x_n with integral form of remainder to show that

$$T_n = \int_{x_n}^{x_{n+1}} G(s)z_n'''(s)\, ds$$

where

$$G(s) = \tfrac{1}{2}[(x_{n+1} - s)^2 - h(x_{n+1} - s)]$$

Show that $G(s)$ has constant sign in $x_n \leq s \leq x_{n+1}$ and therefore, by the second mean-value theorem for integrals,†

$$T_n = z_n'''(\xi)\int_{x_n}^{x_{n+1}} G(s)\, ds = -\tfrac{1}{12}h^3 z_n'''(\xi) \qquad x_n \leq \xi \leq x_{n+1}$$

12. Discuss the behavior of the error E_n in the use of the trapezoidal rule (assume A_n and T_n are constants).

13. Investigate the relative error for the trapezoidal rule using the differential equation $y' = Ay$.

12-4 Parasitic solutions and stability

To illustrate the concept of parasitic solutions, we use Nystrom's method (see Prob. 5 of Sec. 12-3):

$$y_{n+1} = y_{n-1} + 2hy_n' \tag{24}$$

We shall apply this method to the simple differential equation

$$y' = Ay \qquad y(x_0) = y_0 \tag{25}$$

which has for its exact solution $z_n = y(x_n) = y_0 e^{A(x_n - x_0)} = y_0(e^{hA})^n$. Since $y_n' = Ay_n$, (24) becomes

$$y_{n+1} - 2hAy_n - y_{n-1} = 0 \tag{26}$$

The characteristic equation for this difference equation is

$$r^2 - 2hAr - 1 = 0 \tag{27}$$

with roots

$$r_0 = hA + \sqrt{1 + h^2A^2} \qquad r_1 = hA - \sqrt{1 + h^2A^2} \tag{28}$$

The solution of (26) is therefore

$$y_n = c_0 r_0^n + c_1 r_1^n \tag{29}$$

† See Courant [12], vol. 1, p. 256.

The constants c_0 and c_1 are determined from the initial condition y_0 and the value y_1. Since Nystrom's method is not a starting method, y_1 must be determined by some other method, such as the Taylor series (see Prob. 5 of Sec. 12-3):

$$y_1 = y_0 + hy'_0 + \tfrac{1}{2}h^2 y''_0$$

Evaluating c_0 and c_1, we get

$$c_0 = \frac{y_1 - y_0(hA - \sqrt{1 + h^2 A^2})}{2\sqrt{1 + h^2 A^2}}$$

$$c_1 = \frac{y_0(hA + \sqrt{1 + h^2 A^2}) - y_1}{2\sqrt{1 + h^2 A^2}}$$

(30)

We now consider what happens as $h \to 0$ and $n \to \infty$ in such a way that $nh = x - x_0$ remains constant. In Prob. 1 we show that

$$\lim_{\substack{h \to 0 \\ n \to \infty}} r_0^n = \lim_{\substack{h \to 0 \\ n \to \infty}} (hA + \sqrt{1 + h^2 A^2})^n = e^{A(x - x_0)} \tag{31}$$

Since $\lim_{h \to 0} c_0 = y_0$, we see that the first term in (29) approaches the exact solution as $h \to 0$. The second term is called a *parasitic solution*. It arises because the *first-order* differential equation (25) was approximated by a *second-order* difference equation (26). In order to ensure a small error, the parasitic solution must be small or

$$|r_1| \leq 1 \tag{32}$$

If this is the case, the method is called *stable*. For small relative error, the parasitic solution should be small relative to the exact solution or $|r_1/e^{hA}| \leq 1$. Since, for small h, $r_0 \approx e^{hA}$, we require instead that

$$\left| \frac{r_1}{r_0} \right| \leq 1 \tag{33}$$

When this holds, we call the method *relatively stable*.

It is easy to see that Nystrom's method is stable for $A > 0$ and unstable for $A < 0$. It is relatively stable for $A > 0$ and not relatively stable if $A < 0$.

In general we shall consider numerical methods which express y_{n+1} as a linear combination of present and past values of y_n and y'_n and possibly also of y'_{n+1}. For instance, a class of formulas considered in Sec. 12-6 is

$$\begin{aligned} y_{n+1} = {} & a_0 y_n + a_1 y_{n-1} + a_2 y_{n-2} \\ & + h(b_{-1} y'_{n+1} + b_0 y'_n + b_1 y'_{n-1} + b_2 y'_{n-2}) \end{aligned} \tag{34}$$

When this formula is applied to the equation $y' = Ay$, the following difference equation results:

$$(1 - b_{-1}hA)y_{n+1} = (a_0 + hAb_0)y_n + (a_1 + hAb_1)y_{n-1} + (a_2 + hAb_2)y_{n-2}$$

$$(35)$$

The characteristic equation for this difference equation is

$$(1 - b_{-1}hA)r^3 = (a_0 + hAb_0)r^2 + (a_1 + hAb_1)r + (a_2 + hAb_2) \qquad (36)$$

Calling the roots of this equation r_0, r_1, r_2, the general solution (assuming the roots are distinct) is

$$y_n = c_0 r_0^n + c_1 r_1^n + c_2 r_2^n$$

In all the methods we consider, one of the roots, which we denote by r_0, will have the property that r_0^n approximates the solution to the differential equation, that is,

$$\lim_{\substack{h \to 0 \\ n \to \infty}} (r_0)^n = e^{A(x-x_0)} \qquad (37)$$

The roots r_1, r_2 then give rise to parasitic solutions. Thus we are led to the following definition.

Definition 1   *The numerical method (35) is called stable if r_0 satisfies* (37) *and r_1, r_2 satisfy*

$$|r_1| \leq 1 \qquad |r_2| \leq 1 \qquad (38)$$

It is called relatively stable if r_0 satisfies (37) and

$$\left| \frac{r_1}{r_0} \right| \leq 1 \qquad \left| \frac{r_2}{r_0} \right| \leq 1 \qquad (39)$$

We emphasize that stability has been discussed only for the special equation $y' = Ay$. However, for a general differential equation $y' = f(x,y)$, if f_y changes slowly from step to step, and if A is a typical value of f_y, then we can expect approximately the same behavior for the error as for the simple differential equation $y' = Ay$.

Thus for stable methods the error can be expected to grow slowly, and for relatively stable methods the relative error can be expected to grow slowly. If only a few steps are needed in the calculation, stability is not particularly important and any method having low enough truncation error to ensure the desired accuracy may be used. However, when a large number of steps are needed in the calculation, it is important to use stable numerical methods.

Problems 12-4

1. Using L'Hospital's rule, verify Eq. (31).
2. If $|hA| < 1$, show that r_0 and r_1 given in Eq. (28) can be written

$$r_0 = 1 + hA + \frac{h^2 A^2}{2!} - \tfrac{1}{8}(h^4 A^4) + \cdots$$

$$r_1 = -\left(1 - hA + \frac{h^2 A^2}{2} + \tfrac{1}{8}h^4 + \cdots\right)$$

Note that for small h, $r_0 \approx e^{hA}$, $r_1 \approx -e^{-hA}$. Show therefore that for small h Eq. (29) can be written

$$y_n \approx c_0 e^{A(x_n - x_0)} + c_1(-1)^n e^{-A(x_n - x_0)}$$

Thus the parasitic solution oscillates with decreasing amplitude if $A > 0$ and increasing amplitude if $A < 0$.

3. Show that Euler's method is relatively stable. [Since there is only one root for the difference equation, it is necessary only to verify Eq. (37).]
4. Show that the trapezoidal rule

$$y_{n+1} = y_n + \tfrac{1}{2}h(y'_n + y'_{n+1})$$

is relatively stable. (See Prob. 3.)

Investigate the following methods for stability and relative stability:

5. $y_{n+1} = y_n + \tfrac{1}{2}h(3y'_n + y'_{n-1})$
6. $y_{n+1} = y_{n-1} + \tfrac{1}{3}h(y'_{n+1} + 4y'_n + y'_{n-1})$
7. $y_{n+1} = \tfrac{1}{8}(9y_n - y_{n-2}) + \tfrac{3}{8}h(y'_{n+1} + 2y'_n - y'_{n-1})$

12-5   A second-order predictor-corrector method

Milne† has suggested a clever combination of Nystrom's method

$$y_{n+1} = y_{n-1} + 2hy'_n \qquad [T_n = \tfrac{1}{3}h^3 z'''(\xi)] \tag{40}$$

and the trapezoidal rule

$$y_{n+1} = y_n + \tfrac{1}{2}h(y'_n + y'_{n+1}) \qquad [T_n = -\tfrac{1}{12}h^3 z'''(\xi)] \tag{41}$$

(See Sec. 12-3, Probs. 5, 6, 9, 10, 11.) Nystrom's method is simple to compute but is not relatively stable if $f_y < 0$ (see Sec. 12-4). On the other hand, the trapezoidal rule has a lower truncation error and is relatively stable but is more difficult to compute. Since y_{n+1} also appears on the right side of (41) in the term $y'_{n+1} = f(x_{n+1}, y_{n+1})$, we must use an iteration procedure to solve for y_{n+1} (see Sec. 12-3, Prob. 9). Therefore Milne has proposed that we use Nystrom's method to *predict* a value $y^{(0)}_{n+1}$:

$$\text{Predictor: } y^{(0)}_{n+1} = y_{n-1} + 2hy'_n \tag{42}$$

† See Milne [27].

and use the trapezoidal rule to *correct* the value using $y_{n+1}^{\{0\}}$ as an initial approximation for the iteration

Corrector:

$$y_{n+1}^{\{i+1\}} = y_n + \tfrac{1}{2}h[f(x_n,y_n) + f(x_{n+1},y_{n+1}^{\{i\}})] \qquad i = 0, 1, 2, \ldots \qquad (43)$$

We have seen in Sec. 12-3, Prob. 10, that this iteration converges if $\theta = |hA/2| < 1$ where $A \geq |f_y|$. In general we pick h small enough so that only one or two applications of (43) are needed to give the required accuracy.

Example 1   DE: $y' = 1 + y^2$

  IC: $y(0) = 0$

We use $h = 0.2$ and approximate $y(0.2)$, $y(0.4)$. To get started, we need $y_1 \approx y(0.2)$, which we calculate by the Taylor series

$$y_1 = y_0 + hy_0' + \frac{h^2}{2!}y_0'' + \frac{h^3}{3!}y_0'''$$

From the differential equation we get $y_0 = 0$, $y_0' = 1$, $y_0'' = 0$, $y_0''' = 2$, so that

$$y_1 = 0.2 + \tfrac{0.008}{3} = 0.2027$$

Now

$$y_1' = f(x_1,y_1) = 1 + (0.2027)^2 = 1.0411$$

From Eq. (42)

$$y_2^{\{0\}} = y_0 + 2hy_1' = 0 + 0.4(1.0411) = 0.4164$$

Then

$$(y_2^{\{0\}})' = f(x_2,y_2^{\{0\}}) = 1 + (y_2^{\{0\}})^2 = 1.1734$$

Finally, from Eq. (43) we obtain

$$y(0.4) \approx y_2^{\{1\}} = y_1 + \tfrac{1}{2}h[y_1' + (y_2^{\{0\}})'] = 0.4241$$

A second application of Eq. (43) yields

$$y_2^{\{2\}} = 0.4248$$

A third application yields $y_2^{\{3\}} = 0.4249$; therefore $y_2 = 0.425$, rounded to three decimal places.

An important feature of predictor-corrector methods is that an estimation of the truncation error can be obtained from the calculation itself. From Eqs. (42) and (43) and the definition of truncation error, the following approximations hold:

$$z_{n+1} \approx y_{n+1}^{\{0\}} + \tfrac{1}{3}h^3 z'''(\xi_1) \qquad x_n \leq \xi_1 \leq x_{n+1} \qquad (44)$$

$$z_{n+1} \approx y_{n+1}^{\{1\}} - \tfrac{1}{12}h^3 z'''(\xi_2) \qquad x_n \leq \xi_2 \leq x_{n+1} \qquad (45)$$

Therefore

$$c_0 \equiv y_{n+1}^{\{0\}} - y_{n+1}^{\{1\}} \approx \tfrac{1}{3}h^3 z'''(\xi_1) + \tfrac{1}{12}h^3 z'''(\xi_2) \tag{46}$$

If $z'''(\xi)$ is approximately constant in the interval, then

$$c_0 \approx \tfrac{5}{12}h^3 z'''(\xi) \qquad x_n \leq \xi \leq x_{n+1} \tag{47}$$

Therefore the truncation error is

$$T \approx \tfrac{1}{5}c_0 \tag{48}$$

It is useful to tabulate $c_0/5$, so that an approximation to the truncation error is available as the calculation proceeds.† If $\tfrac{1}{5}c_0$ is larger than the desired accuracy, the step size can be reduced. If $A = f_y$, and $\theta = |\tfrac{1}{2}hA|$ is also tabulated, an estimate of the number of iterations that are needed is also available.

In Tables 4, 5, and 6 we show the computations, using the predictor-corrector method, for the three sample problems, Eqs. (6), (7), and (8). We note in Tables 4 and 5 a considerable increase in accuracy over Euler's method (Tables 1 and 2). In Table 6 the rapid growth of c_0, T, A gives a warning that a large error is to be expected. We have carried the calculation several steps beyond where one could reasonably expect three-place accuracy in order to show what happens.

Table 4 *Second-order Predictor-corrector Method for $y' = y$, $y(0) = 1$ (when $h = 0.1$)*

x	y_n	$y_n^{\{0\}}$	c_0	$T = \dfrac{c_0}{5}$	$\theta = \lvert\tfrac{1}{2}hA\rvert$	Error	Percent relative error
0.0	1.000	—	—	—	—	—	—
0.1	1.105	—	—	—	—	—	—
0.2	1.221	1.221	—	—	0.06	—	—
0.3	1.350	1.350	—	—	0.07	—	—
0.4	1.492	1.491	−0.001	—	0.07	—	—
0.5	1.649	1.648	−0.001	—	0.08	—	—
0.6	1.823	1.822	−0.001	—	0.09	0.001	0.06
0.7	2.014	2.013	−0.001	—	0.10	—	—
0.8	2.226	2.225	−0.001	—	0.11	—	—
0.9	2.461	2.460	−0.001	—	0.12	0.001	0.04
1.0	2.720	2.718	−0.002	—	0.14	0.002	0.07

† In general, c_0 changes slowly from step to step. If a sudden change occurs in c_0, it often indicates an arithmetical mistake by the person (or machine) doing the computation. Therefore the immediately preceding computations should be checked. This is another reason why c_0 should be tabulated.

Table 5 *Second-order Predictor-corrector Method for* $y' = -y$, $y(0) = 1$ (when $h = 0.1$)

x	y_n	$y_n^{\{0\}}$	c_0	$T = \dfrac{c_0}{5}$	$\theta = \lvert \tfrac{1}{2}hA \rvert$	Error	Percent relative error
0.0	1.000	—	—	—	—	—	—
0.1	0.905	—	—	—	—	—	—
0.2	0.819	0.819	—	—	0.04	—	—
0.3	0.741	0.741	—	—	0.04	—	—
0.4	0.670	0.671	$+0.001$	—	0.03	—	—
0.5	0.607	0.607	—	—	0.03	—	—
0.6	0.549	0.549	—	—	0.03	—	—
0.7	0.497	0.497	—	—	0.03	—	—
0.8	0.449	0.449	—	—	0.02	—	—
0.9	0.407	0.407	—	—	0.02	—	—
1.0	0.368	0.368	—	—	0.02	—	—

Table 6 *Second-order Predictor-corrector Method for* $y' = 2xy^2$, $y(0) = 1$ (when $h = 0.1$)

x	y_n	$y_n^{\{0\}}$	c_0	$T = \dfrac{c_0}{5}$	$\theta = \lvert \tfrac{1}{2}hA \rvert$	Error	Percent relative error
0.0	1.000	—	—	—	—	—	—
0.1	1.010	—	—	—	0.02	—	—
0.2	1.042	1.041	-0.001	—	0.04	—	—
0.3	1.100	1.097	-0.003	—	0.07	-0.001	-0.09
0.4	1.193	1.187	-0.006	-0.001	0.10	-0.003	-0.25
0.5	1.338	1.328	-0.010	-0.002	0.13	-0.005	-0.38
0.6	1.572	1.551	-0.021	-0.004	0.19	-0.011	-0.70
0.7	1.981	1.931	-0.050	-0.010	0.27	-0.021	-1.1
0.8	2.826	2.671	-0.155	-0.031	0.45	-0.048	-1.7
0.9	5.317	4.536	-0.781	-0.156	0.96	-0.054	-1.0
1.0	28.021	13.003	-15.018	—	—	(∞)	(∞)

Problems 12-5

1. Find approximations of $y(0.1)$, $y(0.2)$, $y(0.3)$, $y(0.4)$ for $y' = 1 + y^2$, $y(0) = 0$. Use $h = 0.1$ and three-decimal-place accuracy.

2. Check the calculations in Tables 4, 5, and 6.

3. Solve to three-place accuracy for $y(0.5)$ for $y' = -2xy^2$, $y(0) = 1$.

4. From (47) and (44) we see that the truncation error in the predictor is

$$\tfrac{1}{3}h^3 z'''(\xi_1) \approx \tfrac{12}{15}(y_{n+1}^{\{0\}} - y_{n+1}^{\{1\}})$$

Furthermore the difference between the predicted and corrected values changes slowly from step to step so that

$$\tfrac{1}{3}h^3z'''(\xi_1) \approx \tfrac{12}{15}(y_n^{\{0\}} - y_n)$$

where $y_n \approx y_n^{\{1\}}$ is the final calculated value at x_n. Thus

$$\bar{y}_{n+1}^{\{0\}} = y_{n+1}^{\{0\}} + \tfrac{12}{15}(y_{n+1}^{\{0\}} - y_n)$$

will in general be an improved value of the prediction. Thus we are led to the method:

    Predictor: $y_{n+1}^{\{0\}} = y_{n-1} + 2hy_n'$

    Modifier: $\bar{y}_{n+1}^{\{0\}} = y_{n+1}^{\{0\}} + \tfrac{4}{5}(y_n^{\{0\}} - y_n)$

    Corrector: $y_{n+1}^{\{i+1\}} = y_n + \tfrac{1}{2}h[f(x_n,y_n) + f(x_{n-1},y_{n+1}^{\{i\}})]$

with $\bar{y}_{n+1}^{\{0\}}$ being used in the corrector initially. Use this method to solve Prob. 1. (The modifier is omitted at the first step since $y_1^{\{0\}}$ is not available.)

12-6   *Fourth-order predictor-corrector methods*

We now consider the derivation of fourth-order predictors and correctors. Fourth-order methods are often used on digital computers because they give sufficient accuracy for most problems and are relatively simple to use and analyze. The class of formulas we shall consider is

$$y_{n+1} = a_0y_n + a_1y_{n-1} + a_2y_{n-2} + h(b_{-1}y_{n+1}' + b_0y_n' + b_1y_{n-1}' + b_2y_{n-2}') \quad (49)$$

If $b_{-1} = 0$, the formula expresses y_{n+1} explicitly in terms of known quantities. Such formulas are called *open-type* formulas. Because of their simplicity, open-type formulas are often used for predictors. If $b_{-1} \neq 0$, the formula is called a *closed-type formula*. Such a formula is an implicit equation for y_{n+1} [since $y_{n+1}' = f(x_{n+1},y_{n+1})$ appears on the right-hand side], which must be solved by iteration. Closed-type formulas are often used for correctors because they generally have better stability properties and smaller truncation errors than open-type formulas.

    The truncation error T_n for Eq. (49) is defined by

$$z_{n+1} = a_0z_n + a_1z_{n-1} + a_2z_{n-2}$$
$$+ h(b_{-1}z_{n+1}' + b_0z_n' + b_1z_{n-1}' + b_2z_{n-2}') + T_n \quad (50)$$

To get methods with fourth-order accuracy, we determine the coefficients in (50) so that the Taylor-series expansions about x_n on both sides agree up to the fourth-derivative terms. This will determine five of the seven coefficients in (49). The remaining two coefficients can then be picked so that the method has good stability properties or other desirable properties such as possession of zero coefficients.

In Prob. 1 we show that the above process yields the following equations:

$$a_0 = 24b_2 - 9a_2 \qquad a_1 = 1 + 8a_2 - 24b_2$$
$$b_{-1} = \tfrac{1}{3}(1 - a_2 + 3b_2) \qquad b_0 = \tfrac{1}{3}(4 + 14a_2 - 39b_2)$$
$$b_1 = \tfrac{1}{3}(1 + 17a_2 - 39b_2) \qquad (51)$$

where a_2 and b_2 can be picked arbitrarily. The truncation error T_n is obtained from the remainder terms of the Taylor series (see Prob. 2). The general expression for arbitrary a_2 and b_2 is fairly complicated. However, in the particular cases in common use the truncation error can be expressed simply as

$$T_n = kh^5 z^{(5)}(\xi) \qquad x_{n-2} \le \xi \le x_{n+1} \qquad (52)$$

where k is a suitable constant.

Milne's corrector is obtained by setting $a_2 = b_2 = 0$ in Eq. (49). This gives $a_0 = 0$, $a_1 = 1$, $b_{-1} = b_1 = \tfrac{1}{3}$, $b_0 = \tfrac{4}{3}$, so that Milne's corrector is

$$y_{n+1} = y_{n-1} + \tfrac{1}{3}h(y'_{n+1} + 4y'_n + y'_{n-1}) \qquad (53)$$

This has a truncation error of $-\tfrac{1}{90}h^5 z^{(5)}(\xi)$ (see Prob. 3). This corrector is often used because of its simplicity and its low truncation error. It is relatively stable if $f_y > 0$ but is unstable for $f_y < 0$ (see Sec. 12-4, Prob. 6). Therefore this method must be used with caution if $f_y < 0$ and a large number of steps are needed in the calculation.

Hamming's corrector is obtained by setting $b_2 = 0$ and $a_2 = -\tfrac{1}{8}$. This yields

$$y_{n+1} = \tfrac{1}{8}(9y_n - y_{n-2}) + \tfrac{3}{8}h(y'_{n+1} + 2y'_n - y'_{n-1}) \qquad (54)$$

and a truncation error of $T_n = -\tfrac{1}{40}h^5 z^{(5)}(\xi)$ (see Prob. 4).

Hamming's corrector has a larger truncation error than Milne's corrector. However, it is relatively stable regardless of the sign of f_y, provided the step size h is small enough (see Sec. 12-4, Prob. 7.). Thus Hamming's corrector is to be preferred to Milne's corrector if $f_y < 0$ or if a large number of steps are needed in the calculation.

We now discuss suitable fourth-order predictors to go with the above fourth-order correctors. For predictors we set $b_{-1} = 0$ so that Eq. (49) becomes

$$y_{n+1} = a_0 y_n + a_1 y_{n-1} + a_2 y_{n-2} + h(b_0 y'_n + b_1 y'_{n-1} + b_2 y'_{n-2}) \qquad (55)$$

Evaluating the coefficients as above, we obtain (see Prob. 5)

$$a_0 = -8 - a_2 \qquad a_1 = 9 \qquad b_0 = \tfrac{1}{3}(17 + a_2)$$
$$b_1 = \tfrac{1}{3}(14 + 4a_2) \qquad b_2 = \tfrac{1}{3}(-1 + a_2) \qquad (56)$$

where a_2 is arbitrary. It turns out that these methods are not relatively stable. However, stability is not particularly important for predictors since the final values satisfy the corrector formula. It is more important to have a simple formula with low truncation error. Another factor to consider is the round-off error. As a general rule[†] large values of the coefficients a_i tend to amplify round-off errors. Since this class of formulas all have $a_1 = 9$, a relatively large value, we turn to other predictors.

In order to get one more degree of freedom to seek better behavior of the round-off error, we can add an extra value y_{n-3} to get the class of formulas

$$y_{n+1} = a_0 y_n + a_1 y_{n-1} + a_2 y_{n-2} + a_3 y_{n-3} + h(b_0 y'_n + b_1 y'_{n-1} + b_2 y'_{n-2}) \quad (57)$$

These include the Milne predictor, Eq. (59), and are called Milne-type predictors.

For a fourth-order formula we get the following equations (see Prob. 7):

$$a_0 = -8 - a_2 + 8a_3 \qquad a_1 = 9 - 9a_3$$
$$b_0 = \tfrac{1}{3}(17 + a_2 - 9a_3) \qquad b_1 = \tfrac{1}{3}(14 + 4a_2 - 18a_3)$$
$$b_2 = \tfrac{1}{3}(-1 + a_2 + 9a_3) \qquad (58)$$

where a_2 and a_3 are arbitrary. Various choices are possible.[‡] The choice $a_3 = 1, a_2 = 0$ yields the popular Milne predictor

$$y_{n+1} = y_{n-3} + \tfrac{4}{3}h(2y'_n - y'_{n-1} + 2y'_{n-2}) \quad (59)$$

with a truncation error (Prob. 7)

$$T_n = \tfrac{14}{45}h^5 z^{(5)}(\xi) \quad (60)$$

This gives a simple formula with only four terms; it has low truncation error and good round-off properties.

Thus a good fourth-order predictor-corrector method is:

$$\text{Predictor: } y_{n+1} = y_{n-3} + \tfrac{4}{3}h(2y'_n - y'_{n-1} + 2y'_{n-2})$$
$$\text{Corrector: } y_{n+1} = \tfrac{1}{8}(9y_n - y_{n-2}) + \tfrac{3}{8}h(y'_{n+1} + 2y'_n - y'_{n-1}) \quad (61)$$

An estimate of the truncation error is given (see Prob. 8) by

$$T_n \approx \tfrac{9}{121}(y_{n+1} - y_{n+1}^{\{0\}}) \quad (62)$$

The corrector formula must be solved by iterations. The iterations will converge if the *convergence factor* $\theta < 1$, where

$$\theta = \tfrac{3}{8}hA \quad (62a)$$

and $|f_y| \le A$ (see Prob. 11). In general, h is picked small enough so that only one or two iterations are necessary to obtain the desired accuracy.

---

† See Hamming [17], pp. 200, 201.
‡ See Hamming [17], pp. 201–204, or Ralston [30], pp. 179–189.

Table 7   Fourth-order Predictor-corrector Method for $y' = y$, $y(0) = 1$ (when $h = 0.1$)

x	y_n	$y_n^{\{0\}}$	$T = \frac{9}{121}[y_{n+1} - y_{n+1}^{\{0\}}]$	$\theta = \frac{3}{8}hA$	Error
0.0	1.0000000	—	—	—	—
0.1	1.1051708	—	—	—	1.0×10^{-7}
0.2	1.2214026	—	—	—	2.0×10^{-7}
0.3	1.3498585	—	—	—	3.0×10^{-7}
0.4	1.4918245	1.4918208	2.8×10^{-8}	0.06	2.0×10^{-7}
0.5	1.6487213	1.6487169	3.2×10^{-8}	0.06	0.0×10^{-7}
0.6	1.8221191	1.8221139	3.8×10^{-8}	0.07	-3.0×10^{-7}
0.7	2.0137533	2.0137473	4.5×10^{-8}	0.08	-6.0×10^{-7}
0.8	2.2255419	2.2255352	5.0×10^{-8}	0.08	-1.0×10^{-6}
0.9	2.4596046	2.4595971	5.6×10^{-8}	0.09	-1.5×10^{-6}
1.0	2.7182838	2.7182756	6.1×10^{-8}	0.10	-2.0×10^{-6}

We illustrate the use of (61) on the sample problems, Eqs. (6) and (7).   The results are shown in Tables 7 and 8.   The calculations were performed with ten significant digits and the results rounded to eight significant digits.   We note the considerable increase in accuracy over the second-order method shown in Tables 4 and 5.   The starting values were computed by the fourth-order Runge-Kutta method, Eq. (71), discussed in the next section.

Table 8   Fourth-order Predictor-corrector Method for $y' = -y$, $y(0) = 1$ (when $h = 0.1$)

x	y_n	$y_n^{\{0\}}$	$T = \frac{9}{121}[y_{n+1} - y_{n+1}^{\{0\}}]$	$\theta = \frac{3}{8}hA$	Error
0.0	1.00000000	—	—	—	—
0.1	0.90483750	—	—	—	-8.0×10^{-8}
0.2	0.81873090	—	—	—	-1.5×10^{-7}
0.3	0.74081842	—	—	—	-2.0×10^{-7}
0.4	0.67031997	0.67032254	-2.1×10^{-7}	0.03	8.0×10^{-8}
0.5	0.60653041	0.60653306	-2.0×10^{-7}	0.02	2.5×10^{-7}
0.6	0.54881120	0.54881388	-2.0×10^{-7}	0.02	4.4×10^{-7}
0.7	0.49658473	0.49658750	-2.1×10^{-7}	0.02	5.8×10^{-7}
0.8	0.44932827	0.44933076	-1.9×10^{-7}	0.02	6.9×10^{-7}
0.9	0.40656888	0.40657118	-1.7×10^{-7}	0.02	7.8×10^{-7}
1.0	0.36787860	0.36788068	-1.6×10^{-7}	0.01	8.4×10^{-7}

Problems 12-6

1. Expand z_{n+1}, z'_{n+1}, z_{n-1}, z'_{n-1}, z_{n-2}, z'_{n-2} in a Taylor series up to fourth-derivative terms plus an integral form of the remainder.   Substitute in Eq. (50) and derive Eq. (51).

2. Using the remainder terms in Prob. 1, show that the truncation error in (50) can be written as

$$T_n = \int_{x_{n-2}}^{x_{n+1}} G(s) z^{(5)}(s)\, ds$$

where

$$4!\, G(s) = \overline{(x_{n+1} - s)}^4 - 4hb_{-1}\overline{(x_{n+1} - s)}^3 + a_1\overline{(x_{n-1} - s)}^4$$
$$+ 4hb_1\overline{(x_{n-1} - s)}^3 + a_2\overline{(x_{n-2} - s)}^4 + 4hb_2\overline{(x_{n-2} - s)}^3$$

where

$$\overline{(x_{n-i} - s)} = \begin{cases} x_{n-i} - s & \begin{cases} x_{n-i} \leq s \leq x_n & i \neq -1 \\ x_n \leq s \leq x_{n+1} & i = -1 \end{cases} \\ 0 & \text{otherwise} \end{cases}$$

3. Derive the truncation error for Milne's corrector, Eq. (53). Substitute a_i and b_i in Prob. 2, show $G(s)$ has constant sign, and use the second law of the mean for integrals to write the truncation error in the form

$$T_n = z^{(5)}(\xi) \int_{x_{n-2}}^{x_{n+1}} G(s)\, ds \qquad x_{n-2} \leq \xi \leq x_{n+1}$$

4. Derive the truncation error for Hamming's corrector, Eq. (54).
5. Derive Eqs. (56).
6. Derive Eqs. (58).
7. Derive the truncation error for Milne's predictor, Eq. (59).
8. Derive Eq. (62) (see Sec. 12-5).
9. Derive a predict-modify-correct method based on Eq. (61) (see Sec. 12-5, Prob. 4).
10. Use (61) on $y' = 1 + y^2$, $y(0) = 0$. Use $h = 0.2$.
11. Show that the iterations for the corrector formula (61) converge if the convergence factor satisfies (62a) (see Sec. 12-3, Prob. 10).

12-7   *Starting methods and Runge-Kutta methods*

All the methods we have discussed other than the simple Euler method are not starting methods. Some other method must be used for the first few points. A straightforward method of starting the solution is to use a truncated Taylor series up to rth-derivative terms

$$y_{n+1} = y_n + hy'_n + \frac{h^2}{2} y''_n + \cdots + \frac{h^r}{r!} y_n^{(r)} \tag{63}$$

The truncation error is $h^{r+1}z^{(r+1)}(\xi)/(r+1)!$. The main disadvantage of this approach is that it requires the computation of derivatives of $f(x,y)$. This may be complicated and time-consuming if f is not a simple function.

Runge-Kutta methods are starting methods which have the same order of truncation error as Taylor series but which involve computation of only values of $f(x,y)$ and not the derivatives of $f(x,y)$.

The Euler method $y_{n+1} = y_n + hf(x_n,y_n)$ involves the computation of only values of $f(x,y)$ and is a first-order Runge-Kutta method. The second-order Runge-Kutta methods are of the form

$$y_{n+1} = y_n + h\Phi(x_n,y_n,h) \tag{64}$$

where $\Phi(x_n,y_n,h)$ is an *average slope* obtained by weighting the slopes of the calculated solution at two points in the interval $x_n \leq x \leq x_{n+1}$. That is,

$$\Phi(x_n,y_n,h) = w_1 f(x_n,y_n) + w_2 f(x_n + \alpha_1 h, y_n + \beta_1 h y_n')$$

where w_1, w_2, α_1, β_1 are picked so that the Taylor-series expansions about x_n of both sides of (64) agree up through second-derivative terms. The expansion of the right-hand side yields

$$y_n + hw_1 f(x_n,y_n) + hw_2 f(x_n,y_n)$$
$$+ h^2 w_2 [\alpha_1 f_x(x_n,y_n) + \beta_1 y_n' f_y(x_n,y_n)] \tag{65}$$

The expansion of the left-hand side of (64) gives

$$y_n + hy_n' + \frac{h^2}{2} y_n'' = y_n + hf(x_n,y_n) + \frac{h^2}{2} [f_x(x_n,y_n) + y_n' f_y(x_n,y_n)]$$

Comparing terms, we get the equations

$$w_1 + w_2 = 1 \qquad w_2\alpha_1 = \tfrac{1}{2} \qquad w_2\beta_1 = \tfrac{1}{2} \tag{66}$$

We may take $w_2 = \omega$ as arbitrary and obtain

$$w_1 = 1 - \omega \qquad \alpha_1 = \beta_1 = \frac{1}{2\omega} \qquad \omega \neq 0 \tag{67}$$

Thus the second-order Runge-Kutta methods are

$$y_{n+1} = y_n + h\left\{(1 - \omega)f(x_n,y_n) + \omega\left[f\left(x_n + \frac{h}{2\omega}, y_n + \frac{h}{2\omega} y_n'\right)\right]\right\} \tag{68}$$

If $\omega = \tfrac{1}{2}$, we obtain the *improved Euler method* or *Heun's method*

$$y_{n+1} = y_n + \frac{h}{2} [f(x_n,y_n) + f(x_n + h, y_n + hy_n')] \tag{69}$$

For $\omega = 1$ we get the *modified Euler method*

$$y_{n+1} = y_n + hf\left(x_n + \frac{h}{2}, y_n + \frac{h}{2} y_n'\right) \tag{70}$$

These are two commonly used methods either for obtaining the whole solution or for starting second-order predictor-corrector methods.

In order to start fourth-order predictor-corrector methods, we need a fourth-order Runge-Kutta method. The derivation of such formulas follows the same procedure as for the second-order case, but the algebra is considerably more complicated. We refer the reader to textbooks on numerical analysis for the derivations.†

The most commonly used fourth-order Runge-Kutta method is

$$y_{n+1} = y_n + \tfrac{1}{6}(k_1 + 2k_2 + 2k_3 + k_4) \tag{71a}$$

where

$$k_1 = hf(x_n, y_n)$$

$$k_2 = hf\left(x_n + \frac{h}{2}, y_n + \frac{k_1}{2}\right)$$

$$k_3 = hf\left(x_n + \frac{h}{2}, y_n + \frac{k_2}{2}\right) \tag{71b}$$

$$k_4 = hf(x_n + h, y_n + k_3)$$

The second term on the right of (71a) represents an average slope over various points in the interval $x_n \leq x \leq x_{n+1}$. In Prob. 4 we verify that the expansions of both sides of the first equation in a Taylor series agree up to fourth-derivative terms.

Example 1 We illustrate the fourth-order Runge-Kutta method for $y' = 1 + y^2$, $y(0) = 0$, using $h = 0.2$ and working to six decimal places. The results are tabulated below:

x	y	k_1	k_2	k_3	k_4	Error
0.0	0.000000	0.200000	0.202000	0.202040	0.2081640	—
0.2	0.2027074	0.2082181	0.2188273	0.2194849	0.2356491	3×10^{-6}
0.4	0.4277890	—	—	—	—	4×10^{-6}

The Runge-Kutta methods are useful for starting predictor-corrector methods. They can be used also for the entire solutions; however, they are less attractive than the predictor-corrector methods for two reasons: (1) It is difficult to analyze the error in the Runge-Kutta method. (2) The fourth-order Runge-Kutta method requires four evaluations of $f(x,y)$ at each step whereas a fourth-order predictor-corrector usually requires only two evaluations. Since evaluation of $f(x,y)$ is often the most time-consuming part of the solution, the predictor-corrector method is more efficient.

† See Ralston [30], pp. 191–201.

Problems 12-7

1. Use Heun's method, Eq. (69), on the differential equation $y' = Ay$, $y(0) = y_0$. Solve explicitly for y_n and show that y_n approaches the exact solution as $h \to 0$.

2. Solve $y' = y$, $y(0) = 1$, $0 \le x \le 1$ by the modified Euler method, Eq. (70). Use $h = 0.1$ and three decimal places.

3. Verify the computations in Example 1. Also find an approximation to $y(0.6)$.

4. Verify that the Taylor-series expansions of both sides of (71a) agree up to fourth-derivative terms.

12-8 Higher-order equations and systems of equations

It is quite easy to extend the methods of this chapter to systems of first-order equations. For illustration we shall consider a system of two first-order equations

$$\frac{dy}{dx} = f(x,y,z) \qquad \frac{dz}{dx} = g(x,y,z) \tag{72}$$

$$y(x_0) = y_0 \qquad z(x_0) = z_0$$

Euler's method for this system is simply

$$\begin{aligned} y_{n+1} &= y_n + hy'_n \\ z_{n+1} &= z_n + hz'_n \end{aligned} \tag{73}$$

where $y'_n = f(x_n,y_n,z_n)$ and $z'_n = g(x_n,y_n,z_n)$. Since y_0, z_0 are known, $y_1, z_1, y_2, z_2, \ldots$ can be successively calculated.

The predictor-corrector methods can also be used on systems. For example, the fourth-order method given in Eq. (61) yields:

Predictor: $y_{n+1} = y_{n-3} + \frac{4}{3}h(2y'_n - y'_{n-1} + 2y'_{n-2})$

$$z_{n+1} = z_{n-3} + \frac{4}{3}h(2z'_n - z'_{n-1} + 2z'_{n-2}) \tag{74}$$

Corrector: $y_{n+1} = \frac{1}{8}(9y_n - y_{n-2}) + \frac{3}{8}h(y'_{n+1} + 2y'_n - y'_{n-1})$

$$z_{n+1} = \frac{1}{8}(9z_n - z_{n-2}) + \frac{3}{8}h(z'_{n+1} + 2z'_n - z'_{n-1}) \tag{75}$$

The starting values may be obtained by the Taylor series or by the Runge-Kutta method, Eqs. (84) and (85).

A second-order equation

$$y'' = f(x,y,y') \qquad y(x_0) = y_0 \qquad y'(x_0) = y'_0 \tag{76}$$

may be reduced to a system of first-order equations by letting $y' = z$ to obtain

$$y' = z \qquad z' = f(x,y,z) \tag{77}$$

and the methods considered above can be applied.

Some simplifications can be made when the second-order equation has no y' term:

$$y'' = f(x,y) \qquad y(x_0) = y_0 \qquad y'(x_0) = y_0' \tag{78}$$

Instead of reducing to a system, we look for direct methods of the form†

$$y_{n+1} = a_0 y_n + a_1 y_{n-1} + h^2(b_{-1} y_{n+1}'' + b_0 y_n'' + b_1 y_{n-1}'') \tag{79}$$

We determine the coefficients so that the Taylor-series expansions agree up to fourth-derivative terms (see Prob. 1). Thus we obtain a useful corrector formula

$$y_{n+1} = 2y_n - y_{n-1} + \tfrac{1}{12}h^2(y_{n+1}'' + 10y_n'' + y_{n-1}'') \tag{80}$$

This has a truncation error of

$$-\tfrac{1}{240}h^6 z^{(6)}(\xi) \tag{81}$$

A suitable predictor‡ to go with (80) is

$$y_{n+1} = 2y_{n-1} - y_{n-3} + \tfrac{4}{3}h^2(y_n'' + y_{n-1}'' + y_{n-2}'') \tag{82}$$

with a truncation error of

$$\tfrac{16}{240}h^6 z^{(6)}(\xi) \tag{83}$$

The Runge-Kutta method can also be extended to systems. For the second-order system, Eq. (72), the Runge-Kutta method is given below for reference purposes.

$$\begin{aligned} y_{n+1} &= y_n + \tfrac{1}{6}(k_1 + 2k_2 + 2k_3 + k_4) \\ z_{n+1} &= z_n + \tfrac{1}{6}(l_1 + 2l_2 + 2l_3 + l_4) \end{aligned} \tag{84}$$

where

$$\begin{aligned} k_1 &= hf(x_n, y_n, z_n) \\ l_1 &= hg(x_n, y_n, z_n) \\ k_2 &= hf(x_n + \tfrac{1}{2}h, y_n + \tfrac{1}{2}k_1, z_n + \tfrac{1}{2}l_1) \\ l_2 &= hg(x_n + \tfrac{1}{2}h, y_n + \tfrac{1}{2}k_1, z_n + \tfrac{1}{2}l_1) \\ k_3 &= hf(x_n + \tfrac{1}{2}h, y_n + \tfrac{1}{2}k_2, z_n + \tfrac{1}{2}l_2) \\ l_3 &= hg(x_n + \tfrac{1}{2}h, y_n + \tfrac{1}{2}k_2, z_n + \tfrac{1}{2}l_2) \\ k_4 &= hf(x_n + h, y_n + k_3, z_n + l_3) \\ l_4 &= hg(x_n + h, y_n + k_3, z_n + l_3) \end{aligned} \tag{85}$$

† See Hamming [17], pp. 213, 214, or Ralston [30], pp. 210, 211.
‡ See Ralston [30], p. 211, or Hamming [17], p. 214.

Problems 12-8

1. Derive Eqs. (80) and (81).

2. Consider $y'' = y$, $y(0) = 0$, $y'(0) = 1$. Find an approximate value of $y(0.3)$, using Euler's method. Use $h = 0.1$ and carry three decimal places.

3. Find approximate values of $y(0.2)$, $y(0.4)$, $y(0.6)$ for the equation in Prob. 2, using the Runge-Kutta method. Use $h = 0.2$ and carry four decimal places.

4. Find approximate values for $y(0.8)$, $y(1.0)$ for the equation in Prob. 2, using the predictor-corrector method, Eqs. (74) and (75). Use the starting values obtained in Prob. 3.

5. Do Prob. 4 using Eqs. (80) and (82).

Boundary-value Problems

13

13-1 Introduction

Up to this point we have been mainly concerned with initial-value problems for differential equations, that is, problems in which the subsidiary conditions are given at one point. We now take up boundary-value problems with the subsidiary conditions given at two points. There are considerable differences between initial- and boundary-value problems. For initial-value problems it is rather generally true that a unique solution exists (see Chap. 4), whereas for boundary-value problems, even *linear* ones, there may be no solution, many solutions, or a unique solution.

Our study of boundary-value problems will lead us naturally to many new and interesting concepts: eigenvalues and eigenfunctions, orthogonal functions, Fourier series, Sturm-Liouville problems, eigenfunction expansions, and the Green's functions. All of these topics are useful in many parts of mathematics, science, and engineering. We cannot cover all of these subjects thoroughly;† in several cases it will be necessary to state theorems without proof. However, many illustrations will be given to aid in the understanding, appreciation, and use of the theorems.

13-2 Homogeneous boundary-value problems

We consider the homogeneous boundary-value problem

$$\text{DE: } Ly = 0 \qquad a \leq x \leq b$$
$$\text{BC: } \alpha_1 y(a) + \beta_1 y'(a) = 0 \tag{1}$$
$$\alpha_2 y(b) + \beta_2 y'(b) = 0$$

---

† For more information see Churchill [8], Sagan [31], or Courant and Hilbert [13].

where L is the second-order linear operator defined by

$$Ly = a_0(x)y'' + a_1(x)y' + a_2(x)y \tag{2}$$

with $a_0(x) \neq 0$ and a_0, a_1, a_2 continuous in $a \leq x \leq b$. We assume that α_1, β_1 are not both zero and also that α_2, β_2 are not both zero. The boundary conditions in (1) are called *linear* and *homogeneous*, since if $u(x)$ and $v(x)$ satisfy the boundary conditions, so do any linear combination of $u(x)$ and $v(x)$. The boundary conditions in (1) are also called *unmixed* boundary conditions since each boundary condition involves the value of y and y' at only *one* value of x.

It is easy to see that the homogeneous boundary-value problem (1) always possesses the trivial solution $y(x) \equiv 0$. We ask whether or not nontrivial solutions exist. A condition for the existence of nontrivial solutions can easily be obtained. Let $y_1(x)$ and $y_2(x)$ be two linearly independent solutions of the differential equation $Ly = 0$. Then the general solution is

$$y = c_1 y_1 + c_2 y_2 \tag{3}$$

In order to satisfy the boundary conditions, we must have

$$\alpha_1 y(a) + \beta_1 y'(a) = 0$$
$$\alpha_2 y(b) + \beta_2 y'(b) = 0 \tag{4}$$

or, using (3), we obtain

$$\alpha_1[c_1 y_1(a) + c_2 y_2(a)] + \beta_1[c_1 y_1'(a) + c_2 y_2'(a)] = 0$$
$$\alpha_2[c_1 y_1(b) + c_2 y_2(b)] + \beta_2[c_1 y_1'(b) + c_2 y_2'(b)] = 0 \tag{5}$$

Collecting coefficients of c_1 and c_2, we obtain

$$c_1 B_a(y_1) + c_2 B_a(y_2) = 0$$
$$c_1 B_b(y_1) + c_2 B_b(y_2) = 0 \tag{6}$$

where we have used the abbreviations†

$$B_a(u) = \alpha_1 u(a) + \beta_1 u'(a)$$
$$B_b(u) = \alpha_2 u(b) + \beta_2 u'(b) \tag{7}$$

We see that the two algebraic equations (6) for c_1 and c_2 possess nontrivial solutions if and only if

$$\begin{vmatrix} B_a(y_1) & B_a(y_2) \\ B_b(y_1) & B_b(y_2) \end{vmatrix} = 0 \tag{8}$$

Therefore nontrivial solutions of the boundary-value problem (1) *exist if and only if* (8) *holds.*

† Note that with these abbreviations the boundary conditions in (1) can be written in the compact form $B_a(y) = 0$, $B_b(y) = 0$. (See Sec. 13-12.)

Example 1 DE: $y'' - 4y = 0$

BC: $y(0) = y(1) = 0$

The general solution of the differential equation is

$$y = c_1 e^{2x} + c_2 e^{-2x}$$

Substituting the boundary conditions, we obtain

$$0 = c_1 + c_2$$
$$0 = c_1 e^2 + c_2 e^{-2}$$

Since

$$\begin{vmatrix} 1 & 1 \\ e^2 & e^{-2} \end{vmatrix} = e^{-2} - e^2 \neq 0$$

the only solution is $c_1 = c_2 = 0$, which yields $y \equiv 0$.

Example 2 DE: $y'' + 4y = 0$

BC: $y(0) = y(\pi) = 0$

$$y = c_1 \sin 2x + c_2 \cos 2x$$

Substituting the boundary conditions, we obtain

$$0 = c_1 \cdot 0 + c_2 \cdot 1 \quad \text{or} \quad c_2 = 0$$
$$0 = c_1 \cdot 0 + c_2 \cdot 1 \quad \text{or} \quad c_2 = 0$$

Clearly the solution of these equations is $c_2 = 0$, c_1 arbitrary. Therefore $y = c_1 \sin 2x$ is a solution of the boundary-value problem for any number c_1.

In Example 2 above, all solutions of the boundary-value problem were given by the one-parameter family $y = cu(x)$ where $u(x)$ was a particular nontrivial solution. This result holds generally for the boundary-value problem (1) (with unmixed boundary conditions):

Theorem 1 *If $u(x)$ is a particular nontrivial solution of the boundary-value problem (1), then all solutions are given by $y = cu(x)$ where c is an arbitrary constant.*

Proof Let $v(x)$ be any solution and $u(x)$ a particular nontrivial solution of the boundary-value problem. Then we have

$$\alpha_1 u(a) + \beta_1 u'(a) = 0$$
$$\alpha_1 v(a) + \beta_1 v'(a) = 0$$

since both u and v satisfy the first boundary condition (the second boundary condition is not needed). These are two equations for α_1 and β_1 which have a nontrivial solution. Therefore the determinant of the coefficients must vanish:

$$\begin{vmatrix} u(a) & u'(a) \\ v(a) & v'(a) \end{vmatrix} = 0$$

This determinant is the Wronskian of u and v evaluated at $x = a$. We have seen (Sec. 5-2) that if the Wronskian is zero at one point, it is identically zero. Therefore u and v are linearly dependent and $v(x) \equiv cu(x)$.

Problems 13-2

1. Find the nontrivial solutions, if any, of the following:

a. $y'' = 0$
   $y(0) = y(1) = 0$
b. $y'' = 0$
   $y'(0) = y'(1) = 0$
c. $y'' - \lambda y = 0 \qquad \lambda > 0$
   $y(0) = y(1) = 0$
d. $y'' + 4y = 0$
   $y(0) = y(1)$
   $y'(0) = y'(1)$

2. *Nonhomogeneous boundary conditions.* Consider the boundary-value problem with nonhomogeneous boundary conditions

DE: $Ly = 0 \qquad a \le x \le b$
BC: $B_a(y) \equiv \alpha_1 y(a) + \beta_1 y'(a) = A$
$\qquad B_b(y) = \alpha_1 y(b) + \beta_1 y'(b) = B$

where L is given in Eq. (2).

a. Prove that this problem has a unique solution if and only if

$$\Delta = \begin{vmatrix} B_a(y_1) & B_a(y_2) \\ B_b(y_1) & B_b(y_2) \end{vmatrix} \ne 0$$

where y_1 and y_2 are linearly independent solutions of $Ly = 0$.

This result can be stated: *The nonhomogeneous problem has a unique solution if and only if the corresponding homogeneous problem ($A = B = 0$) has only the trivial solution.*

b. If the homogeneous problem has a nontrivial solution, i.e., if $\Delta = 0$, show that the nonhomogeneous problem either has no solution or else an infinity of solutions given by

$$y = y_p(x) + cu(x)$$

where y_p is a particular solution of the nonhomogeneous problem and $u(x)$ is a nontrivial solution of the homogeneous problem.

3. Find solutions, if they exist, for:

a. $y'' + y = 0$
$\quad y(0) = 1,\ y(1) = 0$
b. $y'' + \pi^2 y = 0$
$\quad y(0) = 2,\ y(1) = -2$
c. $y'' + \pi^2 y = 0$
$\quad y(0) = 1,\ y(1) = 2$

4. *Nonhomogeneous differential equations and boundary conditions.* Consider, using the notation of Prob. 2,

DE: $Ly = f(x)$
BC: $B_a(y) = A$
$\quad\ B_b(y) = B$

where $f(x)$ is continuous.

a. Prove that this problem has a unique solution if and only if $\Delta \neq 0$ (see Prob. 2). Does the italicized statement in Prob. 2a still hold?
b. Prove the statement in Prob. 2b for the above problem.

5. Find solutions, if they exist:

a. $y'' + y = 1$
$\quad y(0) = 1,\ y(1) = 2$
b. $y'' + 9y = x$
$\quad y(0) = 0,\ y(\pi) = 0$
c. $y'' + 9y = \sin x$
$\quad y(0) = 0,\ y(\pi) = 1$
d. $y'' + 9y = \sin x$
$\quad y(0) = 1,\ y(\pi) = -1$

6. Show that $y'' + \pi^2 y = f(x)$, $y(0) = y(1) = 0$ has a solution if and only if $\int_0^1 f(x) \sin \pi x\, dx = 0$ and, when the solution exists, it is given by $y = \frac{1}{\pi}\int_0^x f(t) \sin \pi(x - t)\, dt + c \sin \pi x$. Hint: A particular solution of the differential equation is

$$y_p = \frac{1}{\pi}\int_0^x f(t) \sin \pi(x - t)\, dt$$

(See Sec. 5-6, Prob. 6.)

7. Assume $Ly = f(x)$, $B_a(y) = A$, $B_b(y) = B$ has a unique solution. Show that this solution is given by $y = u + v$ where u satisfies $Lu = f(x)$, $B_a(u) = B_b(u) = 0$ and v satisfies $Lv = 0$, $B_a(v) = A$, $B_b(v) = B$.

13-3   Eigenvalue problems

Many homogeneous boundary-value problems arising in physical problems involve a parameter.   An important type of problem is

DE: $L(y) + \lambda y = 0$ $a \leq x \leq b$

BC: $\alpha_1 y(a) + \beta_1 y'(a) = 0$ (9)

$\quad \alpha_2 y(b) + \beta_2 y'(b) = 0$

where λ is a parameter (independent of x) and L is the operator defined by

$$Ly = a_0(x)y'' + a_1(x)y' + a_2(x)y \tag{10}$$

and a_0, a_1, a_2 are continuous and $a_0 \neq 0$ in $a \leq x \leq b$.

The trivial solution $y \equiv 0$ exists for all values of the parameter λ. However, it may be that nontrivial solutions exist for certain values of the parameter λ and not for other values.   If a nontrivial solution exists for a value $\lambda = \lambda_i$, then this value is called an *eigenvalue* of the operator L (relevant to the boundary conditions) and the corresponding nontrivial solution $y_i(x)$ is called an *eigenfunction*.   Eigenfunctions are determined only up to a multiplicative constant.   For if $y_i(x)$ is an eigenfunction corresponding to the eigenvalue λ_i, then so is $c y_i(x)$† where c is any number.

When the operator L is applied to a function, it often changes the function drastically.   However, eigenfunctions enjoy the special property that L applied to an eigenfunction merely multiplies the eigenfunction by a constant (the negative of the eigenvalue).‡

The importance of eigenfunctions and eigenvalues in mathematics and science cannot be overemphasized.   Countless pages have been written on this subject.   In physical problems the eigenvalues and eigenfunctions often have important physical interpretations.   In vibration problems the eigenvalues (actually the square of the eigenvalues) are proportional to the natural frequencies of the system and the eigenfunctions are the shapes of the normal-mode vibrations.   In the problem of buckling of a column, the eigenvalues determine the loads at which the column buckles.   In quantum mechanics the eigenvalues represent the only possible measurable values of the energy of a physical system.

---

† Because the boundary conditions in (9) are unmixed, this is the most general eigenfunction.   This was proved in Sec. 13-2.

‡ It is customary to define the eigenvalues of a differential operator L using the equation $Ly = -\lambda y$, as we have done.   Of course the equation $Ly = \lambda y$ could have been used instead.   Our choice makes the eigenvalues come out positive in the most common cases (see Example 1).   For a matrix A, the eigenvalues are defined using the equation $Ax = \lambda x$ (see Secs. 8-6 and 8-7).

From a mathematical point of view, knowledge of eigenvalues and eigenfunctions of an operator L allows us to solve certain partial differential equations as will be seen in Chap. 14 (see also Secs. 8-6, 8-7 and 11-7).

To give one illustration of the use of eigenfunctions, we consider briefly the nonhomogeneous problem

DE: $Ly = f(x)$

BC: $\alpha_1 y(a) + \beta_1 y'(a) = 0$ (11)

$\quad\quad \alpha_2 y(b) + \beta_2 y'(b) = 0$

where L is the operator (10). We assume that $\lambda_1, \ldots, \lambda_n$ are eigenvalues of L and $y_1, \ldots, y_n$ are the corresponding eigenfunctions. That is

$$Ly_i = -\lambda_i y_i \quad\quad i = 1, 2, \ldots, n \tag{12}$$

and the $y_i(x)$ satisfy the boundary conditions. Suppose also that $f(x)$ is a given linear combination of the eigenfunctions

$$f(x) = A_1 y_1 + A_2 y_2 + \cdots + A_n y_n \tag{13}$$

To determine a solution of the differential equation

$$Ly = A_1 y_1 + A_2 y_2 + \cdots + A_n y_n \tag{14}$$

we assume that the solution y is a linear combination of the eigenfunctions

$$y = c_1 y_1 + c_2 y_2 + \cdots + c_n y_n \tag{15}$$

where the c_i are to be determined so as to satisfy the differential equation. Substituting (15) into (14), we obtain

$$L(c_1 y_1 + \cdots + c_n y_n) = A_1 y_1 + \cdots + A_n y_n \tag{16}$$

However,

$$L(c_1 y_1 + \cdots + c_n y_n) = c_1 L(y_1) + \cdots + c_n L(y_n)$$
$$= c_1(-\lambda_1) y_1 + \cdots + c_n(-\lambda_n) y_n \tag{17}$$

where we have used the linearity of L and the fact that the y_i are eigenfunctions. Therefore we must have

$$-(c_1 \lambda_1 y_1 + \cdots + c_n \lambda_n y_n) = A_1 y_1 + \cdots + A_n y_n \tag{18}$$

Assuming the y_i are linearly independent functions, we must have

$$c_1 = \frac{-A_1}{\lambda_1}, \; c_2 = \frac{-A_2}{\lambda_2}, \ldots, \; c_n = \frac{-A_n}{\lambda_n} \tag{19}$$

assuming also that none of the eigenvalues is zero. Therefore the solution to the problem (11) is

$$y = -\left(\frac{A_1}{\lambda_1} y_1 + \cdots + \frac{A_n}{\lambda_n} y_n \right) \tag{20}$$

We note that (20) satisfies not only the differential equation but also the boundary conditions.

Although we have solved the nonhomogeneous problem under the apparently severe restriction that the right-hand side f was a linear combination of eigenfunctions, we shall see in the sequel that the same method can be used for a wide class of functions f. For we shall see that it is often possible to expand a wide class of functions f into a convergent infinite series of eigenfunctions. The solution for y can then be obtained by expanding y in an infinite series of eigenfunctions with undetermined coefficients. These coefficients can then be determined in exactly the same manner as above.

Example 1   Find the eigenvalues and eigenfunctions for

DE: $y'' + \lambda y = 0$ $0 \le x \le L$ $(L > 0)$
BC: $y(0) = 0$ $y(L) = 0$

Since the form of the solution of the differential equation depends on whether $\lambda = 0$, $\lambda < 0$, or $\lambda > 0$, we consider these cases separately.

a. $\lambda = 0$. The differential equation is simply $y'' = 0$, which has for its general solution

$$y = Ax + B$$

Using the boundary conditions, we find

$$0 = B$$

and

$$0 = A \cdot L \qquad \text{or} \qquad A = 0$$

which means that $y \equiv 0$ is the only solution. Therefore $\lambda = 0$ is not an eigenvalue.

b. $\lambda < 0$. The general solution of the differential equation is

$$y = Ae^{\sqrt{-\lambda}x} + Be^{-\sqrt{-\lambda}x}$$

where $\sqrt{-\lambda}$ is a real number since $\lambda < 0$. Using the boundary conditions, we find

$$0 = A + B$$
$$0 = Ae^{\sqrt{-\lambda}L} + Be^{-\sqrt{-\lambda}L}$$

Solving these equations, we find that $A = B = 0$. Therefore the only solution for $\lambda < 0$ is the trivial solution and the numbers $\lambda < 0$ are not eigenvalues.

c. $\lambda > 0$. In this case the general solution of the differential equation is

$$y = A \cos \sqrt{\lambda}x + B \sin \sqrt{\lambda}x$$

Using the first boundary condition, we find

$$0 = A$$

Using the second boundary condition, we find

$$0 = B \sin \sqrt{\lambda}L$$

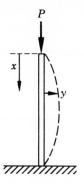

figure 1. *Buckling of a column.*

Therefore either $B = 0$ or $\sin \sqrt{\lambda} L = 0$. If $B = 0$, we obtain the trivial solution. However, if we determine λ so that

$$\sin \sqrt{\lambda} L = 0$$

then B can be taken different from zero and nontrivial solutions will exist. We see that the above equation will hold if and only if

$$\sqrt{\lambda} L = n\pi \qquad n = 1, 2, 3, \ldots$$

or $\qquad \lambda = \lambda_n = \dfrac{n^2 \pi^2}{L^2}$

The numbers $\lambda_n = n^2\pi^2/L^2$ are therefore the eigenvalues of our problem and the corresponding eigenfunctions are

$$y_n = B_n \sin \frac{n\pi x}{L} \qquad n = 1, 2, \ldots$$

where the B_n are arbitrary real numbers.

This example has important physical interpretations. The problem of the buckling of a slender column (see Fig. 1) leads to the differential equation†

$$y'' + \frac{P}{EI} y = 0$$

where P is the applied load, E is Young's modulus, I is the second moment of area about the neutral axis, and x is the distance from the top of the column. If the column is hinged at the bottom and constrained from rotating at the top, the boundary conditions are $y(0) = y(L) = 0$ where L is the length of the column. Letting $\lambda = P/EI$, we have the eigenvalue problem

$$y'' + \lambda y = 0$$
$$y(0) = y(L) = 0$$

We know that this problem has only the trivial solution unless $\lambda = n^2\pi^2/L^2$. Physically this means as we increase the load from zero, the column does not

† See Ziegler [39], vol. 1, p. 200.

buckle until the load reaches the critical value (often called the Euler load)

$$P_1 = \lambda_1 EI = \frac{\pi^2 EI}{L^2}$$

The shape of the column at this first *buckling mode* is the half-sine wave

$$y_1 = B_1 \sin \frac{\pi x}{L}$$

This lowest mode seems to be the only one which can be experimentally obtained for most columns.

In Chap. 14 we shall study the vibrating string. It will be shown there that the *normal modes of vibration* must satisfy the same boundary-value problem as above. The squares of the eigenvalues are proportional to the natural frequencies of the vibration, and the eigenfunctions are the shapes of the normal modes.

Problems 13-3

Find eigenvalues and eigenfunctions:

1. $y'' + \lambda y = 0$ $y'(0) = y(L) = 0$
2. $y'' + \lambda y = 0$ $y(0) = y'(L) = 0$
3. $y'' + \lambda y = 0$ $y'(0) = y'(L) = 0$
4. $y'' + \lambda y = 0$ $y(-L) = y(L), y'(-L) = y'(L)$
5. $(xy')' + \dfrac{\lambda y}{x} = 0$ $y(1) = y(e^\pi) = 0$

6. Assume λ is real and consider $y'' + \lambda^2 y = 0$, $y(0) + y'(0) = 0$, $y(1) = 0$.

a. Show that $\lambda = 0$ is an eigenvalue.

b. If $\lambda > 0$, show that eigenvalues must satisfy the transcendental equation $\tan \lambda = \lambda$.

c. Show graphically that $\tan \lambda = \lambda$ has an infinite number of real roots $\lambda_1 < \lambda_2 < \lambda_3 < \cdots$ and that, for large n, $\lambda_n \approx \frac{1}{4}(2n + 1)^2 \pi^2$.

d. Find the eigenfunctions.

7. Let λ_i be the eigenvalues and y_i be the corresponding eigenfunctions of an operator L. Show that a solution of

$$Ly + \lambda y = \sum_{k=1}^{n} A_k y_k$$

where the A_i are given constants is

$$y = \sum_{k=1}^{n} \frac{A_k y_k}{\lambda - \lambda_k}$$

provided λ is not an eigenvalue.

8. Solve $y'' + \lambda y = \sum_{k=0}^{n} a_k \sin k\pi x$, $y(0) = y(1) = 0$, if $\lambda \neq k^2 \pi^2$.

9. $x(xy')' + 5y = 3 \sin (5 \ln x)$, $y(1) = y(e^{\pi}) = 0$. Hint: Problems 5 and 7 are to be used.

10. Consider $Ly + \lambda y = f(x)$ with the homogeneous boundary condition $B_a(y) = 0$, $B_b(y) = 0$, where L is the operator of Eq. (10) and $f(x)$ is continuous. Show that this problem has a unique solution if and only if λ is not an eigenvalue. Hint: See Sec. 13-2, Prob. 4.

13-4   Orthogonal functions

Before proceeding with out study of eigenvalue problems, we must consider some elementary aspects of the theory of orthogonal functions. We shall develop the concept of orthogonal functions by generalizing orthogonal vectors.

We recall that a vector **v** in three-dimensional space is an order triplet of numbers

$$\mathbf{v} = (v_1, v_2, v_3) \tag{21}$$

where the numbers v_k are called the components of **v**. These components can be thought of as the values of a function $v(x)$ which is defined for only the three values $x = 1, 2, 3$. That is,

$$v(1) = v_1 \qquad v(2) = v_2 \qquad v(3) = v_3 \tag{22}$$

The graph of the function $v(x)$ consists merely of the three points shown in Fig. 2. Each point represents a component of **v**, and the collection of the three points in the natural order is a graphical representation of the vector **v**. (This representation can easily be extended to vectors in N dimensions, as was done in Sec. 11-6.) We see also that the operations of addition of vectors and multiplication by a scalar are the usual operations with functions. For if $v(x)$ and $w(x)$ are two functions defined for $x = 1, 2, 3$, then the usual meaning for the sum of $v(x)$ and $w(x)$ is the function

$$u(x) = v(x) + w(x)$$

whose functional values $u(1) = v(1) + w(1)$, $u(2) = v(2) + w(2)$, and $u(3) = v(3) + w(3)$ are simply the three components of the sum of the vectors v and w defined by the functions $v(x)$ and $w(x)$. Similar remarks hold for multiplication by a scalar. These considerations suggest that functions f

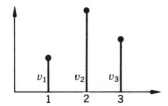

figure 2.   *The vector* **v.**

defined on an interval $a \leq x \leq b$ can be considered as vectors, and the functional values $f(x)$ can be thought of as the components of the *function vector f*.   One fundamental difference between function vectors and ordinary vectors is that the number of components of a function vector is not only infinite but uncountable.†   This will necessitate some change in our definition of inner product of functions.   We recall that for vectors in three dimensions the inner product of v and w was defined by

$$\langle v,w \rangle = \sum_{k=1}^{3} v_k w_k = \sum_{k=1}^{3} v(k)w(k)$$

For functions we define an inner product by replacing the summation with an integral.   That is, we define the inner product of two continuous functions f and g defined in an interval $a \leq x \leq b$ by

$$\langle f,g \rangle = \int_a^b f(x)g(x)\, dx \tag{23}$$

We note that the inner product satisfies the following:

$$\begin{aligned}
&\langle f,g \rangle = \langle g,f \rangle \\
&\langle \alpha f + \beta g, h \rangle = \alpha \langle f,h \rangle + \beta \langle g,h \rangle \\
&\langle f,f \rangle = 0 \quad \text{if } f = 0 \\
&\langle f,f \rangle > 0 \quad\quad \text{if} \quad f \neq 0
\end{aligned} \tag{24}$$

These in fact are the same as the properties of the inner product of ordinary vectors (Sec. 11-6).   Analogous to the length of a vector in three dimensions, the *norm* of a function f, denoted by $\|f\|$, is defined by

$$\|f\| = \sqrt{\langle f,f \rangle}$$

or

$$\|f\| = \sqrt{\int_a^b f^2(x)\, dx} \tag{25}$$

The norm of a function is a measure‡ of how large the function is on the whole interval $a \leq x \leq b$.   If $\|f\| = 1$, we say that f is *normalized* and call f a *unit function*.

---

† A *countable set* is one whose elements can be put into one-to-one correspondence with the positive integers.   An *uncountable set* is one that is not countable. Interesting facts are that the set of all rational numbers is countable, whereas the set of real numbers in $0 \leq x \leq 1$ is uncountable.   See Courant and Robbins [14], p. 81.

‡ This measure, often used in electrical engineering, is called the root-mean-square value of a current or voltage.

The concept of orthogonality of vectors can also be extended to functions. Two functions f and g are said to be orthogonal on the interval $a \leq x \leq b$ if $\langle f,g \rangle = \int_a^b f(x)g(x)\,dx = 0$. Note that the zero function is orthogonal to all functions.

Example 1 The functions 1 and x are orthogonal on the interval $[-1,1]$ since

$$\langle 1,x \rangle = \int_{-1}^1 1 \cdot x\,dx = \tfrac{1}{2}\,x^2 \Big|_{-1}^{1} = 0$$

Example 2 The function 1 and x^2 are not orthogonal on $[-1,1]$ since

$$\langle 1,x^2 \rangle = \int_{-1}^1 1 \cdot x^2\,dx = \tfrac{1}{3}\,x^3 \Big|_{-1}^{1} = \tfrac{2}{3}$$

It often happens in connection with eigenvalue problems that we encounter an infinite set of functions

$$\{\varphi_1, \varphi_2, \ldots\} \tag{26}$$

that are mutually orthogonal on some interval $a \leq x \leq b$, that is,

$$\langle \varphi_i, \varphi_j \rangle = \int_a^b \varphi_i(x)\varphi_j(x)\,dx = 0 \qquad i \neq j \tag{27}$$

Any set of functions satisfying (27) is called an *orthogonal set* of functions on the interval $a \leq x \leq b$. If each of the functions of an orthogonal set has a norm of unity, then we call the set an *orthonormal set*. An orthonormal set is defined by

$$\langle \varphi_i, \varphi_j \rangle = \begin{cases} 0 & i \neq j \\ 1 & i = j \end{cases} \tag{28}$$

It is convenient to introduce a symbol for the right-hand side of (29) called the *Kronecker delta*

$$\delta_{ij} \equiv \begin{cases} 0 & i \neq j \\ 1 & i = j \end{cases} \tag{29}$$

Thus we define an orthonormal set simply by

$$\langle \varphi_i, \varphi_j \rangle = \delta_{ij} \qquad i, j = 1, 2, \ldots \tag{30}$$

If an orthogonal set of nonzero functions $\psi_1, \psi_2, \ldots$ is known, it is easy to construct an orthonormal set by dividing each function by its norm.

That is, the functions

$$\varphi_i = \frac{\psi_i}{\|\psi_i\|} = \frac{\psi_i}{\sqrt{\langle \psi_i, \psi_i \rangle}} = \frac{\psi_i}{\sqrt{\displaystyle\int_a^b \psi_i^2(x)\, dx}} \tag{31}$$

form an orthonormal set (see Prob. 3).

If an orthonormal set has the property that the *only continuous function orthogonal to all members of the set is the zero function*, then we call the set a *complete*† *orthonormal set*. It is easy to see that with three-dimensional vectors a complete orthonormal set must have three members. In dealing with functions, we cannot define completeness in terms of the number of members of an orthonormal set, since there may be an infinite number. Instead we use the above definition to guarantee that no orthonormal functions have been left out.

Example 3   The set of functions

$$\left\{ \sin \frac{n\pi x}{L} \right\} \qquad n = 1, 2, 3, \ldots$$

form an orthogonal set on the interval $[0, L]$. These are the eigenfunctions obtained in the previous section. To show this, we need to prove that

$$\left\langle \sin \frac{m\pi x}{L}, \sin \frac{n\pi x}{L} \right\rangle = \int_0^L \sin \frac{m\pi x}{L} \sin \frac{n\pi x}{L}\, dx = 0 \qquad n \neq m$$

We have

$$\int_0^L \sin \frac{m\pi x}{L} \sin \frac{n\pi x}{L}\, dx = \frac{1}{2} \int_0^L \left[\cos \frac{(m-n)\pi x}{L} - \cos \frac{(m+n)\pi x}{L} \right] dx$$

$$= \frac{1}{2} \left[\frac{\sin \left[(m-n)\pi x/L\right]}{(m-n)\pi/L} - \frac{\sin \left[(m+n)\pi x/L\right]}{(m+n)\pi/L} \right]_0^L \qquad m \neq n$$

$$= 0$$

where we have used a familiar trigonometric identity in the first step above.

To construct an orthonormal set, we evaluate

$$\int_0^L \sin^2 \frac{n\pi x}{L}\, dx = \frac{1}{2} \int_0^L \left(1 - \cos \frac{2n\pi x}{L} \right) dx$$

$$= \frac{1}{2} \left(x - \frac{\sin (2n\pi x/L)}{2n\pi/L} \right) \Big|_0^L = \frac{L}{2}$$

† Sometimes the word *closed* is used instead of *complete*. We have followed the terminology of Churchill [7] and the definitive work *Linear Operators; Part I, General Theory* by N. Dunford and J. T. Schwartz, Interscience Publishers, Inc., New York, 1958. Courant and Hilbert [13] use *closed*.

The norm of $\sin(n\pi x/L)$ is therefore $\sqrt{L/2}$ and the set

$$\left\{\frac{\sin(n\pi x/L)}{\sqrt{\frac{1}{2}L}}\right\} = \left\{\sqrt{\frac{2}{L}}\sin\frac{n\pi x}{L}\right\} \qquad n = 1, 2, \ldots$$

is an orthonormal set.

Problems 13-4

1. Prove that the zero function is orthogonal to all functions.

2. Prove the properties (24) above.

3. Prove that an orthogonal set of nonzero functions becomes an orthonormal set when each function is divided by its norm.

4. Verify the following properties of the Kronecker delta:

a. $\displaystyle\sum_{i=1}^{n} \delta_{ii} = n$

b. $\displaystyle\sum_{i=1}^{n} \alpha_i \delta_{ij} = \alpha_j$

5. Prove that the set of functions

$$\left\{1, \cos\frac{\pi x}{L}, \cos\frac{2\pi x}{L}, \ldots\right\}$$

forms an orthogonal set on the interval $0 \leq x \leq L$. Find the corresponding orthonormal set.

6. *Gram-Schmidt orthogonalization process.* Let $\langle f_1, f_2, \ldots\rangle$ be a set of continuous linearly independent functions on the interval $a \leq x \leq b$. Show that the following set of functions φ_k formed as linear combinations of the f_k form an orthonormal set:

$$\varphi_1 = \frac{f_1}{\|f_1\|}$$

$$\varphi_2 = \frac{f_2 - \langle f_2, \varphi_1\rangle\varphi_1}{\|f_2 - \langle f_2, \varphi_1\rangle\varphi_1\|}$$

$$\varphi_3 = \frac{f_3 - \langle f_3, \varphi_1\rangle\varphi_1 - \langle f_3, \varphi_2\rangle\varphi_2}{\|f_3 - \langle f_3, \varphi_1\rangle\varphi_1 - \langle f_3, \varphi_2\rangle\varphi_2\|}$$

$$\cdot$$
$$\cdot$$
$$\cdot$$

$$\varphi_n = \frac{f_n - \displaystyle\sum_{k=1}^{n-1}\langle f_n, \varphi_k\rangle\varphi_k}{\left\|f_n - \displaystyle\sum_{k=1}^{n-1}\langle f_n, \varphi_k\rangle\varphi_k\right\|}$$

The same process works for ordinary vectors in 3-space or N-space. Give a geometrical interpretation of the above process in 3-space.

7. Using the Gram-Schmidt process, find the first four members of an orthonormal set in the interval $-1 \le x \le 1$, starting with the linearly independent set of functions

$$\langle 1, x, x^2, x^3, \ldots \rangle$$

(The functions thus found are proportional to the Legendre polynomials.)

8. Let λ_n and λ_m be eigenvalues and y_n, y_m be corresponding eigenfunctions for $y'' + \lambda y = 0$, $y(0) = y(L) = 0$. Show, without solving for the eigenvalues and eigenfunctions, that eigenfunctions corresponding to different eigenvalues are orthogonal on $[0,L]$. That is, prove

$$\int_0^L y_n(x) y_m(x)\, dx = 0 \qquad m \ne n$$

Hint: $y_n'' + \lambda_n y_n = 0$, $y_m'' + \lambda_m y_m = 0$. Multiply the first equation by y_m and the second by y_n, subtract, and integrate to obtain

$$(\lambda_n - \lambda_m) \int_0^L y_n y_m\, dx = \int_0^L (y_m y_n'' - y_n y_m'')\, dx$$

Show that the right-hand side is zero.

13-5   *Generalized Fourier series*

We recall that an arbitrary vector $\mathbf{v}$ in 3-space can always be expressed as a linear combination of three orthonormal vectors (a complete orthonormal set) $\{\mathbf{u}^1, \mathbf{u}^2, \mathbf{u}^3\}$. That is,

$$\mathbf{v} = \alpha_1 \mathbf{u}^1 + \alpha_2 \mathbf{u}^2 + \alpha_3 \mathbf{u}^3 \tag{32}$$

Further, the coefficients α_i are easily obtained by taking the inner product of both sides with $\mathbf{u}^i$. For instance, to find α_1, we have

$$\langle \mathbf{v}, \mathbf{u}^1 \rangle = \alpha_1 \langle \mathbf{u}^1, \mathbf{u}^1 \rangle + \alpha_2 \langle \mathbf{u}^2, \mathbf{u}^1 \rangle + \alpha_3 \langle \mathbf{u}^3, \mathbf{u}^1 \rangle \tag{33}$$

but

$$\langle \mathbf{u}^2, \mathbf{u}^1 \rangle = \langle \mathbf{u}^3, \mathbf{u}^1 \rangle = 0 \qquad \text{and} \qquad \langle \mathbf{u}^1, \mathbf{u}^1 \rangle = 1$$

Therefore

$$\alpha_1 = \langle \mathbf{v}, \mathbf{u}^1 \rangle$$

Similarly

$$\alpha_i = \langle \mathbf{v}, \mathbf{u}^i \rangle \qquad i = 1, 2, 3 \tag{34}$$

The expansion (32) with the coefficients given by (34) is sometimes called a *finite Fourier series*. We now extend these ideas to functions defined on an interval $a \le x \le b$.

Suppose that

$$\{\varphi_1, \varphi_2, \ldots\} \tag{35}$$

is a complete orthonormal set of continuous functions on the interval $a \leq x \leq b$, that is,

$$\langle \varphi_i, \varphi_j \rangle = \int_a^b \varphi_i(x)\varphi_j(x)\, dx = \delta_{ij} \tag{36}$$

and let f be an arbitrary continuous function defined on $a \leq x \leq b$. We seek an expansion of the form

$$f(x) = \sum_{i=1}^{\infty} \alpha_i \varphi_i(x) \tag{37}$$

which is analogous to Eq. (32). However, whereas it is easy to prove that such an expansion is possible for vectors in three dimensions, it is not at all obvious that the expansion (37) is possible. It is not within the scope of this introductory treatment to go into any details concerning the validity of this expansion. Later we shall quote some important theorems which will guarantee the existence of such expansions. For the moment we assume that the expansion (37) is valid and consider how the coefficients can be found. We assume not only that the infinite series converges to $f(x)$ in $a \leq x \leq b$ but also that the series obtained by multiplying both sides by $\varphi_j(x)$ can be integrated termwise, and the result converges to the proper sum.

In spite of these remarks, the formal manipulations are easy. Multiplying both sides of $f(x)$ by $\varphi_j(x)$ and integrating, we obtain

$$\int_a^b f(x)\varphi_j(x)\, dx = \sum_{k=1}^{\infty} \alpha_i \int_a^b \varphi_i(x)\varphi_j(x)\, dx$$

$$= \sum_{i=1}^{\infty} \alpha_i \delta_{ij}$$

$$= \alpha_j$$

or

$$\alpha_j = \langle f, \varphi_j \rangle = \int_a^b f(x)\varphi_j(x)\, dx \tag{38}$$

Therefore the coefficient α_j is simply the inner product of $f(x)$ with φ_j, analogous to Eq. (34) for three-dimensional vectors. The expansion

$$f(x) = \sum_{i=1}^{\infty} \alpha_i \varphi_i(x) \tag{39}$$

with the coefficients

$$\alpha_i = \int_a^b f(x)\varphi_i(x)\, dx \tag{40}$$

is often called a *generalized Fourier series*† and the coefficients α_i are called the *Fourier coefficients*.

Since the orthonormal set was assumed to be complete, we see from Eq. (40) that all $\alpha_i = 0$ if and only if $f(x) \equiv 0$. That is, the only continuous function having a Fourier series with all zero coefficients is the zero function. If the orthonormal set were not complete, this property would not hold.

We will often encounter sets which are not complete orthonormal sets but which are complete orthogonal sets.‡

If the complete orthogonal set is

$$\{\psi_1, \psi_2, \ldots\} \tag{41}$$

the Fourier series is

$$f(x) = \sum_{i=1}^{\infty} \alpha_i \psi_i(x) \tag{42}$$

and, by a process similar to above, the coefficients α_i are given by

$$\alpha_i = \frac{\langle f, \psi_i \rangle}{\langle \psi_i, \psi_i \rangle} = \frac{\displaystyle\int_a^b f(x)\psi_i(x)\,dx}{\displaystyle\int_a^b \psi_i^2(x)\,dx} \tag{43}$$

The following example will illustrate a Fourier-series expansion for a particular function and a particular complete orthogonal set.

Example 1   Expand the function

$$f(x) = \begin{cases} -1 & 0 \le x < \tfrac{1}{2}\pi \\ 1 & \tfrac{1}{2}\pi \le x \le \pi \end{cases}$$

into a Fourier series using the orthogonal set

$$\{\sin nx\} \qquad n = 1, 2, 3, \ldots$$

For the moment, we ignore the fact that $f(x)$ is discontinuous at $x = \tfrac{1}{2}\pi$. We have

$$f(x) = \sum_{n=1}^{\infty} \alpha_n \sin nx$$

† In fact the right-hand side of (39) with the coefficients given by (40) is usually called the Fourier series corresponding to $f(x)$ regardless of whether or not the series converges to $f(x)$. It turns out that the series can often be used for certain purposes even when it does not converge.

‡ A complete orthogonal set is one with all nonzero elements such that the corresponding orthonormal set is complete.

with the coefficients given by

$$\alpha_n = \frac{\int_0^\pi f(x)\sin nx\, dx}{\int_0^\pi \sin^2 nx\, dx}$$

Since $\int_0^\pi \sin^2 nx\, dx = \frac{1}{2}\pi$, we have

$$\alpha_n = \frac{2}{\pi}\int_0^\pi f(x)\sin nx\, dx$$

$$= \frac{2}{\pi}\left(-\int_0^{\frac{1}{2}\pi}\sin nx\, dx + \int_{\frac{1}{2}\pi}^{\pi}\sin nx\, dx\right)$$

$$= \frac{2}{\pi}\left(\frac{\cos nx}{n}\Big|_0^{\frac{1}{2}\pi} - \frac{\cos nx}{n}\Big|_{\frac{1}{2}\pi}^{\pi}\right)$$

$$= \frac{2}{n\pi}(-1 + 2\cos\tfrac{1}{2}n\pi - \cos n\pi)$$

Therefore the Fourier series is

$$f(x) = \frac{2}{\pi}\sum_{n=1}^{\infty}\frac{1}{n}(-1 + 2\cos\tfrac{1}{2}n\pi - \cos n\pi)\sin nx$$

$$= -\frac{2}{\pi}(\tfrac{4}{2}\sin 2x + \tfrac{4}{6}\sin 6x + \tfrac{4}{10}\sin 10x + \cdots) \tag{i}$$

In Fig. 3 we show the function $f(x)$ and the first few partial sums of the Fourier series. From this figure it appears that the partial sums approximate $f(x)$ at all points in the open interval $0 < x < \pi$ except at $x = \frac{1}{2}\pi$. At $x = \frac{1}{2}\pi$, which is a point of discontinuity of $f(x)$, the partial sums do not approximate $f(\frac{1}{2}\pi) = 1$. It is easily seen that at $x = \frac{1}{2}\pi$ the Fourier series converges to zero, and therefore Eq. (i) does not hold at $x = \frac{1}{2}\pi$. We note that zero is the average value of the right- and left-hand limits of $f(x)$ at $x = \frac{1}{2}\pi$:

$$0 = \tfrac{1}{2}[f(\tfrac{1}{2}\pi + 0) + f(\tfrac{1}{2}\pi - 0)] = \tfrac{1}{2}[1 + (-1)]$$

It will be pointed out later that this type of behavior always occurs at a jump discontinuity.

A few remarks about the boundary points $x = 0$ and $x = \pi$. At these points we have each member of the orthogonal set $\sin nx$ equaling zero.† Therefore the Fourier series converges to zero and Eq. (i) does not hold for these points.‡ In general, if the function $f(x)$ satisfies the same boundary conditions as each member of the orthogonal set, the Fourier series will hold for the boundary points, otherwise not.

† Recall that these eigenfunctions arose as solutions of the boundary-value problem $y'' + \lambda y = 0$, $y(0) = y(\pi) = 0$.

‡ See also Sec. 13-9 for an explanation of what happens at the boundary points.

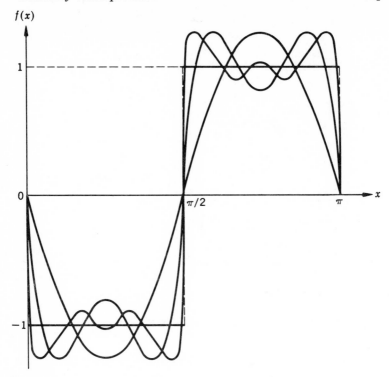

figure 3

In summary, the Fourier series in (i) above converges to the function

$$F(x) = \begin{cases} 0 & x = 0 \\ -1 & 0 < x < \frac{1}{2}\pi \\ 0 & x = \frac{1}{2}\pi \\ 1 & \frac{1}{2}\pi < x < \pi \\ 0 & x = \pi \end{cases}$$

We note that the only difference between $F(x)$ and $f(x)$ is at the boundary points $x = 0$ and $x = \pi$ and at the point of discontinuity $x = \frac{1}{2}\pi$. In other words, by redefining the original $f(x)$ at points of discontinuity and at boundary points, the Fourier series will converge to the function for all x in $0 \leq x \leq \pi$. This simple example illustrates what happens in the general case to be discussed later.

Problems 13-5

1. Derive Eq. (43).

2. Find the Fourier series for $f(x) \equiv x$ in $0 \leq x \leq \pi$ using the orthogonal set

$\{\cos nx\}$ $n = 0, 1, 2, \ldots$

Draw a graph of $f(x)$ and the first few partial sums.

3. Assuming that the necessary operations are legitimate, start with the Fourier series for an *orthonormal* set $\{\varphi_i\}$:

$$f(x) = \sum_{i=1}^{\infty} \alpha_i \varphi_i(x)$$

and show that

$$\int_a^b f^2(x)\, dx = \sum_{i=1}^{\infty} \alpha_i^2$$

where α_i are the Fourier coefficients.

4. *Least-squares approximation.* Let $\varphi_n(x)$ be an orthonormal set on $[a,b]$. Find the value of the coefficients A_k in the finite linear combination

$$\sum_{k=1}^{N} A_k \varphi_k(x) \qquad N \text{ fixed}$$

so that the value of the integral

$$J = \int_a^b [f(x) - \sum_{k=1}^{N} A_k \varphi_k(x)]^2\, dx$$

is a minimum. (This is called a *least-squares approximation*.) Hint: Expand J and show

$$J = \int_a^b f^2(x)\, dx + \sum_{k=1}^{N} A_k^2 - 2 \sum_{k=1}^{N} A_k \alpha_k$$

where α_k are the Fourier coefficients of f given by

$$\alpha_k = \int_a^b f(x)\varphi_k(x)\, dx$$

Then add and subtract $\sum_{k=1}^{N} \alpha_k^2$, complete the square, and obtain

$$J = \int_a^b f^2(x)\, dx - \sum_{k=1}^{N} \alpha_k^2 + \sum_{k=1}^{N} (A_k - \alpha_k)^2$$

From this last expression it is clear that the minimum occurs for $A_k = \alpha_k$. In other words, *the Fourier coefficients give the best approximation to $f(x)$, in the least-squares sense, compared to all other linear combinations.*

5. Using Prob. 4, show that

$$\sum_{k=1}^{N} \alpha_k^2 \leq \int_a^b f^2(x)\, dx$$

and prove that $\sum_{k=1}^{\infty} \alpha_k^2$ converges and that $\lim_{k \to \infty} \alpha_k = 0$. This shows that the sum of the squares of the Fourier coefficients *always converges*; and also that the Fourier coefficients must be arbitrarily small for large enough k.

13-6 Weight functions

For future use we generalize the concept of inner product and orthogonality introduced in Sec. 13-4. We say that two continuous functions $f(x)$ and $g(x)$ are orthogonal in the interval $a \leq x \leq b$ with respect to a continuous *weight function* $w(x)$ provided that

$$\int_a^b w(x)f(x)g(x)\,dx = 0 \tag{44}$$

We assume that $w(x) \geq 0$ in $a \leq x \leq b$ and that $w(x)$ is not identically zero. This is equivalent to defining a new inner product

$$\langle f,g \rangle_w = \int_a^b w(x)f(x)g(x)\,dx \tag{45}$$

This inner product has all the properties of the old inner product (see Prob. 1 below). Using this new inner product, we can easily introduce notions of orthogonal sets and Fourier-series expansions.

A set of continuous functions $\{\varphi_k(x)\}$ is called an orthogonal set in $a \leq x \leq b$ with respect to the weight function $w(x)$ provided that

$$\langle \varphi_k, \varphi_j \rangle_w = \int_a^b w(x)\varphi_k(x)\varphi_j(x)\,dx = 0 \qquad k \neq j \tag{46}$$

and $\{\varphi_k\}$ is called an orthonormal set provided that

$$\langle \varphi_k, \varphi_j \rangle_w = \int_a^b w(x)\varphi_k(x)\varphi_j(x)\,dx = \delta_{kj} \tag{47}$$

A complete orthonormal set is one having the property that the zero function is the only function orthogonal to all members of the set.

The Fourier series for a complete orthogonal set is

$$f(x) = \sum_{k=1}^{\infty} \alpha_k \varphi_k(x) \tag{48}$$

If we assume the validity of such an expansion, we can easily show that the coefficients α_k are given by

$$\alpha_k = \frac{\langle f, \varphi_k \rangle_w}{\langle \varphi_k, \varphi_k \rangle_w} = \frac{\displaystyle\int_a^b w(x)f(x)\varphi_k(x)\,dx}{\displaystyle\int_a^b w(x)\varphi_k^2(x)\,dx} \tag{49}$$

(See Prob. 2 below.)

Problems 13-6

1. Verify that the inner product defined by Eq. (45) satisfies:

a. $\langle f, f \rangle_w > 0 \qquad f \neq 0$
b. $\langle f, f \rangle_w = 0 \qquad f = 0$
c. $\langle f, g \rangle_w = \langle g, f \rangle_w$
d. $\langle \alpha f + \beta g, h \rangle_w = \alpha \langle f, h \rangle_w + \beta \langle g, h \rangle_w$

2. Provide the formal derivation of (49) from (48).
3. Show that the Tchebysheff polynomials defined by

$$T_0(x) = 1$$

$$T_n(x) = \frac{1}{2^{n-1}} \cos (n \arccos x) \qquad n = 1, 2, \ldots$$

form an orthogonal set with respect to the weight function $w(x) = (1 - x^2)^{-\frac{1}{2}}$ in the interval $-1 \leq x \leq 1$.

4. If $\{\varphi_k(x)\}$ form an orthogonal set with respect to a weight function $w(x)$, show that $\{\sqrt{w(x)}\varphi_k(x)\}$ form an orthogonal set in the ordinary sense (with weight function $= 1$).

5. *Complex inner product.* Let $f(x)$ and $g(x)$ be complex-valued functions of a real variable. Define an inner product as follows:

$$\langle f, g \rangle = \int_a^b f(x)\overline{g(x)} \, dx$$

where $\overline{g(x)}$ is the complex conjugate of g. Show that this inner product satisfies

$\langle f, f \rangle > 0 \qquad f \neq 0$
$\langle f, f \rangle = 0 \qquad f = 0$
$\langle f, g \rangle = \overline{\langle g, f \rangle}$
$\langle \alpha f + \beta g, h \rangle = \alpha \langle f, h \rangle + \beta \langle g, h \rangle$

where α and β are complex constants.

6. Show that the set of complex-valued functions

$$\{e^{inx}\} \qquad n = 0, \pm 1, \pm 2, \ldots$$

form an orthogonal set in $0 \leq x \leq 2\pi$, using the inner product of Prob. 5.

7. Referring to Probs. 5 and 6, derive formulas for the coefficients of Fourier series for complex-valued functions.

13-7   The Sturm-Liouville problem

Under certain conditions the set of eigenfunctions for an eigenvalue problem form an orthogonal set. For example, we have seen that the eigenfunctions for

$$y'' + \lambda y = 0$$
$$y(0) = y(L) = 0 \tag{50}$$

are the set of functions

$$\left\{ \sin \frac{n\pi x}{L} \right\} \qquad n = 1, 2, \ldots \tag{51}$$

which we have shown to be an orthogonal set. We now consider a general class of eigenvalue problems for which the eigenfunctions form an orthogonal set.

First we define a *self-adjoint*† operator as the operator defined by

$$L(y) = [p(x)y']' + q(x)y \tag{52}$$

where $p(x)$, $p'(x)$, and $q(x)$ are continuous, and $p(x) \neq 0$ in some interval $a \leq x \leq b$. We consider the eigenvalue problem with unmixed boundary conditions

$$\text{DE: } L(y) + \lambda w(x)y = 0 \qquad a \leq x \leq b$$
$$\text{BC: } \alpha_1 y(a) + \beta_1 y'(a) = 0 \qquad \alpha_1^2 + \beta_1^2 \neq 0 \tag{53}$$
$$\alpha_2 y(b) + \beta_2 y'(b) = 0 \qquad \alpha_2^2 + \beta_2^2 \neq 0$$

where $w(x) \geq 0$ is a continuous function, not the zero function, and L is the self-adjoint operator (52). The problem (53) is called a *Sturm-Liouville problem.*

Theorem 1 *Eigenfunctions of the problem (53) corresponding to different eigenvalues are orthogonal with respect to the weight function $w(x)$.*

Proof If $\lambda_n \neq \lambda_m$ are eigenvalues and y_n, y_m are the corresponding eigenfunctions (assuming they exist), we have

$$L(y_m) + \lambda_m w(x)y_m = 0$$
$$L(y_n) + \lambda_n w(x)y_n = 0$$

Multiplying the first equation by y_n and the second equation by $(-y_m)$ and adding, we obtain

$$y_n L(y_m) - y_m L(y_n) = (\lambda_n - \lambda_m)w(x)y_n y_m$$

† Self-adjoint operators have special properties (see Theorem 1 below). However, every second-order linear operator can be made self-adjoint by multiplying by a suitable factor (see Prob. 4 of this section).

Integrating both sides, we have

$$\int_a^b [y_n L(y_m) - y_m L(y_n)]\, dx = (\lambda_n - \lambda_m) \int_a^b w(x) y_n y_m\, dx \tag{54}$$

In Prob. 1 we show that

$$\int_a^b [y_n L(y_m) - y_m L(y_n)]\, dx = p(x)(y_n y'_m - y_m y'_n) \Big|_a^b \tag{55}$$

Equation (55) for a self-adjoint operator is known as the *Lagrange identity*.†
Using the boundary conditions in (53), we find (see Prob. 2)

$$p(x)(y_n y'_m - y_m y'_n) \Big|_a^b = 0 \tag{56}$$

Therefore the left-hand side of (54) vanishes and, if $\lambda_n \neq \lambda_m$,

$$\int_a^b w(x) y_n y_m\, dx = 0$$

which completes the proof.

We now discuss some extensions of Theorem 1 which are important in certain applications.

Periodic boundary conditions   Consider the boundary-value problem (53) with the boundary conditions replaced by the periodic (mixed) boundary conditions

1. $y(a) = y(b)$
2. $y'(a) = y'(b)$ $\tag{57}$

If we make the additional assumption that $p(a) = p(b)$, then the proof of Theorem 1 goes through in this case (see Prob. 3) so that again eigenfunctions corresponding to different eigenvalues are orthogonal with respect to the weight function $w(x)$.

Singular end points   If in the Sturm-Liouville problem (53) we have $p(a) = 0$, the proof of Theorem 1 will hold with the first boundary condition in (53) removed.   However, in this case it is possible that a solution $y(x)$ of the differential equation may approach infinity as $x \to a$ and that the integrals involving $y(x)$ may not exist.   Therefore for the case $p(a) = 0$ we require that the *solution $y(x)$ be bounded at $x = a$*.   With this condition we have again that eigenfunctions corresponding to different eigenvalues are orthogonal.   The condition that $y(x)$ be bounded at $x = a$ can be thought of as replacing the first boundary condition in (53).

† This identity does not hold if L is not self-adjoint (see Sec. 5-8).   Therefore Theorem 1 is not true if L is not self-adjoint.

In a similar way, if $p(b) = 0$, we can replace the second boundary condition in (53) by the requirement that $y(x)$ be bounded at $x = b$, and the conclusions of Theorem 1 will hold.

If both $p(a) = 0$ and $p(b) = 0$, then both boundary conditions in (53) can be omitted and replaced by the condition that the solution $y(x)$ be bounded at $x = a$ and $x = b$ and the conclusions of Theorem 1 will hold.

Problems 13-7

1. If L is the self-adjoint operator (52) and u and v are twice differentiable functions [not necessarily eigenfunctions of (53)], prove the Lagrange identity.

$$\int_{x_0}^{x_1} [uL(v) - vL(u)]\, dx = p(x)(uv' - vu') \bigg|_{x_0}^{x_1}$$

for any x_0, x_1.   Hint: Integrate the left side by parts.

2. Prove Eq. (56).

3. Prove Theorem 1 for the periodic boundary conditions (57) under the additional assumption that $p(a) = p(b)$.

4. a. Show that the operator

$$L(y) = a_0(x)y'' + a_1(x)y' + a_2(x)y$$

is self-adjoint if $a_1 = a_0'$.

b. Show that if L is not self-adjoint, it can be made self-adjoint by multiplying by the nonzero function

$$\frac{1}{a_0(x)} \exp\left[\int \frac{a_1(x)}{a_0(x)}\, dx \right], \qquad (\text{if } a_0(x) \neq 0)$$

5. Write the following differential equations in self-adjoint form:

a. $x^2 y'' + xy' + (x^2 - n^2)y = 0$

b. $(1 - x^2)y'' - 2xy' + n(n + 1)y = 0$

13-8   Theorems on eigenvalues and eigenfunctions

In this section we quote without proof some important theorems concerning the eigenvalues and eigenfunctions for the Sturm-Liouville problem:

$$\text{DE: } [p(x)y']' + [q(x) + \lambda w(x)]y = 0 \qquad a \leq x \leq b$$

$$\text{BC: } \alpha_1 y(a) + \beta_1 y'(a) = 0 \qquad \alpha_1^2 + \beta_1^2 \neq 0 \tag{58}$$

$$\alpha_2 y(b) + \beta_2 y'(b) = 0 \qquad \alpha_2^2 + \beta_2^2 \neq 0$$

We assume that, in the interval $a \leq x \leq b$, $p(x)$, $p'(x)$, $q(x)$, $w(x)$ are continuous, $p(x) \neq 0$, $w(x) \geq 0$, and $w(x) \not\equiv 0$.

Theorem 1 *The above Sturm-Liouville problem possesses an infinite number of real, nonnegative eigenvalues. The set of all eigenvalues can be ordered in an increasing sequence,*

$$0 \leq \lambda_1 < \lambda_2 < \lambda_3 < \cdots$$

such that $\lambda_n \to \infty$ as $n \to \infty$.

The fact that the eigenvalues are real is proved in Prob. 1.   However, the proof of the existence of an increasing sequence of eigenvalues tending to infinity requires more advanced methods.†

Theorem 2 *For each eigenvalue there exists only one eigenfunction* (up to a multiplicative constant).

The proof of this theorem was given in Sec. 13-2.

Theorem 3 *The set of eigenfunctions $\{\varphi_1, \varphi_2, \ldots\}$ corresponding to the eigenvalues form a complete orthogonal set in $a \leq x \leq b$ with respect to the weight function $w(x)$.*

Theorem 4 *If $f(x)$ is any piecewise smooth‡ function in $a \leq x \leq b$, then $f(x)$ can be expanded in a uniformly convergent Fourier series which converges to $f(x)$ at any point in the open interval $a < x < b$ where $f(x)$ is continuous and which converges to the average value of the right- and left-hand limits of $f(x)$ at any point of discontinuity in $a < x < b$. That is,*

$$\tfrac{1}{2}[f(x+0) + f(x-0)] = \sum_{n=1}^{\infty} \gamma_n \varphi_n(x) \qquad a < x < b \tag{59}$$

where

$$\gamma_n = \frac{\displaystyle\int_a^b w(x)f(x)\varphi_n(x)\,dx}{\displaystyle\int_a^b w(x)\varphi_n^2(x)\,dx} \tag{60}$$

The fact that the eigenfunctions form an orthogonal set was proved in Sec. 13-7.   However, the completeness of this orthogonal set and the convergence of the Fourier expansion are much more difficult and require more advanced methods. §

† See Sagan [31], chap. 5, Courant and Hilbert [13], chap. 5, or Birkhoff and Rota [4], chap. 11.

‡ $f(x)$ is piecewise smooth if $f(x)$ is piecewise continuous (continuous except for a finite number of jump discontinuities) and if $f'(x)$ is piecewise continuous.

§ See Sagan [31], chap. 4, Courant and Hilbert [13], chaps. 2 and 3, or Birkhoff and Rota [4], chap. 11.

Periodic boundary conditions   If all the conditions of problem (58) hold except that the boundary conditions are replaced by the periodic boundary conditions

$$y(a) = y(b)$$
$$y'(a) = y'(b)$$

$$(61)$$

and $p(a) = p(b)$, then Theorems 1, 3, and 4 remain true.   However, Theorem 2 does not hold.   It is possible that two linearly independent eigenfunctions correspond to the same eigenvalue.   An important case of this will be illustrated in Sec. 13-9.

Singular end points   If in the Sturm-Liouville problem (58) the coefficient $p(x)$ vanishes at the boundary point $x = a$, we may omit the first boundary condition in (58) and replace it by the condition that $y(x)$ be bounded at $x = a$ and the conclusions of Theorems 1, 2, 3, and 4 still hold.

Similarly if $p(b) = 0$, we may replace the second boundary condition by the requirement that $y(x)$ be bounded at $x = b$ and the conclusions of Theorems 1, 2, 3, and 4 still hold.   If both $p(a) = 0$ and $p(b) = 0$, then neither boundary condition in (58) is needed and the requirement that $y(x)$ be bounded at $x = a$ and $x = b$ will ensure that the theorems hold.

In the next three sections we shall discuss special cases of the Sturm-Liouville problem and illustrate the above theorems.

Problem 13-8

1. Prove that the eigenvalues of the Sturm-Liouville problem (58) are all real.   Hint: Assume $\lambda = \alpha + i\beta$ is an eigenvalue and $y = u + iv$ is the corresponding eigenfunction.   Separate the differential equation into real and imaginary parts.   By appropriate manipulation show that $\int_a^b (u^2 + v^2)p \, dx = 0$.

13-9   Ordinary Fourier series

The ordinary Fourier series are those in which the orthogonal sets are sinusoids.   We discuss the three most common cases.

Fourier sine series   The problem

$$\text{DE: } y'' + \lambda y = 0 \qquad 0 \le x \le L$$
$$\text{BC: } y(0) = y(L) = 0$$

$$(62)$$

is already in self-adjoint form with weight function $w(x) = 1$.   The theorems of Sec. 13-8 ensure the existence of eigenvalues and eigenfunctions.   However, these are easy to find directly as we have already done.   The eigenvalues and eigenfunctions are

$$\lambda_n = \frac{n^2\pi^2}{L^2} \quad \text{and} \quad \varphi_n = \sin\frac{n\pi x}{L} \quad n = 1, 2, \ldots \tag{63}$$

The normalizing factor needed in the Fourier-series expansion is

$$\int_0^L \sin^2\frac{n\pi x}{L}\,dx = \frac{L}{2} \tag{64}$$

The Fourier-series expansion for a piecewise smooth function $f(x)$ in $0 \le x \le L$ is

$$f(x) = \sum_{n=1}^\infty \alpha_n \sin\frac{n\pi x}{L} \tag{65}$$

where

$$\alpha_n = \frac{2}{L}\int_0^L f(x)\sin\frac{n\pi x}{L}\,dx \tag{66}$$

For a particular example of this expansion see Sec. 13-7.

We note that the right-hand side of (65) is a periodic function of period $2L$ and also that the right-hand side is an odd function† of x. Therefore, if this Fourier series converges to $f(x)$ in $0 < x < L$, then the Fourier series will converge for all x to the function $F(x)$ defined by

$$F(x) = \begin{cases} f(x) & 0 < x < L \\ -f(-x) & -L < x < 0 \end{cases} \tag{67}$$

$$F(x + 2L) = F(x)$$

The function $F(x)$ is called the *odd periodic extension* of $f(x)$ and is illustrated in Fig. 4. If $f(0) \ne 0$, then $F(x)$ will be discontinuous at $x = 0$, $\pm L$, $\pm 2L$, $\ldots$. At all of these points the Fourier series converges to zero. Also at all of these points the average of the right-hand and left-hand limits

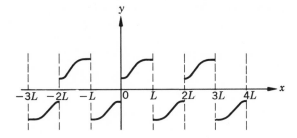

figure 4. *Odd periodic extension.*

† An odd function is a function satisfying $f(-x) = -f(x)$, i.e., a function symmetric about the origin.

of $F(x)$ is also zero.   Therefore the Fourier expansion

$$F(x) = \Sigma \alpha_n \sin \frac{n \pi x}{L} \tag{68}$$

$$\alpha_n = \frac{2}{L} \int_0^L f(x) \sin \frac{n \pi x}{L} \, dx \tag{69}$$

holds for all x with the usual convention regarding points of discontinuity.

Fourier cosine series   Consider the problem

DE: $y'' + \lambda y = 0$
BC: $y'(0) = y'(L) = 0$ $\qquad\qquad\qquad\qquad\qquad\qquad$ (70)

Again the theorems of Sec. 13-8 hold with weight function $w(x) = 1$.   It is easy to show that the eigenvalues and eigenfunctions are

$$\lambda_n = \frac{n^2 \pi^2}{L^2} \qquad \varphi_n = \cos \frac{n \pi x}{L} \qquad n = 0, 1, 2, \ldots \tag{71}$$

We note in particular that $\lambda_0 = 0$ is an eigenvalue and $\varphi_0 = 1$ is the corresponding eigenfunction.   The Fourier expansion is

$$f(x) = \alpha_0 + \sum_{n=1}^{\infty} \alpha_n \cos \frac{n \pi x}{L} \tag{72}$$

where by the usual process we find

$$\alpha_0 = \frac{1}{L} \int_0^L f(x) \, dx \tag{73}$$

$$\alpha_n = \frac{2}{L} \int_0^L f(x) \cos \frac{n \pi x}{L} \, dx \qquad n = 1, 2, \ldots \tag{74}$$

Note that α_0 *cannot* be obtained from α_n by setting $n = 0$.   The coefficient α_0 has a simple interpretation: it is the *average value of $f(x)$* in $0 \le x \le L$.

The right-hand side of (72) is a periodic function of period $2L$ and is also an even function.† Therefore the Fourier series will converge for all x (with the usual convention regarding points of discontinuity) to the *even periodic extension of $f(x)$* defined by

$$F(x) = \begin{cases} f(x) & 0 < x < L \\ f(-x) & -L < x < 0 \end{cases} \tag{75}$$
$$F(x + 2L) = F(x)$$

which is illustrated in Fig. 5.

† An even function is defined by $f(-x) = f(x)$, that is, a function symmetric about the y axis.

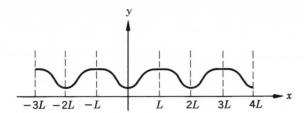

figure 5.   *Even periodic extension.*

Example 1   The Fourier cosine series of period 2π for

$$f(x) = x \qquad 0 \le x \le \pi$$

has coefficients

$$\alpha_0 = \frac{1}{\pi} \int_0^\pi x \, dx = \frac{\pi}{2}$$

For $n \ne 0$

$$\alpha_n = \frac{2}{\pi} \int_0^\pi x \cos nx \, dx = \frac{2}{\pi}\left(\frac{1}{n} x \sin nx \Big|_0^\pi - \frac{1}{n} \int_0^\pi \sin nx \, dx\right) = \begin{cases} 0 & n \text{ even} \\ -\dfrac{4}{n^2\pi} & n \text{ odd} \end{cases}$$

Therefore

$$f(x) = \frac{\pi}{2} - \frac{4}{\pi}\left(\cos x + \frac{\cos 3x}{3^2} + \frac{\cos 5x}{5^2} + \cdots\right)$$

For $x = 0$ we obtain

$$\frac{\pi^{2\bullet}}{8} = 1 + \frac{1}{3^2} + \frac{1}{5^2} + \cdots$$

a remarkably simple expression for π^2.

Full Fourier series   The boundary-value problem

$$\text{DE: } y'' + \lambda y = 0 \qquad 0 \le x \le 2L$$
$$\text{BC: } y(0) = y(2L) \tag{76}$$
$$\qquad y'(0) = y'(2L)$$

also generates an orthogonal set of sinusoids.

It is easy to show that the eigenvalues are

$$\lambda_n = \frac{n^2\pi^2}{L^2} \qquad n = 0, 1, 2, \ldots \tag{77}$$

and the eigenfunctions are

$$y_n = a_n \cos \frac{n\pi x}{L} + b_n \sin \frac{n\pi x}{L} \qquad n = 0, 1, 2, \ldots \tag{78}$$

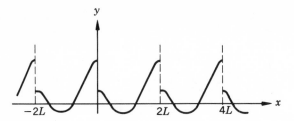

figure 6.   *Periodic extension of $f(x)$.*

We note that $\lambda_0 = 0$ is an eigenvalue with the eigenfunction $y_0 = a_0$, a constant.   For the other eigenvalues λ_n for $n > 0$ we note that there are two linearly independent eigenfunctions corresponding to the single eigenvalue λ_n.   This is due to the fact that the periodic boundary conditions in (76) are *mixed* boundary conditions.

The Fourier expansion in this case is

$$f(x) = a_0 + \sum_{n=1}^{\infty} \left(a_n \cos \frac{n\pi x}{L} + b_n \sin \frac{n\pi x}{L} \right) \tag{79}$$

where the coefficients are given by

$$a_0 = \frac{1}{2L} \int_0^{2L} f(x)\, dx \tag{80}$$

$$a_n = \frac{1}{L} \int_0^{2L} f(x) \cos \frac{n\pi x}{L}\, dx \tag{81}$$

$$b_n = \frac{1}{L} \int_0^{2L} f(x) \sin \frac{n\pi x}{L}\, dx \tag{82}$$

Again the constant term a_0 is the average value of $f(x)$ in the interval $0 \le x \le 2L$.   The right-hand side of (79) is a periodic function of period $2L$. Therefore the Fourier series (79) will converge for all x with the usual convention regarding points of discontinuity to the *periodic extension* of $f(x)$ defined by

$$\begin{aligned} F(x) &= f(x) \qquad 0 < x < L \\ F(x + 2L) &= f(x) \qquad \text{all } x \end{aligned} \tag{83}$$

which is illustrated in Fig. 6.

Example 2   Expand the function

$$f(x) = x \qquad 0 < x < 4$$

into a full Fourier series of period 4.

The coefficients are given by Eqs. (80), (81), and (82) with $2L = 4$ or $L = 2$:

$$a_0 = \tfrac{1}{4} \int_0^4 x \, dx = 2$$

$$a_n = \tfrac{1}{2} \int_0^4 x \cos \frac{n\pi x}{2} \, dx = 0 \qquad n > 0$$

$$b_n = \tfrac{1}{2} \int_0^4 x \sin \frac{n\pi x}{2} \, dx = -\frac{8}{n\pi}$$

Therefore the Fourier series is

$$f(x) = 2 - \frac{8}{\pi} \sum_1^\infty \frac{1}{n} \sin \frac{n\pi x}{2}$$

Problems 13-9

In Probs. 1 through 3 find the required Fourier series for the given functions. Sketch the graph of the function to which the Fourier series converges for two periods.

1. $f(x) = \begin{cases} 1 & 0 < x < \frac{1}{2}\pi \\ 0 & \frac{1}{2}\pi < x < \pi \end{cases}$

   a. cosine series of period 2π
   b. sine series of period 2π
   c. full Fourier series of period π

2. $f(x) = 1 \qquad 0 < x < 4$

   a. cosine series of period 8
   b. sine series of period 8
   c. full Fourier series of period 4

3. $f(x) = x^2 \qquad 0 < x < \pi$

   a. cosine series of period 2π
   b. sine series of period 2π

4. If $f(x)$ is periodic of period $2L$, show that the Fourier expansion is given by

$$f(x) = a_0 + \sum_{n=1}^\infty \left(a_n \cos \frac{n\pi x}{L} + b_n \sin \frac{n\pi x}{L} \right)$$

when

$$a_0 = \frac{1}{2L} \int_c^{c+2L} f(x) \, dx \qquad a_n = \frac{1}{L} \int_c^{c+2L} f(x) \cos \frac{n\pi x}{L} \, dx$$

$$b_n = \frac{1}{L} \int_c^{c+2L} f(x) \sin \frac{n\pi x}{L} \, dx$$

where c is any number. In other words the integrals may be evaluated over any interval of length $2L$.

5. Let $f(x)$ be defined in $-L \leq x \leq L$ and consider the periodic extension $F(x)$, of period $2L$.

a. Show that $F(x)$ has a Fourier series given by

$$F(x) = a_0 + \sum_{n=1}^{\infty} \left(a_n \cos \frac{n\pi x}{L} + b_n \sin \frac{n\pi x}{L} \right)$$

where

$$a_0 = \frac{1}{2L} \int_{-L}^{L} f(x)\, dx \qquad a_n = \frac{1}{L} \int_{-L}^{L} f(x) \cos \frac{n\pi x}{L}\, dx$$

$$b_n = \frac{1}{L} \int_{-L}^{L} f(x) \sin \frac{n\pi x}{L}\, dx$$

b. Specialize part a in case $f(x)$ is an odd function.
c. Specialize part b in case $f(x)$ is an even function.

In Probs. 6 through 10, find Fourier series of period 2π:

6. $f(x) = |x|$ $-\pi \leq x \leq \pi$
7. $f(x) = |\sin x|$ $-\pi < x \leq \pi$
8. $f(x) = x$ $-\pi < x < \pi$
9. $f(x) = 1$ $-\pi < x < \pi$
10. $f(x) = 2 + 3\cos x + 4\sin 10x$ $-\pi < x < \pi$

In Probs. 11 through 14 find the Fourier series of the lowest possible period for the indicated periodic functions.

11.

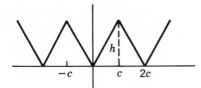

12.

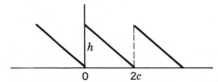

13.

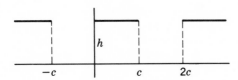

14.

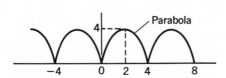

13-10   *Fourier-Bessel series*†

An important boundary-value problem involving the Bessel differential equation is

DE: $x^2 y'' + xy' + (\lambda^2 x^2 - n^2)y = 0 \qquad 0 \le x \le c$

BC: $y(c) = 0$ $\qquad\qquad\qquad\qquad\qquad\qquad\qquad\qquad$ (84)

$\qquad y(0)$ is finite

where n is a nonnegative integer. Since the differential equation can be written in the self-adjoint form

$$(xy')' + \frac{(\lambda^2 x^2 - n^2)}{x} y = 0 \qquad\qquad (85)$$

we see that problem (84) is a special case of the Sturm-Liouville problem where $p(x) = x$ and the weight function $w(x) = x$. Since $p(0) = 0$, no boundary condition is required at $x = 0$; the condition that $y(0)$ be finite can be considered as a hidden boundary condition.

The general solution of the equation can be written in terms of the Bessel functions of order n of the first and second kinds‡

$$y = AJ_n(\lambda x) + BN_n(\lambda x) \qquad\qquad (86)$$

Since N_n is unbounded at the origin,† we must set $B = 0$ to satisfy the second boundary condition in (84). To satisfy the first boundary condition, we must have

$$y(c) = AJ_n(\lambda c) = 0 \qquad\qquad (87)$$

Let α_i, $i = 1, 2, \ldots$ denote the positive roots of $J_n(\alpha) = 0$.§ Then the eigenvalues are given by

$$\lambda c = \alpha_i \qquad \text{or} \qquad \lambda = \lambda_i = \frac{\alpha_i}{c} \qquad i = 1, 2, \ldots \qquad (88)$$

and the corresponding eigenfunctions are

$$J_n(\lambda_i x) \qquad i = 1, 2, \ldots \qquad\qquad (89)$$

The set of functions

$$\{J_n(\lambda_i x)\} \qquad i = 1, 2, \ldots \qquad\qquad (90)$$

for fixed n form an orthogonal set in the interval $0 \le x \le c$ with respect to the weight functions x.

† See Sec. 15-3.
‡ See Sec. 7-4.
§ See Sec. 7-4.

The Fourier expansion in terms of this orthogonal set is

$$f(x) = \sum_{i=1}^{\infty} A_i J_n(\lambda_i x) \tag{91}$$

where the coefficients A_i are found by the usual process to be

$$A_i = \frac{\displaystyle\int_0^c x J_n(\lambda_i x) f(x)\, dx}{\displaystyle\int_0^c x[J_n(\lambda_i x)]^2\, dx} \tag{92}$$

In Prob. 1 below the normalizing factor in the denominator of (92) is evaluated to be

$$\int_0^c x[J_n(\lambda_i x)]^2\, dx = \frac{c^2}{2}[J_{n+1}(\lambda_i c)]^2 \tag{93}$$

Therefore the coefficients A_i are

$$A_i = \frac{2\displaystyle\int_0^c x J_n(\lambda_i x) f(x)\, dx}{c^2[J_{n+1}(\lambda_i c)]^2} \tag{94}$$

Example 1   Expand the function $f(x) = 1$ into a Fourier-Bessel series of the zeroth order in $0 < x < 1$.

The coefficients A_i are given by

$$A_i = \frac{2}{J_1^2(\lambda_i)} \int_0^1 x J_0(\lambda_i x)\, dx \tag{95}$$

where λ_i are roots of $J_0(\lambda_i) = 0$.   The above integral can be easily evaluated using the identity (see Sec. 7-4)

$$\frac{d}{dx}(x J_1(x)) = x J_0(x)$$

or

$$\int_0^x t J_0(t)\, dt = x J_1(x)$$

Letting $\lambda_i x = t$ in (95), we find

$$A_i = \frac{2}{\lambda_i J_1(\lambda_i)}$$

Therefore the Fourier-Bessel series is

$$f(x) = 2\sum_{i=1}^{\infty} \frac{J_0(\lambda_i x)}{\lambda_i J_1(\lambda_i)} \tag{96}$$

Problems 13-10

1. Derive (93) by performing the following steps. Note that $y(x) = J_n(\lambda_i x)$ satisfies

$$(xy')' + \left(\lambda_i^2 x - \frac{n^2}{x}\right) y = 0$$

Multiply both sides of the equation by $2xy'$ to find

$$\frac{d}{dx}(xy')^2 + (\lambda_i^2 x^2 - n^2)\frac{d}{dx}(y^2) = 0$$

Integrate both terms and use integration by parts to find

$$[(xy')^2 + (\lambda_i^2 x^2 - n^2)y^2]\Big|_0^c = 2\lambda_i^2 \int_0^c xy^2 \, dx$$

Noting that $y(c) = J_n(\lambda_i c) = 0$ for all i, obtain

$$\int_0^c x[J_n(\lambda_i x)]^2 \, dx = \tfrac{1}{2}c^2[J_n'(\lambda_i c)]^2$$

Use the relation (see Sec. 7.4) $zJ_n'(z) = nJ_n(z) - zJ_{n+1}(z)$ to obtain the desired result.

2. Derive the formula

$$x = 2\sum_{i=0}^{\infty} \frac{J_1(\lambda_i x)}{\lambda_i J_2(\lambda_i)} \qquad 0 \leq x < 1$$

where λ_i are the positive roots of $J_1(\lambda) = 0$.

13-11   Fourier-Legendre series

An important eigenvalue problem that occurs in connection with potential problems is

$$\text{DE: } [(1 - x^2)y']' + \lambda y = 0 \qquad -1 \leq x \leq +1$$
$$\text{BC: } y \text{ and } y' \text{ finite at } x = \pm 1 \tag{97}$$

This is a special case of the Sturm-Liouville problem (53) with $p(x) = 1 - x^2$, $q(x) = 0$, $w(x) = 1$. In Sec. 7-6 it was pointed out that the *only values of λ for which y and y' are finite at $x = \pm 1$ are*

$$\lambda = \lambda_n = n(n+1) \qquad n = 0, 1, 2, \ldots \tag{98}$$

Thus the eigenvalues are given by (98). The corresponding eigenfunctions are the Legendre polynomials $P_n(x)$, which were derived in Sec. 7-6.

According to the theorems of Sec. 13-8, the Legendre polynomials form a complete orthogonal set in the interval $-1 \leq x \leq 1$ with respect to the weight function $w(x) = 1$.   If $f(x)$ is a piecewise smooth function, then the following Fourier-Legendre expansion holds:

$$f(x) = \sum_{n=0}^{\infty} A_n P_n(x) \tag{99}$$

where we find by the usual process that

$$A_n = \frac{\displaystyle\int_{-1}^{1} f(x) P_n(x)\, dx}{\displaystyle\int_{-1}^{1} P_n^2(x)\, dx} \tag{100}$$

In Probs. 1 and 2 below we find that

$$\int_{-1}^{1} P_n^2(x)\, dx = \frac{2}{2n+1} \tag{101}$$

Therefore the coefficients A_n are given by

$$A_n = \frac{2n+1}{2} \int_{-1}^{1} f(x) P_n(x)\, dx \tag{102}$$

A useful alternative form of Eq. (102) can be derived using Rodrigues's formula (see Sec. 7-6)

$$P_n(x) = \frac{1}{2^n n!} \frac{d^n}{dx^n} [(x^2 - 1)^n] \tag{103}$$

Substituting (103) into (102), we obtain

$$A_n = \frac{2n+1}{2^{n+1} n!} \int_{-1}^{1} f(x) \frac{d^n}{dx^n} [(x^2 - 1)^n]\, dx \tag{104}$$

It is natural to integrate by parts to obtain [assuming $f'(x)$ is continuous]

$$A_n = \frac{2n+1}{2^{n+1} n!} \left\{ f(x) \frac{d^{n-1}}{dx^{n-1}} [(x^2 - 1)^n] \Big|_{-1}^{1} - \int_{-1}^{1} f'(x) \frac{d^{n-1}}{dx^{n-1}} [(x^2 - 1)^n]\, dx \right\} \tag{105}$$

The first term on the right side vanishes since the $(n-1)$st derivative of $(1 - x^2)^n$ will still contain a factor of $(1 - x^2)$.   Therefore we have

$$A_n = -\frac{2n+1}{2^{n+1} n!} \int_{-1}^{1} f'(x) \frac{d^{n-1}}{dx^{n-1}} [(1 - x^2)^n]\, dx \tag{106}$$

If $f(x)$ has n continuous derivatives, we can repeat this process and obtain

$$A_n = \frac{2n+1}{2^{n+1} n!} (-1)^n \int_{-1}^{1} f^{(n)}(x)(x^2 - 1)^n\, dx$$

or finally

$$A_n = \frac{2n+1}{2^{n+1}n!} \int_{-1}^{1} f^{(n)}(x)(1-x^2)^n \, dx \tag{107}$$

Example 1   Expand x^3 into a Fourier-Legendre series. Using Eq. (107), we have

$$A = \frac{2n+1}{2^{n+1}n!} \int_{-1}^{1} \left(\frac{d^n}{dx^n} x^3 \right)(1-x^2)^n \, dx$$

Therefore

$$A_0 = \tfrac{1}{2} \int_{-1}^{1} x^3 \, dx = 0$$

$$A_1 = \tfrac{3}{4} \int_{-1}^{1} 3x^2(1-x^2) \, dx = \tfrac{3}{5}$$

$$A_2 = 0$$

$$A_3 = \frac{7}{2^4 \cdot 4} \int_{-1}^{1} 6(1-x^2)^3 \, dx = \tfrac{2}{5}$$

$$A_n = 0 \qquad n > 3$$

Therefore

$$x^3 = \tfrac{2}{5}P_3(x) + \tfrac{3}{5}P_1(x)$$

Problems 13-11

1. Let $u = (x^2-1)^n$; show by successive integration by parts that:

a. $\displaystyle \int_{-1}^{1} u^{(n)}u^{(n)} \, dx = -\int_{-1}^{1} u^{(n-1)}u^{(n+1)} \, dx = \int_{-1}^{1} u^{(n-2)}u^{(n+2)} \, dx = \cdots$

$$= (-1)^n \int_{-1}^{1} uu^{(2n)} \, dx = (2n)! \int_{-1}^{1} (1-x)^n(1+x)^n \, dx$$

b. $\displaystyle \int_{-1}^{1} (1-x)^n(1+x)^n \, dx = \frac{n(n-1)\cdots 2 \cdot 1}{(n+1)(n+2)\cdots(2n)}$

$$\times \int_{-1}^{1} (1+x)^{2n} \, dx = \frac{(n!)^2}{(2n)!(2n+1)} 2^{2n+1}$$

2. Prove Eq. (101). Hint: Use Prob. 1 and Rodrigues's formula.

3. Show that $x^2 = \tfrac{1}{3}P_0(x) + \tfrac{2}{3}P_2(x)$.

4. If $f(x) = \begin{cases} 0 & -1 < x < 0 \\ 1 & 0 < x < 1 \end{cases}$

show that

$$f(x) = \tfrac{1}{2} + \tfrac{3}{4}x + \sum_{n=1}^{\infty} (-1)^n \frac{4n+3}{4n+4} \frac{(2n)!}{2^n(n!)^2} P_{2n+1}(x)$$

Hint: Use Sec. 7-6, Probs. 8 and 9.

13-12   *Nonhomogeneous boundary-value problems*

Most of this chapter has been concerned with homogeneous boundary-value problems. We now consider the nonhomogeneous problem

DE: $Ly = f(x)$ $a \leq x \leq b$

BC: $\alpha_1 y(a) + \beta_1 y'(a) = 0$ $\alpha_1^2 + \beta_1^2 \neq 0$ (108)

  $\alpha_2 y(b) + \beta_2 y'(b) = 0$ $\alpha_2^2 + \beta_2^2 \neq 0$

where $f(x)$ is continuous; L is the operator

$$Ly = p(x)y'' + q(x)y' + r(x)y$$

where p, q, r are continuous and $p \neq 0$ in $a \leq x \leq b$.

It is convenient to introduce the *boundary operators*†

$$B_a(y) \equiv \alpha_1 y(a) + \beta_1 y'(a) \tag{109a}$$

$$B_b(y) \equiv \alpha_2 y(b) + \beta_2 y'(b) \tag{109b}$$

For every differentiable function, the operator $B_a(y)$ assigns a unique number determined by the right-hand side of (109a), and similarly for $B_b(y)$. We note that these operators are *linear*; that is (see Prob. 1)

$$B_a(c_1 y_1 + c_2 y_2) = c_1 B_a(y_1) + c_2 B_a(y_2) \tag{110a}$$

$$B_b(c_1 y_1 + c_2 y_2) = c_1 B_b(y_1) + c_2 B_b(y_2) \tag{110b}$$

The boundary problem (108) can now be written in the compact form

$$Ly = f   B_a(y) = B_b(y) = 0 \tag{111}$$

We proceed to solve this problem. The general solution of $Ly = f$ is

$$y = c_1 y_1 + c_2 y_2 + y_p$$

where y_1 and y_2 are linearly independent solutions of $Ly = 0$ and y_p is a particular solution of $Ly = f$. A convenient y_p has been derived in Sec. 5-8 (see also Sec. 5-6, Prob. 7):

$$y_p = \int_a^x \frac{y_1(t)y_2(x) - y_1(x)y_2(t)}{p(t)W[y_1(t),y_2(t)]} f(t)\, dt \tag{112}$$

where $W[y_1,y_2]$ is the Wronskian. In case L is self-adjoint ($q = p'$), then $p(t)W[y_1(t),y_2(t)]$ is a constant (see Sec. 5-8, Prob. 4). We note also (see Prob. 2) that

$$y_p(a) = y_p'(a) = 0 \tag{113}$$

† Operators of this type which map functions into numbers are sometimes called *functionals*.

Now to satisfy the boundary conditions, we must have

$$B_a(y) = c_1 B_a(y_1) + c_2 B_a(y_2) + B_a(y_p) = 0$$
$$B_b(y) = c_1 B_b(y_1) + c_2 B_b(y_2) + B_b(y_p) = 0 \tag{114}$$

If we can find c_1, c_2 so that these equations are satisfied, then problem (111) *has a solution; if not, then the problem has no solution.*

We know that (114) has a unique solution if and only if

$$\Delta = \begin{vmatrix} B_a(y_1) & B_a(y_2) \\ B_b(y_1) & B_b(y_2) \end{vmatrix} \neq 0 \tag{115}$$

However, (115) holds if and only if the homogeneous problem $Ly = 0$, $B_a(y) = B_b(y) = 0$, has only the trivial solution (see Sec. 13-2). Therefore we have the following theorem.

Theorem 1 $Ly = f$, $B_a(y) = B_b(y) = 0$ *has a unique solution if and only if the corresponding homogeneous problem* $Ly = 0$, $B_a(y) = B_b(y) = 0$ *has only the trivial solution.*

In the case when the homogeneous problem has only the trivial solution, we shall obtain an elegant formula for the solution. We note first that, because of (113), we have

$$B_a(y_p) = 0 \tag{116}$$

It is a straightforward calculation to show (see Prob. 3) that

$$B_b(y_p) = \int_a^b \frac{B_b(y_2)y_1(t) - B_b(y_1)y_2(t)}{p(t)W[y_1(t),y_2(t)]} f(t)\, dt \tag{117}$$

Now in order to obtain a simple formula, we let $y_1(x)$ be a (nontrivial) solution of $L(y) = 0$ which satisfies the left-hand boundary condition $B_a(y_1) = 0$; we let $y_2(x)$ be a linearly independent solution of $Ly = 0$ which satisfies the right-hand boundary condition $B_b(y_2) = 0$. Equations (114) now simplify to

$$c_2 B_a(y_2) = 0$$
$$c_1 B_b(y_1) + B_b(y_p) = 0 \tag{118}$$

Now $B_a(y_2) \neq 0$; otherwise y_2 would be a nontrivial solution of $Ly = 0$, $B_a(y) = B_b(y) = 0$. Therefore $c_2 = 0$ and

$$c_1 = \frac{-B_b(y_p)}{B_b(y_1)} = \int_a^b \frac{y_2(t)f(t)\, dt}{p(t)W[y_1(t),y_2(t)]} \tag{119}$$

where we have used (117) and the fact that $B_b(y_2) = 0$.   We also note that $B_b(y_1) \neq 0$.   Therefore the solution of (111) is

$$y = c_1 y_1(x) + y_p = \int_a^b \frac{y_2(t)y_1(x)}{p(t)W[y_1(t),y_2(t)]}\, f(t)\, dt$$

$$+ \int_a^x \frac{y_1(t)y_2(x) - y_2(t)y_1(x)}{p(t)W[y_1(t),y_2(t)]} f(t)\, dt \quad (120)$$

We break the first integral into two parts to obtain

$$y = \int_a^x \frac{y_2(t)y_1(x)}{p(t)W[y_1(t),y_2(t)]}\, f(t)\, dt + \int_x^b \frac{y_2(t)y_1(x)}{p(t)W[y_1(t),y_2(t)]}\, f(t)\, dt$$

$$+ \int_a^x \frac{y_1(t)y_2(x) - y_2(t)y_1(x)}{p(t)W[y_1(t),y_2(t)]} f(t)\, dt \quad (121)$$

which simplifies to

$$y = \int_x^b \frac{y_2(t)y_1(x)}{p(t)W[y_1(t),y_2(t)]}\, f(t)\, dt + \int_a^x \frac{y_1(t)y_2(x)}{p(t)W[y_1(t),y_2(t)]}\, f(t)\, dt \quad (122)$$

We now introduce the *Green's function*†

$$g(x,t) = \begin{cases} \dfrac{y_2(t)y_1(x)}{p(t)W[y_1(t),y_2(t)]} & x \leq t \\[4mm] \dfrac{y_1(t)y_2(x)}{p(t)W[y_1(t),y_2(t)]} & t \leq x \end{cases} \quad (123)$$

so that (122) can be written in the compact form

$$y(x) = \int_a^b g(x,t)f(t)\, dt \quad (124)$$

Example 1 $y'' = f(x)$

$$y(0) = y(1) = 0$$

The homogeneous problem $y'' = 0$, $y(0) = y(1) = 0$ has only the trivial solution so that (124) applies.   We may take $y_1 = x$, $y_2 = x - 1$.   Since $p(x) \equiv 1$,

$$pW[y_1,y_2] = 1[x \cdot 1 - (x - 1) \cdot 1] = 1$$

† In general any function $g(x,t)$ with the property that a solution to the nonhomogeneous problem can be expressed as a definite integral like Eq. (124) is called a Green's function.   For the physical significance of the Green's functions and their extension to partial differential equations, see Sagan [31], chap. 9.

Therefore

$$g(x,t) = \begin{cases} x(t-1) & x \le t \\ t(x-1) & t \le x \end{cases}$$

and the solution to the problem is

$$y(x) = \int_0^1 g(x,t)f(t)\,dt = \int_0^x t(x-1)f(t)\,dt + \int_x^1 x(t-1)f(t)\,dt$$

This obviously satisfies the boundary conditions. It may easily be checked by differentiation that the differential equation is satisfied. (See Prob. 3.)

When the homogeneous problem has a nontrivial solution We now consider the solution of (111) if the corresponding homogeneous problem has a nontrivial solution. Let $y_1(x)$ be such a nontrivial solution; the most general solution is $cy_1(x)$ (see Sec. 12-2). Then $y_1(x)$ satisfies $B_a(y_1) = B_b(y_1) = 0$. Equations (114) become

$$\begin{aligned} c_2 B_a(y_2) &= 0 \\ c_2 B_b(y_2) + B_b(y_p) &= 0 \end{aligned} \tag{125}$$

Now since y_1 and y_2 are linearly independent, $B_a(y_2) \ne 0$ (see Prob. 4). Therefore $c_2 = 0$, and a solution of (125), and therefore of (111), exists if and only if $B_b(y_p) = 0$. Using (117) and $B_b(y_1) = 0$, we have

$$B_b(y_p) = B_b(y_2) \int_a^b \frac{y_1(t)f(t)\,dt}{p(t)W[y_1(t),y_2(t)]} = 0 \tag{126}$$

A solution of $Ly = f$, $B_a(y) = B_b(y) = 0$ exists if and only if (126) holds. If L is self-adjoint, we know that $p(t)W[y_1,y_2]$ is a constant. Therefore a solution of the self-adjoint problem $Ly = f$, $B_a(y) = B_b(y) = 0$ exists if and only if f satisfies

$$\int_a^b f(t)y_1(t)\,dt = 0 \tag{127}$$

To find an explicit formula for the solution when (126) holds, we have that the solution of (114) is $c_2 = 0$, c_1 arbitrary, so that the general solution of the boundary-value problem is

$$y = c_1 y_1(x) + y_p = c_1 y_1(x) + \int_a^x \frac{y_1(t)y_2(x) - y_2(t)y_1(x)}{p(t)W[y_1(t),y_2(t)]}f(t)\,dt \tag{128}$$

where c_1 is arbitrary. We now use an artifice; we note that

$$\begin{aligned} y = c_1 y_1(x) + \int_a^b \frac{y_1(x)y_2(t)}{p(t)W[y_1(t),y_2(t)]}\,f(t)\,dt \\ + \int_a^x \frac{y_1(t)y_2(x) - y_2(t)y_1(x)}{p(t)W[y_1(t),y_2(t)]}f(t)\,dt \end{aligned} \tag{129}$$

is still a solution since the middle term on the right is simply a constant times $y_1(x)$. We now perform the same steps as in getting Eq. (124) to obtain the solution

$$y = c_1 y_1(x) + \int_a^b g(x,t)f(t)\,dt \tag{130}$$

where the Green's function is given by

$$g(x,t) = \begin{cases} \dfrac{y_2(t)y_1(x)}{p(t)W[y_1(t),y_2(t)]} & x \le t \\[3mm] \dfrac{y_1(t)y_2(x)}{p(t)W[y_1(t),y_2(t)]} & x \ge t \end{cases} \tag{131}$$

Here $y_1(x)$ is a nontrivial solution of the homogeneous problem and y_2 is a linearly independent solution of $Ly = 0$.

Example 2 $y'' = f(x)$
$$y(0) = 0 \qquad y(1) - y'(1) = 0$$

The homogeneous problem has a nontrivial solution $y_1 = x$. We may take $y_2 = 1$. We find that $pW[y_1,y_2] = 1 \cdot (x \cdot 0 - 1 \cdot 1) = -1$. Therefore

$$g(x,t) = \begin{cases} -x & x \le t \\ -t & t \le x \end{cases}$$

and the solution is

$$y = c_1 x + \int_0^1 g(x,t)f(t)\,dt = \int_0^x (-t)f(t)\,dt + \int_x^1 (-x)f(t)\,dt + c_1 x$$

provided that f satisfies $\displaystyle\int_0^1 xf(x)\,dx = 0$. It may be easily checked that this satisfies the differential equation and the boundary condition.

Problems 13-12

1. Prove Eqs. (110a) and (110b).
2. Verify Eq. (113).   Hint: See Sec. 1-7, Eq. (42).
3. Verify directly that the solution of Example 1 satisfies the differential equation and the boundary condition.
4. Let $y_1(x)$ be a nontrivial solution of $Ly = 0$, $B_a(y) = B_b(y) = 0$.   Show that if y_1, y_2 are linearly independent, then $B_a(y_2) \ne 0$.
5. Fill in the steps in the derivation of (130) and (131).

In Probs. 6 through 9 find the Green's function.

6. $y'' = f$ $y(0) = 0$, $y'(1) = 0$

7. $y'' = f$ $y(-1) = 0$, $y(1) = 0$

8. $y'' + 4y = f$ $y(0) = y'(1) = 0$

9. $y'' + \lambda^2 y = f$ $y(0) = y(1) = 0$, $\lambda \neq$ eigenvalue

In Probs. 10 through 12 find the condition that f must satisfy in order to have a solution and find the solution.

10. $y'' = f(x)$ $y'(0) = y'(1) = 0$

11. $y'' + \pi^2 y = f$ $y(0) = y(1) = 0$

12. $y'' + \lambda_n^2 y = f$ $y(0) = y(1) = 0$, $\lambda_n = n\pi$ (n a positive integer)

Partial Differential Equations
of Mathematical Physics

14

14-1 Introduction

All of the differential equations which have been studied in the previous chapters have had one feature in common. They all described the state or the motion of a quantity which depended on only *one independent variable*. Hence any statement about a rate of change of this quantity involved *ordinary derivatives*, and the resulting differential equations were *ordinary differential equations*.

In this chapter we shall study equations which describe the motion or change of state of quantities which depend on *more than one independent variable*. In this case, statements of change necessarily involve *partial derivatives* and the resulting equations are called *partial differential equations*.

The study of partial differential equations has been one of the major projects of mathematicians for two centuries, and practically every field of modern science depends on these equations for the foundation of the quantitative aspects of its theory. The subject is vast and, in fact, many basic problems are still unsolved, but a few important equations have been solved completely and many general methods of solution have been developed.

In this chapter we shall discuss a few special equations that have far-reaching implications in mathematical physics. In particular, we examine the following equations:

$$y_{xx} = \frac{1}{a^2} y_{tt} \qquad \text{wave equation}$$

$$u_{xx} = \frac{1}{k} u_t \qquad \text{heat equation}$$

$$u_{xx} + u_{yy} = 0 \qquad \text{Laplace's equation}$$

The reader may wonder, and naturally so, why the study of partial differential equations should begin with such equations rather than with the presumably simpler *first-order* equations.

The choice is based on the following facts:

1. The second-order equations are far more important than the first-order equations in applications.

2. The methods of solving second-order equations are in many ways simpler and more closely allied to the techniques already developed in this book for the solution of *ordinary* differential equations.

3. Although the *theory* of first-order equations is the more fundamental, the authors feel that it would not be possible to treat it adequately in a book at this level.

Thus we shall limit ourselves to an investigation of the *methods* of *solving* second-order equations and a brief discussion of some of the physical implications of our solutions.†

14-2   *The vibrating string*‡

In this section we shall consider a simple physical problem which leads to a partial differential equation. The solution of the resulting equation will illustrate an important general method of solution, the method of separation of variables.

Derivation of the equation Suppose that a thin flexible string of mass m and length L is tightly stretched horizontally between two supports (see Fig. 1). If the string is given a small vertical displacement and released, it will vibrate up and down (dotted line in Fig. 1). A point P on the string, at a distance x from the left end of the string, will vibrate vertically. That is, its vertical displacement y will vary with the time t. Thus, in general, the displacement y will depend on both the position x and the time t. That is,

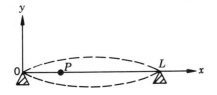

figure 1. *Vibrating string.*

† The methods and results in this chapter (and in Chap. 15) will be mostly heuristic. For a more rigorous development see, for example, Churchill [8].

‡ Compare the discussion in Sec. 11-8.

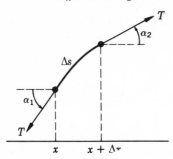

figure 2

y will be a function of x and t which we shall designate by $y(x,t)$. We shall usually assume that $y(x,t)$ has continuous second partial derivatives.

In order to derive the differential equation for $y(x,t)$, we shall make a few simplifying assumptions. We assume that the only force that is exerted on the string is the constant tension T lb. That is, we neglect the effects of gravity, friction, etc. We assume that the mass per unit length δ is constant along the string ($\delta = m/L$). Finally, we assume that the displacement $y(x,t)$ is small in such a way that the angle of inclination α of the displacement curve is always small (see Fig. 2).

Consider a typical element, of length Δs, of the string between x and $x + \Delta x$ (see Fig. 2). The vertical component F of the force on the element is

$$F = T \sin \alpha_2 - T \sin \alpha_1$$

Since we have assumed α to be small, $\sin \alpha \sim \tan \alpha$ and we have approximately

$$F = T(\tan \alpha_2 - \tan \alpha_1)$$

or, by the definition of the partial derivative y_x,

$$F = T[y_x(x + \Delta x, t) - y_x(x,t)] \tag{1}$$

The mass of the element is $\delta \, \Delta s$, where Δs is the length. Our assumptions ensure that $\Delta s \sim \Delta x$, and hence the mass is approximately $\delta \, \Delta x$. Applying Newton's law, we have

$$T[y_x(x + \Delta x, t) - y_x(x,t)] = (\delta \, \Delta x)y_{tt}(x,t) \tag{2}$$

where we assume that the element is small enough so that the vertical acceleration at all points of the element is approximately $y_{tt}(x,t)$.

Dividing Eq. (2) by $\delta \, \Delta x$, we have

$$\frac{T}{\delta}\left[\frac{y_x(x + \Delta x, t) - y_x(x,t)}{\Delta x}\right] = y_{tt}(x,t) \tag{3}$$

If we now take the limit as $\Delta x \to 0$, the expression in brackets becomes the partial derivative of the function y_x, that is, y_{xx}, and we have

$$\frac{T}{\delta}\, y_{xx}(x,t) = y_{tt}(x,t) \tag{4}$$

Since T/δ is positive, we denote it by

$$a^2 = \frac{T}{\delta} \tag{5}$$

and we have finally the partial differential equation for the displacement of a vibrating string:

$$y_{xx}(x,t) = \frac{1}{a^2}\, y_{tt}(x,t) \tag{6}$$

This equation is also referred to as the *one-dimensional wave equation* and is used to describe a wide variety of physical phenomena.† The quantity a has the dimensions of a velocity, and it will be shown below that it represents the velocity of propagation of a disturbance in the medium in which Eq. (6) holds. For example, a small disturbance at a point x on the string will travel along the string with velocity a. See page 456.

The mathematical significance of Eq. (6) as a typical *hyperbolic equation* is discussed in Sec. 14-5.

In order to be a solution of the problem which we have stated, the displacement $y(x,t)$ must satisfy several conditions in addition to Eq. (6). First, the ends of the string must remain fixed; that is, y must be zero at $x = 0$ and $x = L$ for all t. Hence we have the *boundary conditions*

$$y(0,t) = 0 \quad\text{for}\quad t \geq 0 \tag{7}$$

and

$$y(L,t) = 0 \quad\text{for}\quad t \geq 0 \tag{8}$$

The displacement will also depend on the manner in which the string has been set in motion. If we assume that the string has an initial displacement curve $y = f(x)$ and that it is released from rest,‡ we have the *initial conditions*

$$y(x,0) = f(x) \quad\text{for}\quad 0 \leq x \leq L \tag{9}$$

and

$$y_t(x,0) = 0 \quad\text{for}\quad 0 \leq x \leq L \tag{10}$$

† E.g., see Budak, Samarskii, and Tikhonov [6].

‡ The more general situation, in which the string is given an initial velocity $g(x)$, is easily treated by a slight modification of the method which is developed in this section. See Probs. 3 and 4.

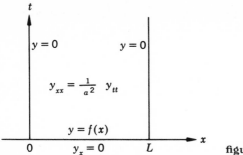

figure 3

Examination of these conditions reveals that the function $f(x)$ is not entirely arbitrary. For instance, we must have $f(0) = f(L) = 0$. Further restrictions on $f(x)$ will be imposed as we develop the solution. These restrictions on $f(x)$ will then ensure that there exists a unique solution to Eq. (6) which satisfies the initial conditions (7) and (8) and the boundary conditions (9) and (10).

The problem which is posed in Eqs. (6) through (10) is represented diagrammatically in Fig. 3.

Solution by separation of variables   In order to solve the mathematical problem which we have stated above, we shall appeal to physical intuition for some indication as to a method of procedure.

The simplest possible motion of the string would be a periodic vibration in which each point on the string moves in simple harmonic motion and in which all points on the string have the same period and frequency and are in phase throughout the motion.† If the initial displacement is given by a function $y = X(x)$, then the particle at x will execute harmonic vibrations with amplitude $X(x)$; that is, the displacement would be represented by a function of the form

$$y(x,t) = X(x) \cos \omega t \tag{11}$$

Such a motion is shown schematically in Fig. 4, page 452.

In general, then, we might expect the simpler motions of the string to be represented by periodic functions $T(t)$ of time with the amplitude $X(x)$, depending on the position x; i.e., we expect solutions of the form

$$y(x,t) = X(x)T(t) \tag{12}$$

If we can find solutions of this form, then, since the differential equation (6) is *linear and homogeneous*, we can combine such solutions to produce more complicated motions.

† Such motions are called *normal modes*. See Secs. 8-3 and 11-3 and also Eq. (37) below.

In order to find solutions of (6) in the form of (12), we substitute into (6) and obtain

$$X''(x)T(t) = \frac{1}{a^2} X(x)T''(t) \tag{13}$$

where primes denote *ordinary* derivatives. Dividing both sides of Eq. (13) by $X(x)T(t)$, we have

$$\frac{X''(x)}{X(x)} = \frac{1}{a^2} \frac{T''(t)}{T(t)} \tag{14}$$

The left-hand side of this equation depends only on x and *not* on t, whereas the right-hand side depends only on t and not on x. But the only condition under which two such functions can be equal, as x and t vary independently, is that both functions are equal to a constant k. That is, we have

$$\frac{X''(x)}{X(x)} = k \qquad \frac{1}{a^2} \frac{T''(t)}{T(t)} = k \tag{15}$$

or

$$X''(x) - kX(x) = 0 \tag{16}$$

and

$$T''(t) - a^2 k T(t) = 0 \tag{17}$$

Thus the problem of finding a solution to the *partial* differential equation (6) has been reduced to the problem of solving a pair of *ordinary* differential equations (16) and (17). That is, we have *separated the variables* x and t. If we can find solutions $X(x)$ and $T(t)$ to Eqs. (16) and (17), then the function $y = XT$ will be a solution to (6).†

If the product $y = XT$ is to be a solution to our problem, it must also satisfy the boundary conditions (7) and (8). That is, XT must vanish for all t when $x = 0$ and $x = L$. Hence we must have

$$X(0) = 0 \tag{18}$$
$$X(L) = 0 \tag{19}$$

since the only other alternative is the trivial case $T(t) \equiv 0$. These relations are *boundary conditions* on X and, together with the differential equation (16), form a *boundary-value problem* for $X(x)$. (See Secs. 13-2 and 13-9.)

Equation (16) is a linear homogeneous differential equation with constant coefficients, and we could write down the solution immediately. However, the form of the solution depends on whether k is positive or negative. If k

---

† The converse is not necessarily true. That is, not all solutions of (6) can be written in the form $y = XT$.

were positive, the solution would be

$$X(x) = c_1 \exp(\sqrt{k}x) + c_2 \exp(-\sqrt{k}x) \tag{20}$$

But a little algebra shows that conditions (18) and (19) then yield the trivial solution $X(x) \equiv 0$.

On the other hand, *if k is negative,*† that is, if

$$k = -\lambda^2 \tag{21}$$

where λ is real, then the differential equation (16) for $X(x)$ becomes

$$X''(x) + \lambda^2 X(x) = 0 \tag{22}$$

and the solution is

$$X(x) = c_1 \cos \lambda x + c_2 \sin \lambda x \tag{23}$$

Applying condition (18), we have $c_1 = 0$ and hence

$$X(x) = c_2 \sin \lambda x \tag{24}$$

In order to satisfy the other boundary condition (19), we must have

$$c_2 \sin \lambda L = 0$$

Thus either $c_2 = 0$, which again implies $X(x) \equiv 0$, or

$$\sin \lambda L = 0 \tag{25}$$

Equation (25) will be satisfied *if and only if*

$$\lambda L = n\pi \qquad n = 0, \pm 1, \pm 2, \ldots \tag{26}$$

that is, if

$$\lambda = \frac{n\pi}{L} \qquad n = 0, \pm 1, \pm 2, \ldots \tag{27}$$

Hence we see that the boundary-value problem [Eq. (22) together with the conditions (18) and (19)] for $X(x)$ can have a nontrivial solution *only* for a special set of values of the parameter λ. These values of λ, satisfying (27), are called *eigenvalues* and the resulting solutions

$$X_n(x) = c \sin \frac{n\pi x}{L} \tag{28}$$

are called eigensolutions or *eigenfunctions*. From Eq. (28) we see that the case $n = 0$ leads to the trivial result $X(x) \equiv 0$, whereas the values $n = -1$, $-2, \ldots$ simply give the negatives of the solutions for $n = +1, +2, \ldots$.

† The case $k = 0$ is treated in Prob. 1.

Hence we shall retain only those values $\lambda = \lambda_n$ where

$$\lambda_n = \frac{n\pi}{L} \qquad n = 1, 2, \ldots \tag{29}$$

Equation (17) for $T(t)$ now becomes

$$T''(t) + a^2\lambda_n^2 T(t) = 0 \tag{30}$$

with solutions

$$T(t) = c_3 \cos a\lambda_n t + c_4 \sin a\lambda_n t \tag{31}$$

or

$$T(t) = c_3 \cos \frac{an\pi t}{L} + c_4 \sin \frac{an\pi t}{L} \tag{32}$$

Applying the second initial condition (10) to the function $y = X(x)T(t)$, we have $X(x)T'(0) = 0$ for all x in $0 \le x \le L$, and hence we must have

$$T'(0) = 0 \tag{33}$$

Therefore $c_4 = 0$ and $T(t)$ reduces to

$$T(t) = c_3 \cos \frac{an\pi t}{L} \tag{34}$$

We have determined the following set of functions which are solutions of (6), (7), (8), and (10):

$$y_n(x,t) = c_n \sin \frac{n\pi x}{L} \cos \frac{an\pi t}{L} \qquad n = 1, 2, \ldots \tag{35}$$

where the constants c_n can be chosen independently for each value of n.

The only condition which still has to be fulfilled is the initial condition (9). Before examining the method of solution for a general $f(x)$, some interesting special cases will be considered.

Suppose the initial displacement of the string is given by

$$y(x,0) = f(x) = A \sin \frac{\pi x}{L} \tag{36}$$

This condition can be satisfied by an expression of the form of Eq. (35) if we take $n = 1$ and $c_1 = A$. This yields the function

$$y(x,t) = A \sin \frac{\pi x}{L} \cos \frac{a\pi t}{L} \tag{37}$$

This function satisfies all the conditions (6), (7), (8), and (10) in addition to the special initial condition (36). The appearance of the vibrating string

figure 4. $y(x,t) = A \sin \dfrac{\pi x}{L} \cos \dfrac{a\pi t}{L}$

in this case is shown in Fig. 4 for various times t. This type of motion is called the *first normal mode of vibration*.

In fact, the solutions given by Eq. (35) are *normal modes* for every value of n. The normal modes for $n = 2$ and $n = 3$ are sketched in Fig. 5. Thus the functions given by Eq. (35) are solutions to our problem in the special case in which the initial condition (9) is of the form

$$y(x,0) = f(x) = c_n \sin \frac{n\pi x}{L}$$

for any $n = 0, \pm 1, \pm 2, \ldots .$†

The motions described by Eq. (35) are periodic in t with circular frequency

$$\omega_n = \frac{an\pi}{L} \tag{38}$$

These frequencies ω_n are called the *natural frequencies* of the system.

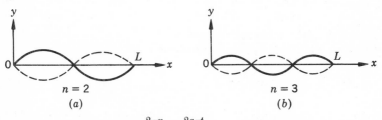

figure 5.   (a) $y(x,t) = B \sin \dfrac{2\pi x}{L} \cos \dfrac{2a\pi t}{L}$

(b) $y(x,t) = C \sin \dfrac{3\pi x}{L} \cos \dfrac{3a\pi t}{L}$

† Note that the case $n = 0$ implies the trivial solution $y \equiv 0$ which would result if the string is given no initial displacement or initial velocity.

Fourier-series solution   If the initial displacement $y = f(x)$ of the string is not exactly in the form of one of the eigenfunctions $\sin n\pi x/L$, then the problem cannot be solved with a single function of the form (35).   However, if $f(x)$ can be represented by a linear combination of eigenfunctions, that is, if

$$y(x,0) = f(x) = \sum_{n=1}^{N} A_n \sin \frac{n\pi x}{L} \tag{39}$$

then, because the problem is linear, the solution may be written as

$$y(x,t) = \sum_{n=1}^{N} A_n \sin \frac{n\pi x}{L} \cos \frac{an\pi t}{L} \tag{40}$$

The reader should check that $y(x,t)$ in Eq. (40) actually satisfies (6), (7), (8), (9), and (10) for the function $f(x)$ in (39).   See Prob. 8.

In general the function $f(x)$ cannot be expressed as a finite sum of the form (39).   However, if f is continuous and piecewise smooth† in the interval $0 \le x \le L$, and if $f(0) = f(L) = 0$, then f can be represented by an *infinite Fourier sine series*‡ of the form

$$f(x) = \sum_{n=1}^{\infty} A_n \sin \frac{n\pi x}{L} \tag{41}$$

The coefficients A_n are given by

$$A_n = \frac{2}{L} \int_{0}^{L} f(x) \sin \frac{n\pi x}{L} dx \tag{42}$$

Now, the initial condition (9) requires that $y(x,0) = f(x)$ and hence, if $f(x)$ is given by the infinite series (41), it is natural to extend the finite summation in Eq. (40) to express the displacement $y(x,t)$ in the form of an infinite series

$$y(x,t) = \sum_{n=1}^{\infty} A_n \sin \frac{n\pi x}{L} \cos \frac{an\pi t}{L} \tag{43}$$

If we substitute $t = 0$ in (43) we obtain

$$y(x,0) = \sum_{n=1}^{\infty} A_n \sin \frac{n\pi x}{L}$$

and hence the initial condition (9) will be satisfied if the coefficients A_n are chosen to be the Fourier coefficients of $f(x)$ as given by Eq. (42).   Note that $y_t(x,0) = 0$ and hence the second initial condition (10) is also satisfied.

The expression for $y(x,t)$ in Eq. (43), with A_n given by (42), is a *formal* solution to the problem posed in Eqs. (6) through (10).   A rigorous proof

† See Sec. 13-8, Theorem 4, p. 425.
‡ See Sec. 13-9 for a discussion of Fourier sine series.

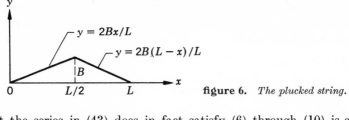

figure 6.   *The plucked string.*

that the series in (43) does in fact satisfy (6) through (10) is given, for example, in Churchill [8]. The convergence of (43) is discussed further at the end of this section. See the footnote on p. 457.

Example 1   *The plucked string.* Suppose the string is given the initial displacement shown in Fig. 6. That is,

$$f(x) = \begin{cases} \dfrac{2Bx}{L} & 0 \le x \le L/2 \\ \dfrac{2B(L-x)}{L} & L/2 \le x \le L \end{cases}$$

The displacement $y(x,t)$ is given by (43) with the coefficients

$$A_n = \frac{2}{L} \int_0^{L/2} \frac{2Bx}{L} \sin \frac{n\pi x}{L} \, dx + \frac{2}{L} \int_{L/2}^{L} \frac{2B(L-x)}{L} \sin \frac{n\pi x}{L} \, dx$$

or

$$A_n = \frac{8B}{n^2 \pi^2} \sin \frac{n\pi}{2}$$

But $\sin \frac{1}{2} n\pi = 0$ when $n = 2k$ $(k = 1, 2, \ldots)$, and $\sin \frac{1}{2} n\pi = (-1)^{k-1}$ when $n = 2k - 1$. Hence

$$A_n = \begin{cases} 0 & \text{if } n = 2k \\ \dfrac{8B(-1)^{k-1}}{\pi^2 (2k-1)^2} & \text{if } n = 2k - 1 \end{cases}$$

and the displacement $y(x,t)$ can be written in the form

$$y(x,t) = \frac{8B}{\pi^2} \sum_{k=1}^{\infty} \frac{(-1)^{k-1}}{(2k-1)^2} \sin \frac{(2k-1)\pi x}{L} \cos \frac{(2k-1)a\pi t}{L} \tag{44}$$

It should be noted that for $t = 0$ this expression becomes the Fourier expansion of $f(x)$. Also, in (44) the factor $1/(2k-1)^2$ ensures absolute and uniform convergence of the series in the interval $0 \le x \le L$.

D'Alembert's solution   It is easy to verify that the functions $y = F(x + at)$ and $y = G(x - at)$ satisfy the wave equation

$$y_{xx} = \frac{1}{a^2} y_{tt} \tag{45}$$

if F and G are any functions with two continuous derivatives. This fact suggests that the wave equation (45) might be simplified by introducing the new independent variables†

$$r = x + at \qquad \text{and} \qquad s = x - at \tag{46}$$

In fact, if we set $y(x,t) = Y(r,s)$, we have

$$y_x = Y_r r_x + Y_s s_x = Y_r + Y_s$$
$$y_{xx} = (Y_r + Y_s)_r r_x + (Y_r + Y_s)_s s_x = Y_{rr} + 2Y_{rs} + Y_{ss}$$

and similarly

$$y_{tt} = a^2(Y_{rr} - 2Y_{rs} + Y_{ss})$$

Substitution into (45) yields the equation

$$Y_{rs} = 0 \tag{47}$$

But this equation can be solved directly since it is equivalent to the requirement that $(Y_r)_s = 0$. Hence we must have

$$Y_r = f(r) \tag{48}$$

where f is an arbitrary continuously differentiable function. Equation (48) in turn implies that

$$Y = \int f(r)\,dr + G(s) \tag{49}$$

The integral of $f(r)$ is an arbitrary twice continuously differentiable function of r, say $F(r)$, and hence we can write the general solution of Eq. (47) in the form

$$Y = F(r) + G(s) \tag{50}$$

But then, returning to the original variables, the general solution (that is, *all* twice continuously differentiable solutions) of the wave equation (45) can be written in the form

$$y(x,t) = F(x + at) + G(x - at) \tag{51}$$

where F and G are arbitrary functions having two continuous derivatives.

This result, Eq. (51), represents one of the very rare instances in which a useful general solution of a partial differential equation can be written down explicitly. The usefulness of the solution depends on the fact that we shall here be able to determine F and G in such a way as to satisfy the other conditions, both boundary and initial.

† See Sec. 14-5 for a further discussion of the mathematical theory behind this change of variables.

The initial conditions (9) and (10) require that $y(x,0) = f(x)$, $y_t(x,0) = 0$†
for $0 \leq x \leq L$. From Eq. (51) we have then

$$F(x) + G(x) = f(x) \tag{52}$$

$$a F'(x) - a G'(x) = 0 \tag{53}$$

On integrating (53), we have

$$F(x) - G(x) = C \tag{54}$$

where C is an arbitrary constant.

From Eqs. (52) and (54) we obtain

$$\begin{aligned} F(x) &= \tfrac{1}{2}[f(x) + C] \\ G(x) &= \tfrac{1}{2}[f(x) - C] \end{aligned} \tag{55}$$

Substituting back into (51), we have

$$y(x,t) = \tfrac{1}{2}[f(x + at) + f(x - at)] \tag{56}$$

This expression gives the general solution to the initial-value problem

$$\text{PDE: } y_{xx} = \frac{1}{a^2} y_{tt} \tag{57}$$

$$\text{IC: } y(x,0) = f(x) \qquad y_t(x,0) = 0$$

It is instructive to examine the solution (56) before attempting to impose the boundary conditions.

In physical terms Eq. (56) states that the displacement y of the string is a combination of the displacements $\tfrac{1}{2}f(x + at)$ and $\tfrac{1}{2}f(x - at)$ where $f(x)$ is the initial displacement of the string. The function $f(x + at)$ can be interpreted physically as a *wave* with shape $f(x)$ traveling to the *left* along the x axis with speed a. This can be seen as follows. Consider some special point (e.g., a maximum) on the initial displacement curve at $x = x_0$. The function $f(x + at)$ reaches this maximum when $x + at = x_0$ or $x = x_0 - at$. Hence the x coordinate of this point changes with time according to the law $dx/dt = -a$. That is, it moves to the left with constant speed a. A similar argument shows that $f(x - at)$ represents a wave with shape $f(x)$ traveling to the right with speed a.

Therefore the resultant displacement y as given by Eq. (56) consists of the sum of two waves traveling in opposite directions. Each wave has the shape of the initial-displacement curve with half the amplitude.

Example 2   Suppose that a long string is given the initial displacement shown in Fig. 7a. The subsequent motion of the string is indicated in Fig. 7b, c, d for time intervals of $\tfrac{1}{2}a$.

† For the case of a general initial velocity $g(x)$, see Probs. 9 and 10.

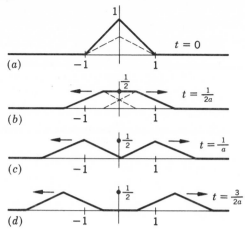

figure 7

In order to complete the solution of the vibrating-string problem, including the boundary conditions (7) and (8), we note that Eq. (56) includes *all* solutions to (6), (9), and (10). Hence it includes the Fourier-series solution (43). To see this, we apply the trigonometric identity

$$(\sin Au)(\cos Av) = \tfrac{1}{2}[\sin A(u + v) + \sin A(u - v)] \tag{58}$$

and (43) becomes

$$y(x,t) = \tfrac{1}{2} \sum_{n=1}^{\infty} A_n \left[\sin \frac{n\pi}{L} (x + at) + \sin \frac{n\pi}{L} (x - at) \right] \tag{59}$$

or

$$y(x,t) = \tfrac{1}{2} \sum_{n=1}^{\infty} A_n \sin \frac{n\pi}{L} (x + at) + \tfrac{1}{2} \sum_{n=1}^{\infty} A_n \sin \frac{n\pi}{L} (x - at) \tag{60}$$

Thus y is represented as the sum of a function of $x + at$ and a function of $x - at$. The expression

$$\sum_{n=1}^{\infty} A_n \sin \frac{n\pi}{L} (x + at) \tag{61}$$

with the coefficients A_n given by (42) is precisely the Fourier expansion of $f(x + at)$ if $0 \le x + at \le L$. For values outside this range (61) represents the odd periodic extension† of $f(x + at)$. A similar argument holds for $f(x - at)$.

Therefore, if we define a function f_1 to be the odd periodic extension of f, then the complete solution of our problem can be written simply

$$y(x,t) = \tfrac{1}{2}[f_1(x + at) + f_1(x - at)]‡ \tag{62}$$

† See Sec. 13-9.

‡ It is important to note that the convergence of the Fourier series for f implies the convergence of the series (60), and hence also (43), for $y(x,t)$.

Problems 14-2

1. Show that the boundary-value problem

$$X'' + k^2 X = 0 \qquad X(0) = X(L) = 0$$

has only the trivial solution $X \equiv 0$ for the case $k = 0$.

2. Suppose that the vibrating string described in the text has length $L = \pi$ and is released from rest with initial displacement $y(x,0) = \sin x + 2 \sin 2x$. Set up the problem and solve it *completely*, showing all steps.

3. Assume that the function $g(x)$ has a Fourier expansion and solve the following:

PDE: $y_{xx}(x,t) = \dfrac{1}{a^2} y_{tt}(x,t)$

BC: $y(0,t) = 0 \qquad y(L,t) = 0$

IC: $y(x,0) = 0 \qquad y_t(x,0) = g(x)$

This represents a vibrating string which starts from its equilibrium position with initial velocity $g(x)$.

4. Show how the solution to Prob. 3 can be combined with the solution to the problem in the text in order to solve the general problem

PDE: $y_{xx}(x,t) = \dfrac{1}{a^2} y_{tt}(x,t)$

BC: $y(0,t) = 0 \qquad y(L,t) = 0$

IC: $y(x,0) = f(x) \qquad y_t(x,0) = g(x)$

5. Solve the problem of the plucked string if the length $L = \pi$ and the initial displacement of the midpoint of the string is $B = \frac{1}{2}$. Carry out *all* steps in the solution and then check your result with Eq. (44) in the text.

6. Suppose that the string described in the text is given the initial displacement $y(x,0) = f(x) = x(L - x)$ and is released from rest. Set up the problem and solve it completely, showing all steps.

7. Use the method of separation of variables to solve the problem

PDE: $u_{xx} - u_{tt} = 0$

BC: $u_x(0,t) = 0 \qquad u(1,t) = 0 \qquad t \geq 0$

IC: $u(x,0) = f(x) \qquad u_t(x,0) = 0 \qquad 0 \leq x \leq 1$

8. Verify by direct computation that Eq. (40) satisfies (6), (7), (8), (10), and the form of (9) as given by (39).

9. In the D'Alembert solution of the wave equation replace the initial conditions by $y(x,0) = 0$, $y_t(x,0) = g(x)$ and obtain the following general solution:

$$y(x,t) = \frac{1}{2a} \int_{x-at}^{x+at} g(s)\, ds$$

10. Using the D'Alembert solution of the text and the result of Prob. 9, solve

PDE: $y_{xx} = \dfrac{1}{a^2}\, y_{tt}$

IC: $y(x,0) = f(x)$ $y_t(x,0) = g(x)$

14-3   Heat conduction

Physical background   Whenever there is a difference in temperature between two parts of a solid object, there will be a flow of heat from the hotter part to the cooler. This fact is essentially the content of the second law of thermodynamics. Implied in the statement is the fact that the temperature u (degrees centigrade) is a function of both position and time. Hence the variations of u will be governed by a partial differential equation.

We shall examine first the simple special case in which the heat conductor is a cylinder (e.g., a bar or a wire) parallel to the x axis. If the lateral surface of the cylinder is insulated, and if the ends are perpendicular to the x axis and are kept at constant temperature, then the temperature is a function of x and t only, i.e., $u = u(x,t)$. We also assume that there is no source of heat inside the cylinder. The differential equation governing the changes in u can be derived from the following two laws of physics, which we assume to be known.†

1. The heat content Q of a solid of mass m and specific heat c is

$$Q = cmu \tag{63}$$

if the entire body is at temperature u.

2. The rate at which heat flows *out* of a body through a plane surface of area A is given by

$$-KA\frac{\partial u}{\partial n} \tag{64}$$

where K is the thermal conductivity (assumed to be a positive constant) and $\partial u/\partial n$ is the outward normal derivative.‡

These two laws will be applied to the thin slice of the cylinder between x and $x + \Delta x$ (see Fig. 8). If the mass density of the solid is ρ, then the mass Δm of the slice is $\Delta m = \rho A \, \Delta x$, where A is the (constant) cross-sectional area.

† For a more complete discussion of these laws see, for example, Churchill [8].

‡ A normal derivative is defined as the rate of change of a quantity in a direction perpendicular to a surface at a point.

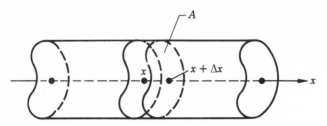

figure 8

Equation (63) implies that the heat content of the slice is

$$Q = c\rho A(\Delta x)u(x,t) \tag{65}$$

where we assume that the temperature at all points in the slice is approximately $u(x,t)$.

At the right-hand boundary of the slice, i.e., at $x + \Delta x$, the outer normal derivative is simply the derivative with respect to x. Hence the rate of heat flow out through the face at $x + \Delta x$ is given by

$$-KAu_x(x + \Delta x, t) \tag{66}$$

Similarly, at the left face, at x the outward normal derivative of u is $-\partial u/\partial x$ and hence the rate of heat loss equals

$$KAu_x(x,t) \tag{67}$$

Expressions (66) and (67) represent the rate at which heat flows *out* of the element. Hence, if we change the signs of both expressions and add, we obtain the rate of *increase* of heat in the slice.

$$KA[u_x(x + \Delta x, t) - u_x(x,t)] \tag{68}$$

From Eq. (65) we can obtain another expression for the time rate of change Q_t of heat content in the slice, i.e.,

$$Q_t(x,t) = c\rho Au_t(x,t)\,\Delta x \tag{69}$$

Combining the results of (68) and (69), we have

$$c\rho Au_t(x,t)\,\Delta x = KA[u_x(x + \Delta x, t) - u_x(x,t)] \tag{70}$$

Dividing by $c\rho A\,\Delta x$ and taking the limit as $\Delta x \to 0$ yields

$$u_t(x,t) = \frac{K}{c\rho}\,u_{xx}(x,t) \tag{71}$$

The positive constant $K/c\rho = k$ is called the *thermal diffusivity*,† and Eq.

† If x is measured in feet and time t in seconds, then k has the units of square feet per second.

(71) can be written in the standard form of

$$u_{xx}(x,t) = \frac{1}{k} u_t(x,t) \tag{72}$$

This is the one-dimensional heat equation† and is the prototype of *parabolic equations* (see Sec. 14-5).

In order to determine the temperature uniquely, additional information is needed.   One set of conditions which leads to a correctly posed problem is the following.‡   Suppose that the ends of our conductor (say at $x = 0$ and $x = L$) are both maintained at a constant temperature of $0°$.   If in addition to these boundary conditions we also know the temperature distribution at some instant, say $t = 0$, then the temperature at all points of the conductor is determined for all later time.   To summarize, the problem we have posed is the following: Find a function $u(x,t)$ which satisfies

$$\text{PDE: } u_{xx} = \frac{1}{k} u_t \qquad \begin{matrix} 0 < x < L \\ t > 0 \end{matrix} \tag{73}$$

$$\text{BC: } u(0,t) = 0 \qquad t \geq 0 \tag{74}$$

$$u(L,t) = 0 \qquad t \geq 0 \tag{75}$$

$$\text{IC: } u(x,0) = f(x) \qquad 0 \leq x \leq L. \tag{76}$$

where $f(x)$ is a given function of x.   In the next subsection we shall solve this problem and examine some of the physical implications of our solution.

Solution by separation of variables   The problem which is posed in Eqs. (73) through (76) can be solved by exactly the same *method* that was used in solving the vibrating-string problem of Sec. 14-2.   That is, we seek solutions of the form

$$u(x,t) = X(x)T(t) \tag{77}$$

Substituting in (73) and dividing by $X(x)T(t)$, we have

$$\frac{X''(x)}{X(x)} = \frac{1}{k}\frac{T'(t)}{T(t)} \tag{78}$$

As in Eq. (14), each side of (78) depends on a different independent variable, and hence (78) can hold identically only if both sides are constant, say $-\lambda^2$.§

---

† This equation is identical in form to the *one-dimensional diffusion equation* as well as the *skin-effect* equation and many other equations which are important in physics.   E.g., see Pipes [29], chap. 18.

‡ For some other possible conditions see Probs. 1 through 5.

§ The constant is chosen to be negative, since otherwise solutions would not exist.   See Sec. 14-2.

This yields the two ordinary differential equations

$$X''(x) + \lambda^2 X(x) = 0 \tag{79}$$

$$T' + \lambda^2 kT = 0 \tag{80}$$

In terms of the function $u = X(x)T(t)$ the boundary conditions (74) and (75) become

$$X(0) = X(L) = 0 \tag{81}$$

Hence the boundary-value problem for $X(x)$ has the eigensolutions

$$X_n = c_n \sin \frac{n\pi}{L} x \qquad n = 1, 2, \ldots \tag{82}$$

Solving (80) for $T(t)$, we have

$$T(t) = ce^{-(n\pi/L)^2 kt} \tag{83}$$

Hence a set of functions satisfying the differential equation (73) and the boundary conditions (74) and (75) is given by

$$u_n(x,t) = c_n e^{-(n\pi/L)^2 kt} \sin \frac{n\pi x}{L} \tag{84}$$

Finally, in order to satisfy the initial condition (76), we are again faced with the necessity of combining solutions of the form of (84). In fact, if the initial heat distribution $f(x)$ can be represented in the Fourier series

$$f(x) = \sum_{n=1}^{\infty} c_n \sin \frac{n\pi}{L} x \tag{85}$$

where

$$c_n = \frac{2}{L} \int_0^L f(x) \sin \frac{n\pi}{L} x \, dx \tag{86}$$

then the formal solution of our problem (73) through (76) is

$$u(x,t) = \sum_{n=1}^{\infty} c_n e^{-(n\pi/L)^2 kt} \sin \frac{n\pi}{L} x \tag{87}$$

where the constants c_n are given by (86) so that the initial condition $u(x,0) = f(x)$ will be satisfied. It can be shown† that the expression (87) actually represents the solution to the problem as long as the initial temperature function is piecewise smooth.‡

† E.g., see Churchill [8].
‡ See Sec. 13-8, Theorem 4.

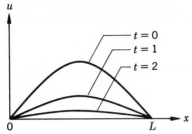

figure 9. *Graph of* $u = e^{-\pi^2 kt/L^2} \sin \dfrac{\pi x}{L}$

In order to investigate the physical meaning of our solution, it will simplify matters somewhat if we examine a single eigensolution, e.g.,

$$u_1(x,t) = c_1 e^{-(\pi^2/L^2)kt} \sin \frac{\pi x}{L}$$

This function represents the temperature in a cylinder which had initial temperature distribution

$$u_1(x,0) = c_1 \sin \frac{\pi}{L} x$$

The exponential time factor indicates that the temperature at each point will decrease steadily toward zero, the temperature at the ends. See Fig. 9.

The general situation exhibits similar behavior. That is, the time factors tend to smooth out any temperature differences and the body approaches a smooth steady-state temperature distribution. See Sec. 14-4.

Problems 14-3

1. Show directly that $u(x,t)$ as given by Eq. (87) satisfies Eqs. (73) through (76). Assume that the series in (87) may be differentiated termwise.

2. Solve the problem of the text if the face at $x = L$ is kept at temperature A. *Hint:* Let $u(x,t) = v(x,t) + (A/L)x$ where $v(x,t)$ can be obtained by using the solution (87) in the text. Show that the constants c_n are obtained by replacing $f(x)$ by $f(x) - Ax/L$ in Eq. (86).

3. If an infinite slab of homogeneous material bounded by the planes $x = 0$ and $x = \pi$ has an initial temperature $f(x)$ which depends on x alone, then the temperature at any time will be of the form $u(x,t)$. This function will satisfy the heat equation $u_{xx} = (1/k)u_t$ and the condition $u(x,0) = f(x)$. Suppose also that the face at $x = 0$ is kept at zero while the face at $x = \pi$ is thermally insulated. From Eq. (64) the condition at $x = \pi$ becomes $u_x(\pi,t) = 0$. Use separation of variables to find $u(x,t)$.

4. Solve Prob. 3 by considering a slab with faces at $x = 0$ and $x = 2\pi$ both kept at temperature zero. Use an initial temperature that is symmetric with respect to $x = \pi$ so that the condition $u_x(\pi,t) = 0$ will be satisfied.

5. Solve the problem of the text if $L = \pi$, $u(\pi,t) = A$, and if the initial temperature is a step function

$$f(x) = \begin{cases} 0 & \text{if } 0 \le x < \frac{1}{2}\pi \\ A & \text{if } \frac{1}{2}\pi \le x \le \pi \end{cases}$$

Make a rough sketch similar to Fig. 9 showing the temperature distribution at several times.   See Prob. 2.

6. The integral function $I(t) = \frac{1}{2} \int_0^\pi [u(x,t)]^2 \, dx$ is called the *energy integral*. If $u(x,t)$ is a solution of

$$u_{xx} = \frac{1}{k} u_t \qquad u(0,t) = u(\pi,t) = 0 \qquad u(x,0) = 0$$

then $I(t)$ can be used to prove that $u(x,t) \equiv 0$ everywhere.   Show that $I(t)$ is a nonincreasing, nonnegative function of t with $I(0) = 0$.   Hence conclude $I(t) \equiv 0$ and therefore $u \equiv 0$.

7. Use the result of Prob. 6 to prove the *uniqueness* of the solution to the problem:

a. $u_{xx} = \dfrac{1}{k} u_t$

b. $u(0,t) = f(t) \qquad u(\pi,t) = g(t)$

c. $u(x,0) = h(x)$

Hint: Assume there are two solutions u and v and show that the function $w = u - v$ satisfies the conditions of Prob. 6.

8. Show that the function

$$u = \frac{1}{2\sqrt{\pi k t}} e^{-(x-s)^2/4kt}$$

satisfies the heat equation $u_{xx} = u_t/k$.   Assume that differentiation under the integral sign is permissible and show that the following function also satisfies $u_{xx} = u_t/k$:

$$u(x,t) = \frac{1}{2\sqrt{\pi k t}} \int_{-\infty}^\infty f(s) e^{-(x-s)^2/4kt} \, ds$$

where $f(s)$ is bounded, i.e., $|f(s)| < M$ for some constant M and all s.   Next, change variables by substituting

$$\sigma = \frac{x - s}{\sqrt{4kt}}$$

and show that

$$u(x,t) = \frac{1}{\sqrt{\pi}} \int_{-\infty}^\infty f(x - 2\sigma\sqrt{kt}) e^{-\sigma^2} \, d\sigma$$

Use the fact that $\displaystyle\int_{-\infty}^\infty e^{-\sigma^2} \, d\sigma = \sqrt{\pi}$ to show that $\displaystyle\lim_{t \to 0^+} u(x,t) = f(x)$.

14-4 Laplace's equation

The elliptic equation†

$$u_{xx} + u_{yy} = 0 \qquad (88)$$

is known as the two-dimensional Laplace equation. The derivation of this equation is more natural in the three-dimensional form, which is given in Sec. 15-2. Equation (88) is also called the *potential equation* since it is satisfied by the gravitational potential, electrostatic potential, velocity potential in hydrodynamics, etc. For example, if a long thin wire is perpendicular to the xy plane and passes through the origin, then the gravitational potential at a point $P(x,y)$ is given by‡

$$u(x,y) = -km \ln r \qquad m, k \text{ const} \qquad (89)$$

where

$$r \equiv \sqrt{x^2 + y^2} \qquad (90)$$

To show that u in Eq. (89) satisfies Laplace's equation (88), we compute

$$u_x = u_r r_x = -\frac{km}{r}\frac{x}{r} = -\frac{kmx}{r^2}$$

$$u_{xx} = -\frac{km}{r^2} + \frac{2kmx}{r^3}\frac{x}{r} = \frac{km}{r^2}\left(-1 + \frac{2x^2}{r^2}\right) \qquad (91)$$

Similarly

$$u_{yy} = \frac{km}{r^2}\left(-1 + \frac{2y^2}{r^2}\right) \qquad (92)$$

On adding (91) and (92), we see that (88) is satisfied, except at the origin.

Laplace's equation is also satisfied by the steady-state temperature in a two-dimensional region (see Chap. 15). For example, suppose that a rectangular plate ($ABCD$ in Fig. 10) is subjected to the following temperature distribution: $u = f(x)$ on the lower edge and $u = 0$ on the other three sides. In Chap. 15 it is shown that the temperature u at any point in the rectangle will approach a steady state $u(x,y)$ as $t \to \infty$. This steady-state temperature $u(x,y)$ satisfies

$$\text{PDE: } u_{xx} + u_{yy} = 0 \qquad \begin{matrix} 0 < x < \pi \\ 0 < y < 1 \end{matrix} \qquad (93)$$

$$\text{BC: } u(0,y) = u(\pi,y) = 0 \qquad 0 \le y \le 1 \qquad (94)$$

$$\text{IC: } u(x,0) = f(x) \qquad u(x,1) = 0 \qquad 0 \le x \le \pi \qquad (95)$$

† See Sec. 14-5 for a definition of the term *elliptic equation.*
‡ See Kellogg [22].

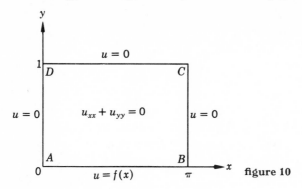

figure 10

Solving by separation of variables, we assume

$$u = X(x) Y(y)$$

Substitution into (93) yields

$$\frac{X''}{X} = -\frac{Y''}{Y} = -\lambda^2 \dagger \tag{96}$$

and hence

$$X'' + \lambda^2 X = 0 \tag{97}$$

$$Y'' - \lambda^2 Y = 0 \tag{98}$$

Equation (97) together with the boundary conditions (94) yields eigenfunctions

$$X_n(x) = \sin nx \qquad n = 1, 2, \ldots \tag{99}$$

with the eigenvalues $\lambda = n$. Equation (98) then has solutions

$$Y_n(y) = \sinh n(y + c) \tag{100}$$

and since $Y_n(1) = 0$, we have $c = -1$ and

$$Y_n(y) = \sinh n(y - 1) \tag{101}$$

Thus we have solutions of the form

$$u_n(x,y) = c_n \sin nx \sinh n(y - 1) \tag{102}$$

and, in order to satisfy the condition $u(x,0) = f(x)$, we can take formally

$$u(x,y) = \sum_{n=1}^{\infty} c_n \sin nx \sinh n(y - 1) \tag{103}$$

† See discussion following Eq. (21) above.

By setting $y = 0$, we see that we must have

$$f(x) = \sum_{n=1}^{\infty} c_n \sinh(-n) \sin nx \tag{104}$$

Thus the quantities $c_n \sinh(-n)$ are the Fourier sine coefficients of $f(x)$. That is,

$$c_n = -\frac{2}{\pi \sinh n} \int_0^\pi f(x) \sin nx \, dx \tag{105}$$

With these coefficients c_n, Eq. (103) then gives the solution to the problem (93), (94), and (95).

The problem (93), (94), and (95), which has just been solved, is a special case of a general problem which is known as the *interior Dirichlet problem*: Find a function $u(x,y)$ which satisfies $u_{xx} + u_{yy} = 0$ in some region R of the xy plane (e.g., a rectangle) and which assumes the value f on the boundary of R, where f is a given continuous function. In the *exterior Dirichlet problem* the function u satisfies $u_{xx} + u_{yy} = 0$ *outside* some region R and equals f on the boundary. Another important type of boundary-value problem is the so-called Neumann problem where the *normal derivative* $\partial u/\partial n$ is prescribed on the boundary.

The Dirichlet problem for a circle   *Poisson's solution.*   The steady-state temperature (or the potential) in a circular disk of radius R satisfies the following conditions:

$$u_{xx} + u_{yy} = 0 \qquad \text{for} \qquad x^2 + y^2 < R^2 \tag{106}$$

$$u(x,y) = f(\theta) \qquad \text{for} \qquad x^2 + y^2 = R^2 \tag{107}$$

where $f(\theta)$ is a given continuous function of the angular position θ of a point on the circumference of the circle. Hence $f(\theta)$ must be a periodic function of θ with period 2π, i.e.,

$$f(\theta) = f(\theta + 2\pi)$$

Since the boundary condition is given on a circle, it is more convenient to phrase the whole problem in a compatible system of coordinates. That is, we shall transform the problem into polar coordinates r and θ, so that the boundary can be expressed simply as $r = R$.† The differential equation can be transformed by means of the chain rule (see Prob. 4), and our problem is now that of finding a function $u(r,\theta)$ which satisfies

$$u_{rr} + \frac{1}{r} u_r + \frac{1}{r^2} u_{\theta\theta} = 0 \qquad \text{if } r < R \tag{108}$$

$$u(R,\theta) = f(\theta) \tag{109}$$

† The fact that the equation of the boundary involves only one of the coordinates is necessary for the success of the method of separation of variables.

In order to separate variables, we seek a solution of the form

$$u(r,\theta) = P(r)T(\theta) \tag{110}$$

Substitution into (108) and division by PT yield

$$\frac{P''}{P} + \frac{1}{r}\frac{P'}{P} + \frac{1}{r^2}\frac{T''}{T} = 0 \tag{111}$$

or

$$r^2\frac{P''}{P} + r\frac{P'}{P} = -\frac{T''}{T} = \lambda^2 \tag{112}$$

where the constant λ^2 is chosen positive to produce periodic functions $T(\theta)$. In fact, we have

$$T(\theta) = A\cos n\theta + B\sin n\theta \tag{113}$$

with $\lambda = n$, a nonnegative integer (why?), and the equation for P becomes

$$r^2 P'' + rP' - n^2 P = 0 \tag{114}$$

This is a Euler equation (see Sec. 5-7) with solution

$$P = r^n\dagger \tag{115}$$

The eigensolutions of our problem are then

$$u_n(r,\theta) = r^n(A_n\cos n\theta + B_n\sin n\theta) \tag{116}$$

Now, suppose $f(\theta)$ has the Fourier expansion

$$f(\theta) = \sum_{n=0}^{\infty} a_n\cos n\theta + b_n\sin n\theta \tag{117}$$

with

$$a_0 = \frac{1}{2\pi}\int_0^{2\pi} f(\varphi)\,d\varphi$$

$$a_n = \frac{1}{\pi}\int_0^{2\pi} f(\varphi)\cos n\varphi\,d\varphi \qquad n = 1, 2, \ldots \tag{118}$$

$$b_n = \frac{1}{\pi}\int_0^{2\pi} f(\varphi)\sin n\varphi\,d\varphi \qquad n = 1, 2, \ldots$$

Then the boundary condition (109) will be satisfied by a function

$$u(r,\theta) = \sum_{n=0}^{\infty} u_n(r,\theta) \tag{119}$$

---

† Equation (114) also has the solution $P = r^{-n}$ when $n > 0$, but this function becomes unbounded at $r = 0$, the center of the circle. For the case $n = 0$ see Prob. 6.

where the coefficients A_n and B_n of u_n in (116) are given by $A_n = R^{-n}a_n$, $B_n = R^{-n}b_n$. Thus the solution can be written†

$$u(r,\theta) = \frac{1}{\pi} \int_0^{2\pi} \left[\tfrac{1}{2} + \sum_{n=1}^{\infty} \left(\frac{r}{R}\right)^n \cos n\theta \cos n\varphi \right.$$
$$\left. + \left(\frac{r}{R}\right)^n \sin n\theta \sin n\varphi \right] f(\varphi)\, d\varphi \qquad (120)$$

or

$$u(r,\theta) = \frac{1}{\pi} \int_0^{2\pi} \left[\tfrac{1}{2} + \sum_{n=1}^{\infty} \left(\frac{r}{R}\right)^n \cos n(\theta - \varphi) \right] f(\varphi)\, d\varphi \qquad (121)$$

Now, using the fact that $\cos x = \tfrac{1}{2}(e^{ix} + e^{-ix})$, we have

$$u(r,\theta) = \frac{1}{\pi} \int_0^{2\pi} \left[\tfrac{1}{2} + \tfrac{1}{2}\sum_{n=1}^{\infty} \left(\frac{r}{R} e^{i(\theta-\varphi)}\right)^n + \tfrac{1}{2}\sum_{n=1}^{\infty} \left(\frac{r}{R} e^{-i(\theta-\varphi)}\right)^n \right] f(\varphi)\, d\varphi \qquad (122)$$

But these infinite series are just geometric series and can be summed by using

$$\sum_{n=1}^{\infty} z^n = \frac{1}{1-z} - 1 \qquad \text{if } |z| < 1‡ \qquad (123)$$

Hence, since

$$\left| \frac{r}{R} e^{i(\theta-\varphi)} \right| = \frac{r}{R} < 1 \qquad \text{if } r < R$$

(122) becomes

$$u(r,\theta) = \frac{1}{\pi} \int_0^{2\pi} \left[-\frac{1}{2} + \frac{1}{1-(r/R)e^{i(\theta-\varphi)}} + \frac{1}{1-(r/R)e^{-i(\theta-\varphi)}} \right] f(\varphi)\, d\varphi \qquad (124)$$

and a little algebra reduces this to

$$u(r,\theta) = \frac{1}{2\pi} \int_0^{2\pi} \frac{R^2 - r^2}{R^2 + r^2 - 2Rr \cos(\theta - \varphi)} f(\varphi)\, d\varphi \qquad (125)$$

This result is known as *Poisson's integral*, and it gives the solution§ to the Dirichlet problem for a circle in a useful form. Since (125) holds for all $r < R$ and u is assumed to be continuous in the circle plus boundary, we have in particular

$$u(0,\theta) = \frac{1}{2\pi} \int_0^{2\pi} f(\varphi)\, d\varphi = \frac{1}{2\pi} \int_0^{2\pi} u(R,\varphi)\, d\varphi \qquad (126)$$

† Assuming that it is permissible to interchange the order of integration and summation.

‡ See Appendix B.

§ We have established (125) only as a formal solution. That it is actually a solution is somewhat difficult to show. E.g., see Petrovsky [28], sec. 29.

Since the integral on the right-hand side of (126) can be interpreted as the mean value of f on the circumference, Eq. (126) yields the following general property of solutions to Laplace's equation.

The mean-value property   If u is a solution of $u_{xx} + u_{yy} = 0$ inside a circle and continuous in the circle plus boundary, then the value of u at the center of the circle equals the mean of the values of u on the circumference.

Another important general property is an immediate corollary of this result.

The maximum principle   If u is a solution of $u_{xx} + u_{yy} = 0$ inside a region G bounded by a closed curve B, and if u is continuous in G plus its boundary (i.e., in $G + B$), then the maximum value and the minimum of u in $G + B$ are attained on the boundary B.†

Proof   First of all the maximum, call it M, must actually be attained since u is continuous in the closed region $G + B$.   Now suppose $u = M$ at some interior point P in G.   Then on the circumference of a circle about P in G the mean value of $u = M$.   But since u cannot exceed M, we must have $u \equiv M$ on every circle about P.   This would imply $u \equiv M$ throughout $G + B$.   Thus, in any case, the maximum is attained on B, and, moreover, if it is also attained inside G, then u is identically constant.   The proof for the minimum is essentially the same.

The maximum principle affords a simple proof of the following important *uniqueness theorem* for the interior Dirichlet problem.

Theorem 1   Uniqueness.   *If $u(x,y)$ satisfies $u_{xx} + u_{yy} = 0$ inside a region G bounded by B, and if $u = f$ on B, and if u is continuous in $G + B$, then u is the only such function.*

Proof   First consider the problem $w_{xx} + w_{yy} = 0$ in G and $w = 0$ on B. The only possible solution to this problem is $w \equiv 0$ since, by the maximum principle, the boundary value, zero, is both maximum and minimum for w. Now if $u_{xx} + u_{yy} = 0$, and $u = f$ on B, and if also $v_{xx} + v_{yy} = 0$ and $v = f$ on B, then the function $u - v$ still satisfies Laplace's equation but vanishes on B.   Hence $u - v \equiv 0$ or $u \equiv v$.

Problems 14-4

1. Solve by separation of variables:

PDE: $u_{xx} + u_{yy} = 0$ $\begin{array}{l} 0 < x < 1 \\ 0 < y < 1 \end{array}$

BC: $u(x,0) = u(x,1) = 0$ $0 \leq x \leq 1$

$u(0,y) = g(y)$ $u(1,y) = 0$ $0 \leq y \leq 1$

† See, e.g., Kellogg [22] for a discussion of the conditions on G and B.

2. Solve: PDE: $u_{xx} + u_{yy} = 0$ $\begin{array}{l} -1 < x < 1 \\ -1 < y < 1 \end{array}$

BC: $u(x,-1) = f_1(x)$ $u(x,1) = f_2(x)$ $-1 \leq x \leq 1$

$u(-1,y) = g_1(y)$ $u(1,y) = g_2(y)$ $-1 \leq y \leq 1$

Hint: Solve four separate problems and then use superposition.

3. Find a solution $u(x,y)$ of Laplace's equation in the semi-infinite strip $0 < x < \pi$, $y > 0$, if $|u(x,y)| < K$ for some constant K (that is, u is bounded in the strip) and if u satisfies:

BC: $u(0,y) = u(\pi,y) = 0$ $y > 0$

$u(x,0) = 1$

4. Derive Laplace's equation in polar coordinates, Eq. (108).   Hint: Change variables in Eq. (88) by using $r = \sqrt{x^2 + y^2}$, $\theta = \tan^{-1}(y/x)$, and the chain rule.   For example, $u_x = u_r r_x + u_\theta \theta_x = u_r(x/r) + u_\theta(-y/r^2)$, etc.

5. Solve: PDE: $u_{xx} + u_{yy} = 0$ $x^2 + y^2 < 1$
BC: $u(x,y) = \cos\theta$ $x^2 + y^2 = 1$

where θ is the angular position of a point on the circumference of the unit circle.

6. Set $n = 0$ in Eq. (114) and show that $P(r) = $ const is the only solution that is bounded at $r = 0$.

14-5   Theory of second-order equations

In this section we give a brief introduction to the mathematical theory of second-order partial differential equations.   We shall restrict our study to equations of the general type

$$A(x,y)u_{xx} + 2B(x,y)u_{xy} + C(x,y)u_{yy} + F(x,y,u,u_x,u_y) = 0\dagger \tag{127}$$

where A, B, and C are functions of x and y with continuous second-order partial derivatives.

In certain common cases it will be possible to change variables in (127) and reduce the equation to a simple *canonical form*.   Let r and s be given as twice differentiable functions of x and y by the following transformation:

$$r = r(x,y) \qquad s = s(x,y) \tag{128}$$

where the functions $r(x,y)$ and $s(x,y)$ will be determined later.   Also let

$$u(x,y) = v(r,s) \tag{129}$$

† Equations of type (127) are called *quasi-linear* since they are linear in the highest derivative although not necessarily linear in all derivatives of u.

and then, by repeated use of the chain rule for partial derivatives (see Prob. 1), Eq. (127) becomes

$$(Ar_x^2 + 2Br_xr_y + Cr_y^2)v_{rr} + 2(Ar_xs_x + Br_xs_y + Br_ys_x + Cr_ys_y)v_{rs}$$
$$+ (As_x^2 + 2Bs_xs_y + Cs_y^2)v_{ss} + f(r,s,v,v_r,v_s) = 0 \qquad (130)$$

This equation would be greatly simplified if we could pick the functions r and s in such a way as to eliminate the coefficients of v_{rr} and v_{ss}. It is easily seen that this would be accomplished if we set $r = w_1(x,y)$ and $s = w_2(x,y)$ where w_1 and w_2 are solutions of the equation

$$Aw_x^2 + 2Bw_xw_y + Cw_y^2 = 0 \qquad (131)$$

But this equation is equivalent to the pair of equations

$$w_x = \lambda_1 w_y \qquad \text{and} \qquad w_x = \lambda_2 w_y \qquad (132)$$

where

$$\lambda_1 = \frac{-B + \sqrt{B^2 - AC}}{A} \qquad \text{and} \qquad \lambda_2 = \frac{-B - \sqrt{B^2 - AC}}{A} \qquad (133)$$

Now, as we shall see, the character of the solutions depends intimately on the character of the quantities λ_1 and λ_2 and hence on the value of the discriminant $B^2 - AC$. In order to keep the discussion as simple as possible, it is necessary that we restrict ourselves to a region in the xy plane in which $B^2 - AC$ does not change sign. In fact, we assume further that $B^2 - AC$ never vanishes unless it is *identically* zero. Also we assume $A \neq 0$. Thus three cases arise.

Case I. $B^2 - AC > 0$. *Hyperbolic case.*† In this case λ_1 and λ_2 are real functions of x which are everywhere distinct. We recall that if the equation $w(x,y) = c$ defines y as a differentiable function of x, then by implicit differentiation

$$w_x + w_y \frac{dy}{dx} = 0 \qquad (134)$$

---

† The general second-degree *algebraic* equation in two variables can be written in the form

$$Ax^2 + 2Bxy + Cy^2 + Dx + Ey + F = 0$$

where A, B, C, D, E, and F are constants. This equation has a graph which is a hyperbola, parabola, or ellipse depending on whether the discriminant $B^2 - AC$ is positive, zero, or negative. In analogy with this situation the three types of partial differential equation under consideration are called hyperbolic ($B^2 - AC > 0$), parabolic ($B^2 - AC \equiv 0$), and elliptic ($B^2 - AC < 0$).

On comparing Eq. (134) with (132), we see that (132) will be satisfied if

$$\frac{dy}{dx} = -\lambda_1 \quad \text{or} \quad \frac{dy}{dx} = -\lambda_2 \tag{135}$$

Thus, if these equations have the solutions

$$w_1(x,y) = c_1 \quad \text{and} \quad w_2(x,y) = c_2\dagger \tag{136}$$

then, clearly, w_1 and w_2 satisfy (132) and hence (131).

Thus, if we set

$$r = w_1(x,y) \quad \text{and} \quad s = w_2(x,y) \tag{137}$$

where $w_1 = c_1$ and $w_2 = c_2$ are solutions of (135), we have the desired transformation which changes (127) into the form

$$pv_{rs} + f = 0 \tag{138}$$

It is easy to show (see Prob. 2) that $p \neq 0$, and hence we can divide through by it and write the equation in the form

$$v_{rs} = g(r,s,v,v_r,v_s) \tag{139}$$

This is the so-called *canonical form* for hyperbolic equations.

Example 1   Reduce the one-dimensional wave equation

$$u_{xx} - \frac{1}{a^2} u_{yy} = 0$$

to canonical form. The discriminant $B^2 - AC = 1/a^2 > 0$ and $\lambda_1 = 1/a$, $\lambda_2 = -1/a$. The ordinary differential equations (135) become

$$\frac{dy}{dx} = -\frac{1}{a} \quad \text{and} \quad \frac{dy}{dx} = \frac{1}{a}$$

with solutions

$$x + ay = c_1 \quad \text{and} \quad x - ay = c_2$$

Thus the transformation (137) is

$$r = x + ay \quad \text{and} \quad s = x - ay$$

yielding the equation

$$v_{rs} = 0$$

This is the canonical form. See Sec. 14-2 for further discussion of this equation and its solution.

---

† The set of curves represented by Eq. (136) are called the *characteristics* of Eq. (127).

Case II. $B^2 - AC \equiv 0$ *Parabolic equation.* In this case $\lambda_1 \equiv \lambda_2 = -B/A$. Let $r(x,y)$ be defined as in Case I, Eq. (137), so that $r_x = \lambda_1 r_y = -(B/A)r_y$. Also, let $s(x,y)$ be a function which is *not* a solution of (132). On substituting in (130) and simplifying,† we have

$$(As_x^2 + 2Bs_x s_y + Cs_y^2)v_{ss} + f(r,s,v,v_r,v_s) = 0 \tag{140}$$

The choice of s ensures that the coefficient of v_{ss} never vanishes, and hence we can write (140) in the form

$$v_{ss} = g(r,s,v,v_r,v_s) \tag{141}$$

This is the canonical form for parabolic equations. For example, the one-dimensional heat equation‡

$$u_{xx} = \frac{1}{a^2} u_t \tag{142}$$

is already in canonical form.

Case III. $B^2 - AC < 0$ *Elliptic equation.* From Eq. (133) we see that in this case λ_1 and λ_2 are complex conjugates. If we proceed as in Case I (assuming that our reasoning remains valid for complex variables), we again obtain the equation

$$v_{rs} = g(r,s,v,v_r,v_s) \tag{143}$$

Here, however, r and s would be complex conjugates, i.e.,

$$r = p + iq \qquad s = p - iq \tag{144}$$

where p and q are real.

To transform (143) into a real canonical form, we set

$$p = \tfrac{1}{2}(r + s) \qquad q = \frac{r - s}{2i} \tag{145}$$

and

$$w(p,q) = v(r,s) \tag{146}$$

Then, since

$$v_{rs} = \tfrac{1}{4}(w_{pp} + w_{qq}) \tag{147}$$

Eq. (143) becomes

$$w_{pp} + w_{qq} = h(p,q,w,w_p,w_q) \tag{148}$$

the canonical form for *elliptic equations.*

† See Prob. 3.
‡ See Sec. 14-3.

The simplest example of an elliptic equation is Laplace's equation†

$$u_{xx} + u_{yy} = 0 \tag{149}$$

Problems 14-5

1. Derive Eq. (130).
2. Prove that $p \neq 0$ in Eq. (138).
3. Prove that the coefficient of v_{rs} is zero in Eq. (140).
4. Classify as to type and reduce to canonical form

$$u_{xx} - y^2 u_{yy} = 0$$

5. Classify, reduce to canonical form, and *solve*:

$$u_{xx} + 2u_{xy} + u_{yy} = 0$$

† See Sec. 14-4.

Further Applications of
Partial Differential Equations

15

15-1 Introduction

In Chap. 14 we studied the three main types of second-order partial differential equations for functions of two independent variables. In this chapter we extend our investigations to equations involving functions of three or more independent variables. This method of classifying partial differential equations is somewhat artificial but convenient nevertheless, since we are not attempting a logical development of the entire subject but merely a brief glimpse at some of the principal techniques of solution.

Three classical problems in physics provide us with the equations that are to be solved in this chapter. In Sec. 15-2 the problem of constructing a mathematical model of gravitational force leads to Laplace's equation in three dimensions. In Sec. 15-3 the temperatures in an infinite cylinder are analyzed by utilizing the heat equation in cylindrical coordinates. Finally, in Sec. 15-4 the vibrations of a membrane are studied with the help of the two-dimensional wave equation.

All three problems are of interest in themselves and, in addition, the techniques developed for solving them have widespread application throughout the broad field of mathematical physics.

15-2 Laplace's equation in three dimensions†

Suppose that a mass m_1 is held fixed at a point P which has coordinates (a,b,c) (see Fig. 1). Then the gravitational force $\mathbf{F}$ exerted *by* m_1 *on* a mass

---

† In this section we assume a working knowledge of vector analysis, e.g., the material in Courant [12], vol. 2, chaps. 2 and 5.

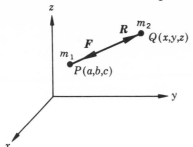

figure 1

m_2 at a point Q with coordinates (x,y,z) is, by Newton's law of gravitation,

$$\mathbf{F} = -\left(\frac{Gm_1m_2}{|\mathbf{R}|^2}\right)\frac{\mathbf{R}}{|\mathbf{R}|} \tag{1}$$

where $\mathbf{R}$ is a vector from P to Q and G is the universal gravitation constant. That is, $\mathbf{F}$ is a vector with magnitude $(Gm_1m_2)/|\mathbf{R}|^2$ directed from Q toward P.

Since

$$\mathbf{R} = (x - a)\mathbf{i} + (y - b)\mathbf{j} + (z - c)\mathbf{k}, \tag{2}$$

if we let

$$r = |\mathbf{R}| = \sqrt{(x - a)^2 + (y - b)^2 + (z - c)^2} \tag{3}$$

then $\mathbf{F}$ can be expressed as

$$\mathbf{F} = -\frac{Gm_1m_2}{r^3}[(x - a)\mathbf{i} + (y - b)\mathbf{j} + (z - c)\mathbf{k}] \tag{4}$$

Since we are ignoring all frictional and dissipative forces which would cause a moving body to lose energy, the force field due solely to gravitational attraction is *conservative*. That is, the sum of the kinetic and potential energy of a body is constant, and the work done in moving through this force field is independent of the path. These properties of the gravitational force field, as determined by Eq. (4), can be established mathematically if we can produce a *potential function* $u(x,y,z)$, which satisfies the equation

$$\mathbf{F} = \nabla u \tag{5}$$

where $\nabla u = \mathbf{grad}\ u$ is the gradient of u. This function u can be obtained by evaluating the line integral

$$u = \int_{Q_1}^{Q} \mathbf{F} \cdot \mathbf{ds} \tag{6}$$

where Q_1 is an arbitrary fixed point in space $(Q_1 \neq P)$ and the integral is taken along any path (not through P) from Q_1 to Q.

Substituting into Eq. (6) the value of **F** from Eq. (4), we see that

$$u = -Gm_1m_2 \int_{Q_1}^{Q} \frac{(x-a)\,dx + (y-b)\,dy + (z-c)\,dz}{[(x-a)^2 + (y-b)^2 + (z-c)^2]^{\frac{3}{2}}} \tag{7}$$

The integrand in (7) is easily recognized as the total differential of $[(x-a)^2 + (y-b)^2 + (z-c)^2]^{-\frac{1}{2}} = 1/r$.   That is,

$$u = +Gm_1m_2 \int_{Q_1}^{Q} d\left(\frac{1}{r}\right)$$

or

$$u = Gm_1m_2\left[\frac{1}{r(Q)} - \frac{1}{r(Q_1)}\right]$$

Since the term $1/r(Q_1)$ is an arbitrary constant, we can obtain a simple form of the potential function u by setting $1/r(Q_1) = 0$.   This is equivalent to taking the initial point of the line integral to be the point at infinity, and thus we are determining the potential u at a point Q by finding the work done in moving a body from infinity to Q.   The result is†

$$u = Gm_1m_2\left[\frac{1}{r(Q)}\right] = \frac{Gm_1m_2}{[(x-a)^2 + (y-b)^2 + (z-c)^2]^{\frac{1}{2}}} \tag{8}$$

That is, the gravitational potential u is proportional to $1/r$.   Thus the quantity $1/r$ is often referred to as the *Newtonian potential*.   It is an easy exercise (see Prob. 1) to verify that this quantity $1/r$ satisfies the partial differential equation

$$\mathbf{\nabla} \cdot \mathbf{\nabla}\left(\frac{1}{r}\right) = 0 \tag{9}$$

i.e.,

$$\frac{\partial^2(1/r)}{\partial x^2} + \frac{\partial^2(1/r)}{\partial y^2} + \frac{\partial^2(1/r)}{\partial z^2} = 0 \tag{10}$$

or, if we set $u = 1/r$, then u satisfies

$$u_{xx} + u_{yy} + u_{zz} = 0 \tag{11}$$

This equation is known as *Laplace's equation* or the *potential equation*.   The left-hand side of the equation is obtained by taking the divergence of the gradient of u, $(\mathbf{\nabla} \cdot \mathbf{\nabla}u)$.   This operation is sometimes abbreviated as $\nabla^2 u$ which is read "del squared u" or "Laplacian of u."

The gravitational attraction produced by a *continuous distribution* of mass can be described by a direct extension of the above technique.   Suppose the mass has a density $\sigma(a,b,c)$ and volume V.   Consider an element Δm

† The reader should verify that $\mathbf{\nabla}u = \mathbf{F}$.

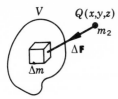

figure 2

of mass at the point (a,b,c) (see Fig. 2)

$$\Delta m = \sigma \, \Delta V \tag{12}$$

where ΔV is the volume of the element. The force exerted on a point mass m_2 situated *outside the body* at Q is given approximately by

$$\Delta \mathbf{F} = - \frac{G m_2}{|\mathbf{R}|^2} \frac{\mathbf{R}}{|\mathbf{R}|} \sigma \, \Delta V \tag{13}$$

where again $\mathbf{R}$ is given by (2). The total force $\mathbf{F}$ is obtained as the limit of the sum of the elements in (13). This gives $\mathbf{F}$ as the volume integral

$$\mathbf{F} = - G m_2 \iiint\limits_V \frac{\mathbf{R}}{|\mathbf{R}|^3} \sigma \, dV \tag{14}$$

Using Eq. (8) as a guide, we can again define a potential function u

$$u = G m_2 \iiint\limits_V \frac{\sigma}{r} \, dV \tag{15}$$

Now since $\sigma(a,b,c)$ is independent of x, y, z and since it is permissible to differentiate inside the integral in (15),† we have once again

$$\nabla u = G m_2 \iiint\limits_V \sigma \, \nabla \left(\frac{1}{r} \right) dV = G m_2 \iiint\limits_V \sigma \, \frac{\mathbf{R}}{|\mathbf{R}|^3} \, dV = \mathbf{F} \tag{16}$$

Also

$$\nabla^2 u = G m_2 \iiint\limits_V \sigma \, \nabla^2 \left(\frac{1}{r} \right) dV \tag{17}$$

and since the quantity $1/r$ satisfies Laplace's equation $\nabla^2(1/r) = 0$, the integrand in (17) is identically zero and we have

$$\nabla^2 u = 0 \tag{18}$$

Thus the gravitational potential (15) of any continuous distribution of mass satisfies Laplace's equation.

In addition to gravitational potential, there are analogous quantities in the theories of electricity and magnetism. These electrostatic and magnetic potentials also satisfy Laplace's equation. In hydrodynamics

† See Courant [12], vol. 2.

and aerodynamics, if the fluid being studied is assumed to be incompressible and irrotational, then the components of the velocity of the fluid are obtained by differentiating a velocity potential. In Sec. 15-3 we shall see that the steady-state temperature of a homogeneous solid also satisfies Laplace's equation. There are many other physical theories in which Laplace's equation plays a fundamental role, and hence the equation and its solutions have received considerable attention from mathematicians in the past two centuries. Of particular interest are those solutions which have continuous second derivatives. These are called *harmonic functions*.

In many ways the harmonic functions in three dimensions behave like the solutions of the two-dimensional Laplace equation studied in Sec. 14-4. For example, the value of a harmonic function at the center of a hollow sphere is equal to the average of its values on the surface of the sphere.† By applying the methods of Sec. 14-4, we could prove (see Probs. 3 and 4) the following consequences of this *mean-value property*. The maximum (minimum) value of a harmonic function in a region is attained on the boundary. Also there can be no more than one harmonic function which takes on a given continuous set of values on the boundary of a region. That is, the *interior Dirichlet problem*‡ has a unique solution in this case.

Potential inside a hollow sphere Suppose that the surface of a hollow sphere of radius a is being maintained at a certain electrical potential. If we wish to determine the resulting potential at points inside the sphere, we seek a function u which satisfies Laplace's equation $\nabla^2 u = 0$ inside the sphere and which assumes the given values on the surface.

Since the boundary condition is to be applied on the surface of a sphere, it is natural to introduce spherical coordinates (ρ,θ,φ). See Fig. 3.§ The

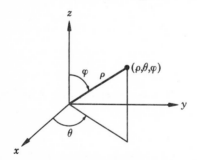

figure 3

† A proof of this important and useful theorem can be found, for example, in Kellogg [22]. See also Prob. 6.

‡ See Sec. 14-4.

§ The reader should be aware of the fact that in many books the angles θ and φ are interchanged.

transformation to these new coordinates is achieved by means of the relations

$$x = \rho \sin \varphi \cos \theta$$
$$y = \rho \sin \varphi \sin \theta \qquad (19)$$
$$z = \rho \cos \varphi$$

Laplace's equation itself must also be expressed in terms of these three new independent variables. The result is *Laplace's equation in spherical coordinates.*†

$$\rho \frac{\partial^2 (\rho u)}{\partial \rho^2} + \frac{1}{\sin \varphi} \frac{\partial}{\partial \varphi}\left(\sin \varphi \frac{\partial u}{\partial \varphi}\right) + \frac{1}{\sin^2 \varphi} \frac{\partial^2 u}{\partial \theta^2} = 0 \qquad (20)$$

In order to simplify the solution of the problem which has been posed, we shall restrict ourselves to the special case where the given surface charge is independent of the longitude θ. That is, we assume a boundary condition of the form

$$\lim_{\rho \to a} u = f(\varphi) \qquad 0 \le \varphi \le \pi \qquad (21)$$

Since the distribution of the potential on the sphere is independent of θ, it is clear that the resulting potential inside the sphere will not depend on θ. Thus the last term in the left-hand side of Eq. (20) will vanish and we shall now seek a solution $u(\rho,\varphi)$ of the equation

$$\rho \frac{\partial^2 (\rho u)}{\partial \rho^2} + \frac{1}{\sin \varphi} \frac{\partial}{\partial \varphi}\left(\sin \varphi \frac{\partial u}{\partial \varphi}\right) = 0 \qquad \begin{array}{l} 0 \le \rho < a \\ 0 \le \varphi \le \pi \end{array} \qquad (22)$$

Using the technique of separation of variables, we attempt to find solutions of the form

$$u(\rho,\varphi) = R(\rho)\Phi(\varphi) \qquad (23)$$

Substitution into (22) and subsequent division by $R\Phi$ yield

$$\frac{\rho}{R} \frac{d^2 (\rho R)}{d\rho^2} = -\frac{1}{\Phi \sin \varphi} \frac{d}{d\varphi}\left(\sin \varphi \frac{d\Phi}{d\varphi}\right) \qquad (24)$$

Since the variables have been separated in Eq. (24), each side must equal a constant λ. Thus we have the pair of equations

$$\rho \frac{d^2 (\rho R)}{d\rho^2} = \lambda R \qquad (25)$$

$$\frac{d}{d\varphi}\left(\sin \varphi \frac{d\Phi}{d\varphi}\right) + \lambda \Phi \sin \varphi = 0 \qquad (26)$$

† The change to spherical coordinates can be effected by a direct application of the chain rule and the transformation equations (19). For a more elegant method see Courant [12], vol. 2, pp. 369 and 391.

Equation (26) can be transformed into a more familiar form by means of the substitution

$$s = \cos \varphi \tag{27}$$

We have

$$\frac{d\Phi}{d\varphi} = \frac{d\Phi}{ds}\frac{ds}{d\varphi} = -\sin \varphi \frac{d\Phi}{ds} \tag{28}$$

and Eq. (26) becomes

$$-\sin \varphi \frac{d}{ds}\left(-\sin^2 \varphi \frac{d\Phi}{ds}\right) + \lambda\Phi \sin \varphi = 0 \tag{29}$$

or

$$\frac{d}{ds}\left[(1 - s^2)\frac{d\Phi}{ds}\right] + \lambda\Phi = 0 \tag{30}$$

This is Legendre's equation, which has been solved in Sec. 7-6 in the form

$$(1 - s^2)\frac{d^2\Phi}{ds^2} - 2s\frac{d\Phi}{ds} + \lambda\Phi = 0 \tag{31}$$

It can be shown† that for this equation to have nonzero solutions which have continuous derivatives of the first order in the interval $-1 \le s \le 1$, we must have

$$\lambda = n(n + 1) \qquad n = 0, 1, 2, \ldots \tag{32}$$

For these values of λ the solutions of (31) are the Legendre polynomials $P_n(s) = P_n (\cos \varphi)$, which are defined in Sec. 7-6.

Returning to Eq. (25), we have

$$\rho^2 \frac{d^2R}{d\rho^2} + 2\rho \frac{dR}{d\rho} - n(n + 1)R = 0 \tag{33}$$

This is Euler's equation, which is discussed in Sec. 5-7, and the solution is obtained by trying $R = \rho^\alpha$. Substituting into (33), we obtain the indicial equation for α

$$\alpha^2 + \alpha - n(n + 1) = 0 \tag{34}$$

with solutions

$$\alpha = n \qquad \alpha = -n - 1 \tag{35}$$

That is, we have two possible solutions for R, ρ^n and ρ^{-n-1}. Of these only the function ρ^n is finite at the center of the sphere, and hence we have obtained the following set of solutions to Eq. (22):

$$u_n(\rho,\varphi) = A_n\rho^n P_n (\cos \varphi) \qquad n = 0, 1, 2, \ldots \tag{36}$$

† E.g., see Churchill [8], p. 203.

In order to satisfy the boundary condition (21), we shall need, in general, an infinite series of such solutions. That is, if the function $f(\varphi) = g(\cos \varphi)$ and its first derivative are piecewise continuous in the interval $0 < \varphi < \pi$, then $f(\varphi)$ can be expanded in a series of Legendre polynomials[†]

$$f(\varphi) = g(\cos \varphi) = \sum_{n=0}^{\infty} A_n a^n P_n(\cos \varphi) \tag{37}$$

where the constants A_n are given by

$$A_n = \frac{1}{a^n} \frac{2n+1}{2} \int_0^{\pi} f(\varphi) P_n(\cos \varphi) \sin \varphi \, d\varphi \tag{38}$$

or

$$A_n = \frac{1}{a^n} \frac{2n+1}{2} \int_{-1}^{1} g(s) P_n(s) \, ds \tag{39}$$

The solution of the problem is then given formally[‡] by

$$u(\rho, \varphi) = \sum_{n=0}^{\infty} A_n \rho^n P_n(\cos \varphi) \tag{40}$$

where the A_n are given by (38) or (39).

Problems 15-2

1. Verify that the Newtonian potential is a solution of Laplace's equation. That is, show that $\nabla^2(1/r) = 0$ where

$$r = \sqrt{(x-a)^2 + (y-b)^2 + (z-c)^2}$$

2. Find the potential inside a hollow sphere of radius a if the upper hemisphere is maintained at the constant potential A and the lower hemisphere is kept at potential zero. In order to evaluate the integrals for the coefficients in the solution, see Probs. 8 and 9 in Sec. 7-6.

3. State and prove a *maximum principle* for solutions of Laplace's equation in three dimensions. See the discussion of the *mean-value property* in this section and in Sec. 14-4.

4. State and prove a *uniqueness theorem* for solutions of Laplace's equation in three dimensions. See Prob. 3 and Sec. 14-4.

5. For the problem which is solved in the text, show that the potential on the positive z axis in the range $0 < \rho \le a$ is given by

$$u(\rho, 0) = \frac{a(a^2 - p^2)}{2} \int_0^{\pi} \frac{f(\varphi') \sin \varphi' \, d\varphi'}{[a^2 - 2a\rho \cos \varphi' + \rho^2]^{\frac{3}{2}}}$$

Hint: Substitute $\varphi = 0$ in Eq. (40) and use the result of Prob. 6, Sev. 7-6.

6. Use the formula in Prob. 5 to show that the potential at the center of the sphere is equal to the mean value of the potential on the surface.

† See Sec. 13-11.
‡ The validity of the formal solution is established, for example, in Churchill [8], p. 195.

15-3   Temperature in an infinite cylinder

Consider an infinitely long, solid, circular cylinder with its axis coinciding with the axis of a system of cylindrical coordinates r, θ, z.   Let the surface of the cylinder at $r = c$ be kept at some fixed temperature, say zero.   If at time $t = 0$ the temperature u at any point in the cylinder is a function of r alone, i.e., if $u(r,\theta,z,0) = F(r)$, then u will always be independent of θ and z.[†]   That is, $u = u(r,t)$.

In Sec. 14-3 the one-dimensional heat equation is derived:

$$u_{xx}(x,t) = \frac{1}{k} u_t(x,t) \tag{41}$$

By an extension of the argument in Sec. 14-3[‡] it can be shown that the temperature $u(x,y,z,t)$ which depends on three space variables satisfies the *three-dimensional heat equation.*

$$u_{xx} + u_{yy} + u_{zz} = \frac{1}{k} u_t \tag{42}$$

or in cylindrical coordinates r, θ, z

$$u_{rr} + \frac{1}{r} u_r + \frac{1}{r^2} u_{\theta\theta} + u_{zz} = \frac{1}{k} u_t \tag{43}$$

The temperature distribution that is approached by a system after a long time, that is, the so-called steady-state distribution, is characterized by the condition $u_t \equiv 0$.   Hence the differential equation describing this situation is Laplace's equation

$$u_{xx} + u_{yy} + u_{zz} = 0$$

In the special problem under consideration in this section, u is independent of θ and z and the heat equation reduces to

$$u_{rr}(r,t) + \frac{1}{r} u_r(r,t) = \frac{1}{k} u_t(r,t) \tag{44}$$

We have assumed the initial condition

$$u(r,0) = f(r) \tag{45}$$

and the boundary condition

$$u(c,t) = 0 \tag{46}$$

† This conclusion and the following discussion would also hold for a *finite* cylinder if the ends were insulated.

‡ E.g., see Churchill [8], sec. 6.

The determination of the function $u(r,t)$ satisfying (44), (45), and (46) will be effected by the technique of separation of variables. Assume a solution of the form

$$u(r,t) = R(r)T(t) \tag{47}$$

Substituting into (44) and dividing by RT, we have

$$\frac{R''(r)}{R(r)} + \frac{R'(r)}{rR(r)} = \frac{1}{k}\frac{T'(t)}{T(t)} \tag{48}$$

By the usual argument this equation is satisfied identically only if both sides equal some constant, say $-\lambda^2$. This yields the pair of equations

$$T' + k\lambda^2 T = 0 \tag{49}$$

$$rR'' + R' + \lambda^2 rR = 0 \tag{50}$$

Solving for T in (49), we have

$$T(t) = T_0 e^{-k\lambda^2 t} \qquad T_0 = \text{const} \tag{51}$$

Equation (50) for R should be recognized as a form of Bessel's equation of order zero (see Prob. 1, Sec. 7-4) with the solution

$$R(r) = c_1 J_0(\lambda r) + c_2 N_0(\lambda r) \tag{52}$$

However, since the Neumann function N_0 becomes infinite at $r = 0$† and since we are seeking a solution $R(r)$ which will be finite at $r = 0$ (i.e., at the axis of the cylinder), we take $c_2 = 0$ and $R(r)$ is then just a multiple of $J_0(\lambda r)$. Thus functions of the form $J_0(\lambda r)e^{-k\lambda^2 t}$ satisfy the heat equation (44) and remain finite at $r = 0$. In order to satisfy the boundary condition (46), we must have

$$J_0(\lambda c) = 0 \tag{53}$$

This equation is satisfied only if $\lambda = \lambda_n$ where $\lambda_n c$ is the nth zero of $J_0(x)$. We have now constructed the set of eigensolutions

$$u_n = A_n J_0(\lambda_n r)e^{-k\lambda_n^2 t} \qquad n = 1, 2, \ldots \tag{54}$$

which satisfy (44) and (46) for any constants A_n.

The initial condition (45) can be satisfied in general by using an infinite sum of eigenfunctions. That is, a formal solution to our problem is given by

$$u(r,t) = \sum_{n=1}^{\infty} A_n J_0(\lambda_n r)e^{-k\lambda_n^2 t} \tag{55}$$

† See Fig. 2, Sec. 7-4.

where the constants A_n are chosen to be the coefficients of the Fourier-Bessel expansion of $f(r)$,[†] i.e.,

$$A_n = \frac{2}{c^2[J_1(\lambda_n c)]^2} \int_0^c rf(r)J_0(\lambda_n r)\, dr \qquad n = 1, 2, \ldots \tag{56}$$

Problems 15-3

1. Find the temperature distribution in an infinite cylinder if the initial temperature is given by $u(r,\theta,z,0) = u_0 + F(r)$ where u_0 is a constant and if the surface $r = c$ is maintained at temperature u_0.

2. Find the temperature distribution in an infinite cylinder if the surface $r = c$ is insulated, i.e., $u_r(c,\theta,z,t) = 0$. Assume an initial temperature

$$u(r,\theta,z,0) = f(r)$$

3. a. Find the steady-state temperature distribution in a finite cylinder of height L and radius $r = 1$ if the surface $r = 1$ is insulated, the base is kept at constant temperature zero, and the top $(z = L)$ is kept at $f(r)$.

b. Solve the problem in part a if $f(r) = u_1 = $ const.

15-4   Vibrating membranes

In this section we shall study two important special cases of the vibratory motions of tightly stretched membranes. We assume the membrane to be thin and flexible and to be attached to a frame in the xy plane. A constant tension T, per unit length across any line, is assumed to act tangential to the membrane at each point. We assume that the mass per unit area, δ, is constant and that the displacement of the membrane at each point is small in such a way as to justify approximations similar to those that were made in deriving the equation of motion of the vibrating string (see Sec. 14-2). If no external forces act on the membrane, the transverse displacement z at each point will be found to satisfy the *two-dimensional wave equation*[‡]

$$z_{xx} + z_{yy} = \frac{1}{a^2} z_{tt} \tag{57}$$

where $a^2 = T/\delta$. We shall solve this equation for the cases of a circular and a rectangular membrane.

The circular membrane   We shall study the vibrations of a circular membrane of radius c with its center at the origin of the xy plane. If the outer edge is to be kept fixed, we have a boundary condition $z \equiv 0$ on the

[†] See Sec. 13-10.
[‡] E.g., see Churchill [8], sec. 5.

circle $x^2 + y^2 = c^2$. This suggests the introduction of polar coordinates r, θ in the xy plane so that if $z = z(r,\theta,t)$, then the boundary condition becomes

$$z(c,\theta,t) = 0 \qquad -\pi < \theta \leq \pi \qquad t \geq 0 \tag{58}$$

On changing variables in the differential equation (57), we have the new equation

$$z_{rr} + \frac{1}{r} z_r + \frac{1}{r^2} z_{\theta\theta} = \frac{1}{a^2} z_{tt} \tag{59}$$

for

$$0 \leq r \leq c \qquad -\pi < \theta < \pi \qquad t > 0$$

In addition to the boundary condition (58) we need initial conditions. If we assume that the membrane is given an initial displacement $f(r,\theta)$ and released from rest, we have

$$z(r,\theta,0) = f(r,\theta) \tag{60}$$

$$z_t(r,\theta,0) = 0 \tag{61}$$

Hence the problem we have posed is to find a solution $z(r,\theta,t)$ of (59) which satisfies (58), (60), and (61).

In order to separate variables, we seek solutions in the form

$$z = R(r)S(\theta)T(t) \tag{62}$$

and (59) becomes

$$\frac{R''(r)}{R(r)} + \frac{R'(r)}{rR(r)} + \frac{1}{r^2}\frac{S''(\theta)}{S(\theta)} = \frac{T''(t)}{a^2 T(t)} \tag{63}$$

Since the left side is independent of t and the right independent of r and θ, both sides must equal a constant. This constant will turn out to be negative, and hence we designate it by $-\lambda^2$. The equation for $T(t)$ becomes

$$T'' + a^2\lambda^2 T = 0 \tag{64}$$

and so (61) implies

$$T(t) = \cos a\lambda t \tag{65}$$

The equation for R and S is

$$\frac{R''}{R} + \frac{R'}{rR} + \frac{1}{r^2}\frac{S''}{S} = -\lambda^2 \tag{66}$$

or

$$\frac{r}{R}(rR'' + R') + \lambda^2 r^2 = -\frac{S''}{S} \tag{67}$$

Once again both sides must be a constant, say μ^2. For S we obtain

$$S(\theta) = c_1 \cos \mu\theta + c_2 \sin \mu\theta \tag{68}$$

But since the displacement z is to be single-valued as a function of x and y, it must be periodic in θ with period 2π. (*Is this always true?*) Thus μ must be an integer n and (68) becomes

$$S(\theta) = c_1 \cos n\theta + c_2 \sin n\theta \qquad n = 0, 1, 2, \ldots \tag{69}$$

Finally, the equation for $R(r)$ is now

$$r^2 R'' + r R' + (\lambda^2 r^2 - n^2)R = 0 \tag{70}$$

This should be recognized as a form of Bessel's equation with the solution

$$R(r) = c_3 J_n(\lambda r) + c_4 Y_n(\lambda r) \tag{71}$$

However, the second solution Y_n to Bessel's equation becomes infinite at $r = 0$,† whereas z and hence R must be finite there.‡ Thus c_4 must equal zero and $R(r)$ can be taken as

$$R(r) = J_n(\lambda r) \tag{72}$$

The boundary condition (58) will be satisfied if $J_n(\lambda c) = 0$. Hence λ must equal λ_{nk}, $k = 1, 2, \ldots$, where $\lambda_{nk}c$ is the kth zero of J_n. That is,

$$J_n(\lambda_{nk}c) = 0 \qquad n = 0, 1, 2, \ldots$$
$$k = 1, 2, \ldots$$

Combining our results, we have a function z which satisfies all but (60), i.e.,

$$z = J_n(\lambda_{nk}r)(c_1 \cos n\theta + c_2 \sin n\theta) \cos a\lambda_{nk}t \tag{73}$$

This is a solution for any integral n and k, and hence by choosing the constants c_1 and c_2 separately for each n and k, we construct a doubly infinite set z_{nk} of solutions (or *eigenfunctions*):

$$z_{nk} = J_n(\lambda_{nk}r)(a_{nk} \cos n\theta + b_{nk} \sin n\theta) \cos a\lambda_{nk}t \tag{74}$$

where a_{nk} and b_{nk} are constants.

In order to satisfy the initial condition (60), suppose that $f(r,\theta)$ has the Fourier expansion

$$f(r,\theta) = \sum_{n=0}^{\infty} [A_n(r) \cos n\theta + B_n(r) \sin n\theta] \tag{75}$$

† See Sec. 7-4, Prob. 15.

‡ The assumption that z is finite at $r = 0$ is actually a second boundary condition to supplement (58).

Then

$$A_0(r) = \frac{1}{2\pi} \int_{-\pi}^{\pi} f(r,\theta) \, d\theta \tag{76}$$

$$A_n(r) = \frac{1}{\pi} \int_{-\pi}^{\pi} f(r,\theta) \cos n\theta \, d\theta \qquad n = 1, 2, \ldots \tag{77}$$

$$B_n(r) = \frac{1}{\pi} \int_{-\pi}^{\pi} f(r,\theta) \sin n\theta \, d\theta \qquad n = 1, 2, \ldots \tag{78}$$

Now the Fourier series (75) will be a sum of terms of the form of (74) (for $t = 0$) only if the functions $A_n(r)$ and $B_n(r)$ can be represented as sums of the Bessel functions $J_n(\lambda_{nk}r)$. That is, if

$$A_n(r) = \sum_{k=1}^{\infty} a_{nk} J_n(\lambda_{nk}r) \tag{79}$$

$$B_n(r) = \sum_{k=1}^{\infty} b_{nk} J_n(\lambda_{nk}r) \tag{80}$$

then a_{nk} and b_{nk} are just the coefficients of the Fourier-Bessel expansion.†
Thus

$$a_{nk} = \frac{2}{c^2[J_{n+1}(\lambda_{nk}c)]^2} \int_0^c r J_n(\lambda_{nk}r) A_n(r) \, dr \tag{81}$$

$$b_{nk} = \frac{2}{c^2[J_{n+1}(\lambda_{nk}c)]^2} \int_0^c r J_n(\lambda_{nk}r) B_n(r) \, dr \tag{82}$$

where $A_n(r)$ and $B_n(r)$ are given by (77) and (78).

Therefore, assuming convergence of the series involved, we have the following solution to the problem (58), (59), (60), and (61):

$$z(\lambda,\theta,t) = \sum_{n=0}^{\infty} \sum_{k=1}^{\infty} J_n(\lambda_{nk}r)(a_{nk} \cos n\theta + b_{nk} \sin n\theta) \cos a\lambda_{nk}t \tag{83}$$

where a_{nk} and b_{nk} are given by (81) and (82).

The general solution (83) yields much qualitative information about the kinds of vibrations that one can expect from a circular membrane under special conditions. For example, if the initial displacement is independent of θ, i.e., if

$$f(r,\theta) = F(r) \tag{84}$$

then

$$A_0(r) = F(r) \qquad A_n(\theta) = B_n(\theta) = 0 \qquad n = 1, 2, \ldots \tag{85}$$

† See Sec. 13-10.

and the solution (83) reduces to

$$z(r,t) = \sum_{k=1}^{\infty} a_{0k} J_0(\lambda_{0k}r) \cos a\lambda_{0k}t \tag{86}$$

and the a_{0k} are the Fourier-Bessel coefficients of $F(r)$.

Note that in either case (83) or case (86) the eigenfunctions or component parts of the vibration involve terms of the form $g(r,\theta) \cos a\lambda_{nk}t$. That is, the component vibrations have frequencies $a\lambda_{nk}$ which are multiples of the irrational numbers λ_{nk} and which have the property that the higher frequencies are not integral multiples of some fundamental. Thus the vibrations will not combine like the harmonic vibrations of a string, for example. That is, total displacement as given by (83) is not, in general, a periodic function of time. This is a mathematical explanation of the reason why a drum usually does not produce a musical note such as that emanating from a vibrating banjo string. However, it is a curious fact that since an individual eigenfunction (74) *is* periodic, a musical note could be produced by giving the drum surface the initial displacement

$$f(r,\theta) = J_n(\lambda_{nk}r)(a_{nk} \cos n\theta + b_{nk} \sin n\theta)$$

In any kind of vibrating system such as the one under study here, it is of interest to examine the so-called nodes or points which remain motionless when the system is executing an eigenvibration. From Eq. (74) we see that the eigenfunctions vanish for all t at those points (r,θ) for which

$$J_n(\lambda_{nk}r)(a_{nk} \cos n\theta + b_{nk} \sin n\theta) = 0 \tag{87}$$

This equation is satisfied if either factor vanishes. For example, if $r = r_1$ is such that $J_n(\lambda_{nk}r_1) = 0$, then the points (r_1,θ) are all nodes for all values of θ. These points form a curve called a *nodal curve* (in this case the circle $r = r_1$). Similarly, the term $a_{nk} \cos n\theta + b_{nk} \sin n\theta$ will vanish for certain θ independent of r. Thus the nodal curves for the eigenfunctions will be a discrete set of circles concentric with the membrane and straight lines emanating from the center.

The rectangular membrane   Consider a membrane supported by a rectangular frame formed by the straight lines $x = 0$, $x = A$, $y = 0$, and $y = B$. Suppose that the membrane is given an initial displacement $f(x,y)$ and is released from rest. The resulting displacement $z = z(x,y,t)$ must satisfy the wave equation

$$z_{xx} + z_{yy} = \frac{1}{a^2} z_{tt} \tag{88}$$

together with the initial conditions

$$z(x,y,0) = f(x,y) \tag{89a}$$

$$z_t(x,y,0) = 0 \tag{89b}$$

and the boundary conditions

$$z(0,y,t) = z(A,y,t) = 0 \tag{90a}$$

$$z(x,0,t) = z(x,B,t) = 0 \tag{90b}$$

Assuming a solution of the form

$$z = X(x)Y(y)T(t) \tag{91}$$

we obtain from (88)

$$\frac{X''(x)}{X(x)} + \frac{Y''(y)}{Y(y)} = \frac{T''(t)}{a^2 T(t)} \tag{92}$$

Each term in (92) is a function of only one of the independent variables, and by the usual argument each must be constant if (92) is to hold identically. Set

$$\frac{X''}{X} = -\lambda^2 \qquad \frac{Y''}{Y} = -\mu^2 \tag{93}$$

and then we have

$$\frac{T''}{a^2 T} = -\lambda^2 - \mu^2 \tag{94}$$

Solving these equations and applying (89b), (90a), and (90b), we find

$$X = \sin\frac{m\pi x}{A} \qquad Y = \sin\frac{n\pi y}{B} \qquad T = \cos\left(a\sqrt{\frac{m^2\pi^2}{A^2} + \frac{n^2\pi^2}{B^2}}\,t\right) \tag{95}$$

where $m, n = 1, 2, \ldots$.

Thus the solution to the entire problem (88), (89), and (90) is

$$z = \sum_{m=1}^{\infty} \sum_{n=1}^{\infty} a_{mn} \cos\left(\pi a t\sqrt{\frac{m^2}{A^2} + \frac{n^2}{B^2}}\right) \sin\frac{m\pi x}{A} \sin\frac{n\pi y}{B} \tag{96}$$

where the a_{mn} are the coefficients of the double Fourier sine series for $f(x,y)$,† i.e.,

$$a_{mn} = \frac{2}{B}\int_0^B g(y) \sin\frac{n\pi y}{B}\, dy \tag{97}$$

where

$$g(y) = \frac{2}{A}\int_0^A f(x,y) \sin\frac{m\pi x}{A}\, dx \tag{98}$$

† See Prob. 1.

Since the frequencies of the individual terms (eigenfunctions) in (96) are given by the expressions $\pi a (m^2/A^2 + n^2/B^2)^{\frac{1}{2}}$ and since these numbers are not all integral multiples of some fixed fundamental frequency, the displacement z will not be periodic, in general. Of course, as with the circular membrane, with special initial conditions the expansion (96) can be reduced to a single (periodic) term.

Problem 15-4

1. Derive Eq. (97).

Appendix A: Infinite Series

Infinite Series.   In this appendix we shall summarize those results in the theory of infinite series which are most important in the study of differential equations. Theorems are stated without proof and very few illustrative examples are given. For proofs of the theorems and a more thorough discussion of the theory, the reader should consult a good calculus book.†

If a function f is defined for all the positive integers, then the set of values $f(1), f(2), f(3), \ldots$ is called an *infinite sequence* and is designated by the symbol $\{f(k)\}$.   Frequently we let $a_k = f(k)$ and discuss the sequence $\{a_k\}$.

The sequence $\{a_k\}$ is said to *converge to the limit* L if, for any $\epsilon > 0$, there exists a number N such that $|a_k - L| < \epsilon$ for all $k > N$.   In this case we write $\lim_{k \to \infty} a_k = L$. If $\lim_{k \to \infty} a_k$ does not exist, the sequence $\{a_k\}$ is said to *diverge*.

If, in the sequence $\{a_k\}$, the elements satisfy the inequality $a_k \leq a_{k+1}$ for all k, then the sequence is called *nondecreasing*; if $a_k \geq a_{k+1}$, *nonincreasing*.   If there is a constant M such that $a_k \leq M$ for all k, M is called an *upper bound* for $\{a_k\}$, and similarly N is a *lower bound* if $a_k \geq N$ for all k.   The following theorem is fundamental for such sequences.

Theorem 1   *If a nondecreasing sequence has an upper bound, then it must converge.   If a nonincreasing sequence has a lower bound, it must converge.*‡

In many important cases the terms a_k in a sequence are functions $a_k(x)$ of a variable x in some interval.   For example, in the sequence $1, x, x^2, x^3, \ldots$, $x^k, \ldots$ we have $a_k = a_k(x) = x^k$ for $k = 0, 1, 2, \ldots$.   If, for each fixed value of x in some interval, the sequence $\{a_k(x)\}$ converges to a limit L, then this limit will in general be a function of x and we write $\lim_{k \to \infty} a_k(x) = L(x)$.

If we begin with a sequence $\{a_k\}$ and form a new sequence $\{S_n\}$ by setting $S_1 = a_1, S_2 = a_1 + a_2, \ldots, S_n = a_1 + a_2 + \cdots + a_n, \ldots$ this new sequence of

---

† E.g., see Courant [12], vol. 1, or Buck [5].

‡ For a proof see, for example, Courant [12], vol. 1.

partial sums $\{S_n\}$ is called an *infinite series*. The term S_n is called the nth *partial sum* of the series. If $\lim_{n \to \infty} S_n = S$, then we say that the series *converges* to the *sum* S and we write $a_1 + a_2 + \cdots + a_n + \cdots = S$. If the $\lim_{n \to \infty} S_n$ does not exist, the series† $a_1 + a_2 + \cdots$ is said to *diverge*.

If the series $\sum_{n=1}^{\infty} |a_n|$ of absolute values converges, then $\sum_{n=1}^{\infty} a_n$ is said to be *absolutely convergent*. If Σa_n converges but $\Sigma |a_n|$ diverges, then Σa_n is called *conditionally convergent*. For example, $\sum_{n=1}^{\infty} (-1)^n/n^2$ is absolutely convergent, but $\sum_{n=1}^{\infty} (-1)^n/n$ is only conditionally convergent.

Alternating Series. If a series can be written in the form $\sum_{k=0}^{\infty} (-1)^k p_k$ with $p_k \geq 0$, then the series is called an *alternating series*.

Theorem 2   *Given the alternating series* $\sum_0^{\infty} (-1)^k p_k$, $p_k \geq 0$, *if* $p_{k+1} \leq p_k$, $k = 0, 1, 2, \ldots$, *and* $\lim_{k \to \infty} p_k = 0$, *then the series converges to some number* S. *Furthermore* $|S - \sum_{k=0}^{n} (-1)^k p_k| \leq p_{n+1}$.

This theorem can be proved by writing the partial sums of the series in the form $p_0 - (p_1 - p_2) - (p_3 - p_4) - \cdots$ to show that the sums are bounded above, and then in the form $(p_0 - p_1) + (p_2 - p_3) + \cdots$ to see that the sums are *nondecreasing*. Theorem 1 then implies the first part of the result. A similar argument proves the second part.

One of the most useful tests for absolute convergence is the so-called ratio test.

Ratio test.‡   If $\lim_{n \to \infty} |a_{n+1}|/|a_n| = L$ and if $L < 1$, then Σa_n is absolutely convergent. If $L > 1$, then Σa_n diverges, whereas if $L = 1$, the test fails.

Power Series. The ratio test is especially well adapted to *power series*, that is, series of the form $\sum_{n=0}^{\infty} c_n (x - a)^n$. For example, if $c_n = (-1)^n/n$ and $a = 0$, then the power series is $\sum_{n=0}^{\infty} (-1)^n x^n/n$ and, to apply the ratio test, we compute

$$\lim_{n \to \infty} \left| \frac{(-1)^{n+1} x^{n+1}}{n+1} \cdot \frac{n}{(-1)^n x^n} \right| = \lim_{n \to \infty} \frac{n}{n+1} |x| = |x|$$

Hence $\Sigma (-1)^n x^n/n$ is absolutely convergent if $|x| < 1$ and divergent if $|x| > 1$. The open interval $|x| < 1$ or $-1 < x < 1$ is called the *interval* of *convergence*, whereas the set of *all* points where the series converges is called the *domain* of *convergence*. For the series $\Sigma (-1)^n x^n/n$, one finds convergence for $x = 1$ but divergence for $x = -1$, and hence the domain of convergence is $-1 < x \leq 1$.

For values of x *inside* the interval of convergence the following important theorems hold.

† Whenever it will not cause confusion, we will abuse the notation by speaking of the "series" $a_1 + a_2 + \cdots$. We shall also use $\sum_{n=1}^{\infty} a_n$ or even Σa_n to represent both the series and its sum.

‡ See Courant [12], vol. 1, for a proof.

Theorem 3 *A power series may be differentiated or integrated term by term inside its interval of convergence.*

That is, if a power series $\Sigma c_n(x - a)^n$ converges to $f(x)$ for $|x - a| < R$, then $f'(x) = \Sigma n c_n(x - a)^{n-1}$ and

$$\int_a^x f(t)\, dt = \Sigma c_n \int_a^x (t - a)^n\, dt = \Sigma c_n \frac{(x - a)^{n+1}}{n + 1}$$

for $|x - a| < R$.

For example, the geometric series $\sum_{n=0}^{\infty} x^n$ converges to $(1 - x)^{-1}$ for $|x| < 1$, i.e., $(1 - x)^{-1} = \sum_{n=0}^{\infty} x^n$. Hence, on differentiating, we have $(1 - x)^{-2} = \sum_{n=1}^{\infty} n x^{n-1} = \sum_{n=0}^{\infty} (n + 1)x^n$. On integrating the original series, we have

$$-\ln|1 - x| = \int_0^x (1 - t)^{-1}\, dt = \sum_{n=0}^{\infty} \int_0^x t^n\, dt = \sum_{n=0}^{\infty} \frac{x^{n+1}}{n + 1}$$

These results hold at least for $|x| < 1$.

Theorem 4 *Identity theorem. If two power series $\Sigma a_n(x - a)^n$ and $\Sigma b_n(x - a)^n$ both converge to the same function throughout some open interval containing $x = a$, then the two series are identical, i.e., $a_n = b_n$ for $n = 0, 1, 2, \ldots$. In particular, if $\Sigma a_n(x - a)^n = 0$ for all x satisfying $|x - a| < b$, then $a_n = 0$ for $n = 0, 1, 2, \ldots$.*

One of the most important results in calculus, and one which is fundamental in the theory of power series, is the following theorem.

Theorem 5 *Taylor's formula. If f is continuous, together with its first $n + 1$ derivatives, in an open interval I containing $x = a$, then for each x in the interval I*

$$f(x) = f(a) + \frac{f'(a)}{1!}(x - a) + \frac{f''(a)}{2!}(x - a)^2 + \cdots + \frac{f^{(n)}(a)}{n!}(x - a)^n + R_n(x)$$

where

$$R_n(x) = \frac{f^{(n+1)}(c)}{(n + 1)!}(x - a)^{n+1}$$

and c lies between x and a.[†]

For example, if $f(x) = e^x$, then $f^{(n)}(x) = e^x$ and

$$e^x = e^a + \frac{e^a}{1!}(x - a) + \frac{e^a}{2!}(x - a)^2 + \cdots + \frac{e^a}{n!}(x - a)^n + R_n(x)$$

[†] The remainder $R_n(x)$ may also be written in the form

$$R_n(x) = \frac{1}{n!} \int_a^x (x - t)^n f^{(n+1)}(t)\, dt$$

E.g., see Courant [12], vol. 1.

and

$$R_n(x) = \frac{e^c}{(n+1)!} (x - a)^{n+1}$$

or

$$e^x = e^a \left[1 + \frac{(x-a)}{1!} + \frac{(x-a)^2}{2!} + \cdots + \frac{(x-a)^n}{n!} \right] + \frac{e^c}{(n+1)!} (x - a)^{n+1}$$

In particular, if $a = 0$, we have

$$e^x = 1 + x + \frac{x^2}{2!} + \cdots + \frac{x^n}{n!} + e^c \frac{x^{n+1}}{(n+1)!}$$

where c lies between x and 0.

If the function f has continuous derivatives of all orders, then $R_n(x)$ is defined for all n. If it can also be shown that $\lim_{n \to \infty} R_n(x) = 0$, then it is clear that $f(x)$ is represented by the resulting infinite series

$$f(x) = f(a) + \frac{f'(a)}{1!} (x - a) + \cdots + \frac{f^{(n)}(a)}{n!} (x - a)^n + \cdots$$

This is the so-called *Taylor-series* expansion of $f(x)$ about the point $x = a$. Using the convention that $f^{(0)}(x) \equiv f(x)$, we can write the result in the form

$$f(x) = \sum_{n=0}^{\infty} \frac{f^{(n)}(a)}{n!} (x - a)^n$$

The identity theorem shows that the coefficients in the Taylor expansion are unique.

Uniform Convergence.   The properties of differentiability and integrability, which are ensured for power series by Theorem 3, are shared by certain other series that converge "strongly" enough. The type of convergence needed is *uniform convergence.*

Definition 1   *A series $\sum_{k=0}^{\infty} f_k(x)$ is said to converge uniformly to a function $f(x)$ on an interval I if and only if, given any $\epsilon > 0$, there exists a number $N(\epsilon)$ which depends on ϵ, but not on x, such that $|f(x) - \sum_{k=0}^{n} f_k(x)| < \epsilon$, for all $n > N$ and for all x in I.*

An example of a series that converges, but *not* uniformly, is $s = \sum_0^{\infty} x^k$ on $0 < x < 1$. Since

$$s - s_n = \frac{1}{1-x} - \frac{1 - x^{n+1}}{1 - x} = \frac{x^{n+1}}{1 - x}$$

it is clear that, for any fixed n, $|s - s_n|$ can be made $> \epsilon$ by taking x close enough to $x = 1$.

A useful test for uniform convergence is the following theorem.

Theorem 6 *Weierstrass M test. Let $\{f_k\}$ be a sequence of functions defined on an interval I. Let $\{M_k\}$ be a sequence of positive constants such that $|f_k(x)| \leq M_k$ for all x in I and for all k. If $\sum_{k=0}^{\infty} M_k$ converges then $\sum_{k=0}^{\infty} f_k(x)$ converges uniformly on I.*

One of the most important results for uniformly convergent series is the following theorem.

Theorem 7 *Given the series $\sum_{k=0}^{\infty} f_k(x)$ where the functions f_k are defined on some common interval I:*

1. *If all the functions f_k are continuous on I and if $\Sigma f_k(x)$ converges uniformly to $f(x)$ on I, then $f(x)$ is continuous and, furthermore,*

$$\int_a^b f(x)\,dx = \int_a^b \left[\sum_{k=0}^{\infty} f_k(x) \right] dx = \sum_{k=0}^{\infty} \left[\int_a^b f_k(x)\,dx \right]$$

where a and b are any two points in I.

2. *If all the functions f_k have continuous derivatives on I and if the series of derivatives $\sum_{k=0}^{\infty} f_k'(x)$ converges uniformly to a function $g(x)$ on I, then, if $\Sigma f_k(x)$ converges to $f(x)$, we have $g(x) = f'(x)$.*

A useful, although somewhat loose, interpretation of this theorem is that (1) a uniformly convergent series of continuous functions may be integrated termwise and (2) a convergent series may be differentiated termwise, provided the differentiated series converges uniformly.

Theorem 3 is a corollary of Theorem 7 since power series can be shown to converge uniformly in any closed interval contained in their (open) interval of convergence.

Appendix B: Functions of a Complex Variable

Complex Numbers. The reader is expected to be familiar with the elementary algebra of complex numbers of the form $x + iy$ where $i = \sqrt{-1}$. This algebra will be redeveloped in this appendix, from an axiomatic point of view that has certain aesthetic advantages—in particular, the fact that the properties of complex numbers are defined in terms of the familiar real numbers. The axioms are of course motivated by the properties that we desire the complex numbers to have.

A complex number z is defined as an ordered pair (x,y) of real numbers satisfying the following axioms.

If $z_1 = (x_1,y_1)$ and $z_2 = (x_2,y_2)$ are two complex numbers, then:

1. $z_1 = z_2$ if and only if $x_1 = x_2$, $y_1 = y_2$
2. $z_1 + z_2 = (x_1 + x_2, y_1 + y_2)$
3. $rz_1 = (rx_1, ry_1)$ if r is real
4. $z_1 z_2 = (x_1 x_2 - y_1 y_2, x_1 y_2 + x_2 y_1)$†

Some immediate consequences of these axioms are the facts that:

1. The number $(0,0)$ is the additive identity or zero of the system, since $(x,y) + (0,0) = (x,y)$.
2. The numbers $(x_1,0)$ and $(x_2,0)$ have the property that sums and products are again of the same form. That is,

$$(x_1,0) + (x_2,0) = (x_1 + x_2,\ 0)$$

$$(x_1,0) \cdot (x_2,0) = (x_1 x_2, 0)$$

In fact, these numbers behave exactly like the real numbers x_1 and x_2, and hence we can identify the number $(x,0)$ with the real number x.

---

† The reader should check these axioms against the properties of $x_1 + iy_1$ and $x_2 + iy_2$.

498

3. The number $(0,1)$ has the property that $(0,1) \cdot (0,1) = (-1,0)$. (This follows immediately from axiom 4.) Hence $(0,1)$ is a number whose square is identified with -1. Thus we can define i to be $(0,1)$. Then $-i$ becomes $(0,-1)$.

4. Any complex number $z = (x,y)$ can be written in the form $(x,y) = x(1,0) + y(0,1)$ or, equivalently, $(x,y) = x + iy$ where the meaning of this symbol is now established by the above axioms and remarks. The numbers x and y are called the *real* and *imaginary* parts of z respectively, and we write $\text{Re } z = x$, $\text{Im } z = y$.

Complex Plane. The fact that complex numbers are defined as ordered pairs (x,y) of reals suggests a geometric interpretation in terms of the coordinates of a point in the plane. In fact, if we identify the complex number $(x,y) = x + iy$ with the corresponding point (x,y) in the xy plane, we obtain a geometrical picture of any set of complex numbers. The complex number (x,y) may also be interpreted as a vector with components x and y.

The *magnitude* or *absolute value* of a complex number $z = x + iy$ is given by

$$|z| = \sqrt{x^2 + y^2}$$

Note that this is just the distance from the origin to the point (x,y), i.e., the r of polar coordinates. The polar coordinate θ is called the argument of z or $\arg z$. For example, the set of all complex numbers such that $|z| \leq 1$ is represented geometrically by the unit circle and its interior, whereas the inequality $0 < \arg z < \frac{1}{2}\pi$ represents the open first quadrant.

Functions of a Complex Variable.† A *function* of a complex variable z can be defined as a set of ordered pairs of *complex numbers* (z,w) such that no two different pairs have the same first entry z. That is, w is determined by z, and we say w is given by a *function f* of z and write $w = f(z)$. For example, the set of all pairs (z,z^2) is the *square* function, i.e., $w = z^2$, or $f(z) = z^2$. According to our definition, a *function* is necessarily *single-valued*. If we represent the complex number w in terms of its real and imaginary parts, $w = u + iv$ where $u = \text{Re } w$, $v = \text{Im } w$ then, since $w = f(z) = f(x + iy)$, both u and v are functions of x and y. That is, $u = u(x,y)$, $v = v(x,y)$. For example, if $w = z^2$, then $u = x^2 - y^2$ and $v = 2xy$.

A complex function f is said to approach a complex number L as a limit as z approaches z_0 if, given any real number $\epsilon > 0$, there exists a real number $\delta > 0$ such that $|f(z) - L| < \epsilon$, for all z satisfying $0 < |z - z_0| < \delta$. Note that this is the same definition that is usually given for real functions except that the symbol $|\cdots|$ is here the magnitude of a complex number. Geometrically, if $f(z)$ has the limit L as $z \to z_0$, the numbers $f(z)$ must lie inside a circle of radius ϵ about L for all z inside a circle of radius δ about z_0 (except for $z = z_0$). All the elementary properties of limits of real functions still hold for complex functions. For example, the limit of a sum is the sum of the limits, etc. Continuity is

† For a more thorough treatment of complex functions see, for example, Churchill [8] or Courant [12], vol. 2, chap. 8.

defined as for real functions; i.e., f is continuous at z_0 if $\lim\limits_{z \to z_0} f(z) = f(z_0)$ and polynomials in z, for example, are continuous everywhere.

Example 1 $\lim\limits_{z \to i} z^3 - 5z = i^3 - 5i = -6i$

As with real functions, a *sequence*† of complex functions $f_1(z), f_2(z), \ldots, f_n(z), \ldots$ is said to approach a limit function $f(z)$, i.e., $\lim\limits_{n \to \infty} f_n(z) = f(z)$, if and only if $|f_n(z) - f(z)| < \epsilon$ for all n greater than some N. Similarly, *infinite series* are defined as sequences of partial sums and we write

$$f_1(z) + f_2(z) + \cdots + f_n(z) + \cdots = \sum_{k=1}^{\infty} f_k(z) = \lim_{n \to \infty} \sum_{k=1}^{n} f_k(z)$$

The derivative of $f(z)$ is defined by

$$f'(z) = \lim_{h \to 0} \frac{f(z + h) - f(z)}{h}$$

where h is a complex number. [Note that $f'(z)$ exists only if the limit is the same, no matter how h approaches zero.]

Example 2   If $f(z) = z^2$

$$f'(z) = \lim_{h \to 0} \frac{(z + h)^2 - z^2}{h} = \lim_{h \to 0} \frac{2zh + h^2}{h} = \lim_{h \to 0} 2z + h = 2z$$

From this example one might infer (correctly) that derivatives of *rational functions* of z can be computed by the *differentiation formulas* of elementary calculus.‡ However, transcendental functions like $\sin z$ or e^z need to be approached carefully.

Definition 1   *If a complex function f has a derivative in a certain region of the complex plane, then f is said to be analytic (or holomorphic, or regular) there.*

The study of *analytic functions* of a complex variable has been one of the most fruitful undertakings of mathematicians for two centuries, but we can mention here only a few of the results which are of immediate interest and use to us in the study of differential equations.

Theorem 1   *If a function f has a derivative f' in some region containing a point z_0 in its interior, then (1) all derivatives (i.e., f'', f''', ...) of f exist at $z = z_0$, and (2) f can be represented by its Taylor series in some circle around z_0, i.e., $f(z) = \sum_{k=0}^{\infty} (1/k!)f^{(k)}(z_0)(z - z_0)^k$, for $|z - z_0| < \delta$.*

† See Appendix A for a brief review of sequences and series.

‡ It also follows that if $f(t) = u(t) + iv(t)$ where u and v are real differentiable functions of the real variable t, then $f'(t) = u'(t) + iv'(t)$.

Example 3   Given $f(z) = 1/(1 - z)$, it is easily seen that $f'(z)$ exists except at $z = 1$ and in fact $f'(z) = 1/(1 - z)^2$. In a circle about $z = 0$ of radius less than 1, f' exists and hence so do all higher derivatives, and in fact, the *Taylor series* for f is the direct analog of the real series. That is,

$$\frac{1}{1 - z} = 1 + z + z^2 + \cdots + z^n + \cdots = \sum_{0}^{\infty} z^k \qquad |z| < 1$$

Theorem 2   *If a power series $\sum_{k=0}^{\infty} a_k(z - z_0)^k$ converges to some function $f(z)$ in some circle c about z_0, then* (1) $a_k = (1/k!) f^{(k)}(z_0)$, *i.e., the series is the Taylor series for $f(z)$, and* (2) *the series can be differentiated termwise in the circle c, i.e.,* $f'(z) = \sum_{k=1}^{\infty} k a_k(z - z_0)^{k-1}$.

Example 4   The series $1 + z + z^2/2! + \cdots + z^k/k! + \cdots = \sum_{0}^{\infty} z^k/k!$ converges to some limit function for all z.† Since, for real z, the series is the familiar exponential function, it is natural to use this series to *define* the *complex exponential function* e^z by

$$e^z = \sum_{k=0}^{\infty} \frac{z^k}{k!} = 1 + z + \frac{z^2}{2!} + \cdots + \frac{z^k}{k!} + \cdots \tag{B-1}$$

On differentiating the series termwise, we see immediately that

$$\frac{d}{dz} e^z = e^z$$

and thus the complex exponential retains the fundamental differentiation property that gave the real exponential central importance in the study of linear differential equations. In fact we have

$$\frac{d}{dz} c e^{\lambda z} = c \lambda e^{\lambda z}$$

for any complex constants c and λ.

Since formula (B-1) holds for all complex numbers z, it holds in particular for $z = i\theta$ and $z = -i\theta$ where θ is real. Substituting into (B-1), we have

$$e^{i\theta} = 1 + i\theta - \frac{\theta^2}{2!} + i\frac{\theta^3}{3!} + \cdots + \frac{i^k}{k!} \theta^k + \cdots$$

and

$$e^{-i\theta} = 1 - i\theta - \frac{\theta^2}{2!} - i\frac{\theta^3}{3!} + \cdots + \frac{(-1)^k i^k}{k!} \theta^k + \cdots$$

By adding these series termwise, we have

$$e^{i\theta} + e^{-i\theta} = 2\left[1 - \frac{\theta^2}{2!} + \frac{\theta^4}{4!} - \cdots + \frac{(-1)^n \theta^{2n}}{(2n)!} + \cdots\right]$$

† This is shown by the use of the ratio test in the same way that one shows convergence of the series for the real function e^x.

or, recognizing the series for $\cos \theta$, we see that $e^{i\theta} + e^{-i\theta} = 2 \cos \theta$. By subtracting the series we obtain

$$e^{i\theta} - e^{-i\theta} = 2i \sin \theta$$

We have then the famous *Euler relations*

$$\cos \theta = \tfrac{1}{2}(e^{i\theta} + e^{-i\theta}) \qquad \sin \theta = \frac{1}{2i}(e^{i\theta} - e^{-i\theta})$$

It immediately follows that

$$e^{i\theta} = \cos \theta + i \sin \theta$$

Finally, by applying the identities for the cosine and sine of $\theta_1 + \theta_2$, it is easily seen that

$$e^{i(\theta_1 + \theta_1)} = e^{i\theta_2} \cdot e^{i\theta_2}$$

Recall that if x is real and $e^x = y$, then $x = \ln y$. We can extend the definition of the logarithm to complex numbers by defining the function $\log z$ as follows.

Definition 2   *For any real number θ in $0 \leq \theta < 2\pi$*

$$\log (e^{i\theta}) = i\theta$$

and for any real $r > 0$ define

$$\log (re^{i\theta}) = \ln r + i\theta \qquad 0 \leq \theta < 2\pi$$

Now, since any complex number z can be written in the form $re^{i\theta}$ with $0 < r$, $0 \leq \theta \leq 2\pi$, we have a definition of $\log z = \log (re^{i\theta}) = \ln r + i\theta$.

Note that if $z_1 = r_1 e^{i\theta_1}$ and $z_2 = r_2 e^{i\theta_2}$, then

$$\begin{aligned}
\log (z_1 z_2) &= \log (r_1 e^{i\theta_1} r_2 e^{i\theta_2}) \\
&= \log (r_1 r_2 e^{i(\theta_1 + \theta_2)}) \\
&= \ln r_1 r_2 + i(\theta_1 + \theta_2)\dagger \\
&= \ln r_1 + i\theta_1 + \ln r_2 + i\theta_2
\end{aligned}$$

Finally, the general power function w^z can be defined by

$$w^z = e^{z \log w}$$

† According to our definition of $\log z$, this step is valid only if $0 \leq \theta_1 + \theta_2 < 2\pi$. The extension to other angles requires the introduction of a many-valued function and should be attempted only by those who are well versed in the theory of functions of a complex variable.

References

1. Agnew, R. P.: *Differential Equations*, 2d ed., McGraw-Hill Book Company, New York, 1960.
2. Bailey, T. J.: *The Mathematical Theory of Epidemics*, Charles Griffin & Company, Ltd., London, 1957.
3. Birkhoff, G. D., and S. MacLane: *A Survey of Modern Algebra*, 3d ed., The Macmillan Company, New York, 1965.
4. Birkhoff, G. D., and G. Rota: *Ordinary Differential Equations*, Ginn and Company, Boston, 1962.
5. Buck, R. C.: *Advanced Calculus*, 2d ed., McGraw-Hill Book Company, New York, 1965.
6. Budak, B. M., A. A. Samarskii, and A. N. Tikhonov: *A Collection of Problems on Mathematical Physics*, D. M. Brink (ed.), The Macmillan Company, New York, 1964.
7. Churchill, R. V.: *Complex Variables and Applications*, 2d ed., McGraw-Hill Book Company, New York, 1960.
8. Churchill, R. V.: *Fourier Series and Boundary Value Problems*, 2d ed., McGraw-Hill Book Company, New York, 1963.
9. Churchill, R. V.: *Operational Mathematics*, 2d ed., McGraw-Hill Book Company, New York, 1958.
10. Clement, P. R., and W. C. Johnson: *Electrical Engineering Science*, McGraw-Hill Book Company, New York, 1960.
11. Coddington, E. A., and N. Levinson: *Theory of Ordinary Differential Equations*, McGraw-Hill Book Company, New York, 1955.
12. Courant, R.: *Differential and Integral Calculus*, vols. 1 and 2, Interscience Publishers, Inc., New York, 1937.
13. Courant, R., and D. Hilbert: *Methods of Mathematical Physics*, vol. 1, Interscience Publishers, Inc., New York, 1953.
14. Courant, R., and H. Robbins: *What Is Mathematics?* Oxford University Press, Fair Lawn, N.J., 1958.
15. Finkbeiner, D.: *Introduction to Matrices and Linear Transformations*, W. H. Freeman and Company, San Francisco, 1960.

16. Friedman, B.: *Principles and Techniques of Applied Mathematics*, John Wiley & Sons, Inc., New York, 1956.
17. Hamming, R. W.: *Numerical Methods for Scientists and Engineers*, McGraw-Hill Book Company, New York, 1962.
18. Hurewicz, W.: *Lectures on Ordinary Differential Equations*, The M.I.T. Press, Cambridge, Mass., 1958.
19. Ince, E. L.: *Ordinary Differential Equations*, First American Edition, Dover Publications, Inc., New York, 1956.
20. Jahnke, E., F. Emde, and F. Lösch: *Tables of Higher Functions*, 6th ed., McGraw-Hill Book Company, New York, 1960.
21. Kaplan, W.: *Ordinary Differential Equations*, Addison-Wesley Publishing Company, Inc., Reading, Mass., 1958.
22. Kellogg, O. D.: *Foundations of Potential Theory*, Frederick Ungar Publishing Company, New York, 1929.
23. La Salle, J., and S. Lefschetz: *Stability by Liapunov's Direct Method*, Academic Press Inc., New York, 1961.
24. Lighthill, M. J.: *Introduction to Fourier Analysis and Generalized Functions*, Cambridge University Press, New York, 1958.
25. Lotka, A. J.: *Elements of Mathematical Biology*, Dover Publications, Inc., New York, 1956.
26. Martin, W. T., and E. Reissner.: *Elementary Differential Equations*, 2d ed., Addison-Wesley Publishing Company, Inc., Reading, Mass., 1961.
27. Milne, W. E.: *Numerical Solution of Differential Equations*, John Wiley & Sons, Inc., New York, 1953.
28. Petrovsky, I. G.: *Lectures on Partial Differential Equations*, Interscience Publishers, Inc., New York, 1954.
29. Pipes, L. A.: *Applied Mathematics for Engineers and Physicists*, 2d ed., McGraw-Hill Book Company, New York, 1958.
30. Ralston, A.: *A First Course in Numerical Analysis*, McGraw-Hill Book Company, New York, 1965.
31. Sagan, H.: *Boundary and Eigenvalue Problems in Mathematical Physics*, John Wiley & Sons, Inc., New York, 1961.
32. Sears, F. W.: *Principles of Physics*, 2d ed., vol. I, Addison-Wesley Publishing Company, Inc., Reading, Mass., 1950.
33. Stoker, J. J.: *Nonlinear Vibrations in Mechanical and Electrical Systems*, Interscience Publishers, Inc., New York, 1950.
34. Synge, J. L., and B. A. Griffith: *Principles of Mechanics*, 3d ed., McGraw-Hill Book Company, New York, 1959.
35. Thomas, G. B., Jr.: *Calculus and Analytic Geometry*, 3d ed., Addison-Wesley Publishing Company, Inc., Reading, Mass., 1960.
36. Watson, G. N.: *A Treatise on the Theory of Bessel Functions*, Cambridge University Press, New York, 1922.
37. Widder, D. V.: *The Laplace Transform*, Princeton University Press, Princeton, N.J., 1941.
38. Wylie, C. R., Jr.: *Advanced Engineering Mathematics*, 3d ed., McGraw-Hill Book Company, New York, 1966.
39. Ziegler, H.: *Mechanics*, vol. I, Addison-Wesley Publishing Company, Inc., Reading, Mass., 1965.

Answers and Hints

Chapter 1

Problems 1–2, p. 4

1. a. First order, first degree, linear, independent variable x, unknown function $y(x)$.
   b. Second order, first degree, linear, unknown function y.
   c. Third order, second degree, nonlinear, independent variable t, unknown function $s(t)$.
   d. Second order, first degree, linear, independent variable x, unknown function $y(x)$.
   e. nth order, first degree, linear, independent variable x, unknown function $y(x)$.
   f. Second order, partial, linear, independent variables x and y, unknown function $u(x,y)$.
   g. Second order, nonlinear system, independent variable x, unknown functions $y(x)$ and $z(x)$.
   h. First order, first degree, linear, independent variable y, unknown function $x(y)$.

2. $a(x,y)u_{xx} + b(x,y)u_{xy} + c(x,y)u_{yx} + d(x,y)u_{yy} + e(x,y)u_x + f(x,y)u_y$
   $$+ g(x,y)u = h(x,y)$$

4. Show that if y_1 is a solution of $y'' + y^2 = f_1$ and y_2 is a solution of $y'' + y^2 = f_2$, then $y_1 + y_2$ is not necessarily a solution of $y'' + y^2 = f_1 + f_2$.

Problems 1–3, p. 8

1. Evaluate k by noting that at the surface of the earth the force of attraction is mg.

Problems 1–4, p. 14

3. $x > \frac{1}{2}$

4. a. $a = 0$ b.   No value.

   c. $a = \frac{1}{29}$

5. $-2, -3$

8. $w = 0$ a, b arbitrary

   $w = \pm 2$ $a = b$

Problems 1–5, p. 18

1. $y = x + \frac{1}{24} x^4$

2. $y = \frac{1}{4}(e^{-2x} + 3e^{2x})$

3. a. $x = x_0 \cos \lambda t$ b. $x = \dfrac{v_0 \sin \lambda t}{\lambda}$

   c. $x = x_0 \cos \lambda t + v_0 \dfrac{\sin \lambda t}{\lambda}$

4. $y'(0) = 1$ $y''(0) = 2$

6. Applying the boundary condition, we find a solution exists if and only if λ satisfies $\sin \lambda L = 0$, but this implies that $\lambda L = n\pi$ for integer n.

Problems 1–6, p. 22

1. a. $y = 4 - e^{-x}$

   b. $y = \frac{1}{2}(5 - x^2)$ for $x \le 0$

      and

      $y = \frac{1}{2}(5 + x^2)$ for $x \ge 0$

2. $y = 2$ $x < 0$

   $y = 2 + x$ $x > 0$

6. $f(x) = 4x^3$

Problems 1–7, p. 26

1. a. $F'(x) = -\dfrac{1}{2x^2} + \frac{1}{2}e^{-x^3}\left(3x + \dfrac{1}{x^2}\right)$

   b. $F'(x) = x^2 - (x - x^2)^2(1 - 2x)$

5. Change the order of integration on the right-hand side.

Problems 1–8, p. 28

1. $\mathrm{Erf}(-x) = \dfrac{2}{\sqrt{\pi}} \displaystyle\int_0^{-x} e^{-t^2}\, dt$

   Change variable of integration, i.e., let $t = -\tau$.

3. $1 = 2 \dfrac{dK}{d(\theta/2)} \dfrac{d(\theta/2)}{dt}$

Problems 1–9, p. 30

1. $\Gamma(5.6) = (4.6)(3.6)(2.6)(1.6)(0.8935)$

    $$\Gamma(-2.4) = \frac{(0.8935)}{(-2.4)(-1.4)(-0.4)(0.6)}$$

2. $\Gamma(\tfrac{1}{2}) = \sqrt{\pi}$

4, 6. Make an appropriate substitution in Eq. (43).

6. Show that $\Gamma(p) = \displaystyle\int_0^1 \ln (1/y)^{p-1} \, dy$.   Hint: Let $x = -\ln y$ in Eq. (43).

Chapter 2

Problems 2–2, p. 39

1. a. $\tfrac{1}{3}x^3 + \tfrac{1}{2}y^2 = c$
    b. $(1 + x^2)(1 + y^2) = c$
    c. $u = c(1 + v^2)^2$
    d. $2 \sin y - \cos 2x = c$
    e. $y = \ln (1 + x) - x + c$
    f. $(3 + s^2) = c(1 + t^2)$

2. $y = (2e^{-x^3/3} - 1)^{\frac{1}{2}}$ $x < (3 \ln 2)^{\frac{1}{3}}$

3. $xy = \tfrac{1}{2}y^2 + c$

4. No. It is not an equation involving derivatives.

Problems 2–3, p. 46

1. a. $y = x - 1 + ce^{-x}$
    b. $(x + 1)y = \tfrac{1}{4}x^4 + \tfrac{1}{3}x^3 + c$
    c. $s = -\tfrac{5}{2}t + ct^3$
    d. $y = -2 \cos^2 x + c \cos x$
    e. Regard y as independent variable, $x = -y + \tfrac{1}{5}y^3 + c\sqrt{y}$
    f. $ye^{-\cos x} = k\int xe^{-\cos x} \, dx + c$

2. a. $y = (1 + 3e^{-1})e^{-x} + xe^{-x}$
    b. $ye^{-\cos x} = k \displaystyle\int_0^x te^{-\cos t} \, dt$

3. $y = x^2 - \tfrac{3}{5}(x + 3)$

6. a. $y = -5 + ce^{2x}$
    b. Look for a particular solution in form $y = Ax + B$. $y = -\tfrac{1}{2}x - \tfrac{11}{4} + ce^{2x}$
    c. Treat each term on the right separately.
        $y = -\tfrac{1}{2}x + e^{3x} + ce^{2x}$
    d. Look for a particular solution in form $y = Axe^{2x}$.
        $y = ce^{2x} + xe^{2x}$

7. $p = ae^{r^3/3}$

8. $y = c_1e^{2x} - \tfrac{5}{2}x + c^2$

9. $y = c_0e^{2x} - \dfrac{5}{2}\dfrac{x^n}{n!} + c_1x^{n-1} + c_2x^{n-2} + \cdots + c_{n-1}x + c_n$

Problems 2–4, p. 52

1. a. $xy - \frac{1}{3}x^3 + \frac{1}{3}y^3 = c$ b. $e^x \cos y = c$

   c. $\frac{1}{2}a(x^2 + y^2) + bxy = c$ d. $\ln (xy\sqrt{4y^2 - x^2}) = c$

2. $(x - 1)^2 y^3 + y^2 = 9$

5. $f(x) = c - 2 \cos x$

Problems 2–5, p. 56

1. $\dfrac{x}{y} + y = c$

2. $x^2 + y^2 = ce^{-2y}$

3. $xy^3 - x^{-1}y^2 = c$

4. $3x + y = c\sqrt{x}$

5. $f(x) = \frac{1}{2}x + \dfrac{c}{x}$

6. a. $x^2 + \dfrac{2x}{y} = c$ b. $y^2 + e^{y/x} = c$

9. $(Iy)^{1-n} = (1 - n)\int I^{(1-n)}Q \, dx + c$

Problems 2–6, p. 61

1. $x = ce^{-y^2/x^2}$

2. $x^2 - y^2 = cx$

3. $\ln (x^2 + y^2) + 2 \arctan \dfrac{y}{x} = c$

4. $e^y = \dfrac{x}{2} + \dfrac{c}{x}$

5. $y^5 = \frac{5}{6}x + \dfrac{c}{x^5}$

6. $ay = e^{-2\sqrt{x/y}}$

7. a. Let $v = y/x^n$.   b. Let $v = x^2 + y^2$.

   c. Let $y = x^a v$.

8. Let $x = vy$, $x + ye^{x/y} = c$.

Problems 2–7, p. 64

1. $y = c_1 e^{-3x} + c_2 + \frac{2}{3}x$

2. $y = [\sqrt{2}(3x + 2)]^{\frac{1}{3}}$

3. $y = c_1 e^{ax} + c_2 e^{-ax}$ $a \neq 0$

   $y = c_1 + c_2 x$ $a = 0$

4. $y = \dfrac{30x^{n+3}}{(n+3)!} + c_0 x^{n+2} + c_1 x^{n-1} + c_2 x^{n-2} + \cdots + c_n$

5. $x = \left(x_0 - \dfrac{mg}{k}\right) \cos \sqrt{\dfrac{k}{m}}\, t + v_0 \sqrt{\dfrac{m}{k}} \sin \sqrt{\dfrac{k}{m}}\, t + \dfrac{mg}{k}$

Supplementary problems for chapter 2, p. 65

1. $x^4 + 3x^2 y^2 + y^3 = c$

2. $u = \dfrac{v + c}{1 - cv}$

3. $xy^2 - 2x^2 y - cx - 2 = 0$

4. $y \cos x = 2(x \sin x + \cos x) + c$

5. $x^2 y + 1 = c\sqrt{x}$

6. $e^{-y} = (c - x)e^x$

7. $y = \dfrac{x}{cx^2 - 1}$

8. $N = \dfrac{c}{1 + (c/N_0 - 1)e^{-kct}}$

9. $x = \dfrac{ab(1 - e^{-(b-a)t})}{b - ae^{-(b-a)t}}$

10. $x^2 + y^2 = (x + c)^2$

11. $x^2 - y^2 = ce^{x^2}$

12. $\dfrac{1}{x^2} \cdot \dfrac{1}{F(y/x)}$

14. $\dfrac{-e^{-x^2}}{y - x} + \displaystyle\int_0^x e^{-t^2}\, dt = c$

16. $y = x + \dfrac{3}{1 + ce^{-x^3}}$

Chapter 3

Problems 3–2, p. 70

1. $v = 400(1 - e^{-\frac{2}{25}t})$

   $x = 100 + 400t - 5{,}000(1 - e^{-\frac{2}{25}t})$

   $v = 200$ when $t = 25 \ln \sqrt{2}$

2. $v_l = \frac{1}{2}$ ft/sec

3. $v = \dfrac{\sqrt{mg/k}\,(1 - e^{-2\sqrt{gk/m}\,t})}{1 + e^{-2\sqrt{gk/m}\,t}}$

4. $v = \dfrac{400(1 - e^{-25,600t})}{1 + e^{-25,600t}}$

$x = \frac{1}{64} \ln\left[(1 + e^{-25,600t})(1 + e^{+25,600t})\right] + (100 - \frac{1}{64} \ln 4)$

$v = 200$ when $t = \ln \frac{3}{25,600}$

5. DE: $-mg - kv = m \dfrac{dv}{dt}$

IC: $v(0) = v_0$

$v = \dfrac{-mg}{k} + \left(v_0 + \dfrac{mg}{k}\right)e^{-kt/m}$

$v = 0$ when $t = \dfrac{m}{k} \ln\left(\dfrac{v_0 k}{mg} + 1\right)$ max height

6. $v = \pm\left(\dfrac{2kt}{m} + v_0^{-2}\right)^{-\frac{1}{2}}$

$x = \pm\dfrac{m}{k}\left(\dfrac{2kt}{m} + v_0^{-2}\right)^{\frac{1}{2}} + \left(x_0 \mp \dfrac{m}{kv_0}\right)$

Problems 3–3, p. 74

1. $v = \dfrac{-t}{\sqrt{1 - t^2}}$

$x = \sqrt{1 - t^2}$
$x = 0$ when $t = 1$

2. $x = 16\left(\arctan e^{t/2} - \dfrac{\pi}{4}\right)$

as $t \to \infty$, $x \to 4\pi$

3. $\dfrac{v_0^2}{2g}$

Problems 3–4, p. 77

1. $F = 6w$, assuming the portion of chain on the floor has zero momentum.
    (w is specific weight.)

2. $v = \left(\dfrac{gF}{w} - \frac{2}{3}gx\right)^{\frac{1}{2}}$

where F is the force, w the specific weight of the chain.

3. $v = \dfrac{g(m_0 - at)}{k - a} + \left(v_0 - \dfrac{gm_0}{k - a}\right)\left(\dfrac{m_0 - at}{m_0}\right)^{k/a}$

4. $v = -\dfrac{\rho}{k}\dfrac{g}{4}\left(r_0 - \dfrac{k}{\rho}t\right) + \dfrac{\rho}{k}\dfrac{r_0 g}{4}\left(\dfrac{r_0}{r_0 - (k/\rho)t}\right)^3$

5. a. $v = b \ln\left(\dfrac{m_0 - at}{m_0}\right)$

    b. $v = -\dfrac{ab}{k} + \dfrac{ab}{k}\left(\dfrac{m_0 - at}{m_0}\right)^{k/a}$

Problems 3–5, p. 79

1. $80°$

2. $T = \frac{200}{3} + \frac{100}{3}e^{-\frac{3}{2}kt}$ where $k = -\frac{1}{15}\ln(0.4)$

   $T_s = 5° + \dfrac{100 - T}{2}$

   $T_1 = \frac{200}{3}$

3. $\dfrac{dT_0}{dt} = -\dfrac{1}{A}\dfrac{dT}{dt}$ and use Eq. (44).

Problems 3–6, p. 80

1. $y = \dfrac{e^{-3t/100}}{10}$ $t = \frac{3}{100}\ln 2$

2. $y_1 = 2(1 - e^{-3t/100})$

   $y_2 = 2(1 - e^{-3t/200}) - \frac{3}{50}te^{-3t/100}$

3. $x = 300\left(1 - \dfrac{t}{100}\right) - 300\left(1 - \dfrac{t}{100}\right)^5$

4. $\frac{1}{200}$ lb/gal

5. $y_1 = (1 - e^{-\frac{3}{50}t})$

   $y_2 = 2 - y_1$

   Equilibrium concentration $= 1$ lb/gal

Problems 3–7, p. 82

1. $m_0(\frac{1}{2})^{\frac{1}{10}}$

4. $x = \dfrac{a^2kt}{1 + akt}$ $\lim x(t) = a$

5. $x = \sqrt{\dfrac{k_1a}{k_2}}\left[\dfrac{\exp(2\sqrt{k_1k_2a}\,t) + 1}{\exp(2\sqrt{k_1k_2a}\,t) - 1}\right]$

   $\lim_{t\to\infty} x = \sqrt{\dfrac{k_1a}{k_2}}$

Problems 3–9, p. 90

1. $I = \dfrac{E_0}{\alpha L + R}$ yes

2. $I = I_0e^{-Rt/L}$

3. $I = \dfrac{E_b}{R}(1 - e^{-Rt/L})$ $0 \le t \le t_1$

   $I = \dfrac{E_b}{R}(1 - e^{-Rt_1/L})e^{-Rt/L}$ $t \ge t_1$

5. b. $I_p = \dfrac{E_0 \cos (\omega t + \delta)}{\sqrt{R^2 + 1/(\omega^2 c^2)}}$ $\delta = \arctan (R\omega c)^{-1}$

$Z = R - \dfrac{i}{\omega c}$

Problems 3–10, p. 93

1. a. $y = kx$ b. $y = \dfrac{k}{x}$

   c. $2y^2 + x^2 = K$ d. $y^2 + 2x = K$
   e. $x^2 + y^2 = cy$

Supplementary problems for chapter 3, p. 93

1. b. $x = 72\sqrt{(3\pi)/(2g)}$

3. $x = \dfrac{m}{k(1 - e^{-kt/m})v_0 \cos \alpha}$ $y = \dfrac{mg}{k}t + \dfrac{m}{k}\left(v_0 \sin \alpha - \dfrac{mg}{k}\right)(1 - e^{-kt/m})$

4. Assume origin is at B, y axis along river, x axis perpendicular to river.

   a. $y = \dfrac{w}{2}\left[1 - \left(\dfrac{x}{w}\right)^2\right]$ b. $y = \tfrac{1}{2}\left[\left(\dfrac{x}{w}\right)^{-\frac{1}{2}} - \left(\dfrac{x}{w}\right)^{\frac{1}{2}}\right]$

5. Let the man start at $(0,0)$ and move along the y axis with speed v_1 and let
   the dog start at $(0,a)$ and move with speed v_2. At (x,y) on the curve of
   pursuit we have

   $\dfrac{d}{dt}(y - px) = \dfrac{v_1}{v_2}\dfrac{ds}{dt}$

   where $p = y'$ and s is arc length. Solution:

   $2y = \dfrac{2ka}{1 - k^2} + \dfrac{x}{1 + k}\left(\dfrac{x}{a}\right)^k - \dfrac{x}{1 - k}\left(\dfrac{x}{a}\right)^{-k}$ $k = \dfrac{v_1}{v_2} \neq 1$

   $4ay = \ln a - a^2 + x^2 - 2 \ln x$ $\dfrac{v_1}{v_2} = 1$

6. One possibility: Take position of submarine as origin of polar coordinates.
   Let a = speed of submarine, $3a$ = speed of destroyer. Wait $4/a$ hours
   until submarine is 4 miles away from origin. At time t after this submarine
   will be at $r_s = 4 + at$ miles from D. Move destroyer in a spiral-like
   course so that distance of destroyer from origin is also $4 + at$. If path of
   destroyer is $r = f(t)$, $\theta = y(t)$, then $ds/dt = (\dot{r} + r^2\dot{\theta}^2)^{\frac{1}{2}} = 3a$. If $\theta = 0$
   when $t = 0$, solution is $r = 4 + at$, $\theta = \sqrt{8}\ln \tfrac{1}{4}(4 + at)$ or $r = 4\exp (\theta/\sqrt{8})$.

Chapter 4

Problems 4–2, p. 98

1. $y = mx$
2. a. $y = 2x - 2$ b. None.

5.   a. $v^2 + x^2 = c$ b. $x + v = c$

     c. $x = \dfrac{v^2}{2} + c$ d. $v^2 - x^2 = c$

6.   For $v = \dot{x} > 0$ the phase trajectories are circles with center at $(-1,0)$. For $v = \dot{x} < 0$ the phase trajectories are circles with center at $(+1,0)$.

7.   a.   For $x > 0$ the phase trajectories are parabolas $x = -\frac{1}{2}v^2 + c$, and for $x < 0$, $x = \frac{1}{2}v^2 + c$.
     b.   Above the line $x + y = 0$ the phase curves are $x = -\frac{1}{2}v^2 + c$, and below $x + y = 0$ they are $x = \frac{1}{2}v^2 + c$.

Problems 4–3, p. 103

1.   Change the independent variable by $x = \alpha z$ for $\alpha > 1$.

3. $|x| \le \dfrac{b}{M}$

5.   Yes.
6. $y = 0$ is a solution.   For a second solution separate variables.
7. $\alpha \ge 1$
8.   a.   The half plane $x > 0$ or the half plane $x < 0$.
     b.   The half plane $y > 0$ or the half plane $y < 0$.
     c.   Same as part b.
     d.   Any R not including $(0,0)$.
     e.   Same as part b.
     f.   Any R.
     g.   Any R in the strip $a \le x \le b$, $-\infty < y < \infty$.

Problems 4–4, p. 106

2.   a. $y_0 = 0$ $y_1 = x$ $y_2 = x + \frac{1}{3}x^3$ $y_3 = x + \frac{1}{3}x^3 + \frac{2}{15}x^5 + \frac{1}{63}x^7$
     b. $y_0 = 1$ $y_1 = 1 + x + \frac{1}{3}x^3$
         $y_2 = 1 + x + x^2 + \frac{2}{3}x^3 + \frac{1}{6}x^4 + \frac{2}{15}x^5 + \frac{1}{63}x^7$

Problems 4–5, p. 109

2.   a. $y_0 = 1$ $y_1 \approx 0.90$ $y_2 \approx 0.83$ $y_3 \approx 0.78$ $y_4 \approx 0.75$
         $y_5 \approx 0.73$
     b. $y_0 = 0$ $y_1 \approx 0.000$ $y_2 \approx 0.001$ $y_3 \approx 0.005$ $y_4 \approx 0.014$
         $y_5 \approx 0.030$

Problems 4–6, p. 111

1.   The actual solutions are:

     a. $y = 2e^{2x}$ b. $y \equiv 0$
     c. $y = e^x - 1 - x$ d. $y = 6e^{(x-1)} - 1 - x$
     e.   Use initial condition $y(0) = a$, $y = ce^x - x^2 - 2x - 2$, where $c = a + 2$.
     f. $y = \sin x$

2. a. $y = x + \frac{1}{3}x^3 + \frac{1}{15}x^5 + \cdots$

    b. $y = 3 + 10(x - 1) + 31(x - 1)^2 + \frac{287}{3}(x - 1)^3 + \cdots$

    c. $y = 1 - x - \dfrac{x^3}{3!} + \dfrac{2x^4}{4!} + \cdots$

    d. $y = 1 + \dfrac{x^2}{2!} + \dfrac{x^4}{3!} + \cdots$

Problems 4–7, p. 113

1. If $n = 2$, write $f(x,y_1,y_2) - f(x,z_1,z_2) = [f(x,y_1,y_2) - f(x,z_1,y_2)] + [f(x,z_1,y_2) - f(x,z_1,z_2)]$ and use the triangle inequality. Proceed in a similar way for $n > 2$.

2. a. R: $a_1 \leq x \leq a_2$, $b_1 \leq y \leq b_2$, $c_1 \leq y' \leq c_2$ for arbitrary values of a_i, b_i, c_i.

    b. x and y are as in part a; z can be in any interval not containing $z = 0$.

    c. y and y' are as in part a; x can be in any interval not containing $x = 0$.

Problems 4–8, p. 115

1. $y \equiv 0$
2. $-\infty < x < \infty$
3. No; $\tan x$ does not exist for all x.
4. Reduce the nth-order equation to a system of first-order equations.
5. $n > 3$
6. R can be any set of points $(x, y, y', \ldots, y^{(n-1)})$ such that $x \in I$, $c_i \leq y^{(i)} \leq d_i$ for arbitrary values of c_i, d_i.

Chapter 5

Problems 5–2, p. 126

1. Part d is LD; all others are LI.
2. $y = c_1 e^x + c_2 + x$

Problems 5–3, p. 128

4. $e^x \cos x \qquad e^x \sin x$
6. $x \cos (\ln x) \qquad x \sin (\ln x)$
7. No; the differential equation does not have real coefficients.

Problems 5–4, p. 134

1. a. $y = c_1 x + c_2$ $\qquad\qquad\qquad$ b. $y = c_1 + c_2 e^{2x}$

    c. $y = c_1 e^{ax} + c_2 e^{-ax} \qquad a \neq 0$

     $\quad\ y = c_1 + c_2 x \qquad a = 0$

    d. $y = c_1 \sin ax + c_2 \cos ax \qquad a \neq 0$

     $\quad\ y = c_1 + c_2 x \qquad a = 0$

    e. $y = c_1 + c_2 e^{-x}$ $\qquad\qquad\qquad$ f. $y = (c_1 + c_2 x)e^{-x}$

    g. $y = c_1 e^{-4x} + c_2 e^{-\frac{3}{2}x}$

    h. $y = e^{-\frac{1}{2}x}(c_1 \cos \frac{1}{2}\sqrt{3}x + c_2 \sin \frac{1}{2}\sqrt{3}x)$

4. a. $y'' = 0$ b. $y'' + 4y' - 5y = 0$
    c. $y'' + 16y = 0$ d. $y'' + 2y' + 2y = 0$
    e. $y'' - 4y = 0$ f. $y'' - 6y' + 9y = 0$
    g. $y'' - 2y' + 2y = 0$ h. No such equation exists.

5. Yes, since the coefficients are continuous for all x.

6. $\ln x$ is not defined for $x \le 0$.

7. $y = \frac{1}{5}(4e^{-3x} + e^{2x})$

8. a. $y = c_1 + c_2 x + c_3 x^2$
    b. $y = c_1 e^x + e^{-\frac{1}{2}x}(c_2 \cos \frac{1}{2}\sqrt{3}x + c_3 \sin \frac{1}{2}\sqrt{3}x)$
    c. $y = c_1 e^{3x} + (c_2 + c_3 x)e^{2x}$ d. $y = (c_1 + c_2 x + c_3 x^2)e^x$

Problems 5–5, p. 141

1. $y = \dfrac{4e^{2x}}{5} - \frac{6}{5} e^{-3x}$

2. $y = \frac{1}{97}(-8 \sin 3x - 18 \cos 3x)$

3. $y = \frac{2}{3}x \sin 3x$

4. $y = -\frac{2}{5}xe^{-4x} - \frac{5}{4}$

5. $y = \frac{1}{2}(3x^2 e^{-2x}) + e^{-x}$

6. $y = \frac{1}{4}x^3 + \frac{9}{16}x^2 + \frac{39}{32}x + \frac{81}{128}$

7. $y = \frac{3}{5}e^x(-2 \sin x + \cos x)$

8. $y = -\frac{3}{2}e^x - \frac{2}{9}x^3 - \frac{2}{9}x^2 - \frac{4}{27}x$

9. $y = \frac{1}{4}e^x (\cos x + \sin x)$

10. $y = (\frac{3}{8}x^2 - \frac{3}{16}x)e^{2x}$

Problems 5–6, p. 143

1. $y = -\cos x \ln (\sec x + \tan x) + c_1 \sin x + c_2 \cos x$

2. $y = c_1 e^{3x} + c_2 e^{2x} + e^{4x}$

3. $y = c_1 e^x + c_2 e^{-x} - \frac{1}{2} + \frac{1}{10} \cos 2x$

4. $y = c_1 e^x + c_2 e^{-x} + xe^{-x} - x$

5. $y = c_1 e^{-2x} + c_2 xe^{-2x} - e^{-2x} \ln x$

8. $y = c_1 x^2 + c_2 x^{-1} + \frac{1}{9}x^5$

Problems 5–7, p. 146

1. a. $c_1 x + c_2 x^{-\frac{1}{2}} - x^2 + x \ln x$ b. $c_1 x^{-5} + c_2 x^{-1} + 2 - \dfrac{\ln x}{x}$

2. b. $y = c_1 x + c_2 x^2 + 2 + x^2 \ln x$

Problems 5–8, p. 149

3. $y = \displaystyle\int_0^x \cosh (x - s)f(s)\, ds$ 5. $W = \dfrac{c}{x}$.

Problems 5–9, p. 155

5. a. $c_1 e^x + c_2 e^{-x} + c_3 e^{2x}$

   b. $c_1 e^{-x} + c_2 e^{\frac{1}{2}x} \cos \left(\dfrac{\sqrt{3}x}{2} + \delta \right)$

   c. $A_1 e^{\sqrt{2}x/2} \cos \left(\dfrac{\sqrt{2}x}{2} + \delta_1 \right)$
   $+ A_2 \cos \left(\dfrac{\sqrt{2}x}{2} + \delta_2 \right)$

   d. $c_1 \cosh (x + \delta_1) + c_2 \cos (x + \delta_2)$

   e. $c_4 +$ sol. of b.

   f. $(c_1 + c_2 x) \cos x + (c_3 + c_4 x) \sin x$

   g. $(c_1 + c_2 x + c_3 x^2) e^x$

   h. $c_1 e^x + c_2 e^{-x} + c_3 e^{\sqrt{2}x} + c_4 e^{-\sqrt{2}x}$

7. a. $-1 + e^{2x} + \frac{1}{6}x^3 e^x$

   b. $-\frac{1}{4}x \sin x + \frac{1}{4}x \cos x$

   c. $-\frac{1}{8}x^2 \sin x$

   d. $\frac{2}{3}x^3 - 4x^2 + 13x$

   e. $-\frac{1}{5}e^x(\sin x + 2 \cos x)$

Chapter 6

Problems 6–2, p. 164

1. $x = \frac{3}{2} \sin 4t \qquad f_n = \dfrac{2}{\pi}, \qquad x = 0$ when $t = \dfrac{n\pi}{4}$, $n = 0, 1, 2, \ldots$

2. $\dot{x} = \dfrac{3k}{\sqrt{m(k+m)}} + 3\sqrt{\dfrac{m}{k+m}}$ when $x = 2\sqrt{3}$

3. a. Let $v = d\theta/dt$ and obtain $Lv\, dv/d\theta + g \sin \theta = 0$.   Thus
   $$v = \pm\sqrt{v_0^2 + (2g/L)(\cos \theta - \cos \theta_0)}$$
   See Sec. 10-2.

   b. $\theta = A \cos \sqrt{\dfrac{g}{l}}\, t + B \sin \sqrt{\dfrac{g}{l}}\, t$

4. $\dfrac{d^2x}{dt^2} + \dfrac{64}{3}x = 0 \qquad x = 0.5 \cos \frac{8}{3}\sqrt{3}\, t \qquad T = \frac{1}{4}\pi\sqrt{3}$

5. $\dfrac{d^2y}{dt^2} + \dfrac{2T}{mL}y = 0 \qquad y = s_0 \cos \sqrt{\dfrac{2T}{mL}}\, t \qquad f = \dfrac{1}{2\pi}\sqrt{\dfrac{2T}{mL}}$

6. $x = \sqrt{153}e^{-\frac{3}{4}t} \cos (\frac{1}{3}t - d) \qquad d = \tan^{-1}(-4)$

7. $r = 33.3$ lb-sec/ft.

8. $\ddot{x} + \dfrac{2\mu g}{d}x = 0 \qquad T = 2\pi \sqrt{\dfrac{d}{2\mu g}}$

Problems 6–3, p. 171

1. $\ddot{x} + 10\dot{x} + 100x = 200 \sin 10t$
   $x = \frac{5}{2}e^{-5t}(\cos 5\sqrt{3}t + \frac{1}{3}\sqrt{3} \sin 5\sqrt{3}t) - 2 \cos 10t$
   as $t \to \infty$, $x \to -2 \cos 10t$

2. $x = 4e^{-t} - 2e^{-2t} - 2$

3. When $\omega = 5$, $c = F_0/\sqrt{339}$.   When $\omega = \sqrt{7}$, $c = F_0/\sqrt{15}$.
   $\omega = \sqrt{7}$ is the resonant frequency.

4. $\omega = \sqrt{\dfrac{2mk - r^2}{2m^2}}$

5. Exact solution: $x = \frac{900}{62} \sin 0.1t \sin 3.1t$.

   Approximate solution: $x = \frac{900}{64} \sin 0.1t \sin 3.2t$.

Problems 6–4, p. 179

1. $Q = -\frac{1}{85,000} E_0 \cos 100t + \frac{1}{21,250} E_0 \sin 100t$

   $I = \frac{1}{850} E_0 \sin 100t + \frac{1}{212.5} E_0 \cos 100t$

2. $I = 3\sqrt{2} \times 10^{-2} e^{-100t} \sin 50\sqrt{2}t$. The steady-state current is zero.

3. $Q_T = \frac{1}{200} A_0 e^{-100t} \sin 200t$

   $I_T = -\frac{1}{2} A_0 e^{-100t} (\sin 200t + 2 \cos 200t)$

6. $p = \dfrac{1}{2\,|Z|} E_0^2 \cos \theta$

Problems 6–5, p. 183

1. $\ddot{r} - r(\dot{\theta})^2 = \dfrac{P(r,\theta)}{m} \qquad r\ddot{\theta} + 2\dot{\theta}\dot{r} = 0$

2. Assume $\mathbf{F} = P(r,\theta)\mathbf{u}_r$ and derive the equation

   $$\frac{d^2u}{d\theta^2} + u = -\frac{P(r,\theta)}{mh^2u^2}$$

   for $u = 1/r$. Substituting

   $$u = \frac{1 + e \cos \theta}{ep}$$

   get $P(r,\theta) = -Km/r^2$.

3. $\ddot{r} - r\dot{\theta}^2 = -\dfrac{K}{r^3} \qquad r\ddot{\theta} + 2\dot{\theta}\dot{r} = 0$

   $r = A_1 \sec\left(\sqrt{1 - \dfrac{K}{h^2}}\,\theta - \delta_1\right) \qquad$ if $h^2 > K$

   $r = A_2 \operatorname{sech}\left(\sqrt{\dfrac{K}{h^2} - 1}\,\theta - \delta_2\right) \qquad$ if $h^2 < K$

   $r = (a\theta + b)^{-1} \qquad$ if $h^2 = K$

5. $r_p = \dfrac{ep}{1 + e} \qquad r_a = \dfrac{ep}{1 - e}$

6. $2a = r_p + r_a \qquad b = a\sqrt{1 - e^2}$

8. Determine K from $|F| = mg = Km/r^2$. Using $g = 32$ ft/sec^2, obtain

   $T = 110$ min, approximately.

9. $\ddot{r} = \dfrac{h^2}{r} - \dfrac{K}{r^2} \qquad \dot{r} = \pm\sqrt{\dfrac{2K}{r} + 2h^2 \log r + C}$

Chapter 7

Problems 7–2, p. 194

1. a. $x = 0$ $x = 1$
    b. $x = 0$ $x = 1$
    c. $x = 0$.   ($x = 1$ is removable.)
    d. $x = -1$ $x = \frac{1}{2} \pm \frac{1}{2}i\sqrt{3}$
    e. $x = -1$ $x = (n + \frac{1}{2})\pi$ $n = 0, 1, 2, \ldots$
    f. $x = 0$. The series for f about $x = 0$ vanishes identically and hence does not represent f for any $x \neq 0$.

2. a. No singularities.   b. $x = 0$ $x = 1$ $x = 2$
    c. $x = \pm 1$ d. $x = 1$.   ($x = 0$ is removable.)
    e. None. ($x = 0$ is removable.)   f. $x = 0$ $x = -1 \pm i$

3. $y = a_0\left[1 - \dfrac{(2x)^2}{2!} + \dfrac{(2x)^4}{4!} - \cdots\right] + a_1\left[2x - \dfrac{(2x)^3}{3!} + \dfrac{(2x)^5}{5!} - \cdots\right]$

    $= a_0 \cos 2x + a_1 \sin 2x$

4. a. $y = a_0(1 - x^2) + a_1\left(x - \dfrac{x^3}{3!} - \dfrac{x^5}{5!} - \dfrac{3x^7}{7!} - \cdots\right)$

        $(n + 2)(n + 1)a_{n+2} - (n - 2)a_n = 0$, series converges for all x.
    b. $y = a_0(1 - \frac{1}{6}x^3 + \frac{1}{180}x^6 + \cdots) + a_1(x - \frac{1}{12}x^4 + \frac{1}{504}x^7 + \cdots)$

        $(n + 3)(n + 2)a_{n+3} + a_n = 0$, series converges for all x.

5. a. $y = 12 + 20x - x^4 - x^5 + \frac{1}{56}x^8 + \frac{1}{72}x^9 + \cdots$

        $(n + 4)(n + 3)a_{n+4} + a_n = 0$
    b. $y = 1 - \dfrac{1}{2}x^2 + \dfrac{1}{2^2 2!}x^4 - \dfrac{1}{2^3 3!}x^6 + \dfrac{1}{2^4 4!}x^8 - \dfrac{1}{2^5 5!}x^{10} + \cdots$

        $(n + 2)a_{n+2} + a_n = 0$
    c. $y = 1 + 2x - x^2 - x^3 + \frac{1}{3}x^4 + \frac{1}{4}x^5 + \cdots$

        $(n + 1)a_{n+2} + a_n = 0$

6. $y = 1 + (x - 2) - \frac{1}{24}(x - 2)^3 + \frac{11}{576}(x - 2)^4 + \cdots$ $|x - 2| < 2$

    $4(n + 2)(n + 1)a_{n+2} + (4n^2 + 4n - 2)a_{n+1} + (n^2 - 2n + 2)a_n = 0$

7. a. $y = 1 - \frac{1}{2}x^2 + \frac{1}{8}x^4 - \frac{1}{48}x^6 + \cdots$ all x
    b. $y = 1 + 2x - x^2 - x^3 + \cdots$ all x
    c. $y = x - \frac{1}{4}x^3 + \frac{1}{16}x^4 - \frac{1}{320}x^6 + \cdots$ $|x| < 2$
    d. $y = 1 + \frac{1}{2}x^2 - \frac{1}{6}x^3 - \frac{1}{24}x^4 + \cdots$ all x
    e. $y = 1 - \frac{1}{2}x^2 + \frac{1}{12}x^4 - \frac{1}{72}x^6 + \cdots$
    f. $y = 1 + \frac{3}{2}(x - 1)^2 - \frac{1}{2}(x - 1)^3 + \frac{3}{4}(x - 1)^4 + \cdots$ $|x - 1| < 1$
    g. $y = 1 + 2x - \frac{1}{8}x^2 - \frac{1}{12}x^3 + \cdots$ $|x| < 2$

8. $y = a_0(1 - x^2 - \frac{1}{3}x^3 - \frac{1}{12}x^4 + \cdots) + a_1(x + \frac{1}{2}x^2 - \frac{1}{6}x^3 - \frac{1}{24}x^4 + \cdots)$
    $|x| < 1$

    $y = a_0[1 - (x-1)^2 + \frac{5}{6}(x-1)^4 + \ldots] + a_1[(x-1) - \frac{5}{6}(x-1)^3 + \ldots]$

Problems 7–3, p. 203

1. a. $y = x^{-2} \sum\limits_{n=0}^{\infty} a_n x^n + x^{\frac{1}{4}} \sum\limits_{n=0}^{\infty} b_n x^n$

   b. $y = [\cos(\log x)] \sum\limits_{n=0}^{\infty} a_n x^n + [\sin(\log x)] \sum\limits_{n=0}^{\infty} b_n x^n$

   c. $x = 1$ is an irregular singularity.

   d. $y = x^3 \sum\limits_{n=0}^{\infty} a_n x^n + x^{\frac{1}{2}} \sum\limits_{n=0}^{\infty} b_n x^n$

   e $x = 0$ is an irregular singularity. However, if we substitute $t = \sqrt{x}$, $u(t) = y(x)$, the equation for u becomes $u'' - (1/2t)u' + 4t^3 e^{-t^2} u = 0$. This equation has a regular singularity at $t = 0$ and the general solution is $u = \sum_{n=0}^{\infty} a_n t^n + t^{\frac{3}{2}} \sum_{n=0}^{\infty} b_n t^n$. Hence the original equation has the solution $y = \sum_{n=0}^{\infty} a_n x^{n/2} + x^{\frac{3}{4}} \sum_{n=0}^{\infty} b_n x^{n/2}$ for $x > 0$.

3. a. Irregular singularity at ∞.   b. Ordinary point at ∞.
   c. Regular singularity at ∞.

4. Regular singularity at $x = 1$; irregular singularity at $x = 0$.

Chapter 8

Problems 8–2, p. 228

1. a. $x = c_1 e^t + c_2 e^{5t}$ $y = -3c_1 e^t + c_2 e^{5t}$
   b. $x = c_1 + 2c_2 e^{5t}$ $y = -2c_1 + c_2 e^{5t}$
   c. $x = (c_1 + c_2 t)e^{3t}$ $y = (c_1 - \frac{1}{2}c_2 + c_2 t)e^{3t}$
   d. $x = e^t(c_1 \cos 3t + c_2 \sin 3t)$ $y = e^t(-c_2 \cos 3t + c_1 \sin 3t)$
   e. $x = c_1 e^t + c_2 e^{-2t}$ $y = \frac{1}{2}c_1 e^t - c_2 e^{-2t}$
   f. $x = c_1$ $y = c_1$
   g. $x = c_1 e^t + c_2 e^{-t} + c_3 e^{2t}$ $y = c_1 e^t - c_2 e^{-t} + \frac{1}{2}c_3 e^{2t}$
   $z = c_1 e^t - c_2 e^{-t} + 2c_3 e^{2t}$
   h. $x = c_1 e^{3t}$ $y = c_2 e^{2t}$ $z = c_3 e^{-t}$

3. $x = c_1 e^{2t} + c_2 e^{-t} + 2$ $y = -c_1 e^{2t} + 2c_2 e^{-t} + 3$

5. a. $x = -2e^t + e^{5t}$ $y = 6e^t + e^{5t}$
   b. $x = (1 + 2t)e^{3t}$ $y = 2te^{3t}$

6. $x = c_1 e^t + e^{-\frac{1}{2}t}(c_2 \cos \frac{1}{2}\sqrt{3}t + c_3 \sin \frac{1}{2}\sqrt{3}t)$
   $y = c_1 e^t + e^{-\frac{1}{2}t}[(-\frac{1}{2}c_2 - \frac{1}{2}\sqrt{3}c_3) \cos \frac{1}{2}\sqrt{3}t + (-\frac{1}{2}c_3 + \frac{1}{2}\sqrt{3}c_2) \sin \frac{1}{2}\sqrt{3}t]$

Problems 8–3, p. 232

1. $A_1 = A_2 = 0$ $B_1 = \frac{1}{2}$ $B_2 = -\frac{1}{2}$

4. DE: $m\ddot{y}_1 + \dfrac{2T}{L}y_1 - \dfrac{T}{L}y_2 = 0$ $m\ddot{y}_2 + \dfrac{2T}{L}y_2 - \dfrac{T}{L}y_1 = 0$

$\omega_1 = \sqrt{\dfrac{T}{mL}}$ $y_1 = a_1 \sin(\omega_1 t + \epsilon_1)$ $y_2 = a_1 \sin(\omega_1 t + \epsilon_1)$

$\omega_2 = \sqrt{\dfrac{3T}{mL}}$ $y_1 = a_2 \sin(\omega_2 t + \epsilon_2)$ $y_2 = -a_2 \sin(\omega_2 t + \epsilon_1)$

6. DE: $m\ddot{\theta}_1 + \left(\dfrac{mg}{2L} + \dfrac{k}{4}\right)\theta_1 - \dfrac{k}{4}\theta_2 = 0$

$$m\ddot{\theta}_2 + \left(\dfrac{mg}{2L} + \dfrac{k}{4}\right)\theta_2 - \dfrac{k}{4}\theta_1 = 0$$

$\omega_1 = \sqrt{\dfrac{g}{2L}}$ $\qquad \theta_1 = \theta_2 = a_1 \sin(\omega_1 t + \epsilon_1)$

$\omega_2 = \sqrt{\dfrac{g}{2L} + \dfrac{k}{2m}}$ $\qquad \theta_1 = -\theta_2 = a_2 \sin(\omega_2 t + \epsilon_2)$

Problems 8–4, p. 238

2. a. LI   b. LI
   c. LI   d. LI
   e. LD   f. LI
   g. LD

4. $x = -1$

Problems 8–6, p. 251

1. Hint: first show that if the λ_i are distinct then the c_i are linearly independent.

2. $x = c_1 e^t \begin{bmatrix} 1 \\ 0 \end{bmatrix} + c_2 e^t \begin{bmatrix} t \\ 1 \end{bmatrix}$

Problems 8–7, p. 264

4. If $A = \begin{bmatrix} a & b \\ c & d \end{bmatrix}$, then $A^{-1} = \dfrac{1}{|A|}\begin{bmatrix} d & -b \\ -c & a \end{bmatrix}$

6. $\sin A = \begin{bmatrix} 1 & -\dfrac{4}{\pi} \\ 0 & -1 \end{bmatrix}$

9. $\dfrac{d}{dt}(A^2) = \begin{bmatrix} 0 & 1 + 3t^2 \\ 0 & 4t^3 \end{bmatrix}$ $\qquad 2A\dfrac{dA}{dt} = \begin{bmatrix} 0 & 2 + 4t^2 \\ 0 & 4t^3 \end{bmatrix}$

11. b. $\mathbf{x}(t) = X(t)\left[\mathbf{c}_0 + \displaystyle\int_{t_0}^{t} X^{-1}(s)\mathbf{f}(s)\,ds\right]$

    c. $\mathbf{x}(t) = X(t)\left[X^{-1}(t_0)\mathbf{x}_0 + \displaystyle\int_{t_0}^{t} X^{-1}(s)\mathbf{f}(s)\,ds\right]$

12. $\mathbf{x}(t) = \begin{bmatrix} c_1 e^{2t} + c_2 e^{-t} \\ -c_1 e^{2t} + 2c_2 e^{-t} \end{bmatrix}$

13. $\mathbf{x}(t) = \begin{bmatrix} 4e^{-3t} - 3e^{-7t} \\ 6e^{-3t} - 6e^{-7t} \end{bmatrix}$

17. a. $\begin{bmatrix} 0 & 0 \\ 0 & 0 \end{bmatrix}$ b. $\begin{bmatrix} 0 & 1 \\ 0 & 0 \end{bmatrix}$

c. $\begin{bmatrix} 1 & 1 \\ 0 & 1 \end{bmatrix}$ d. $\begin{bmatrix} e & 0 \\ 0 & 1 \end{bmatrix}$

e. $\begin{bmatrix} e & e-1 \\ 0 & 1 \end{bmatrix}$ f. $\begin{bmatrix} e & 1 \\ 0 & 1 \end{bmatrix}$

g. $\begin{bmatrix} e & e \\ 0 & 1 \end{bmatrix}$

18. $A^5 = 3^4 \times \begin{bmatrix} 8 & -5 \\ 5 & -2 \end{bmatrix}$

Chapter 9

Problems 9–3, p. 272

1. a. $\dfrac{5}{s} + \dfrac{6}{s+1}$ b. $\dfrac{1}{(s-a)^2}$

6. $\frac{13}{10}e^{5x} - \frac{1}{2}e^{3x} - \frac{4}{5}$

9. Solve for y and use Probs. 7 and 8.

10. The homogeneous solution is obviously of exponential order. To show that a particular solution is also, use the formula obtained by variation of parameters in Chap 5 or use the following method. If the roots of the characteristic equation $\lambda^2 + a\lambda + b = 0$ are $\lambda = \alpha$, $\lambda = \beta$, show that solving the differential equation is equivalent to solving successively

$$u' - \alpha u = f(x)$$

$$y' - \beta y = u$$

and use the previous problem.

11. Generalize the approach suggested above.

Problems 9–4, p. 276

3. $\mathscr{L}\{\sinh (x)\} = \dfrac{1}{s^2 - 1}$ $\mathscr{L}\{x \sinh x\} = \dfrac{2s}{(s^2 - 1)^2}$

$\mathscr{L}\{\cosh x\} = \dfrac{s}{s^2 - 1}$ $\mathscr{L}\{x \cosh x\} = \dfrac{s^2 + 1}{(s^2 - 1)^2}$

4. $\dfrac{(-x)^k (\sin ax)}{a}$

5. $(-x)^5 \cos ax$

6. $\dfrac{2b(s - a)}{[(s - a)^2 + b^2]^2}$

9. a. Write $Y(s) = \dfrac{1}{s} \cdot \dfrac{s}{as^2 + bs + c}$

and use Theorem 2.

b. Write $Y(s) = \dfrac{1}{as^2} - \dfrac{bs + c}{as^2(as^2 + bs + c)}$

12. a. $\dfrac{\tanh \frac{1}{2}s}{s}$ b. $\dfrac{1}{s^2} - \dfrac{e^{-s}}{s(1 - e^{-s})}$

Problems 9–5, p. 282

1. $y = \dfrac{(a - 2)e^x + e^{ax}}{a - 1}$ $a \neq 1$

$y = e^x + xe^x$ $a = 1$

2. $y = -\frac{1}{4}e^{-2x} + 2e^{-x} + \frac{1}{2}x - \frac{3}{4}$

3. $y = \frac{1}{2}x^2e^{-2x}$

4. $y = \frac{6}{5}e^{-x} \sin x + \frac{2}{5}e^{-x} \cos x + \frac{1}{5}(\sin x - 2 \cos x)$

5. $y = \frac{1}{3}e^{-x} + \frac{2}{3}e^{x/2} \cos \left(\frac{1}{2}\sqrt{3}x\right)$

6. $y = -\frac{1}{2} \sin x + \frac{1}{2} \sinh x$

7. $y = \cos \left(\frac{1}{2}\sqrt{2}x\right) \cosh \left(\frac{1}{2}\sqrt{2}x\right)$

8. $y = \frac{1}{2}xe^{2x} \sin x$

9. Use arbitrary initial conditions:

$y = c_1 + c_2e^{-x} + \frac{1}{2}x^2$

10. Let $w_n = \sqrt{k/m}$.

a. $x = x_0 \cos w_n t + \dfrac{v_0 \sin (w_n t)}{w_n}$

b. $x = \left(x_0 - \dfrac{1}{k}\right) \cos w_n t + \dfrac{v_0 \sin w_n t}{w_n} + \dfrac{1}{k}$

c. $x = \left(x_0 - \dfrac{F_0}{k - mw^2}\right) \cos w_n t + \dfrac{v_0 \sin w_n t}{w_n} + F_0 \dfrac{\cos wt}{k - mw^2}$ $w \neq w_n$

$x = x_0 \cos w_n t + \dfrac{v_0 \sin w_n t}{w_n} + \dfrac{F_0 t \sin w_n t}{2\sqrt{km}}$ $w = w_n$

11. a. $x(t) = 5e^{-t} - 2e^{4t}$ $y(t) = 5e^{-t} + 3e^{4t}$

b. $x = 1 - e^{-t}$ $y = e^{-t}$

c. $x = \frac{1}{2}(t - 1) + e^t - \frac{1}{4}[(2 + \sqrt{2})e^{\sqrt{2}t} + (2 - \sqrt{2})e^{-\sqrt{2}t}]$

$y = -x + e^t + (t - 1)$

12. $b = 0$ $x = ae^{-2t}$ $y = 0$

13. a. $x_1 = 2e^t - 1$ $x_2 = 0$

b. $x_1 = 2(e^{2t} - e^t)$ $x_2 = e^{2t}$ $x_3 = 0$

c. $x_1 = e^{3t} - e^{2t}$ $x_2 = -\frac{1}{2}e^{3t} + e^{2t} - \frac{1}{2}e^t$

14. $f(x) = \cos x$

Problems 9–6, p. 288

1. a. $\displaystyle \int_0^x e^{a(x-t)}t\, dt = \frac{e^{ax} - ax - 1}{a^2}$

   b. $\frac{1}{2}x \cos 2x + \frac{1}{4} \sin 2x$

2. $y = \displaystyle \int_0^x f(t) \sinh (x - t)\, dt$

3. a. $y = 1 + \frac{1}{2}x^2$

   b. $y = xe^{-x}$

9. $y(x) = a(x - \frac{1}{2}x^2)$

10. a. $12x^2$ 　　　　　　　　　b. 1

   c. g does not exist.

Problems 9–7, p. 292

1. b may have an infinite number of discontinuities in all.

5. a. $\operatorname{sgn} x = u(x) - u(-x) = 2u(x) - 1$

   b. $\delta_\epsilon(x) = \dfrac{u(x) - u(x - \epsilon)}{\epsilon}$

7. $y = (1 - \cos x) - u(x - 5)[1 - \cos (x - 5)]$

Problems 9–8, p. 300

1. a. $Y(s) = \dfrac{1}{s^2 + 1}$ 　　　$w(t) = \sin t$

   $y_0(t) = \displaystyle \int_0^t f(t - \tau)w(\tau)\, d\tau = \int_0^t f(\tau)w(t - \tau)\, d\tau$

   $A(t) = \displaystyle \int_0^t w(\tau)\, d\tau$ 　　$y_1(t) = w(t) + w'(t) = \sin t + \cos t$

   $w(0) = 0$ 　　$w'(0) = 1$

   b. $Y(s) = \dfrac{1}{s^3 + 1}$ 　　$w(t) = \frac{1}{3}e^{-t} + \frac{1}{3}e^{t/2}[-\cos (\frac{1}{2}\sqrt{3}t) + \sqrt{3} \sin (\frac{1}{2}\sqrt{3}t)]$

   $y_1(t) = w'(t)$ 　　$w(0) = w'(0) = 0$ 　　$w''(0) = 1$

   c. $Y(s) = \dfrac{1}{s^4 + 1}$ 　　$w(t) = \sqrt{2}[\sin (\frac{1}{2}\sqrt{2}t) \cosh (\frac{1}{2}\sqrt{2}t)$

   $\quad - \cos (\frac{1}{2}\sqrt{2}t) \sinh (\frac{1}{2}\sqrt{2}t)]$

   $y_1(t) = w'''(t) = \cos (\frac{1}{2}\sqrt{2}t) \cosh (\frac{1}{2}\sqrt{2}t)$ 　　$w(0) = w'(0) = w''(0) = 0$

   $\quad w'''(0) = 1$

   d. $Y(s) = \dfrac{1}{s^3 - s^2 + s - 1}$ 　　$w(t) = \frac{1}{2}(e^t - \cos t - \sin t)$

   $y_1(t) = w(t) - w'(t) + w''(t) = \frac{1}{2}(e^t + \cos t - \sin t)$

   $w(0) = w'(0) = 0$ 　　$w''(0) = 1$

   e. $Y(s) = \dfrac{1}{s}$ 　　$w(t) = 1$ 　　$y_0(t) = \displaystyle \int_0^t f(\tau)\, d\tau$

   $A(t) = t$ 　　$y_1(t) = w(t)$ 　　$w(0) = 1$

7. $w(t) = 12t^2$

9. a. $Y(\omega) = [(6 - \omega^2) + 5i\omega]^{-1}$
    $A(\omega) = [(6 - \omega^2)^2 + 25\omega^2]^{-\frac{1}{2}}$
    steady-state response $= \frac{1}{10}(\sin 4t - 2 \cos 4t)$
    $A(\omega)$ max for $\omega = 0$

    b. $Y(i\omega) = [(1 - \omega^2) + i\omega]^{-1}$
    $A(\omega) = [(1 - \omega^2)^2 + \omega^2]^{-\frac{1}{2}}$
    steady-state response $= -\frac{1}{241}(75 \sin 4t + 20 \cos 4t)$
    $A(\omega)$ max for $\omega = \sqrt{2}$

    c. $Y(i\omega) = (-\omega^2 + i\omega)^{-1}$
    $A(\omega) = (\omega^4 + \omega^2)^{-\frac{1}{2}}$
    steady-state response $= -\frac{1}{68}(20 \sin 4t + 5 \cos 4t)$
    $A(\omega)$ max for $\omega = 0$

14. a. $y = \frac{1}{2}[\sinh 2(t - 1)u(t - 1) + \sinh 2(t - 2)u(t - 2)]$
    b. $y = te^{-2t} + (t - 2)e^{-2(t-2)}u(t - 2)$
    c. $y = \sin t + [1 - \cos (t - 2)]u(t - 2)$
    d. $y = e^6 \sin (t - 3)u(t - 3) + \sin t$

Chapter 10

Problems 10–3, p. 323

1. Stable spirals at $(2n\pi,0)$; saddle points at

    $([2n + 1]\pi, 0)$, $n = 0, \pm1, \pm2, \ldots$

2. The singularities are stable nodes at $(2n\pi,0)$ and saddles at $([2n + 1]\pi, 0)$. Hence if $\dot{x}(0)$ is small, the body will slowly return to equilibrium (without oscillating). For large $\dot{x}(0)$ the body will approach one of the positions $x = 2n\pi$, $n = 1, 2, \ldots$.

3. Center at $(0,0)$, saddle at $(1,0)$.

4. Centers at $(2n\pi,0)$; saddles at $([2n + 1]\pi, 0)$.

5. Center at $(0,0)$; saddles at $(\pm\sqrt{6},0)$.

6. a. Spiral at $(0,0)$.

    b. $\tan^{-1}\frac{y}{x} + \frac{1}{2}\ln (x^2 + y^2) = k$

7. Singularity at $(0,0)$. If $r^2 \geq 4mk$, we have a stable node (overdamped and critically damped cases, i.e., no oscillations). If $r^2 < 4mk$, the singularity is a stable spiral (damped oscillations).

8. b. A linear analysis using Table 1 predicts centers at $(n\pi,0)$ and saddles at $([2n + 1]\pi, 0)$. However, it can be shown that the quadratic damping changes the centers to stable spirals.

c. $\dfrac{v^2}{k} = c_1 e^{-2cx} + \dfrac{2}{1 + 4c^2} (\cos x - 2c \sin x)$ $\qquad v > 0$

$\dfrac{v^2}{k} = c_2 e^{2cx} + \dfrac{2}{1 + 4c^2} (\cos x + 2c \sin x)$ $\qquad v < 0$

d. Let v_0 satisfy $v_{2n-1} < v_0 < v_{2n+1}$, $n = 1, 2, \ldots$, where

$$v_m^2 = \dfrac{2}{1 + 4c^2} (1 + e^{2cm\pi}) \qquad m = 1, 2, \ldots$$

Problems 10–5, p. 329

3. In general the solution curves are spirals which terminate on the segments $v = \pm r$, $-r < x < r$.

4. $t = \frac{3}{2}\pi - \arcsin \frac{1}{3}$

Problems 10–6, p. 333

1. a. Show that the conditions on V assure that V has a minimum at the origin and hence the level curves $V(x,y) = c$ are closed curves around the origin. Then show that since a solution cannot leave the interior of a level curve, it must remain inside a certain circle.

   b. The proof of Theorem 2 follows directly from the argument of Example 1 of the text.

3. The function $V = 3x^2 + 4y^2$ is a Liapunov function and the origin is (asymptotically) stable.

4. a. $A < 0$

5. a. No.

   b. Nothing.

8. $V = \displaystyle\sum_1^n x_i^2$

Chapter 11

Problems 11–2, p. 339

1. $y_n = 5(-5)^n$

2. $y_n = \frac{19}{4} \cdot 3^n - \frac{11}{4} - \frac{1}{2}n$

3. $y_n = (n - 2)! \left(1 + \dfrac{1}{1!} + \dfrac{1}{2!} + \cdots + \dfrac{1}{(n-2)!} \right)$ $\qquad n \geq 2, y_1 = 3$

4. $y_n = \begin{cases} p_{n-2}p_{n-4} \cdots p_0 \cdot a & n \text{ even} \\ p_{n-2}p_{n-4} \cdots p_1 \cdot b & n \text{ odd} \end{cases}$

5. $a_{2m} = \dfrac{p!}{2^{2m} m! (p + m)!} a_0 \qquad m = 0, 1, \ldots$

6. $y_n = 0 \qquad n \le n_0 \qquad y_{n_0+1} \text{ arbitrary}$

$$y_{n_0+k} = \frac{b_{n_0+1}b_{n_0+2}\cdots b_{n_0+k-1}}{a_{n_0+1}a_{n_0+2}\cdots a_{n_0+k-1}} y_{n_0+1} \qquad k = 2, 3, \ldots$$

7. $y_n = \dfrac{a}{(n-1)!} \qquad n \ge 1$

8. $y_n = a^{\frac{1}{2}n(n-1)}$

9. $S_n = \frac{1}{4}n^2(n+1)^2$

10. $y = a^{(2^n-1)}$

Problems 11–3, p. 343

10. b. $\mathbf{x}_n = \left(\displaystyle\prod_{k=0}^{n-1} A_k\right)\mathbf{x}_0 \qquad n > 0$

c. $\mathbf{x}_n = \left(\displaystyle\prod_{k=0}^{n-1} A_k\right)\mathbf{x}_0 + \displaystyle\sum_{j=0}^{n-2}\left(\displaystyle\prod_{k=j+1}^{n-1} A_k\right)\mathbf{F}_j + \mathbf{F}_{n-1}$

11. $y_{n+1} = z_n \qquad z_{n+1} = -\dfrac{b_n}{a_n}z_n - \dfrac{c_n}{a_n}y_n + \dfrac{f_n}{a_n}$

Problems 11–4, p. 346

2. a. $y_n = A\cos\dfrac{n\pi}{2} + B\sin\dfrac{n\pi}{2}$ b. $y_n = c_1(-3)^n + c_2$

c. $y_n = (c_1 + c_2 n)3^n$ d. $y_n = A\cos n\beta + B\sin n\beta$
$\qquad\qquad\qquad\qquad\qquad\qquad\qquad\qquad$ (if $\sin\beta \ne 0$)

e. $y_n = A\cos\frac{2}{3}n\pi + B\sin\frac{2}{3}n\pi$ f. $y_n = 2^n y_0$

3. $y_n = \frac{1}{4}(-3)^n + \frac{3}{4}$

4. a. $y_{n+2} - 2y_{n+1} - 3y_n = 0$ b. $y_{n+2} - 8y_{n+1} + 16y_n = 0$

c. $y_{n+2} - 2\cos\beta\, y_{n+1} + y_n = 0$

5. $y_n = \dfrac{1}{\sqrt{5}}\left(\dfrac{1+\sqrt{5}}{2}\right)^n - \dfrac{1}{\sqrt{5}}\left(\dfrac{1-\sqrt{5}}{2}\right)^n$

7. $y_{n+2} = y_{n+1} + y_n \qquad y_0 = 0 \qquad y_1 = 1$
    (See Prob. 5.)

9. $\beta = \dfrac{k\pi}{N} \qquad k = 1, 2, \ldots, N-1$

10. $|\alpha_1| < 1, |\alpha_2| < 1$ where α_1, α_2 are roots of characteristic equation.

11. a. $y_n = c_1(-1)^n + c_2\cos\frac{1}{3}n\pi + c_3\sin\frac{1}{3}n\pi$

b. $y_n = c_1 2^n + c_2 n 2^n + c_3(3)^n$

c. $y_n = c_1 + c_2 n + c_3 n^2$

13. $y_n = c_1 3^n - 2c_2 2^n$

$z_n = -c_1 3^n + c_2 2^n$

17.   a. $\mathbf{x}_n = \begin{bmatrix} \frac{1}{2}(5^n + 1) & 5^n - 1 \\ \frac{1}{4}(5^n - 1) & \frac{1}{2}(5^n + 1) \end{bmatrix} \mathbf{x_0}$

   b. $\mathbf{x}_n = \begin{bmatrix} \cos n\theta & \sin n\theta \\ -\sin n\theta & \cos n\theta \end{bmatrix} \mathbf{x_0}$

   c. $\mathbf{x}_n = \begin{bmatrix} \cosh na & b \sinh na \\ \dfrac{\sinh na}{b} & \cosh na \end{bmatrix} \mathbf{x_0}$

   d. $\mathbf{x}_n = \begin{bmatrix} 2^{n-1}(2 - n) & n2^{n-1} \\ -n2^{n-1} & 2^{n-1}(n + 2) \end{bmatrix} \mathbf{x_0}$

18.   a. $\mathbf{x}_n = \begin{bmatrix} \frac{1}{2}(-2)^n + \frac{1}{2}(4)^n & -\frac{1}{2}(-2)^n + \frac{1}{2}(4)^n & \frac{3}{2}(2)^n - \frac{3}{2}(4)^n \\ -\frac{1}{2}(-2)^n + \frac{1}{2}(4)^n & \frac{1}{2}(-2)^n + \frac{1}{2}(4)^n & \frac{3}{2}(2)^n - \frac{3}{2}(4)^n \\ 0 & 0 & 2^n \end{bmatrix} \mathbf{x_0}$

   b. $\mathbf{x}_n = \begin{bmatrix} -(2)^n + 8 \cdot 4^{n-1} & -(2)^n + 4 \cdot 4^{n-1} & 2^n - 4 \cdot 4^{n-1} \\ 2^n - 4 \cdot 4^{n-1} & 2^n & -(2)^n + 4 \cdot 4^{n-1} \\ -(2)^n + 4 \cdot 4^{n-1} & -(2)^n + 4 \cdot 4^{n-1} & 2^n \end{bmatrix} \mathbf{x_0}$

   c. $\mathbf{x}_n = \begin{bmatrix} 2^n & \frac{1}{2}n2^n & \frac{1}{8}n(n + 3)2^n \\ 0 & 2^n & \frac{1}{2}n2^n \\ 0 & 0 & 2^n \end{bmatrix} \mathbf{x_0}$

19. $R_n = \dfrac{-4(3)^n + 6(2)^n}{4(3)^n - 3(2)^n}$

20. $R_n = \dfrac{(\sqrt{5} + 1)(3 + \sqrt{5})^n + (\sqrt{5} - 1)(3 - \sqrt{5})^n}{2(3 + \sqrt{5})^n - 2(3 - \sqrt{5})^n}$

   $\lim\limits_{n \to \infty} R'_n = \frac{1}{2}(\sqrt{5} + 1) = \dfrac{1}{\frac{1}{2}(\sqrt{5} - 1)}$

   or reciprocal of golden mean (see Prob. 6).

Problems 11–5, p. 351

1. $y_n = \frac{1}{2}(n + n^2)$

2.   a. $y_n = c_1 + c_2 n - \frac{1}{2}n^2 + \frac{1}{6}n^3$

   b. $y_n = c_1 + c_2 n + \dfrac{a^n}{(a - 1)^2}$

   c. $y_n = c_1 + c_2 n + \dfrac{\sin (n - 1)\alpha}{2 (\cos \alpha - 1)} \qquad \cos \alpha \ne 1$

6.   a. $w_n = 3(-3)^n - 2(-2)^n$ $n \geq 0$

    b. $w_n = 1 + n$ $n \geq 0$

    c. $w_n = \dfrac{\sin (n + 1)\alpha}{\sin \alpha}$ $n \geq 0$ $(\sin \alpha \neq 0)$

8.   a.   Expand $F(z)$ in a power series in z^{-1}; the coefficients are uniquely determined.

    b. $Z^{-1}\left(\dfrac{z}{z^2 - 5z + 6}\right) = 3^n - 2^n$

    c. $Z^{-1}\left(\dfrac{1}{z^2 - 5z + 6}\right) = \begin{cases} 3^{n-1} - 2^{n-1} & n \geq 1 \\ 0 & n = 1 \end{cases}$

    d. $Z^{-1}\left(\dfrac{z}{z^2 + z + 1}\right) = \dfrac{2}{\sqrt{3}} \sin \tfrac{2}{3}\pi n$

12.   d. $\displaystyle\sum_{k=0}^{n} k^4 = \tfrac{1}{30}n(n + 1)(2n + 1)(3n^2 + 3n - 1)$

13.   a. $y_n = \tfrac{1}{6} - \tfrac{13}{42}(-5)^n + \tfrac{1}{7}2^n$

    b. $y_n = \tfrac{23}{36} + \tfrac{13}{36}(-5)^n + \tfrac{1}{6}n$

    c. $y_n = (-\tfrac{1}{9} + \tfrac{1}{6}n)(-2)^n + \tfrac{1}{9}$

    d. $y_n = \sqrt{3} \sin (\tfrac{2}{3}n\pi) + \cos (\tfrac{2}{3}n\pi)$

Problems 11–7, p. 366

2.   c. $\lambda_r = 4 \sin^2 \tfrac{1}{2}\beta_r$ where $\beta_r = \dfrac{(2r - 1)\pi}{2N + 1}$ $r = 1, 2, \ldots, N$

       $a_n^r = B_r \sin (n\beta_r)$ $r = 1, \ldots, N$ $n = 1, \ldots, N$

       $\langle \mathbf{b}^r, \mathbf{b}^r \rangle = \tfrac{1}{4}(2N + 1)B_r^2$

    f. $v_k = \displaystyle\sum_{r=1}^{N} \alpha_r \sin r\beta_k$ $k = 1, 2, \ldots, N$

       $\alpha_r = \dfrac{4}{2N + 1} \displaystyle\sum_{k=1}^{N} v_k \sin k\beta_r$ $r = 1, 2, \ldots, N$

3.   b. $\lambda_r = 4 \sin^2 \tfrac{1}{2}\beta_r$ $\beta_r = \dfrac{r\pi}{N}$ $r = 0, 1, 2, \ldots, N - 1$

       $a_n^r = A_r \cos [\tfrac{1}{2}(2n - 1)\beta_r]$

       $\langle \mathbf{a}^0, \mathbf{a}^0 \rangle = B_0^2 N$ $\langle \mathbf{a}^r, \mathbf{a}^r \rangle = \tfrac{1}{2}B_r^2 N$ $r = 1, \ldots, N - 1$

       $v_k = \displaystyle\sum_{r=0}^{N-1} \alpha_r \cos [\tfrac{1}{2}(2r - 1)\beta_k]$ $k = 1, 2, \ldots, N$

       $\alpha_r = \dfrac{2}{N} \displaystyle\sum_{k=1}^{N} v_k \cos [\tfrac{1}{2}(2k - 1)\beta_r]$ $r = 1, 2, \ldots, N - 1$

       $\alpha_0 = \dfrac{1}{N} \displaystyle\sum_{k=1}^{N} v_k \cos [\tfrac{1}{2}(2k - 1)\beta_r]$

Problems 11–8, p. 371

1. $u_n = \sum_{r=1}^{N} B_r \sin \frac{nr\pi}{N+1} e^{-\mu_r t}$ $n = 1, 2, \ldots, N$

where

$$B_r = \frac{2}{N+1} \sum_{1}^{N} f_n \sin\left(\frac{nr\pi}{N+1}\right) \quad \mu_r = 4\alpha^2 \sin^2 \frac{r\pi}{N+1} \quad r = 1, 2, \ldots, N$$

2. b. $\omega_r = \frac{r\pi}{l} \frac{T}{\rho}$ $r = 1, 2, \ldots$

    c. $u(x,t) = \sum_{r=1}^{\infty} \beta_r \sin \frac{r\pi x}{l} \cos w_r t$

    where

    $$\beta_r = \frac{2}{l} \int_0^l f(x) \sin \frac{r\pi x}{l} \, dx$$

    d. $f(x) = \sum_{r=1}^{\infty} \beta_r \sin \frac{r\pi x}{l}$ (Fourier sine series)

    See Chaps 13 and 14.

Chapter 12

Problems 12–2, p. 377

2. a. $y(0.2) = 0.20$ b. $y(0.2) = 0.21$

Problems 12–3, p. 380

2. It is better to calculate $a(b - c)$.

5. $y(0) = 1.000$ $y(0.1) = 1.105$ $y(0.2) = 1.221$ $y(0.3) = 1.350$
    $y(0.4) = 1.491$ $y(0.5) = 1.647$ $y(0.6) = 1.820$ $y(0.7) = 2.012$
    $y(0.8) = 2.223$ $y(0.9) = 2.456$ $y(1.0) = 2.714$

7. $E_n = c_0 r_0^n + c_1 r_1^n - \dfrac{T}{2Ah}$

    where $r_0 = Ah + \sqrt{A^2 h^2 + 1}$, $r_1 = Ah - \sqrt{A^2 h^2 + 1}$, and c_0, c_1 are constants depending on the initial conditions. If $A > 0$, $r_0 > 1$, whereas if $A < 0$, $|r_1| > 1$. Therefore the errors increase exponentially.

8. $1 - \dfrac{E_n}{z_n} = \dfrac{c_0}{y_0}\left(\dfrac{hA + \sqrt{1 + h^2 A^2}}{e^{hA}}\right)^n + \dfrac{c_1}{y_0}\left(\dfrac{hA - \sqrt{1 + h^2 A^2}}{e^{hA}}\right)^n$

    If $hA > 0$ and small enough, $|E_n/z_n|$ is bounded.
    If $hA < 0$, the second term is unbounded as $n \to \infty$.

9. $y(0.1) = 0.101$ $y(0.2) = 0.203$

11. Use

$$z_{n+1} = z_n + hz'_n + \frac{h^2}{2} z''_n + \frac{1}{2!} \int_{x_n}^{x_{n+1}} (x_{n+1} - s)^2 z^{(3)}(s)\, ds$$

$$z'_{n+1} = z'_n + hz''_n + \int_{x_n}^{x_{n+1}} (x_{n+1} - s) z^{(3)}(s)\, ds$$

12. $(1 - \frac{1}{2}hA)E_{n+1} = (1 + \frac{1}{2}hA)E_n + T$

$$E_n = \left(\frac{1 + \frac{1}{2}hA}{1 - \frac{1}{2}Ah}\right)^n c - \frac{T}{hA}$$

Assume $|\frac{1}{2}Ah| < 1$. If $A < 0$, E_n is bounded as n increases, whereas if $A > 0$, E_n is unbounded.

13. $\dfrac{E_n}{z_n} = 1 - \left[\dfrac{1 + \frac{1}{2}hA}{(1 - \frac{1}{2}hA)e^{hA}}\right]^n$

Assume $|\frac{1}{2}Ah| < 1$. If $A > 0$, E_n/z_n is unbounded; if $A < 0$, E_n/z_n is bounded.

Problems 12–4, p. 385

5. $r_{0,1} = \frac{1}{2}[1 + \frac{3}{2}hA \pm \sqrt{(1 + \frac{3}{2}hA)^2 + 2hA}\,]$

Stable since for $h = 0$, $r_1 = 0$; therefore $|r_1| < 1$ for small h. By the same reasoning, relatively stable for small h.

6. $r_{0,1} = \dfrac{\frac{4}{3}hA \pm \sqrt{4 + \frac{12}{9}(hA)^2}}{2(1 - \frac{1}{3}hA)}$

Let $u = hA$; then $r_1(u)$ has the property that $r_1(0) = -1$, $r'_1(0) = \frac{1}{3}$. Therefore stable for small h if $A > 0$ but unstable if $A < 0$. Relatively stable for $A > 0$ but not relatively stable for $A < 0$.

7. The characteristic equation is
$(1 + \frac{3}{8}hA)r^3 - (\frac{9}{8} + \frac{6}{8}hA)r^2 + \frac{3}{8}hAr + \frac{1}{8} = 0$
For $h = 0$, we have $r^3 - \frac{9}{8}r^2 + \frac{1}{8} = 0$. One root of this equation is $r_0 = 1$; the other roots satisfy $8r^2 - r - 1 = 0$ or $r_{1,2} = \frac{1}{6}(1 \pm \sqrt{33})$. For $h \neq 0$ it can be shown that the root $r_0(h)$ [where $r_0(0) = 1$] approaches the exact solution as $h \to 0$. Since $r_1(h)$, $r_2(h)$ are both less than 1 in magnitude for $h = 0$, there must be an interval about $h = 0$ for which $|r_1(h)| < 1$, $|r_2(h)| < 1$. Therefore the method is stable for small h. By similar reasoning it is relatively stable for small h.

Problems 12–5, p. 388

1. $y(0.1) \approx 0.100$ $y(0.2) \approx 0.206$ $y(0.3) \approx 0.313$ $y(0.4) \approx 0.427$
3. $y(0.5) \approx 0.801$

Problems 12–6, p. 392

10. Predictor: $y_{n+1}^{\{0\}} = y_{n-3} + \frac{4}{3}h(2y'_n - y'_{n-1} + 2y'_{n-2})$
Modifier: $\bar{y}_{n+1}^{\{0\}} = y_{n+1}^{\{0\}} + \frac{112}{121}(y_n - y_n^{\{0\}})$
Corrector: $y_{n+1}^{\{i+1\}} = \frac{1}{8}(9y_n - y_{n-2}) + \frac{3}{8}h[(y_{n+1}^{\{i\}})' + 2y'_n - y'_{n-1}]$

Problems 12–7, p. 396

2. $y(0.1) = 1.105$ $y(0.2) = 1.221$ $y(0.3) = 1.349$ $y(0.4) = 1.491$
   $y(0.5) = 1.648$ $y(0.6) = 1.821$ $y(0.7) = 2.012$ $y(0.8) = 2.223$
   $y(0.9) = 2.456$ $y(1.0) = 2.714$

3. $y(0.6) \approx 0.684133$

Problems 12–8, p. 398

2. $y(0.3) \approx 0.301$

3. $y(0.2) \approx 0.2013$ $y(0.4) \approx 0.4107$ $y(0.6) \approx 0.6366$ $z(0.2) \approx 1.0201$
   $z(0.4) \approx 1.0811$ $z(0.6) \approx 1.1854$ $z = y'$

4. $y(0.8) \approx 0.8881$ $z(0.8) \approx 1.3389$
   $y(1.0) \approx 1.1855$ $z(1.0) \approx 1.5448$

5. $y(0.8) \approx 0.8880$ $y(1.0) = 1.1750$

Chapter 13

Problems 13–2, p. 402

1. Parts a, c, and d have only trivial solutions.
   b. $y = $ const is a nontrivial solution.

3. a. Unique solution: $y = -(\cot 1) \sin x + \cos x$
   b. $y = 2 \cos \pi x + c \sin \pi x$ c arbitrary
   c. No solution.

4. a. The general solution of $Ly = f$ is $y = c_1 y_1 + c_2 y_2 + y_p$ where y_1, y_2 are linearly independent solutions of $Ly = 0$ and y_p is a particular solution of $Ly = f$. Upon substitution of the boundary condition, two linear equations for c_1 and c_2 are obtained with coefficient determinant = Δ. The italicized statement of Prob. 2a does hold.

5. a. $y = 1 + \dfrac{\sin x}{\sin 1}$
   b. No solution.
   c. No solution.
   d. $y = \frac{1}{8} \sin x + \cos 3x + c \sin 3x$

Problems 13–3, p. 408

1. $\lambda_n = \dfrac{(2n + 1)^2 \pi^2}{4L^2}$ $y_n = A_n \cos \left[\dfrac{(2n + 1)\pi x}{2L} \right]$ $n = 0, 1, 2, \ldots$

2. $\lambda_n = \dfrac{(2n + 1)\pi^2}{4L^2}$ $y_n = A_n \sin \left[\dfrac{(2n + 1)\pi x}{2L} \right]$ $n = 0, 1, 2, \ldots$

3. $\lambda^2 = \dfrac{n^2 \pi^2}{L^2}$ $y_n = A_n \cos \dfrac{n\pi x}{L}$ $n = 0, 1, 2, \ldots$

4. $\lambda_n = \dfrac{n^2 \pi^2}{L^2}$ $y_n = A_n \cos \dfrac{n\pi x}{L} + B_n \sin \dfrac{n\pi x}{L}$ $n = 0, 1, 2, \ldots$

5. $\lambda_n = n^2$ $y_n = A_n \sin (n \ln x)$ $n = 1, 2, \ldots$

8. $y = \sum_{k=0}^{n} \dfrac{a_k \sin k\pi x}{\lambda - k^2 \pi^2}$

9. $y = -\frac{3}{20} \sin (5 \ln x)$

Problems 13–4, p. 413

5. Orthonormal set: $\dfrac{1}{\sqrt{L}}, \sqrt{\dfrac{2}{L}} \cos \dfrac{\pi x}{L}, \sqrt{\dfrac{2}{L}} \cos \dfrac{2\pi x}{L}, \ldots$

7. $\dfrac{1}{\sqrt{2}}, \sqrt{\frac{3}{2}}\, x, \frac{1}{2}\sqrt{\frac{5}{2}}\, (3x^2 - 1), \frac{1}{2}\sqrt{\frac{7}{2}}\, (5x^3 - 3x)$

Problems 13–5, p. 418

2. $f(x) = \dfrac{\pi}{2} - \dfrac{4}{\pi}\left(\cos x + \dfrac{\cos 3x}{3^2} + \dfrac{\cos 5x}{5^2} + \cdots \right)$

Problems 13–6, p. 421

7. $f(x) = \sum_{-\infty}^{\infty} \beta_n e^{inx}$

$\beta_n = \dfrac{1}{2\pi} \int_0^{2\pi} f(x) e^{-inx}\, dx$

Note: $\beta_n = \overline{\beta_{-n}}$

Problems 13–7, p. 424

5. a. $(xy')' + \dfrac{(x^2 - n^2)y}{x} = 0$

   b. $[(1 - x^2)y']' + n(n + 1)y = 0$

Problems 13–9, p. 431

1. a. $\dfrac{1}{2} + \dfrac{2}{\pi}(\cos x - \frac{1}{3}\cos 3x + \frac{1}{5}\cos 5x - \cdots)$

   b. $\dfrac{2}{\pi}(\sin x + \frac{2}{2}\sin 2x + \frac{1}{3}\sin 3x + \frac{1}{5}\sin 5x + \frac{2}{6}\sin 6x + \frac{1}{7}\sin 7x$
   $\qquad\qquad\qquad\qquad\qquad\qquad + \frac{1}{9}\sin 9x + \cdots)$

   c. $\dfrac{1}{2} + \dfrac{2}{\pi}(\sin 2x + \frac{1}{3}\sin 6x + \cdots)$

2. a. $f(x) = 1$

   b. $f(x) = \dfrac{4}{\pi}(\sin \frac{1}{4}\pi x + \frac{1}{3}\sin \frac{3}{4}\pi x + \cdots)$

   c. $f(x) = 1$

3. a. $\dfrac{\pi^2}{3} - 4\left(\dfrac{\cos x}{1^2} - \dfrac{\cos 2x}{2^2} + \dfrac{\cos 3x}{3^2} - \cdots \right)$

   b. $\dfrac{2}{\pi}\left[\left(\dfrac{\pi^2}{1} - \dfrac{4}{1^3} \right) \sin x + \left(\dfrac{\pi^2}{3} - \dfrac{4}{3^3} \right) \sin 3x + \cdots \right] - \left(\pi \sin 2x + \dfrac{\pi}{2}\sin 4x \right.$
   $\qquad\qquad\qquad\qquad\qquad\qquad\qquad\qquad\qquad\qquad\qquad\qquad\left. + \cdots \right)$

5. b. $a_0 = a_n = 0 \qquad b_n = \dfrac{2}{L}\displaystyle\int_0^L f(x) \sin \dfrac{n\pi x}{L}\, dx$

   c. $b_n = 0 \qquad a_0 = \dfrac{1}{L}\displaystyle\int_0^L f(x)\, dx \qquad a_n = \dfrac{2}{L}\displaystyle\int_0^L f(x) \cos \dfrac{n\pi x}{L}\, dx$

6. $|x| = \dfrac{\pi}{2} - \dfrac{4}{\pi}\left(\cos x + \dfrac{\cos 3x}{3^2} + \dfrac{\cos 5x}{5^2} + \cdots\right)$

7. $|\sin x| = \dfrac{2}{\pi} - \dfrac{4}{\pi}\displaystyle\sum_{n=1}^{\infty} \dfrac{\cos 2nx}{4n^2 - 1}$

8. $x = 2\left(\sin x - \tfrac{1}{2}\sin 2x + \tfrac{1}{3}\sin 3x - \cdots\right)$

9. $f(x) = 1$

11. $f(x) = \dfrac{h}{2} - \dfrac{4h}{\pi^2}\left(\cos\dfrac{\pi x}{c} + \dfrac{1}{3^2}\cos\dfrac{3\pi x}{c} + \cdots\right)$

$$f(x) = \begin{cases} \dfrac{hx}{c} & 0 \le x \le c \\[2mm] \dfrac{-hx}{c} & -c \le x \le 0 \end{cases}$$

12. $f(x) = \dfrac{h}{2} + \dfrac{h}{\pi}\left(\sin\dfrac{\pi x}{c} + \tfrac{1}{2}\sin\dfrac{2\pi x}{c} + \cdots\right)$

$$f(x) = h\left(1 - \dfrac{x}{2c}\right) \qquad 0 < x < 2c$$

13. $f(x) = \dfrac{h}{2} + \dfrac{2h}{\pi}\left(\sin\dfrac{\pi x}{c} + \tfrac{1}{3}\sin\dfrac{3\pi x}{c} + \cdots\right)$

$$f(x) = \begin{cases} 0 & -c < x < 0 \\ 1 & 0 < x < c \end{cases}$$

14. $x(4 - x) = \dfrac{2}{3} - \dfrac{4}{\pi^2}\left(\cos\dfrac{\pi x}{2} + \dfrac{1}{2^2}\cos\dfrac{2\pi x}{2} + \cdots\right) \qquad 0 \le x \le 4$

Problems 13–12, p. 442

6. $g(x,t) = \begin{cases} -x & x \le t \\ -t & x \ge t \end{cases}$

7. $g(x,t) = \begin{cases} \tfrac{1}{2}(t - x + xt - 1) & x \le t \\ \tfrac{1}{2}(x - t + xt - 1) & x \ge t \end{cases}$

8. $g(x,t) = \begin{cases} -\dfrac{\sin 2x \cos 2(1 - t)}{2\cos 2} & x \le t \\[4mm] -\dfrac{\cos 2(1 - x)\sin 2t}{2\cos 2} & x \ge t \end{cases}$

9. $g(x,t) = \begin{cases} \dfrac{\sin \lambda x \sin \lambda(1-t)}{\lambda \sin \lambda} & x \le t \\[3mm] \dfrac{\sin \lambda t \sin \lambda(1-x)}{\lambda \sin \lambda} & x \ge t \end{cases}$

10. $\displaystyle\int_0^1 f(x)\,dx = 0$

$y = c_1 + \displaystyle\int_0^1 g(x,t)f(t)\,dt \qquad g(x,t) = \begin{cases} t & x \le t \\ x & x \ge t \end{cases}$

11. $\displaystyle\int_0^1 f(x) \sin \pi x\,dx = 0$

$y = c_1 \sin \pi x + \displaystyle\int_0^1 g(x,t)f(t)\,dt \qquad g(x,t) = \begin{cases} \dfrac{\sin \pi x \cos \pi t}{-\pi} & x \le t \\[3mm] \dfrac{\sin \pi t \cos \pi x}{-\pi} & x \ge t \end{cases}$

12. $\displaystyle\int_0^1 f(x) \sin n\pi x\,dx = 0$

$y = c_1 \sin n\pi x + \displaystyle\int_0^1 g(x,t)f(t)\,dt \qquad g(x,t) = \begin{cases} \dfrac{\sin n\pi x \cos n\pi t}{-n\pi} & x \le t \\[3mm] \dfrac{\sin n\pi t \cos n\pi x}{-n\pi} & x \ge t \end{cases}$

Chapter 14

Problems 14–2, p. 458

2. $y(x,t) = \sin x \cos at + 2 \sin 2x \cos 2at$

3. $y(x,t) = \displaystyle\sum_{n=1}^{\infty} c_n \sin \dfrac{n\pi x}{L} \sin \dfrac{an\pi t}{L}$

   where $c_n = \dfrac{2}{an\pi} \displaystyle\int_0^L g(x) \sin \dfrac{n\pi x}{L}\,dx$

5. $y(x,t) = \dfrac{4}{\pi^2} \displaystyle\sum_{k=1}^{\infty} \dfrac{(-1)^{k-1}}{(2k-1)^2} \sin (2k-1)x \cos (2k-1)at$

6. $y(x,t) = \dfrac{L^2}{\pi^3} \displaystyle\sum_{k=1}^{\infty} \dfrac{1}{k^3} \sin \dfrac{2k\pi x}{L} \cos \dfrac{2ak\pi t}{L}$

7. $u(x,t) = \displaystyle\sum_{k=0}^{\infty} A_k \cos (k + \tfrac{1}{2})\pi x \cos (k + \tfrac{1}{2})\pi t$

   $A_k = 2 \displaystyle\int_0^1 f(x) \cos (k + \tfrac{1}{2}) \pi x\,dx$

10. $y(x,t) = \tfrac{1}{2}[f(x + at) + f(x - at)] + \dfrac{1}{2a} \displaystyle\int_{x-at}^{x+at} g(s)\,ds$

Problems 14–3, p. 463

2. $u(x,t) = \dfrac{Ax}{L} + \displaystyle\sum_{n=1}^{\infty} c_n e^{-(n\pi/L)^2 kt} \sin \dfrac{n\pi x}{L}$

    $c_n = (-1)^n \dfrac{2A}{n\pi} + \dfrac{2}{L} \displaystyle\int_0^L f(x) \sin \dfrac{n\pi x}{L}\, dx$

3. $u(x,t) = \displaystyle\sum_{n=0}^{\infty} c_n e^{-(n+\frac{1}{2})^2 kt} \sin\,(n + \tfrac{1}{2})x$ $c_n = \dfrac{2}{\pi} \displaystyle\int_0^{\pi} f(x) \sin\,(n + \tfrac{1}{2})x\, dx$

4.   See solution to Prob. 3.

5. $u(x,t) = \dfrac{Ax}{\pi} + \displaystyle\sum_{n=1}^{\infty} c_n e^{-n^2 kt} \sin nx$ $c_n = \dfrac{2A}{n\pi} \cos \dfrac{n\pi}{2}$

Problems 14–4, p. 470

1. $u(x,y) = 2 \displaystyle\sum_1^{\infty} \dfrac{\sinh\,[n\pi(1-x)]}{\sinh n\pi} \sin n\pi y \displaystyle\int_0^1 g(y') \sin n\pi y'\, dy'$

2.   First solve with boundary condition $u(x,-1) = f_1(x), u(x,1) = 0, u(-1,y) = 0, u(1,y) = 0$. Use symmetry to write the solution for each of the other nonhomogeneous conditions. Sum the four solutions.

3. $u(x,y) = \dfrac{4}{\pi} \displaystyle\sum_{n=1}^{\infty} e^{-(2n-1)y} \dfrac{\sin\,(2n-1)x}{2n-1}$

5. $u(x,y) = x = r \cos \theta$

Problems 14–5, p. 475

4.   Hyperbolic; $r = ye^x$ $s = ye^{-x}$ $v_{rs} = g(r,s,v,v_r,v_s)$

5.   Parabolic; $r = x - y$ $s = x + y$ $v_{ss} = 0$

      $u(x,y) = (x + y)f(x - y) + g(x - y)$

Chapter 15

Problems 15–2, p. 483

2. $u(r,\varphi) = A \left[\dfrac{1}{2} + \dfrac{3}{4} \dfrac{\rho}{a} \cos \varphi - \dfrac{7}{16} \dfrac{\rho^3}{a^3} P_3\,(\cos \varphi) + \cdots \right]$

3.   If u satisfies $u_{xx} + u_{yy} + u_{zz} = 0$ in a region G bounded by a closed surface B, and if u is continuous in $G + B$, then the maximum value (and minimum) of u in $G + B$ is attained on B.

4.   If two functions u_1 and u_2 satisfy Laplace's equation in a region G bounded by a closed surface B, and are both continuous in $G + B$, and if $u_1 = u_2$ at all points of B, then $u_1 = u_2$ in $G + B$.

Problems 15–3, p. 486

1. Let $v = u - u_0$ and then v satisfies the conditions (44), (45), and (46) of the problem that is solved in the text.

2. $u(r,t) = a_1 + \sum\limits_{n=2}^{\infty} a_n J_0(\mu_n r) e^{-\mu_n^2 kt}$ where

$$a_1 = \frac{2}{c^2} \int_0^c sf(s)\, ds$$

$$a_n = \frac{2}{c^2 [J_0(\mu_n c)]^2} \int_0^c s J_0(\mu_n s) f(s)\, ds \qquad n = 2, 3, \ldots$$

and $\mu_2, \mu_3, \ldots$ are the positive roots of the equation $J_1(\mu c) = 0$.

3.   a. $u = \dfrac{2z}{L} \int_0^1 sf(s)\, ds + 2 \sum\limits_{n=2}^{\infty} \dfrac{J_0(\mu_n r) \sinh \mu_n z}{[J_0(\mu_n)]^2 \sinh \mu_n L} \int_0^1 s J_0(\mu_n s) f(s)\, ds$

        where $\mu_2, \mu_3, \ldots$ are the positive roots of $J_1(\mu) = 0$.

   b. $u = \dfrac{z u_1}{L}$

Index

Index